Fish Viruses and Fish Viral Diseases

FISH VIRUSES AND FISH VIRAL DISEASES

KEN WOLF

U.S. FISH AND WILDLIFE SERVICE

Comstock Publishing Associates A DIVISION OF

Cornell University Press | ITHACA AND LONDON

First published 1988 by Cornell University Press.

Library of Congress Cataloging-in-Publication Data

Wolf, Ken.
Fish viruses and fish viral diseases.

Bibliography: p.
Includes indexes.
1. Fishes—Virus diseases. 2. Fishes—Viruses. I. Title.
SH177.V57W65 1988 639.3 87-47968
ISBN 0-8014-1259-5

Printed in the United States of America
The paper in this book is acid-free and meets the guidelines for permanence and durability of the Committee on Production Guidelines for Book Longevity of the Council on Library Resources.

Contents

Preface

My goal in writing this book was to present in one volume comprehensive and detailed information on the virology of fishes—predominantly species of the class Osteichthyes, but including representative members of the classes Chondrichthyes and Myxini. The literature about fish viruses and fish viral diseases has mostly been written by those who have a dominant or, in some cases, an exclusive interest in fish health: researchers, practitioners, culturists, and production managers. Accordingly, the book is intended to be of particular interest to them. A smaller but significant body of literature derives from researchers interested in viruses per se—their biophysical and biochemical properties—and for the virologist, biochemist, and molecular biologist I have attempted to provide highlights of available information on the properties of fish viruses.

The book is divided into six parts, the first four of which appear in descending order of significance to fish health. Pathogenic viruses that have been isolated are presented in Part I, which is subdivided into two sections: a group of chapters on viruses with demonstrably high virulence followed by a smaller group of chapters on viruses commonly regarded as having low virulence. Part II is devoted to isolated viruses whose virulence or pathogenicity remains to be determined. Bona fide viruses seen by electron microscopy but not yet isolated are discussed in Part III, and Part IV presents those disorders of fishes involving viruslike particles. In time, research may demonstrate that some of these viruslike particles are true viruses, whereas others may well be relegated to the category of artifact. Part V, on chlamydia, is included because these etiologic agents, like viruses, are obligate intracellular pathogens. The final part comprises three brief chapters on diseases and agents once thought to be viral but, according to new data, do not actually involve virus.

The chapters within the six parts are an alphabetized mix of common names of fish hosts, internationally agreed upon names of viral diseases, plus generally accepted names of several of the etiologic agents themselves. In defense of such an arrangement, as opposed to categorization by nucleic acid or major viral grouping, the genome and group affiliation of some fish viruses remain to be determined, but common names of hosts are readily recognized by persons concerned with fish health. Also, those interested in major viral groupings or specific nucleic acid will find the index helpful.

The discussions of individual viruses and their infections follow a set format, with some sections omitted for the lesser-known agents. Each chapter begins with a definition and detailed history of the virus or disease. The text then proceeds to describe signs of infection or disease and pathologic changes, etiology, diagnosis, isolation, identification, transmission, source of the virus, host and geographic range, and immunity. Where relevant, a section about control measures is included. Some viruses are well known and have been studied for a long time, and the references cited for them number more than a hundred. Other, newly discovered viruses have been reported only in a single publication; consequently, just a few sections of information are given.

This book materialized as a result of the cooperative effort of many people. G. L. Bullock, Scientific Director of the National Fish Health Research Laboratory, and administrators of the U.S. Fish and Wildlife Service excused me from other duties and authorized support so that the book could be written. A special debt of gratitude is due Paul H. Eschmeyer of the Fish and Wildlife Service Technical Information Service: he provided painstaking and skilled editorial help in preparing and refining the manuscript. Blanche D. Lawson merits special mention for her dedicated conversion of handwritten copy to typescript; for shared proofreading; for seemingly endless manuscript revisions; for noting errors, omissions, unclear text, and inconsistencies; and for her always gracious tolerance of my shortcomings. Joyce A. Mann carried out computer searches of pertinent data bases and furnished printouts; her assistants, Violet J. Catrow and Lora C. McKenzie, acquired the hundreds of references that were used. Photographic services were provided by H. Monte Stuckey. Portions of early drafts of the manuscript were critiqued by F. M. Hetrick, R. M. Malsberger, and J. A. Plumb. The staff at Cornell University Press provided the professional help that resulted in metamorphosis of typescript to a bound book.

Ken Wolf

Kearneysville, West Virginia

Fish Viruses and Fish Viral Diseases

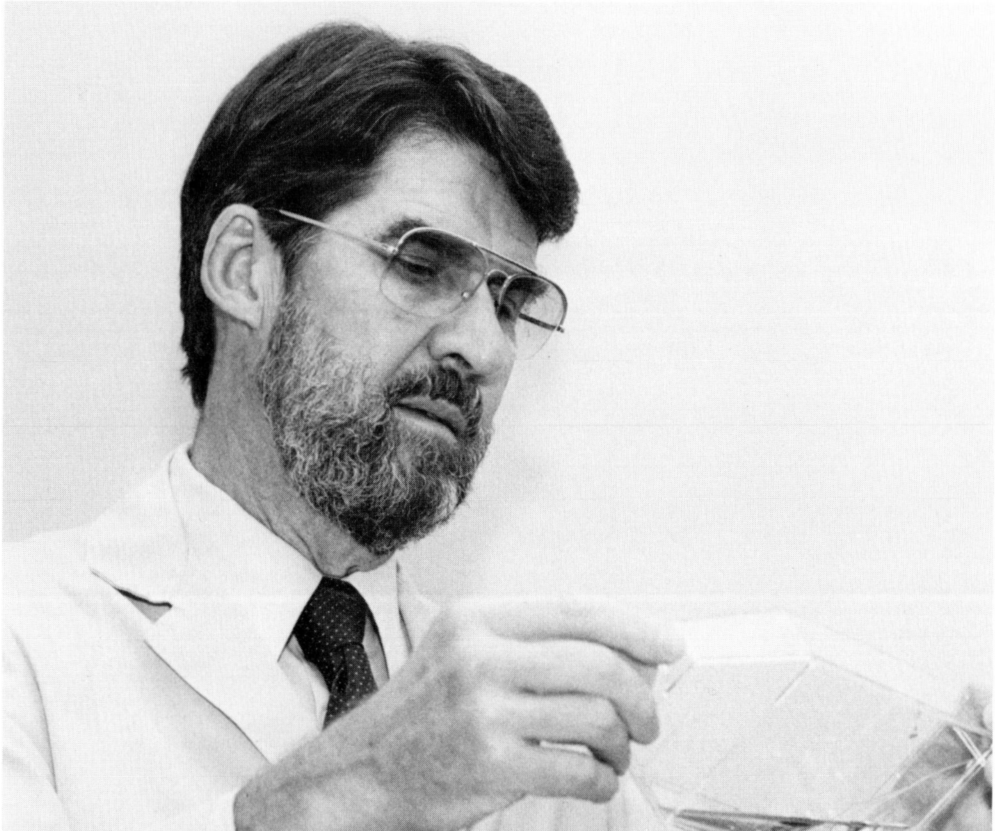

KEN WOLF, 1921–

Introduction

History

Fish husbandry has been practiced for at least 3000 years in China, but we in the West know comparatively little about Oriental fish husbandry and even less about Oriental knowledge and practices in matters of fish health. Accordingly, it seems reasonable to expect that the earliest descriptions of what we now recognize to be viral diseases of cyprinid fishes—carp pox and spring viremia of carp—might be included in ancient literature from the Orient. Until that literature becomes more widely available, however, the West must be content with a history of fish virology that goes back only to Europe and the Middle Ages of Western culture. Bruno Hofer of Germany (Fig. 1), who is generally credited with being the "Father of Fish Pathology," noted in his 1904 text on fish diseases that the famous medieval zoologist K. von Gesner documented the existence of carp pox as early as 1563. Much of the virology of carp pox, however, is yet to be learned. Electron microscopy has shown the presence of an associated herpesvirus-like agent, and Sano and his associates at Tokyo University of Fisheries reported, in 1984, that they had isolated the agent and had shown that it produced tumors experimentally.

In their professional practice, physicians, veterinarians, and fish health specialists all employ basic sciences such as bacteriology, histology, immunology, mycology, parasitology, and virology. Also, the development and maturation of fish health knowledge and its practice have generally followed the pattern first established in human medicine and secondarily in veterinary medicine.

Transmission of foot-and-mouth disease by bacteria-free filtrates—the first demonstration of animal infection by a "filterable virus"—was reported in 1898. Inasmuch as Hofer published his *Handbook of Fish Diseases* a scant 6 years later, he could hardly have included the topic of fish viruses. He did, however, discuss carp pox, but was of the opinion that it represented a secondary effect of myxosporidan parasitism.

In 1914 Weissenberg (Fig. 2) reported from Germany the first of his studies on lymphocystis disease, and at that early date astutely recognized the key viral features of the infected fish cells. He reasoned that virus was the cause of the enormous

Figure 1. Bruno Hofer, 1861–1916.

Figure 2. J. Richard Weissenberg, 1882–1974.

Figure 3. Marianne Plehn, 1863–1946.

Figure 4. Wilhelm Schäperclaus, 1899– .

and specific hypertrophy of lymphocystis cells. Interestingly, the complete proof of the viral nature of lymphocystis was not established until nearly 50 years later.

The second significant text on fish diseases was published in 1924 by Plehn (Fig. 3), a student of Hofer's. In her *Practice of Fish Diseases*, she considered both lymphocystis and pox disease to be of possible viral etiology. In 1938, Schäperclaus (Fig. 4) described a severe disease in Germany that had as its apparent target the kidneys of rainbow trout (*Salmo gairdneri*).[1] Initially the condition was termed "kidney swelling," and next, "infectious kidney swelling and liver degeneration"; it acquired an array of names and ultimately one that was internationally agreed upon—viral hemorrhagic septicemia.

In 1939, the Russian text *Bacterial Diseases of Fish* by Dogiel (Fig. 5) and his associates included material on infectious dropsy and rubella of common carp (*Cyprinus carpio*), but at the time they wrote, the virus had not yet been isolated. Lyaiman (Fig. 6), in his text *A Course in Diseases of Fish* (1949), discussed acute infectious dropsy and lymphocystis and included a brief paragraph on what eventually became known as viral hemorrhagic septicemia. Originally in Russian, the text was later translated into English. In 1954, when Schäperclaus published the third edition of *Fish Diseases*, the number of viral or postulated viral diseases (including acute infectious dropsy, about which opinions still differ) had grown to five.

In 1941 M'Gonigle, a Canadian physician, described a condition in young salmonids that he termed acute catarrhal enteritis—in all probability the first account of what we now recognize to be infectious pancreatic necrosis. It is relevant to note that Emmeline Moore (Fig. 7), the first American fish pathologist, was in the audience when M'Gonigle read his report at a meeting of the American Fisheries Society. In the oral discussion that followed M'Gonigle's paper, Moore affirmed M'Gonigle's diagnosis on the basis of her own unpublished work.

The first review of fish viruses was made in 1952 by Nigrelli (Fig. 8), who included at least six diseases. The viral etiology of infectious pancreatic necrosis, however, had not yet been proposed. *Culture and Diseases of Game Fishes* (1953) by Davis (Fig. 9) was the first American book to include more than a cursory discussion of fish diseases. It included a brief summary of M'Gonigle's work, but lymphocystis was the only viral disease described.

Beginning in 1953, serious epizootics were reported among fry of sockeye salmon (*Oncorhynchus nerka*) in the Pacific Northwest. Although capabilities for isolating virus had not yet been developed, evidence obtained by Watson and his associates (1954) indicated that virus was killing the salmon fry. For a few years at least, the condition was known as "sockeye salmon virus disease" or "Oregon sockeye disease." When Watson reviewed fish viruses in 1954, he described eight diseases—five of which we know today to be viral. In retrospect, the 1950s were marked by an unusual consolidation of personnel, research opportunities, funding, and techniques; a period of mushrooming growth of research in fish virology had begun.

More than any other single factor, the application of cell and tissue culture methods opened fish virology to rapid advance. Although fish tissue culture had its beginnings during the early 1900s, the techniques were largely in the domain of

1. The names of North American fishes used here follow Robins (1980). Classification to family uses follow Jordan (1963).

Figure 5. Valentin Alexandrovich Dogiel, 1882–1955.

Figure 6. E. M. Lyaiman, 1903–1969.

Figure 7. Emmeline Moore, 1872–1963.

Figure 8. Ross F. Nigrelli, 1903– .

anatomists and embryologists; a virologic application had not yet been made. Highly significant advances were made during the 1950s, however, when it was demonstrated that virus could be grown in animal tissue cultures. Rather surprisingly, a mammalian virus, the cause of eastern equine encephalomyelitis, was the first virus to be grown in fish "tissue culture"—embryos of *Gambusia* sp. (Sanders and Soret 1954; Soret and Sanders 1954).

During 1955, histologic examination of trout fry with clinical signs of acute catarrhal enteritis revealed neither parasites nor bacteria, but showed pancreatic necrosis morphologically similar to that produced by coxsackie virus. That early work of Wood and his co-workers provided the basis for what became known as infectious pancreatic necrosis.

Although no virus was found, the first applications of fish tissue culture to investigations of fish diseases were made during the mid-1950s. In America, at what was then the U.S. Fish and Wildlife Service Microbiological Laboratory, Wolf (1956) inoculated explant cultures of trout tissues with blue-sac fluid when he investigated the possibility that virus might be the cause of that disease.

In Germany's Robert Koch Institute, Grützner (1956a, b) described and used explant cultures for the investigation of lymphocystis disease and fish pox. She rightfully deserves credit for having made the first application of fish tissue culture to a fish virus disease. Two years later, she reported what was to be the first preparation of monolayer cultures of fish cells. Meanwhile, the virus of infectious pancreatic necrosis was isolated in America (Wolf et al. 1959; Wolf, Snieszko, et al. 1960).

During the mid-1950s to the mid-1960s, three more texts on fish diseases appeared that included the subject of viral infections. German texts by Amlacher (1961) (Fig. 10) and Reichenbach-Klinke (1966) (Fig. 11) gave then current coverage of viral diseases in books that were later translated into English. A book in English by Van Duijn (1956), however, which was strongly directed toward the aquarist, included only superficial material on viruses.

In 1960, from the Fish and Wildlife Service Laboratory in Seattle, Washington, Ross and co-workers reported that fry of chinook salmon (*Oncorhynchus tshawytscha*) were subject to epizootics that involved a filterable agent. Two years later, an Italian booklet on fish diseases published by Ghittino (Fig. 12) included seven viral diseases, among which was the newly reported chinook salmon virus disease. That infection was eventually shown to be caused by the same rhabdovirus that causes sockeye salmon virus disease. By the early 1980s these diseases had become known as infectious hematopoietic necrosis.

The useful but time-consuming fish tissue culture gave way to monolayer cell culture—first by Grützner (1958) and then by Wolf, Quimby, et al. (1960) and Clem et al. (1961, 1965); Fryer (1964) (Fig. 13) and his associates (1965), and others, soon followed. The effectiveness of the newer techniques was soon evident: In Denmark, Jensen (1963, 1965) isolated Egtved virus, the etiologic agent of what is now termed viral hemorrhagic septicemia; Clem et al. (1965) described an "orphan" virus of a North American marine fish (however, the viral nature of that isolant has not been confirmed); and, as reported first by Parisot et al. (1965), the sockeye salmon disease virus was isolated by J. L. Fryer at Oregon State University.

Wolf described the characteristics of fish viruses known in 1964. The viruses, together with the associated diseases, some 17 items, were further treated in detail

Figure 9. Herbert Spencer Davis, 1875–1958.

Figure 10. Erwin Amlacher, 1922– .

Figure 11. Heinz-Hermann Reichenbach-Klinke, 1914– .

Figure 12. Pietro Ghittino, 1929– .

in a review by Wolf and in a brief overview by Malsberger and Wolf, both in 1966. Also during 1966—some 52 years after Weissenberg had correctly proposed a viral etiology for lymphocystis—Wolf and colleagues from Lehigh University isolated the virus and fulfilled Rivers' postulates. A Danish book on fish diseases by Christensen (Fig. 14), which also appeared in 1966 and was translated into French 2 years later, included material on five viral infections.

During the late 1960s, several noteworthy events occurred in fish virology. In 1968, Fijan, then in the United States as a visiting scientist from Yugoslavia, investigated epizootics among young channel catfish (*Ictalurus punctatus*) and implicated what was soon confirmed to be the channel catfish virus (Fijan et al. 1970).

Amend et al. (1969) published a landmark report describing an infectious condition that involved necrosis of hematopoietic tissue in North American rainbow trout and salmon fry. That work led the way to the eventual determinations that the so-called chinook salmon and sockeye salmon virus diseases were actually the same infection—infectious hematopoietic necrosis.

During 1969, oral papillomas of eels with so-called cauliflower disease were examined by Pfitzner of the Robert Koch Institute. She isolated a virus that Schubert (1969) showed to have an icosahedral form. This was the first of several viruses to be isolated from eels, but it has never been implicated as the cause of the bizarre epidermal ("cauliflower") lesions. At about the same time, Hoffman et al. (1969) isolated a virus from the bluegill (*Lepomis macrochirus*) and described a newly recognized disease, termed epitheliocystis, that was shown to involve a chlamydial organism—the first to be recognized in a fish.

As the 1960s neared a close, a new chapter in fish virology opened. Departing from traditional studies of host, pathogen, and environment, workers began to focus attention on the biophysical aspects of the pathogen, and reports soon appeared on the molecular biology of fish viruses. Lymphocystis and infectious pancreatic necrosis viruses received early attention, but during the 1970s researchers became aware that a rich source of new information lay in the several fish rhabdoviruses.

Lopez et al. (1969) reported on rates of protein and nucleic acid synthesis in cell cultures infected with lymphocystis virus, but for years little else followed—a lapse no doubt reflecting the lengthy growth cycle of the virus. In contrast, when reports of work on the rapidly replicated virus of infectious pancreatic necrosis appeared, the flow of research resumed. The characterization of that virus by Moss and Gravell (1969) included a suggestion that the viral genome was double-stranded RNA. The RNA was later confirmed as being double stranded and consisting of two segments; these features led to the proposal by Dobos et al. (1977) that infectious pancreatic necrosis virus and related agents be regarded as members of a new family, the Birnaviridae.

As a group, the rhabdoviruses had become the most numerous kind of fish virus, and the knowledge of their molecular biology developed rapidly. As examples, Lenoir (1973) undertook a study of the agent of spring viremia of common carp and determined the protein structure of the virus and the size and number of the peptides; Clark and Soriano (1974) found that Egtved virus and the agents of spring viremia of carp and infectious hematopoietic necrosis could be replicated by reptilian and mammalian cells as well as by fish cells, but most notably that the viruses and cells had different temperature optima; and Roy et al. (1975) characterized the pike fry rhabdovirus RNA and proteins and showed that the virion had an RNA-dependent RNA polymerase. In England, Hill et al. (1975) carried out a comparative characterization of the then known fish rhabdoviruses, determined the polypeptides and sedimentation rates of virus RNA, and drew comparisons between the fish rhabdoviruses and prototype viruses of rabies and vesicular stomatitis.

The history of the molecular biology of fish virology has been short, but new advances are being made rapidly, and progress can be expected to continue at an undiminished rate for some years.

In 1969 Schäperclaus published a review of fish virus infections that included a wealth of information developed by European workers—particularly on diseases known solely from that continent.

Fish cell and tissue culture has provided a practical approach toward obtaining most of what has been learned in fish virology. Fish tissue culture had its beginning very early in the development of tissue culture, but it was not until the development of virology during the 1950s that animal tissue culture began thriving. Wolf and Quimby (1969) closed the decade of the 1960s with a history of fish tissue culture and a comprehensive review of the subject.

The pace of discovery in fish virology accelerated during the 1970s. In Yugoslavia, Fijan et al. (1971) firmly established the role of virus in common carp with signs of acute infectious dropsy. The etiology of carp dropsy had long been the subject of divergent opinions. Some workers considered it to be bacterial, others considered it to be viral, and still others postulated that virus and bacteria were jointly responsible. Fijan's work brought the subject of spring viremia of carp definitively into the literature.

Notable compilations and texts on fish diseases also continued during the 1970s. A book on diseases of marine fish and shellfish by Sindermann (Fig. 15) appeared in 1970. Ghittino (1962, 1970) published two well-illustrated companion volumes in Italian—one on fish diseases and one on fish culture. The German text by Amlacher (1961) was translated into English and an enlarged second edition was published in 1972.

In 1972 the Zoological Society of London published a symposium entitled "Diseases of Fish" (Mawdesley-Thomas 1972), which contained an authoritative review of infectious dropsy by N. Fijan; a report on Egtved virus serology and fluorescent antibody technique by P. E. Vestergard Jorgensen; a review by N. O. Christensen of the principal diseases in Danish trout farms, including two viral infections, and a review of progress in fish virology for the period 1966–1971 by K. Wolf, which included an account of the eight distinct viruses that had been isolated.

Reports by Bootsma (1971), Bootsma and van Vorstenbosch (1973), and de Kinkelin and Galimard (1973) brought another rhabdovirus into the literature: epizootics among fry of northern pike (*Esox lucius*) in the Netherlands were shown to be caused by a bullet-shaped virus, and the rhabdoviruses clearly became a major group of fish pathogens.

An entirely new kind of fish virus was reported by Walker (1971), who described intracytoplasmic icosahedra in erythrocytes of cod fishes (Gadidae). His work provided concrete evidence for the viral nature of piscine erythrocytic necrosis that had been described by Laird and Bullock (1969) and postulated as possibly being viral. Confirming reports followed on various Atlantic species and on Pacific salmon.

Figure 13. John L. Fryer, 1929– .

Figure 14. N. O. Christensen, 1914–1983.

Figure 15. Carl J. Sindermann, 1922– .

Figure 16. Syuzo Egusa, 1920– .

A second herpesvirus isolation was made by Sano (1976) in Japan, and Wolf (1976) reported still another herpesvirus in North America; Sano also reported a presumably new icosahedral virus and a rhabdovirus from eels.

As the 1980s began, reports of new viruses became more frequent. Some of the agents were known only from electron microscopy but others were isolated and characterized. Tabulation of all such agents showed that 17 had been isolated and 15 others were known from electron micrographs (Wolf and Mann 1980).

The several reviews of fish viruses published during the late 1970s and early 1980s generally included discussions of the recognized diseases and other agents, but differed somewhat in emphasis: McAllister (1979) emphasized biophysical and biological properties of the agents that had been isolated; Pilcher and Fryer (1980a, b) published two comprehensive and scholarly works, one on 11 diseases of proven viral etiology and one on 19 other diseases in which a viral etiology was suspect but at the time unproven; a review by Ahne and Wolf (1980) followed traditional lines and emphasized the various aspects of diseases where known; and a review by Wolf (1981) stressed the point that the various viral diseases of fishes posed no threat to human health.

Viruses and the viruslike particles that were newly reported during the period 1977–1981 were discussed by Wolf (1982), and a later tabulation (Wolf 1984) showed that the broad subject of fish viruses then included 50 entities or topics.

Landmarks of other fish health literature that dealt partly or entirely with virology merit mention. In 1964, the New York Academy of Sciences sponsored an international conference entitled "Viral Diseases of Poikilothermic Vertebrates." As the first major work of its kind, the collected papers, which included 29 reports on fish viruses and viral diseases, and related topics, were published as a 680-page annal (Whipple 1965). Beginning in 1966, the Japan Research Group of Fish Pathology began publishing the journal *Fish Pathology*, edited by Egusa (Fig. 16). Initially, *Fish Pathology* was printed entirely in Japanese but English titles and abstracts were later added.

The American Fisheries Society published "A Symposium on Diseases of Fishes and Shellfishes," edited by Snieszko (Fig. 17), in 1970. The symposium, a collection of 42 papers by recognized authorities, included 6 on fish virology.

Under the auspices of the Food and Agriculture Organization of the United Nations, the European Inland Fisheries Advisory Commission (EIFAC) held an international "Symposium on the Major Communicable Fish Diseases in Europe and Their Control" in Amsterdam in 1972 that led to the publication of 29 panel reviews and relevant papers, in English or French (Dill 1973). The publication is particularly valuable for its authoritative summaries of a number of problem fish diseases, including four major viral diseases.

In 1972–1985, the U.S. Fish and Wildlife Service produced and distributed *Fish Health News*, a quarterly compilation of abstracts of current literature on the subject. In 1986, that publication merged with *Fisheries Review*, which is available from Superintendent of Documents, U.S. Government Printing Office, Washington, D.C. 20402.

A symposium held in Washington, D.C., in 1972 under the joint sponsorship of The Registry of Comparative Pathology, The American Registry of Pathology, and the

Figure 17. Stanislas F. Snieszko, 1902-1984.

Figure 18. O. N. Bauer, 1915– .

Figure 19. R. J. Roberts, 1941– .

Armed Forces Institute of Pathology, led to the publication of 39 papers, including 6 on fish virology, in a 1004-page volume edited by Ribelin and Migaki (1975).

In 1975, the Japan Research Group of Fish Pathology, under chairman S. Egusa, held an international seminar in Tokyo entitled "Recent Advances in Fish Pathology." The review papers on fish virology in Japan and in North America filled an entire issue of *Fish Pathology* (Egusa 1976).

Tavolek Inc. of Redmond, Washington, a transient entrant in the field of proprietary biologicals for fish, held an international "Symposium on Diseases of Cultured Salmonids" in Seattle in 1977 and published the proceedings as a mixture of abstracts and full papers. Three of the papers were worthwhile reviews of fish virologic topics and a fourth, although only an abstract, was generally agreed to be the most significant contribution of all—Vestergard Jorgensen's account of an 11-year effort during which he and a part-time assistant led a program of eradication of viral hemorrhagic septicemia and infectious pancreatic necrosis from a significant percentage of 380 Danish trout farms.

During autumn 1977, the International Office of Epizootics held the second session of the Cooperative Program of Research on Aquaculture (COPRAQ) at Brest, France, the proceedings of which included summaries of 21 papers on fish virology (COPRAQ 1977).

International awareness of the need to make information on fish diseases more easily accessible became evident with the appearance in 1977 of a Russian text by Bauer (Fig. 18) and associates, and in 1978 of a terse English-language text edited by Roberts (Fig. 19), and a Japanese-language book by Egusa. All three works contain material on fish virus diseases.

Unique in its history, the fourth edition of *Fischkrankheiten* appeared in 1979, again under the name of Schäperclaus, but for the first time with two co-editors (Kulow and Schreckenbach) and seven others listed as collaborators. Representing more than a 50% increase over the third edition, the two-volume work totaled 1089 pages.

The *Journal of Fish Diseases*, the first serial publication in the English language dealing exclusively with reports on that subject, was launched in 1978 under the editorship of R. J. Roberts and R. Wooten, both of Scotland.

Three relevant symposia marked the noteworthy year 1981: (1) The International Association of Biological Standardization held an international symposium in Leetown, West Virginia, USA, entitled "Fish Biologics: Serodiagnostics and Vaccines." The published volume includes 11 papers on aspects of four viral diseases that cause serious mortality among fish under husbandry. (2) The symposium "Microbial Diseases of Fish," was held under the aegis of the Pathogenicity Group of the Society of General Microbiology at the University of Edinburgh, Scotland. Among the 12 contributed papers published in a special publication (Roberts 1982), 3 dealt with fish viruses—one reviewed viruses of warm-water fishes, a second reviewed the virus of infectious pancreatic necrosis and its virulence, and the third reviewed new viruses reported from 1977 to 1981. (3) An international symposium in Talloires, France, "Antigens of Fish Pathogens" (Anderson et al. 1983) included five virus papers and, again, the topic of each concerned an agent that caused significant mortality.

Fish health specialists in Europe banded together in 1980 to form a society that began publication of a new serial in 1981, the *Bulletin of the European Association of Fish Pathologists*. Membership in the association soon spread to other parts of the world, and by the mid-1980s the *Bulletin* was fulfilling a vital role in fostering the rapid exchange of knowledge.

The Sixth International Congress of Virology was convened in Sendai, Japan, in 1984. One of the workshops consisted of 10 papers that dealt with fish viruses—3 concerning newly reported agents. Unfortunately, the literature of that Congress was restricted to abstracts that were not widely distributed.

After the Congress in Sendai, an International Seminar on Fish Pathology was held in Tokyo in 1984 to mark the 20th anniversary of the Japanese Society of Fish Pathology. The 3-day event included a half-day session during which 10 virus papers were presented. Complete papers of the seminar were published in 1985 as a single issue—volume 20 (nos. 2 and 3) of *Fish Pathology*.

References

Ahne, W., and K. Wolf. 1980. Viruserkrankungen der Fische. Pages 56–105 in H. H. Reichenbach-Klinke, ed. Krankheiten und Schädigungen der Fische, 2nd ed. Gustav Fischer Verlag, Stuttgart.

Amend, D. F., W. T. Yasutake, and R. W. Mead. 1969. A hematopoietic virus disease of rainbow trout and sockeye salmon. Trans. Am. Fish. Soc. 98:796–804.

Amlacher, E. 1961. Taschenbuch der Fischkrankheiten. Gustav Fischer Verlag, Jena.

Amlacher, E. 1970. Textbook of fish diseases. D. A. Conroy and R. L. Herman, translators. T.F.H. Publications, Jersey City, New Jersey.

Amlacher, E. 1972. Taschenbuch der Fischkrankheiten, 2nd ed. Gustav Fischer Verlag, Jena.

Anderson, D. P., M. Dorson, and P. Dubourget, editors. 1983. Antigens of fish pathogens. Collection Fondation Marcel Merieux, Lyon, France.

Bauer, O. N., V. A. Musselius, V. M. Nikolaeva, and Y. A. Strelkov. 1977. Ikhtiopatologiya (Ichthyopathology). Izdatel'stvo "Pischchevaya promyshlennost," Moscow.

Bootsma, R. 1971. Hydrocephalus and red-disease in pike fry *Esox lucius*. J. Fish Biol. 3:417–419.

Bootsma, R., and C. J. A. H. van Vorstenbosch. 1973. Detection of a bullet-shaped virus in kidney sections of pike fry (*Esox lucius*). Neth. J. Vet. Sci. 98:86–90.

Christensen, N. O. 1966. Fiskesygdomme Saetryk af Medlemsblad for Den danske Drylaegeforening 49:331–337, 516–520, 541–546, 914–944, 1031–1044. Copenhagen.

Christensen, N. O. 1966. Maladies des poissons. P. Besse, translator (1968). Syndicat des Pisciculteurs Salmoniculteurs de France, Paris.

Clark, H. F., and E. Z. Soriano. 1974. Fish rhabdovirus replication in non-piscine cell culture: new system for the study of rhabdovirus-cell interaction in which the virus and cell have different temperature optima. Infect. Immunol. 10:180–188.

Clem, L. W., L. Moewus, and M. M. Sigel. 1961. Studies with cells from marine fish in tissue culture. Proc. Soc. Exp. Biol. Med. 108:762–766.

Clem, L. W., M. M. Sigel, and R. R. Friis. 1965. An orphan virus isolated in marine fish cell tissue culture. Ann. N.Y. Acad. Sci. 126:343–361.

COPRAQ. 1977. Proceedings of second session of the Cooperative Program of Research on Aquaculture. Bull. Off. Int. Epizoot. 87(5–6):359–522.

Davis, H. S. 1953. Culture and diseases of game fishes. Univ. California Press, Berkeley.

de Kinkelin, P., and B. Galimard. 1973. Isolation and identification of the causative agent of "red disease" of pike (*Esox lucius* L. 1766). Nature 241:465–467.

Dill, W. A., editor. 1973. Symposium on the major communicable fish diseases in Europe and their control. FAO of the United Nations. EIFAC (Eur. Inland Fish. Advis. Comm.), Tech. Paper 17, Suppl. 2.

Dobos, P., R. Hallett, D. T. C. Kells, O. Sorensen, and D. Rowe. 1977. Biophysical studies of infectious pancreatic necrosis virus. J. Virol. 22:150–159.

Dogiel, V. A., M. A. Peshkov, and H. B. Guseva. 1939. Bakterialnye zabolevaniya ryb (Bacterial diseases of fish). Pishchepromizdat (Food Industries), Moscow.

Egusa, S., editor. 1976. Recent advances in fish pathology. In Proceedings of an international seminar on fish diseases. Japan Research Group of Fish Pathology and Committee VI, 3 Nutrition and Production of Fish in the International Union of Nutrition Science, Tokyo, August 11–12, 1975. (Also Fish Pathol. 10:103–259.)

Egusa, S. 1978. The infectious diseases of fish. Koseisha Koseikaku Co., Tokyo. In Japanese.

Fijan, N. 1968. Progress report on acute mortality of channel catfish fingerlings caused by a virus. Bull. Off. Int. Epizoot. 69:1167–1168.

Fijan, N. N., T. L. Wellborn, Jr., and J. P. Naftel. 1970. An acute viral disease of channel catfish. U.S. Fish Wildl. Serv., Tech. Paper 43.

Fijan, N., Z. Petrinec, D. Sulimanovic, and L. O. Zwillenberg. 1971. Isolation of the viral causative agent from the acute form of infectious dropsy of carp. Vet. Arch. 41:125–138.

Fryer, J. L. 1964. Methods for the in vitro cultivation of cells and tissues of salmonid fishes. Ph.D. dissertation. Oregon State Univ., Corvallis.

Fryer, J. L., A. Yusha, and K. S. Pilcher. 1965. The in vitro cultivation of tissue and cells of Pacific salmon and steelhead trout. Ann. N.Y. Acad. Sci. 126:566–586.

Ghittino, P. 1962. Le principali malattie dei pesci Stab. Grafico F. LLI Lega, Faenza, Italy.

Ghittino, P. 1970. Piscicoltura e ittiopatologia: Vol. 2, Ittiopatologia. Edizioni Rivista di Zootecnia Stampa Strada, Sesto S. G. Italy.

Grützner, L. 1956. Versuch zur Züchtung des Gewebes von *Macropodus opercularis* (Linne') und *Lebistes reticulatus* (Peters) in vitro. Zentralbl. Bakteriol. Parasitenkd. Infektionskr. Hyg. Abt. I Orig. 165:8–24.

Grützner, L. 1956. Überprüfung einiger Anwendungsmöglichkeiten der Gewebekultur von *Lebistes reticulatus* (Peters) und *Macropodus opercularis* (Linne') in der Virusforschung. Zentralbl. Bakteriol. Parasitenkd. Infektionskr. Hyg. Abt. I Orig. 165:81–96.

Grützner, L. 1958. In vitro—Züchtung des Leber—und Nierengewebes von Tinca vulgaris Cuv. (Schleie) in trypsinierten Einschichtgewebekulturen. Zentralbl. Bakteriol. Parasitenkd. Infektionskr. Hyg. Abt. I Orig. 173:195–202.

Hill, B. J., B. O. Underwood, C. J. Smale, and F. Brown. 1975. Physico-chemical and serological characterization of five rhabdoviruses infecting fish. J. Gen. Virol. 27:369–378.

Hofer, B. 1904. Handbuch der Fischkrankheiten. Verlag Allg. Fischerei-Zeitung. B. Heller, Munich.

Hoffman, G. L., C. E. Dunbar, K. Wolf, and L. O. Zwillenberg. 1969. Epitheliocystis, a new infectious disease of the bluegill (*Lepomis macrochirus*). Antonie Van Leeuwenhoek 35:146–158.

International Association of Biological Standardization. 1981. International symposium on fish biologics: serodiagnostics and vaccines. Dev. Biol. Stand. 49:1489.

International seminar on fish pathology. 1985. Fish Pathol. 20(2–3).

Jensen, M. 1963. Preparation of fish tissue cultures for virus research. Bull. Off. Int. Epizoot. 59:131–134.

Jensen, M. 1965. Research on the virus of Egtved disease. Ann. N.Y. Acad. Sci. 126:422–426.

Jordan, D. S. 1963. The genera of fishes and a classification of fishes. Stanford Univ. Press, Stanford, California.

Laird, M., and W. L. Bullock. 1969. Marine fish haematozoa from New Brunswick and New England. J. Fish. Res. Board Can. 26:1075–1102.
Lenoir G. 1973. Structural proteins of spring viremia of carp. Biochem. Biophys. Res. Commun. 51: 895–899.
Lopez, D. M., M. M. Sigel, A. E. Beasley, and L. S. Dieterich. 1969. Biochemical and morphologic studies of lymphocystis disease. Page 223–236 in C. J. Dawe and J. C. Harshbarger, eds. A symposium on neoplasms and related disorders of invertebrate and lower vertebrate animals. Natl. Cancer Inst. Monogr. 31.
Lyaiman, E. M. 1949. Kurse bolezni ryb. (A course in fish diseases.) V. O. Pahn, translator. Pishchepromizdat (Food Industries), Moscow.
McAllister, P. E. 1979. Fish viruses and viral infections. In H. Fraenkel-Conrat and R. R. Wagner, eds. Comprehensive virology, Vol. 14, pp. 401–470. Plenum, New York.
M'Gonigle, R. H. 1941. Acute catarrhal enteritis of salmonid fingerlings. Trans. Am. Fish. Soc. 70:297–303.
Malsberger, R. G., and K. Wolf. 1966. Virus diseases of fish. Pages 677–684 in J. E. Prier, ed. Basic medical virology. Williams and Wilkins, Baltimore.
Mawdesley-Thomas, L. E., editor. 1972. Diseases of fish. Symp. Zool. Soc. Lond. No. 30.
Moss, L. H., and M. Gravell. 1969. Ultrastructure and sequential development of infectious pancreatic necrosis virus. J. Virol. 3:52–58.
Nigrelli, R. F. 1952. Virus and tumors in fishes. Ann. N.Y. Acad. Sci. 54:1076–1902.
Parisot, T. J., W. T. Yasutake, and G. W. Klontz. 1965. Virus diseases of the Salmonidae in western United States. I. Etiology and epizootiology. Ann. N.Y. Acad. Sci. 126:502–519.
Pfitzner, I. 1969. Zur Aetiologie der Blumenkohlkrankheit der Aale. Arch. Fischereiwiss. 20:24–35.
Pfitzner, I., and G. Schubert. 1969. Ein Virus aus dem Blut mit Blumenkohlkrankheit behafteter Aale. Z. Naturforsch. 24b:790.
Pilcher, K. S., and J. L. Fryer. 1980a. The viral diseases of fish: a review through 1978. Part 1: Diseases of proven viral etiology. CRC Crit. Rev. Microbiol. 7:287–363.
Pilcher, K. S., and J. L. Fryer. 1980b. The viral diseases of fish: a review through 1978. Part 2: Diseases in which a viral etiology is suspected but unproven. CRC Crit. Rev. Microbiol. 8:1–25.
Plehn, M. 1924. Praktikum der Fischkrankheiten. E. Schweizerbarth Verlag, Stuttgart.
Reichenbach-Klinke, H. H. 1966. Krankheiten und Schädigungen der Fische. Gustav Fischer Verlag, Stuttgart.
Reichenbach-Klinke, H. H. 1973. Fish pathology. C. Ahrens, translator. T.F.H. Publications, Neptune City, N. J. Translated from German by C. Ahrens.
Ribelin, W. E., and G. Migaki, editors. 1975. The pathology of fishes. Univ. Wisconsin Press, Madison.
Roberts, R. J., editor. 1978. Fish pathology. Bailliere Tindall, London.
Roberts, R. J., editor. 1982. Microbial diseases of fish. Soc. Gen. Microbiol., Spec. Publ. 9., Academic Press, London.
Robins, C. R., chairman. 1980. A list of common and scientific names of fishes from the United States and Canada, 4th ed. American Fisheries Society, Bethesda, Maryland.
Ross, A. J., J. Pelnar, and R. R. Rucker. 1960. A virus-like disease of chinook salmon. Trans. Am. Fish. Soc. 89:160–163.
Roy, P., H. F. Clark, H. P. Madore, and D. H. L. Bishop. 1975. RNA polymerase associated with virions of pike fry rhabdovirus. J. Virol. 15:338–347.
Sanders, M., and M. G. Soret. 1954. Cultivation of animal viruses in embryonic teleost cells. Trans. N.Y. Acad. Sci. 17:19–25.
Sano, T. 1976. Viral diseases of cultured fishes in Japan. Fish Pathol. 10:221–226.
Sano, T., H. Fukuda, and M. Furukawa. 1984. *Herpesvirus cyprini*: biological and ocogenic

properties. Page 55 in International seminar on fish pathology for the 20th anniversary of the Japanese Society of Fish Pathology, Tokyo, September 8–10, 1984. Abstract.
Schäperclaus, W. 1938. Die Schädigungen der deutschen Fischerei durch Fischparasiten und Fischkrankheiten. Fisch. Ztg. 41:267–270.
Schäperclaus, W. 1954. Fischkrankheiten, 3rd ed. Akademie-Verlag, Berlin, D.D.R.
Schäperclaus, W. 1969. Virusinfektionen bei Fischen. Pages 1067–1141 in H. Röhrer, ed. Handbuch der Virusinfektionen bei Tieren. Gustav Fischer Verlag, Jena.
Schäperclaus, W., H. Kulow, and K. Schreckenbach, editors. 1979. Fischkrankheiten, 4th ed. Akademie Verlag, Berlin.
Schubert, G. 1969. Elektronenmikroskopische Untersuchungen an der Haut mit Blumenkohlkrankheit behafteter Aale. Arch. Fischereiwiss. 20:36–49.
Sindermann, C. J. 1970. Principal diseases of marine fish and shellfish. Academic Press, New York.
Snieszko, S. F., editor. 1970. A symposium on diseases of fishes and shellfishes. Am. Fish. Soc., Spec. Publ. 5.
Soret, M. G., and M. Sanders. 1954. In vitro method for cultivating eastern equine encephalomyelitis virus in teleost embryos. Proc. Soc. Exp. Biol. Med. 87:526–529.
Symposium on diseases of cultured salmonids. (1977) Tavolek Inc., Redmond, Washington.
Van Duijn, C. 1956. Diseases of fishes, 1st ed. Water Life, Dorset House, London.
Walker, R. 1971. PEN, a viral lesion of fish erythrocytes. Am. Zool. 11:707.
Watson, S. W. 1954. Virus diseases of fish. Trans. Am. Fish. Soc. 83:331–334.
Watson, S. W., R. W. Guenther, and R. R. Rucker. 1954. A virus disease of sockeye salmon: interim report. U.S. Fish Wildl. Serv., Spec. Sci. Rep. Fish. 138.
Weissenberg, R. 1914. Über infektiöse Zellhypertrophie bei Fischen (Lymphocystiserkrankung). Sitzungsber. kgl. preuss. Akad. Wiss. 30:792–804.
Whipple, H. E., editor. 1965. Viral diseases of poikilothermic vertebrates. Ann. N.Y. Acad. Sci. 126(Art. 1).
Wolf, K. E. 1956. The cause and control of blue-sac disease. Ph.D. dissertation. Utah State Univ., Logan.
Wolf, K. 1964. Characteristics of viruses found in fishes. Dev. Ind. Microbiol. 5:139–148.
Wolf, K. 1966. The fish viruses. Adv. Virus Res. 12:35–101.
Wolf, K. 1976. Fish viral diseases in North America, 1971–75, and recent research at the Eastern Fish Disease Laboratory, U.S.A. Fish Pathol. 10:135–154.
Wolf, K. 1981. Viral diseases of fish and their relation to public health. Pages 403–437 in J. H. Steele and G. Beran, eds. Characteristics of viral diseases of fish, Section B. Viral zoonoses, Vol. II. CRC Press, Boca Raton, Florida.
Wolf, K. 1982. Newly discovered viruses and viral diseases of fishes, 1977–1981. Pages 59–90 in R. J. Roberts, ed. Microbial diseases of fish. Soc. Gen. Microbiol., Spec. Publ. 9, Academic Press, London.
Wolf, K. 1984. Fish viruses: their biology, classification, hosts, pathology, and control. Pages 197–215 in E. Kurstak and R. G. Marusyk, eds. Control of virus diseases. Marcel Dekker, New York.
Wolf, K., and J. A. Mann. 1980. Poikilotherm vertebrate cell lines and viruses: a current listing for fishes. In Vitro 16:168–179.
Wolf, K., and M. C. Quimby. 1969. Fish cell tissue culture. Pages 253–305 in W. S. Hoar and D. J. Randall, eds. Fish physiology, Vol. 3. Academic Press, New York.
Wolf, K., S. F. Snieszko, and C. E. Dunbar. 1959. Infectious pancreatic necrosis, a virus-caused disease of fish. Excerpta Med. 13(Sect. 1):228.
Wolf, K., M. C. Quimby, E. A. Pyle, and R. P. Dexter. 1960. Preparation of monolayer cell cultures from tissues of some lower vertebrates. Science 132:1890–1891.

Wolf, K., S. F. Snieszko, C. E. Dunbar, and E. Pyle. 1960. Virus nature of infectious pancreatic necrosis in trout. Proc. Soc. Exp. Biol. Med. 104:105–108.

Wolf, K., M. Gravell, and R. G. Malsberger. 1966. Lymphocystis virus: isolation and propagation in fish cell lines. Science 151:1004–1005.

Wood, E. M., S. F. Snieszko, and W. T. Yasutake. 1955. Infectious pancreatic necrosis in brook trout. Arch. Pathol. 60:26–28.

PART I

ISOLATED VIRUSES AND RESULTING DISEASES

SECTION I DISEASES AND AGENTS OF MODERATE TO HIGH VIRULENCE

1 Channel Catfish Virus Disease

Definition

Channel catfish virus disease (CCVD) is an acute hemorrhagic herpesvirus infection that typically results in high mortality—sometimes nearly 100%—of young-of-the-year *Ictalurus punctatus*. The disease occurs in problem proportions among hatchery stocks of channel catfish, but other ictalurids are susceptible, at least to experimental infection.

History

In the United States during the 1960s, a substantial and rapid expansion of the pond-cultured channel catfish industry was accompanied by reports of serious fish health problems. One problem was the occurrence of severe epizootics among some lots of newly feeding fry. In some instances, handling, high temperature, or hypoxia had preceded the outbreaks, but in others the environmental factors and husbandry practices had been acceptable, if not favorable. Flexibacteria or aeromonads were sometimes prevalent enough to be considered as causal, but an overall pattern of bacterial involvement was not evident. In effect, and in order of decreasing probability, virus, nutritional factors, or toxicants remained as suspected causes for most of the outbreaks.

The breakthrough discovery of the viral nature of the fry mortality was made by N. N. Fijan, then a visiting professor from Yugoslavia at Auburn University, Alabama. In fewer than 200 words he announced transmission of disease by injection of a filterable agent, cell culture isolation, and production of the disease with culture-grown virus (Fijan 1968). In a timely presentation at a conference of fish farmers, the channel catfish virus situation was reviewed by Wellborn (1969), one of Fijan's collaborators, and he was the first to publish the term "channel catfish virus disease."

Fijan et al. (1970) published the first research report on CCVD, which included the results of investigation of seven epizootics. Virus had been isolated from four of five epizootics involving very young catfish, but was not found in subadults or adults.

The investigators gave details of experimental transmission, showed the host specificity of the virus, documented the isolation of the agent, and described some of its characteristics. They illustrated cell culture changes—syncytia and pyknosis—but most significantly, they announced that Rivers' postulates had been fulfilled. At that time, the virus had been found in Alabama, Arkansas, Kentucky, and Texas.

Fijan's research on CCVD ended when he returned to Yugoslavia, but the work was continued by a successor at Auburn University, J. A. Plumb, who began his doctoral studies on CCV in 1969. Elsewhere, the results of viral characterization left no doubt that CCV was a member of the herpesvirus group (Wolf and Darlington 1971). Plumb's work was soon productive; he found that the virus could not be isolated from suspect carriers, as had been done with salmonid viruses, but that it could be quantified in victim tissues (Plumb 1971a). As in the other severe viral diseases of fishes, the kidneys were the prime target organ and produced the most virus (Plumb 1971b). By 1971, five additional states had been added to the list where CCVD had occurred: Georgia, Kansas, Mississippi, Oklahoma, and West Virginia (Plumb 1971c). By that time, too, a total of 23 virus isolations had been documented.

During the following year, McCraren (1972) reviewed the disease; Wolf et al. (1972) reported the histopathologic changes of experimental CCVD; and Plumb (1972) completed his dissertation, in which he reported attempted carrier detection and serologic assessment of suspect carriers, evaluated the immune response, quantified organs for virus, determined temperature effects, and compared the susceptibility of different fish strains.

Plumb showed the sparing effect of temperature reduction on fish with CCV (Plumb 1973a), and found a correlation between virus-neutralizing activity in serum from suspect populations and the absence of such activity in fish with no history of CCV (Plumb 1973b). Plumb et al. (1973) also quantified the survival of CCV in fish tissue under warm decomposition and in cold storage. Gratzek et al. (1973) described microcultures for quantification of CCV and for titration of antibody against CCV. Wolf (1973) compiled a comprehensive review of herpesviruses of poikilotherm vertebrates and included previously unpublished data on the effects of temperature on in vitro replication and on the biophysical properties of CCV.

Details of the growth retardation or stunting phenomenon in CCVD survivors were published by McGlamery and Gratzek (1974), and Plumb (1974) reviewed both the virus and the disease for an aquacultural audience. Plumb et al. (1974) also described and reported on the histopathology and electron microscopy of fingerling catfish with CCVD, and Galla and Hartmann (1974) extended the host range to experimental infection of the walking catfish (*Clarias batrachus*), although that extension has not been confirmed.

Plumb was by now the author with the greatest number of publications on CCV. His epizootiologic data showed that during the 7 years since it had been discovered, CCV had been found in 29 places (Plumb 1975). Comparative testing of catfish strains showed that there were marked differences in susceptibility and that two hybrid strains were the most resistant (Plumb et al. 1975). Plumb and Gaines (1975) described the sequential development of histopathologic changes and virus recovery from organs of experimentally infected fingerling catfish. Major et al. (1975) described the histopathologic changes of naturally infected channel catfish fingerlings, and Heartwell (1975) reported on the kinetics of the immune response of channel catfish to CCV and characterized the resulting antibody. Yasutake (1975) included CCVD in his comparative review of the histopathologic effects of fish viral diseases.

Before 1975, virtually everything that had been done with CCV had been fishery oriented. However, virologists were becoming interested in CCV. Goodheart and Plummer (1975) showed relationships between the DNAs of 28 herpesviruses, including that of CCV. By 1977, CCV had been exported to Honduras in infected fry, and Nebraska, California, and Colorado were added to the list of states with the disease.

A new line of investigation was opened at Florida Atlantic University, where cell lines were developed from the walking catfish and used to attenuate CCV for purposes of immunization (Noga 1977). Research on CCV was also under way at Texas A&M University, where McConnell and Austen (1978) concluded that whenever CCV was accompanied by high mortality, secondary infection and environmental factors were contributory. Koment and Haines (1978), investigators of herpesvirus disease in turtles, determined that CCV was sensitive to phosphonacetic acid, but markedly less so than herpesviruses from homeotherms. Robin and Rodrigue (1978) described a method of purifying CCV, Granoff and Naegele (1978) determined the molecular weights of fish herpesvirus DNAs, and Lee et al. (1978) reported the buoyant density and other properties of CCV. Plumb (1978) discussed the epizootiology of the disease and once again considered secondary infections as contributing significantly to the mortality. Plumb and Chappell (1978) found that the blue catfish (*Ictalurus furcatus*) was susceptible to CCV when the virus was injected.

During 1979, an elegant work published in France included a map of biophysical attributes of the virus DNA (Chousterman et al. 1979) and, in America, Huston completed her doctoral research on attempts to detect the ever-elusive virus in suspect carrier fish. Ten years had elapsed since the virus had been isolated, and no one had yet succeeded in revealing its hiding place. On a different theme, Gerba et al. (1979) described details of photoinactivation of CCV by visible light.

During the late 1970s, investigators interested in CCV could be categorized as those concerned with aspects of fish health and those interested solely in the properties of the virus. The literature of the 1980s reflected significant growth in knowledge of the molecular biology of CCV.

On the basis of Bowser's (1978) dissertation, a trio of publications described establishment of the CCO[1] cell line its rates of growth, and its advantages over BB cells in work with CCV (Bowser and Plumb 1980a, b, c). Interest in fish viruses continued at the University of Sherbrooke, Quebec. There, the response of CCV to several physicochemical factors was reported by Robin and Rodrigue (1980a), who also isolated and made a preliminary characterization and visualization of the DNA (Robin and Rodrigue 1980b). Comparable work was carried out at Harvard University by Dixon and Farber (1980), who determined the biophysical properties of the DNA and the kinetics of its replication, and identified virus-specific polypeptides and the mode of their synthesis.

The establishment of cell lines from the walking catfish and their application in developing an effective attenuated vaccine against CCVD was reported by Noga and

1. The identity and properties of commonly used fish cell lines are given in Table 2 of Appendix 2.

Hartmann (1981) and Walczak et al. (1981). Evidence for the detection of CCV carriers by the fluorescent antibody technique was presented by Plumb et al. (1981). Brady and Ellender (1982) described the effects of pH, temperature, salinity, and soil sediment on the infectivity of CCV.

Limitations in the reliability of using serum neutralizing activity as an indicator of the presence of CCV in catfish populations were noted by Amend and McDowell (1983). Sophisticated molecular biology was reported from French workers who used restriction endonuclease digestions for determination of biophysical properties of the DNA (Cebrian et al. 1983). And France hosted an international symposium on antigens of fish pathogens, the published papers of which included an authoritative and comprehensive review of CCVD (Plumb and Jezek 1983).

In the following year, CCV was included in a review paper on immunization of warm-water fish (Plumb 1984), and the pros and cons of using serum neutralizing activity as an indicator of prior exposure to CCV were evaluated (Amend and McDowell 1984). At the University of Hawaii, various microcarrier beads for growing BB cells and (in turn) CCV were compared by Buck and Loh (1985). They found growth on specific beads to be superior to that in stationary cultures. Plumb et al. (1985) continued to investigate aspects of the disease; they tested the sheatfish or European catfish (*Siluris glanis*) and found it to be resistant to CCV.

The year 1985 was particularly noteworthy in that breakthroughs were made in detection of CCV DNA in adult fish and in isolating the virus. Using a recombinant DNA technique, Wise and Boyle (1985) demonstrated CCV DNA in yearling channel catfish that had been injected with the virus 2 to 4 weeks earlier. Next, virus DNA was detected in two populations of adults—one that had survived an epizootic, and another with no history of the infection (Wise et al. 1985). Bowser et al. (1985) isolated the virus from brood stock. Isolations were made from 3 of 21 adults by inoculating CCO cells with tissue homogenate, but one blind passage was necessary before a cytopathic effect (CPE) was evident. More significant results were obtained when leukocytes were taken from immunosuppressed brood stock and co-cultivated on CCO cultures; with that approach, virus was isolated from all 7 fish examined. However, other investigators have found it difficult to reproduce the induction of virus with corticosteroid (R. P. Hedrick, personal communication, 1986).

Signs and Pathologic Changes

An abrupt rise in mortality of young-of-the-year channel catfish, usually when water temperatures are 25°C or higher, is an early indication of CCVD. The onset of the disease is sudden and its course acute; an outbreak can claim most of a population within 7 to 10 days. The disease itself is best characterized as a hemorrhagic viremia.

Behavior

Victim fish are clearly distressed; some may maintain a head-high posture at the surface, especially at pond or trough edges. Fijan et al. (1970) reported that the head-

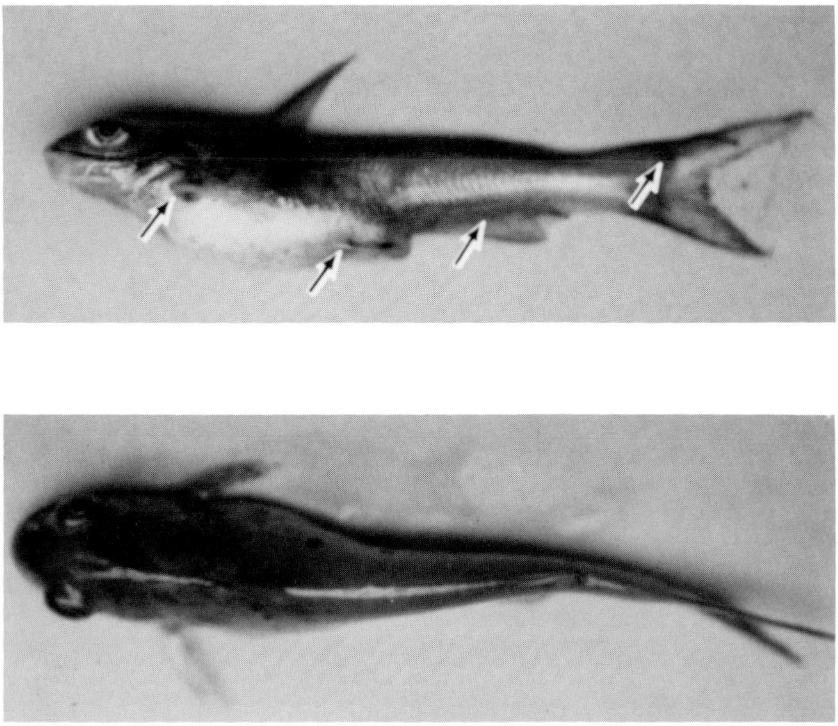

Figure 1. External appearance of channel catfish fingerlings with channel catfish virus disease. *Top.* Lateral view showing distended abdomen and urogenitial vent; arrows indicate areas of hemorrhage at the fin bases. *Bottom.* Dorsal view showing pronounced abdominal distention and exophthalmia. From Fijan et al. (1970).

high or hanging posture occurred in 20 to 50% of the epizootics, and that it was a specific sign. There is reason to question the specificity of the abnormal posture, for it has occurred in epizootics among dying fry from which virus could not be isolated, and it may be a response to stress.

Channel catfish virus infection among young fish is associated with convulsive swimming, often in spirals. Such hyperactivity can be triggered by feeding or alarm reactions. At termination, victims become lethargic and then prostrate.

External Signs

Affected animals show signs of a hemorrhagic disease and kidney dysfunction. The abdomen is swollen and in some the vent may be distended (Fig. 1). Fin bases—especially those of the ventral fins—abdomen, and caudal peduncle are typically hemorrhagic. Such fish develop exophthalmia, and some have shallow, yellowish external lesions indicating secondary invasion by *Flexibacter columnaris* or *Aeromonas hydrophila*. Gills are usually pale, but some victims may show brachial hemorrhages. Gill chambers of moribund fry may also show a high incidence of infestation with aquatic phycomycetes.

Internal Signs

The peritoneal cavity typically contains yellowish fluid, sometimes tinged with red. The visceral mass is generally pale, and that impression is enhanced by the absence of food in the digestive tract. The intestine may contain yellowish mucoid material. The spleen is usually enlarged and dark and the kidneys and liver are hemorrhagic or flecked with petechiae.

Histopathologic Findings

Histologically, the victims of CCVD show a severe hemorrhagic disease, with generalized edema and marked necrotic changes in several vital organs (listed here in the order of decreasing damage): kidneys, liver, gastrointestinal tract, spleen, skeletal muscle, neural tissue, and pancreas. All of five reports on the histopathologic changes of CCVD are in essential agreement. Wolf et al. (1972) described findings in experimentally infected specimens; Plumb et al. (1974) compared changes as detected by light and electron microscopy; Plumb and Gaines (1975) described a sequence of changes that they correlated with virologic quantification; and Major et al. (1975) described the changes in both experimental and natural infections (this work was the first in which neural and pancreatic damage was noted). Hedrick et al. (1987) documented the pathologic changes that occurred in an adult channel catfish that died following waterborne infection with CCV. Yasutake (1975) gave a brief review of CCVD's histopathologic highlights, and a comparison with the histopathology evoked by the salmonid viruses.

Sections of kidneys generally show the effects of systemic edema and widespread necrosis (Fig. 2). Necrosis occurs in hematopoietic tissue, in tubules, and in a few renal corpuscles. Areas of macrophage concentration occur in hematopoietic tissue. Plumb and Gaines (1975) detected increased lymphoid cells as early as 6 hours postinoculation, and a peak at 36 hours. Hemorrhage and pyknotic tubule cell nuclei were seen at 18 hours, and hematopoietic necrosis, first found at 48 hours, was massive by 96 hours. As might be expected, peak virus titers coincided with the appearance of greatest tissue damage.

Changes in the liver consist of edema, necrosis, vascular congestion, and some hemorrhaging; Major et al. (1975) also reported occasional intracytoplasmic eosinophilic inclusions. Plumb and Gaines (1975) found that the progress of damage in the liver was slower than that in the kidneys. Focal necrosis was found at 36 hours and was extensive by 72 hours; hemorrhage was generalized by 96 hours. Infectivity peaked in liver tissue at 120 hours—about 24 hours later than in kidney tissues.

The gastrointestinal tract was uniformly edematous and had focal areas of macrophage concentration; some specimens showed necrosis and sloughing of the mucosa or submucosa or both. Hemorrhaging was absent in the submucosa of

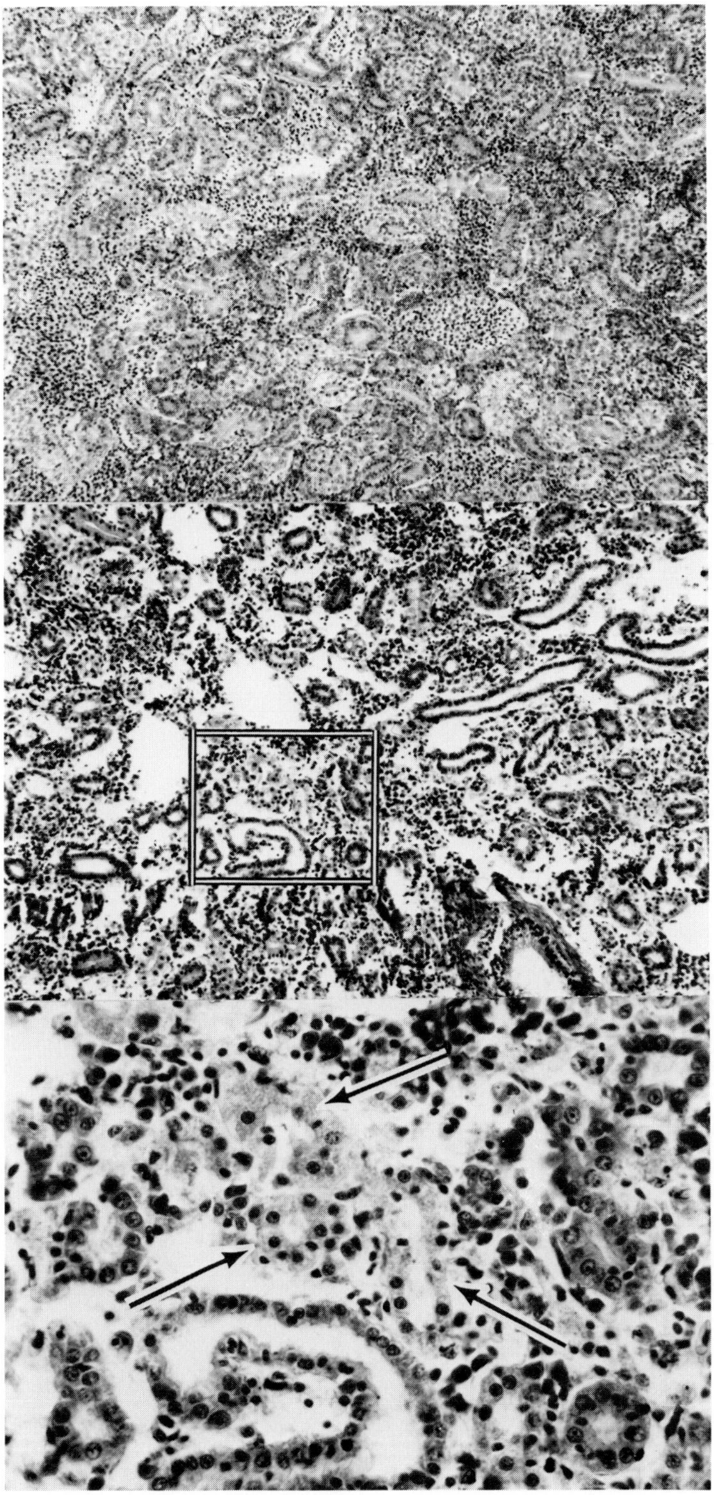

Figure 2. Renal pathology in channel catfish virus disease. *Top.* Normal catfish kidney section showing hematopoietic and excretory tissues. ×65. H & E. *Center.* Kidney of a catfish with channel catfish virus disease showing extensive edema and necrosis of tubules. ×65. H & E. *Bottom* (insert from center panel). Necrotic kidney tubules (arrows) showing pyknotic nuclei and cellular lysis. ×260. H & E. From Wolf et al. (1972). Reprinted with the permission of the Fisheries Research Board of Canada.

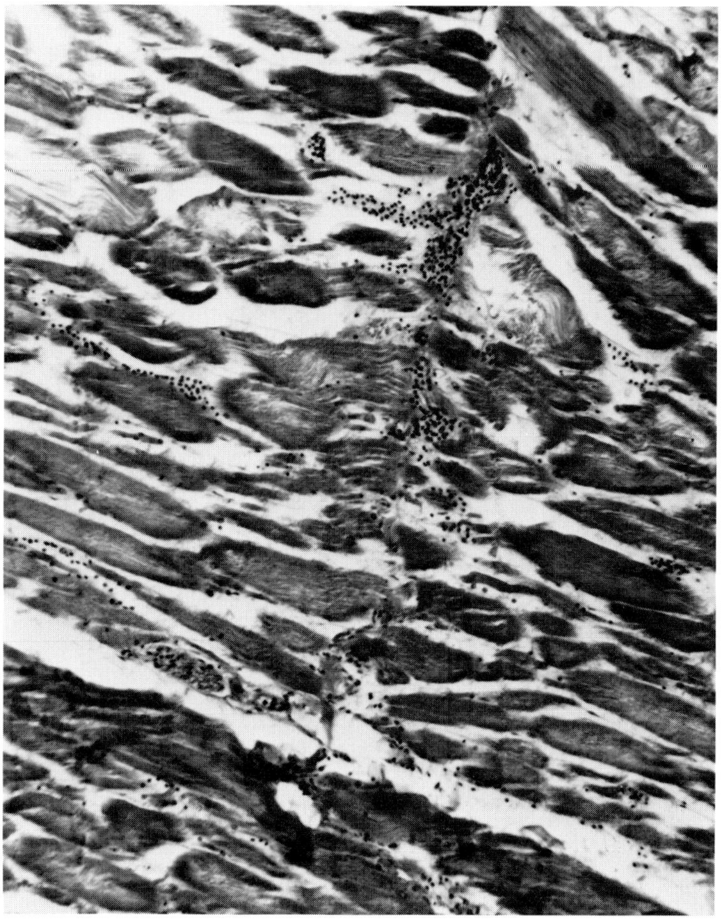

Figure 3. Focal hemorrhage in the skeletal muscle of a catfish with channel catfish virus disease. ×75. H & E. From Wolf et al. (1972). Reprinted with the permission of the Fisheries Research Board of Canada.

some specimens, severe in the submucosa and villi of others (to the point of leaking into the lumen in some), and absent in others. Hemorrhages were found as early as 18 hours after infection and were massive at 72 hours. Viral infectivity peaked at 72 hours, and high titers persisted to 120 hours.

The spleen was congested, and sometimes hemorrhagic, and some specimens showed mild necrosis of hematopoietic tissue. Lymphoid tissue was reduced and macrophages were engorged with the debris of erythrocytes. Viral infectivity peaked at 96 hours postinoculation and again at 168 hours.

Whereas the pattern of disease development in the kidneys, liver, intestine, and spleen was fairly uniform throughout the several studies, other findings were not consistent. Skeletal muscle hemorrhage (Fig. 3) was pronounced in the study by Wolf et al. (1972), seldom evident in the study by Major et al. (1975), and not found by Plumb and Gaines (1975). However, the occurrence of microscopic skeletal hemorrhage is consistent with macroscopic hemorrhages at fin bases. Pancreatic necrosis,

Table 1. Quantification of virus in selected tissues of fingerling channel catfish with experimentally transmitted CCVD

	Condition of fish		
Tissue	Dead[a]	Living[b]	Living[c]
Kidney	5.68	6.25	5.77
Intestine	5.13	5.75	5.20
Liver	3.89	5.50	≥5.71
Spleen	ND	4.50	ND
Blood	ND	4.50	ND
Brain	≥3.00	4.50	≥5.00
Muscle	ND	ND	3.28

Note: Infectivity expressed as exponent of $\log_{10}$ $TCID_{50}$/g; ND, no data.

[a]Adapted from Plumb (1971b).

[b]Peak infectivity values from Plumb (1972) and Plumb and Gaines (1975).

[c]Adapted from Plumb (1971b), based on values for the day on which peak infectivity was found.

neural nuclear vacuolation, and edema surrounding nerve fibers were first reported by Major et al. (1975).

Clinical Findings

Hematologic and other clinical characteristics have not been reported for CCVD; the small size of the usual victim fish and the acute course of the infection undoubtedly discourage the collection of data.

The average amount of virus occurring at peak times in selected tissues of experimentally infected fingerlings has been determined, and for the most part the values coincide with the histopathologic changes that have been noted (Table 1). Brain tissue, in which there was great variability in the amount of virus present and in the amount of alteration that occurred, is an exception.

McGlamery and Gratzek (1974) used controlled procedures to test the observation that survivors of CCV were stunted. They injected 2-g fingerlings with about 1000 plaque-forming units (PFU) CCV; controls received only cell culture medium. Fish surviving to the 18th day were held on a standardized feeding regimen for 6 months; the lengths and weights of 42 fish were then compared with those of 42 randomly chosen controls. Survivors averaged only two-thirds the length and only one-seventh the weight of the controls.

Etiology

The CCV is an acknowledged member of the Herpesviridae, and is provisionally designated *Ictalurid herpesvirus* 1. It is on deposit with the American Type Culture Collection as VR-665.

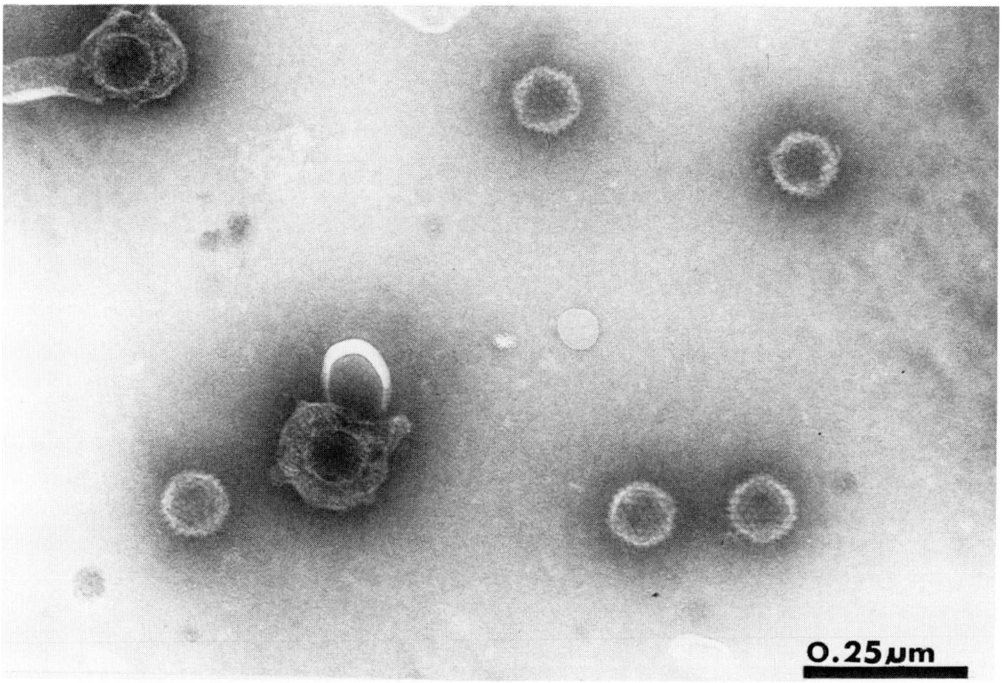

Figure 4. Negatively stained channel catfish virus particles. ×90,000. From Wolf and Darlington (1971). Reprinted with the permission of the American Society for Microbiology.

Size and Shape

The average dimension of enveloped virions is 175 to 200 nm. Negatively stained preparations show a mean capsid diameter of about 100 nm (Fig. 4). Infectivity generally passes membranes having a mean pore diameter of 220 nm but is retained by those with a pore diameter of 100 nm.

Biophysical Properties

The virus is ether sensitive, loses infectivity in glycerol, does not hemagglutinate, requires an envelope for infectivity, and has a high degree of host cell specificity.

Freezing and thawing, commonly used to release other viruses, is contraindicated for CCV. Three cycles of freezing at −20°C each resulted in a loss of one-fourth to one-half of the infectivity.

For storage, infectivity of culture grown virus is maintained longest at −75°C or lower in medium containing 10% serum and having a pH of 7.6 to 8.0. An acid pH is decidedly deleterious (Robin and Rodrigue 1980a). At −20°C, infectivity was gradually lost and virus could not be recovered from victim fry after 1 year of storage. The response of CCV to lyophilization has not been quantified.

As judged from simulation trials, the persistence of CCV under farm pond conditions is brief. Plumb (1978) found that CCV survived only about 2 days in pond water

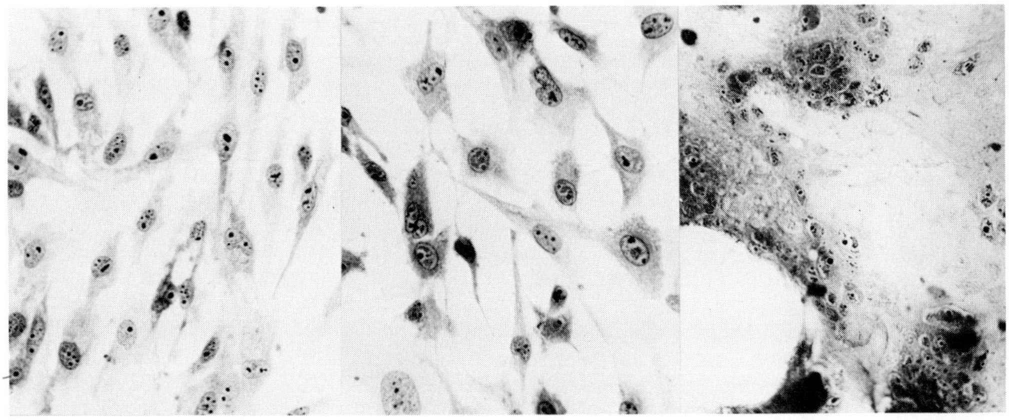

Figure 5. Effects of channel catfish virus in BB cells at 30°C. *Left.* Uninoculated control. *Center.* Margination of chromatin, Cowdry type A intranuclear inclusions, and the beginning of syncytium formation occur several hours after inoculation. *Right.* Karyorrhexis and contraction of syncytia occur after about 12 hours of incubation. ×180. From Wolf (1973). Reprinted with the permission of Acadamic Press, Inc.

at 25°C, but about 11 days in dechlorinated tap water. At 4°C, virus persisted in pond water for nearly 1 month and in tap water for nearly 2 months. Pond mud itself rapidly inactivated the virus, as did drying on concrete and netting. In an investigation of the effects of several environmental factors on CCV, Brady and Ellender (1982) found that soil sediment rapidly and virtually inseparably adsorbed the virus. That finding helps explain the inactivation by pond mud. As could be expected, ultraviolet irradiation inactivates the virus. Lee et al. (1978) found a half-life of only 7 s, but did not cite the energy level. Robin and Rodrigue (1980a) used 24 ergs $\cdot$ mm^{-2} $\cdot$ s^{-1}, found that infectivity was reduced 90% in 20 minutes.

Gerba et al. (1979) reported that CCV was photoinactivated by as little as 0.1 ppm of methylene blue in the presence of visible light; however, some virus was also inactivated in the presence of light without the dye. Farm-scale applications of photoinactivation with methylene blue are not likely.

Host cell specificity is marked: CCV is replicated only in cells of ictalurid and clariid fishes—members of closely related families. All susceptible cultures show syncytium development, followed by pyknosis and lysis (Fig. 5).

The cultures used most for CCV are the ictalurid lines BB and CCO. The clariid line K1K has been used in attenuating CCV, and primary cultures of channel catfish cells find limited use.

Maximal titer (PFU/mL) in conventional stationary culture is 10^7 to $10^{7.5}$ in BB cells, about 10^8 in CCO cells, and 10^5 to $10^{5.5}$ in the K1K line. Buck and Loh (1985), who grew BB cells on microcarrier beads in spinner cultures, realized "modest but distinct advantages" in virus yield over that in their stationary cultures.

With minor variability that reflects temperature tolerances of susceptible cell cultures, CCV is replicated at 10 to 35°C. As judged by rapidity of growth and yield, optimal temperatures are 25 to 30°C.

The time sequence of CCV in the BB line at 30°C was described by Wolf and Darlington (1971): at 2 hours postinoculation, pyknosis was evident, nuclear chro-

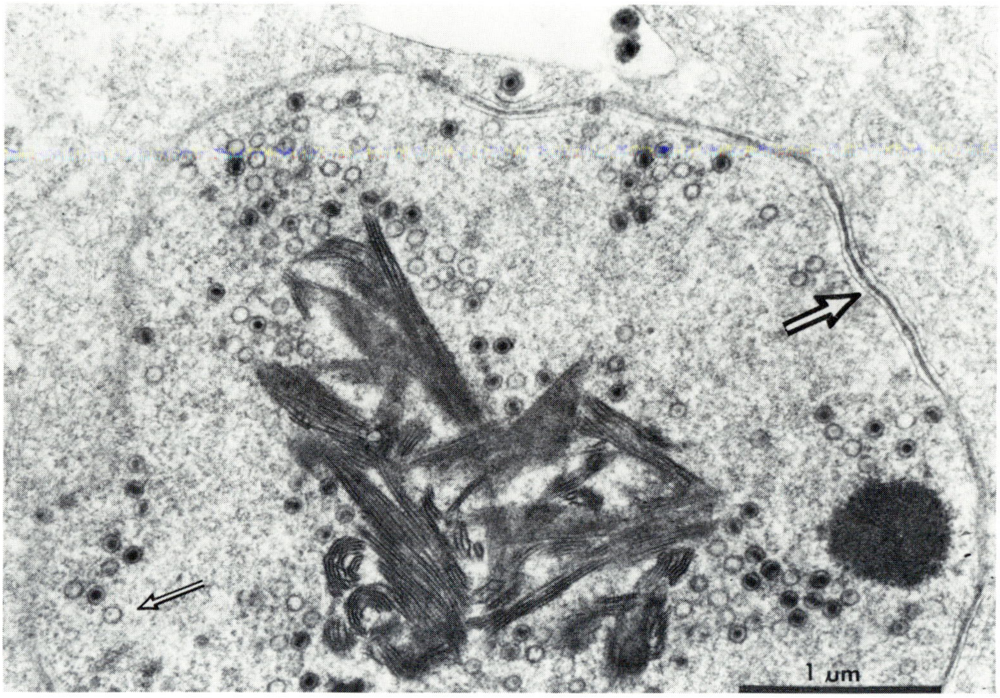

Figure 6. Brown bullhead cell 10 hours after infection with channel catfish virus at 25°C. Extensive reduplication is shown in the nuclear membrane (large arrow), and in some areas the membrane has been disrupted (smaller arrow). Dense lamellar structures occupy a large portion of the nucleus. ×25,800. From Wolf and Darlington (1971). Reprinted with the permission of the American Society for Microbiology.

matin was marginated, and a few cells were coalesced; during hour 3, Cowdry type A intranuclear inclusions appeared and syncytia contained many nuclei; by hour 6, lysis had begun and nuclear contents in the cytoplasm were abnormally basophilic; and about hour 8, the extensive syncytia began to contract, nuclear disintegration was marked, and loss of cytoplasm had begun. Degenerative changes continued for several hours, but nucleoli persisted, apparently unaffected.

While this time sequence was being observed, the authors monitored the changes by electron microscopy. Infected BB cells showed margination of chromatin but no virus until hour 4. Most nuclei then showed virus in various stages of development, particularly in association with a large granular body. Nuclei also showed extremely electron-dense lamellar structures of uncertain nature. Unenveloped particles occurred in the cytoplasm by hour 5, and portions of nuclear membranes had become indistinct. Enveloped virus was evident by hour 10 (Fig. 6). This time coincided with the end of exponential production of titratable viral infectivity.

The age and density of cells seem not to be significant in determining their susceptibility given enough time at favorable temperature, the virus typically destroys the entire cell population. Thus far, the development of resistant cells and an in vitro carrier state has not been reported for CCV.

Growth curve studies have been conducted in BB cells (Wolf and Darlington 1971),

CCO cells (Bowser and Plumb 1980a), and K1K cells (Walczak et al. 1981). The BB and CCO lines supported little growth at 10°C, the average yield from BB cells being less than 10 times the input after 2 weeks. At 16°C in BB cells, a growth curve required about 60 hours and yield was about 50 times the input. Results at 25 to 30°C were similar for BB and CCO cells and the yield was 1000 or more times the input. Replication was most rapid at supraoptimal temperatures for the cells, but yields of virus were reduced. In a typical one-step growth curve in BB cells at 30°C, a 4-hour lag phase preceded the development of new infectivity. Thereafter, replication was exponential and a plateau was approached at about hour 10. Cell-associated virus slightly exceeded released virus in BB and CCO cells, but released virus predominated in K1K cells.

Channel catfish virus has been concentrated and purified by a number of procedures. Culture-grown virus can be pelleted by centrifugation at about 45,000 × g for 30 minutes at 4°C. When the virus is resuspended and put on a 30 to 50% w/v sucrose gradient for 30 minutes at 50,000 × g, the released and cell-associated virus separate. Released virus is present as a finely homogeneous layer of naked and enveloped particles above a coarsely flocculent lower layer of cell-associated virus. Chousterman et al. (1979) sedimented the cell-associated fraction of culture-grown virus, and then pelleted the released fraction and purified it on a cesium chloride step gradient. The virus band was collected and dialyzed. Robin and Rodrigue (1978), using a combination of precipitation with 7% polyethylene glycol 6000 followed by isopycnic centrifugation in a 15 to 45% metrizamide gradient, recovered only about 30% of the initial infectivity, but the purity was 99.9% or higher. Electron microscopy showed almost exclusively intact enveloped virus.

Studies of the nucleic acid of CCV have progressed beyond the presumptive identification of DNA that was based on uptake of thymidine by infected cells and inhibition by base analogues. The buoyant density of CCV DNA of about 1.715 g/mL corresponds to a base composition of 56% (Goodheart and Plummer 1975; Robin and Rodrigue 1980b). General agreement exists that the molecular weight is about 85 $\times 10^6$ (Cebrian et al. 1983; Chousterman et al. 1979; Dixon and Farber 1980; Granoff and Naegele 1978; Robin and Rodrigue 1980b). Details of the molecular structure were reported by Chousterman et al. (1979) who published a physical map of restriction endonuclease sites. Dixon and Farber (1980) confirmed major attributes of the genome and reported the presence of 32 virus-specific polypeptides, of which 18 were structural. Further, they assigned the polypeptides to three classes, in order of their appearance during synthesis. The molecular weight of the polypeptides ranged from 12,000 to 300,000.

Cebrian et al. (1983) used Hite selective extraction and restriction endonuclease digestions of intracellular DNA, and electrophoresed and mapped the products. They found evidence of concatenate fusion that resulted in "endless" viral DNA, which preceded the appearance of unit-length molecules.

Diagnosis

Channel catfish virus disease can be suspected if high mortality, with accompanying behavior and signs of CCVD, occurs among fry or fingerlings during seasonally warm to hot weather. Specimens should be examined carefully to eliminate the possibility that bacteria or parasites are the cause. Accuracy in diagnosis may be increased by histologic examination and the finding of pathologic changes that accompany CCVD.

Viral isolation and identification, either presumptive or serologic, is simple and relatively quick, and recommended to ensure accuracy.

Isolation

Virus is readily isolated from victim fry or fingerlings taken during an epizootic, but at most for only a few days afterward.

The BB or CCO cell lines, or primary cell cultures of channel catfish origin, are suggested for isolation. Inocula are prepared from pooled suspect specimens and decontaminated with antibiotics or by filtration. Reliability is greatest when the inocula are adsorbed on drained preformed cell sheets or are added to cell suspensions before they are seeded. Incubation should be at 22° to 30°C.

Typical epizootic material has a titer of about 10^6 PFU/g of victim fry or fingerlings. The question is, how long will virus be recoverable? Plumb et al. (1973) addressed this problem and obtained results with practical applications. Infected fingerlings were held in water at 22°C, on ice, and frozen at −20°C and at −80°C. Loss of infectivity was 99.99% within 48 hours at 22°C; in iced specimens, 90% had been lost by day 3, and 99.99% by day 14. Loss was least in frozen specimens, in which infectivity was 1% at 50 days and was not completely lost for about 5 months.

Bowser et al. (1985) described the first successful isolation of CCV from channel catfish brood stock. Key factors in the success seem to be, first, that the population had sustained mortality during winter (water temperature was 8 to 14°C) and, second, tissue extracts and leukocytes were co-cultivated and blind passaged. Third, and most important, leukocytes were taken from fish that had been immunosuppressed.

Virus was isolated from 3 of 21 adults when filtrates of organ homogenates or of leukocyte suspensions were inoculated into cultures of CCO cells and incubated at 30°C. However, CPE did not appear until material was in first blind passage, and elapsed time was about 3 weeks. Seven fish were injected intramuscularly with 0.55 mg/kg Dexamethazone, a synthetic analogue of cortisol, and 7 days later leukocytes were separated by gradient centrifugation in Histopaque 1077, washed in phosphate-buffered saline (PBS), and co-cultivated with CCO cells. In incubation at 30°C, CCV was isolated from all seven immunosuppressed adults (three males and four females).

Identification

A presumptive but nevertheless highly accurate identification can be made if suspect inocula produce syncytia in cultures of ictalurid or clariid cells but not in cultures of nonictalurid cells.

Positive identification is obtained either by the serum neutralization test or fluorescent antibody technique. Herpesviruses are generally poorly antigenic, and thus far the best rabbit anti-CCV serum effects only a 50% plaque reduction of 100 PFU at 1:1000 dilution. A nonneutralizable persistent fraction of virus commonly present among herpesviruses probably accounts for the less than complete neutralization, even by serum at low dilutions.

Plumb et al. (1981), who used an indirect fluorescent antibody technique on frozen sections of ovarian tissue from two immunosuppressed catfish, found focal areas of fluorescence. First-transfer cell cultures also showed focal fluorescence. The authors considered the fluorescence to be specific, and their failure to isolate or to visualize virus was attributed to the virus possibly being incomplete.

A recombinant DNA technique has been used to identify the genome of CCV in experimentally infected yearling catfish and in naturally infected brood stock. Wise and Boyle (1985) developed a nucleic acid probe by digesting CCV DNA with the restriction enzyme EcoRI and ligating it with the DNA of an *Escherichia coli* plasmid. The probe identified virus DNA in several tissues of 7 of 11 yearlings that had been injected with CCV 2 weeks earlier. Moreover, it did not bind with DNA from channel catfish or from the channel catfish cell line CCO. Wise et al. (1985) used the DNA probe and identified CCV DNA in the liver of all 22 brood stock catfish from a population with no history of CCVD. From a second population of adult channel catfish, the probe showed the presence of CCV DNA in one or more soft tissues of 7 of 10 fish and in erythrocytes of 4 brood stock that were examined.

Transmission and Incubation

Under farm pond conditions during summer and the usual crowding of young susceptible fish, horizontal transmission and short incubation times are readily demonstrated after contact exposure. Exposed fry begin dying within 3 days and, as in natural outbreaks, most if not all susceptibles die within 7 to 10 days after the first mortality.

When water temperatures are relatively low (20 to 25°C), the fish are favored and contact exposure may result in only transient appearance of disease signs. Incubation time may be a week or more and mortality much lower than at high water temperature.

A wealth of circumstantial evidence exists for vertical transmission.

At 30°C, experimental infections may be transmitted by several means: infective material may be brushed on gills, fed, injected, or sometimes simply added to water. An hour's exposure of susceptibles in water containing culture-grown virus usually results in transmission and at least high—and often complete—mortality. For unknown reasons, some attempts at bath transmission have failed completely, and injection was required (McConnell and Austen 1978; Wolf et al. 1972).

Intramuscular or intraperitoneal injection of virus produces a peracute disease course. Small fingerlings that receive 5×10^5 PFU and are held at 27 to 30°C may die in less than 36 hours, and the last fish seldom live beyond the 5th to 7th day.

Subadult animals with no prior exposure to CCV have sometimes been experimentally infected and killed. Titration of tissues from such victims has shown unequivocal evidence of extensive virus replication. The thriftiness of subadult fish used in such trials, although not quantified, declined. The fish seemed incapable of normal growth and weight gain after the injection of virus.

It has been reported that natural outbreaks of CCVD have followed "some kind of stress"; however, the ease with which the virus can be experimentally transmitted suggests that stress is not an absolute requirement.

Source

Ever since CCV was recognized, some populations of brood stock have been circumstantially implicated as prime sources of the virus and, over the years, considerable supporting evidence has accumulated. Nevertheless, traditional virologic methods have failed to isolate CCV from suspect adult channel catfish. Recently, Bowser et al. (1985), using previously untried methods, isolated CCV from adult channel catfish and demonstrated that indeed brood stock were the prime source of virus. The new findings confirmed items of presumptive evidence, such as observations that epizootics occurred with regularity among progeny of certain populations of brood stock, that such brood stock usually had higher levels of CCV-neutralizing antibody than found in adults with no history of the disease, and that, when indirect fluorescent antibody technique was used, frozen sections of tissues from suspect adults showed apparently specific viral fluorescence (Plumb et al. 1981).

Host and Geographic Range

The channel catfish is the only species known to sustain natural outbreaks of CCVD. Fingerling blue catfish and channel catfish × blue catfish hybrids are susceptible to injected virus, but not to virus delivered orally or by immersion (Plumb and Chappell 1978). Brown bullheads (*Ictalurus nebulosus*) and yellow bullheads (*I. natalis*) are resistant even to injection of the virus. Galla and Hartmann (1974) reported that the walking catfish was susceptible to injected CCV, however, the finding has not been confirmed. The sheatfish has been found to be resistant to injected CCV (Plumb et al. 1985). In typical herpesvirus manner, CCV shows a high degree of host specificity.

The survival of 50 to 60% of some populations of susceptible channel catfish with CCVD prompted speculation about genetic factors of resistance. Plumb et al (1975) found marked differences in survival between six inbred and two outbred strains of catfish. The fish were held at about 28°C, fed virus on 3 consecutive days, and observed for an additional 10 days. Survival ranged from a high of 90% for two outbred strains to a low of only 29% for a wild, inbred (Falcon) strain from Texas (Table 2).

The geographic distribution of virologically documented CCVD coincides with that of principal catfish production in the United States: northwest to Colorado, southwest to California, southeast to Florida, and northeast to West Virginia. The virus was accidentally introduced to Honduras, apparently with infected fry. How-

Table 2. Comparative susceptibility of eight strains of channel catfish fingerlings to challenge with channel catfish virus disease

Strain designation	Origin	Percent survival[a]
Falcon	Wild, Texas	29
Kentucky	Wild, Kentucky	56[b]*
Marion	Domestic, Alabama	68*†
Warrior	Wild, Alabama	72*†
ennessee	Wild, Kentucky	88†‡
Yazoo	Semidomestic	88‡
Warrior × Yazoo	Hybrid	90‡
Yazoo × Tennessee	Hybrid	90‡

Source: Adapted from Plumb et al. (1975).

[a]The difference between the percent mortality of strains sharing the same symbol is not statistically significant.

[b]Extrapolated.

ever, similar exportation elsewhere has not occurred because health histories of source brood stock were carefully checked beforehand.

Immunity

The humoral immune response to CCV varies considerably in both young and adult catfish. Some young that survive an infection produce measurable amounts of serum neutralizing activity, and such activity can be found in serum from some but not all adults that have a history of the disease. Hedrick and McDowell (1987) have passively immunized fingerling catfish by injecting adult serum that contains CCV neutralizing antibody.

Some survivors of CCVD seem not to produce serum neutralizing activity—or do so only in amounts that are below present threshold levels of detection. Amend and McDowell (1983) were unable to show neutralizing activity in serum from one catfish population that had survived an epizootic of low severity (0.1% mortality) 3 months earlier; however, neutralizing activity was demonstrable in serum from a second population consisting of fish that had sustained a 5% mortality. Although the nature of the response—humoral or cellular—is not known, fish that are given adequate initial and follow-up exposures to attenuated vaccine are protected against virulent CCV (Walczak et al. 1981).

Variability in the immune response of fish to CCV is clearly evident in results obtained from work with adults. In studies by Plumb (1973b) and Heartwell (1975), conducted at about 28°C, the primary response of adult fish to an initial injection of live CCV was rapid but highly variable. As in the young, some adults produced little or no virus-neutralizing activity, whereas others produced significant amounts. In both studies, activity peaked at about 9 weeks and then began to decline. At that time, a second injection evoked a secondary response, but it was short-lived, lasting only about 4 weeks. The fish studied by Heartwell, who used a small amount of virus, gave a weak secondary response; however, those studied by Plumb, who used a much larger quantity of virus, yielded a strong secondary response. None of Plumb's fish produced a primary response to heat-killed virus.

Variability in the response of adults to experimental CCV infection helps explain the wide range of virus-neutralizing activity found in hatchery brood stocks that are implicated in CCVD outbreaks (Plumb 1973b). The opposite, unfortunately, is not true; one cannot infer that absence of neutralizing activity indicates that exposure to CCV has not occurred. Moreover, Huston (1979) found CCV-neutralizing activity in some individuals from a stock whose health history included no known exposure to the virus.

Temperature change can affect the immune response to CCV. Hedrick et al. (1987) moved adults from 18°C to 27°C for 40 days and measured the change in CCV-neutralizing activity. They found a dramatic drop in activity level; in fact, a few adults no longer showed measurable activity.

Amend and McDowell (1984) addressed the need to detect neutralizing activity if it is present. To that end, they suggested testing serum at a 1:20 dilution and reacting it for 1 hour at 25°C against 100 $TCID_{50}$ virus. In addition, they recommended confirmation by immunodiffusion or some other method.

Heartwell (1975) used immunodiffusion, but virus concentration was required and the lines of precipitation were weak. Huston (1979) found that sera from some adults with no history of CCVD were reactive in immunodiffusion tests.

Control

Existing measures that are practical for control of CCVD include avoidance, the use of resistant strains, and possibly the use of resistant species hybrids. Additional measures that might apply, but which now have practical limitations, are vaccination, temperature manipulation, and chemotherapy.

Avoidance is the most practical approach, but it requires that sources of the virus be known. Regional fish health personnel should know of virus-free stock. Serologic examinations can be conducted, but existing limitations discussed in the preceding section on immunity should be kept in mind.

In situations where CCV is likely to be present or cannot be avoided, resistant strains of channel catfish should be considered—e.g., the Tennessee, Yazoo, Warrior × Yazoo, or Yazoo-Tennessee strains tested by Plumb et al. (1975) and listed in Table 2. Plumb and Chappell (1978) found hybrids of blue catfish × channel catfish to be resistant to waterborne infection with CCV. That finding is useful for application at production locations where virus is present; however, the market acceptability of such hybrids is a factor to be considered.

Vaccination holds great promise for providing protection against CCVD mortality, and an attenuated live virus has been developed and granted U.S. Patent No. 040,108 (Noga and Hartmann 1981). Under limited experimental conditions—wherein hyperosmotic infiltration was used and fry were given a booster exposure 3 weeks later—the vaccine provided significant protection against challenge with 5 LD_{50} administered by injection. Walczak et al. (1981) reported a survival of 97% among vaccinated fish but only 20% among nonvaccinated controls. The vaccine is not available commercially and has not been licensed by the U.S. Department of Agriculture. Requirements for such licensing are considerable. Basic questions need to be answered: Is the vaccine strain consistently recognizable from wild virus by plaque size and appearance? How stable is the low virulence characteristic of the vaccine strain? What immune response is evoked? Will the vaccine protect the progeny of adults suspected of being carriers in enzootic areas?

Temperature manipulation has proved effective under experimental conditions, but its practicality in large-scale commercial operations is questionable. Plumb (1973a) noted that mortality decreased significantly when water temperature was lowered from 28°C to 19°C or less; mortality was 98% at 28°C, but only 14% at 19°C, and it was lowered to 24% in infected fish that were moved from 28°C to 19°C.

Viral chemotherapy is a subject of growing interest in human and veterinary medicine, but clinical applications are still limited. Accordingly, realistic prospects for applications in fish are exceptionally dim. Nevertheless, in vitro work with CCV has been reported. Koment and Haines (1978), who tested phosphonacetic acid, found that suppression of CCV required 10 to 20 times more drug than was needed to suppress herpes simplex. Kimura et al. (1983) also found that the amount of phosphonacetic acid required for suppression of CCV in vitro (300 μg/mL) was larger than expected.

When epizootics occur, hygienic precautions may be helpful in preventing spread. In several instances, fish with CCV were killed and ponds treated with 20 to 50 ppm chlorine. Chlorine was allowed to dissipate and, after several days, susceptible small fingerlings were placed in a live car as a test for residual infectivity. There was no further evidence of CCV. It is to be understood, however, that water supplies to such ponds did not and should not harbor wild or feral fishes that might be a source of the virus. All indications are that thorough drying of ponds that have held infected fishes would be effective in inactivating CCV.

References

Amend, D. F., and T. McDowell. 1983. Current problems in the control of channel catfish virus. J. World Maricult. Soc. 14:261–267.

Amend, D. F., and T. McDowell. 1984. Comparison of various procedures to detect neutralizing antibody to the channel catfish virus in California brood channel catfish. Prog. Fish-Cult. 46:6–12.

Bowser, P. R. 1978. Development and evaluation of a new cell line from the channel catfish, *Ictalurus punctatus*. Ph.D. dissertation. Auburn Univ., Alabama.

Bowser, P. R., and J. A. Plumb. 1980a. Growth rates of a new cell line from channel catfish ovary and channel catfish virus replication at different temperatures. Can. J. Fish. Aquat. Sci. 37: 871–873.

Bowser, P. R., and J. A. Plumb. 1980b. Fish cell lines: establishment of a line from ovaries of channel catfish. In Vitro 16:365–368.

Bowser, P. R., and J. A. Plumb. 1980c. Channel catfish virus: comparative replication and sensitivity of cell lines from channel catfish ovary and the brown bullhead. J. Wildl. Dis. 16:451–454.

Bowser, P. R., A. D. Munson, H. H. Jarboe, R. Francis-Floyd, and P. R. Waterstrat. 1985. Isolation of channel catfish virus from channel catfish, *Ictalurus punctatus* (Rafinesque), brood stock. J. Fish Dis. 8:557–561.

Brady, Y. J., and R. D. Ellender. 1982. The role of sediment in transmission of channel catfish virus disease. Miss. Ala. Sea Grant Consortium Univ. S. Miss. 67:111.

Buck, C. D., and P. C. Loh. 1985. Growth of brown bullhead and other fish cell lines on microcarriers and the production of channel catfish virus. J. Virol. Methods 10:171–184.

Cebrian, J., D. Buccini, and P. Sheldrick. 1983. Endless viral DNA in cells infected with channel catfish virus. J. Virol. 46:405–412.

Chousterman, S., M. Lacasa, and P. Sheldrick. 1979. Physical map of the channel catfish virus genome: location of sites for restriction endonucleases *Eco*RI, *Hind*III, *Hpa*I, and *Xba*I. J. Virol. 31:73–85.

Dixon, R. A. F., and F. E. Farber. 1980. Channel catfish virus: physicochemical properties of the viral genome and identification of viral polypeptides. Virology 103:267–278.

Fijan, N. 1968. Progress report on acute mortality of channel catfish fingerlings caused by a virus. Bull. Off. Int. Epizoot. 69: 1167–1168.

Fijan, N. N., T. L. Wellborn, Jr., and J. P. Naftel. 1970. An acute viral disease of channel catfish. U.S. Fish Wildl. Serv., Tech. Paper 43.

Galla, J. D., and J. X. Hartmann. 1974. Extension of the host range of channel catfish virus (CCV) to the walking catfish, (*Clarias batrachus* L.). Fla. Sci. 37(Suppl. 1).

Gerba, C. P., R. A. F. Dixon, F. E. Farber, C. Wallis, and J. L. Melnick. 1979. Photodynamic inactivation of fish pathogens. Dev. Ind. Microbiol. 20:647–651.

Goodheart, C. R., and G. Plummer. 1975. The densities of herpesviral DNAs. Prog. Med. Virol. 19:324–352.

Granoff, A., and R. F. Naegele. 1978. The Lucke' tumor: a model for persistent virus infection and oncogenesis. Pages 15–25 in J. Stevens, G. J. Todaro, and C. F. Fox, eds. Persistent viruses: proceedings of the 1978 ICN-UCLA symposia on molecular and cellular biology,Keystone, Colorado, February 1978, Vol. 11. Academic Press, New York.

Gratzek, J. B., M. H. McGlamery, D. L. Dawe, and T. Scott. 1973. Microcultures of brown bullhead (*Ictalurus nebulosus*) cells: their use in quantitation of channel catfish (*Ictalurus punctatus*) virus and antibody. J. Fish. Res. Board Can. 30:1641–1645.

Heartwell, C. M., III. 1975. Immune response and antibody characterization of the channel catfish (*Ictalurus punctatus*) to a naturally pathogenic bacterium and virus. U.S. Fish Wildl. Serv., Tech. Paper 85.

Hedrick, R. P., and T. McDowell. 1987. Passive transfer of sera with antivirus neutralizing activity from adult channel catfish protects juveniles from channel catfish virus disease. Trans. Am. Fish. Soc. 116:277–281.

Hedrick, R. P., J. M. Groff, and T. McDowell. 1987. Response of adult channel catfish to waterborne exposures to channel catfish virus. Prog. Fish-Cult. 49:181–187.

Huston, M. 1979. Channel catfish virus disease: detection of latent virus and correlation with serology. Ph.D. dissertation. Texas A&M Univ., College Station.

Kimura, T., S. Suzuki, and M. Yoshimizu. 1983. In vitro antiviral effect of 9–(2–hydroxyethoxymethyl) guanine on the fish herpesvirus *Oncorhynchus masou* virus (OMV). Antiviral Res. 3:93–101.

Koment, R. W., and H. Haines. 1978. Decreased antiviral effect of phosphonacetic acid on the poikilothermic herpesvirus of channel catfish disease. Proc. Soc. Exp. Biol. Med. 159:21–24.

Lee, M. H., K. M. Tenno, and P. C. Loh. 1978. Some properties of the herpesvirus of channel catfish. Abstr. Annu. Meet. Am. Soc. Microbiol. S262:256.

McConnell, S., and J. D. Austen. 1978. Serologic screening of channel catfish virus. Mar. Fish. Rev. 40:30–32.

McCraren, J. P. 1972. Channel catfish virus disease (CCVD): a current review. Proc. Annu. Conf. West. Assoc. Game Fish Comm. 52:528–537.

McGlamery, M. H., Jr., and J. B. Gratzek. 1974. Stunting syndrome associated with young channel catfish that survived exposure to channel catfish virus. Prog. Fish-Cult. 36:38–41.

Major, R. D., J. P. McCraren, and C. E. Smith. 1975. Histopathological changes in channel catfish (*Ictalurus punctatus*) experimentally and naturally infected with channel catfish virus disease. J. Fish. Res. Board Can. 32: 563–567.

Noga, E. J. 1977. Studies on channel catfish virus passaged in cell cultures of the walking catfish. Ms.Sc. thesis. Florida Atlantic Univ., Boca Raton.

Noga, E. J., and J. X. Hartmann. 1981. Establishment of walking catfish (*Clarias batrachus*) cell lines and development of a channel catfish (*Ictalurus punctatus*) virus vaccine. Can. J. Fish. Aquat. Sci. 38:925–930.

Plumb, J. A. 1971a. Channel catfish virus research at Auburn University. Prog. Rep. Series No. 95, Agric. Exp. Station, Auburn Univ., Alabama.

Plumb, J. A. 1971b. Tissue distribution of channel catfish virus. J. Wildl. Dis. 7:213–216.

Plumb, J. A. 1971c. Channel catfish virus disease in southern United States. Proc. Annu. Conf. Southeast. Assoc. Game Fish Comm. 25:489–493.

Plumb, J. A. 1972. Some biological aspects of channel catfish virus disease. Ph.D. dissertation. Auburn Univ., Alabama.

Plumb, J. A. 1973a. Effects of temperature on mortality of fingerling channel catfish (*Ictalurus punctatus*) experimentally infected with channel catfish virus. J. Fish. Res. Board Can. 30:568–570.

Plumb, J. A. 1973b. Neutralization of channel catfish virus by serum of channel catfish. J. Wildl. Dis. 9:324–330.

Plumb, J. A. 1974. Channel catfish virus. Catfish Farmer World Aquacult. News 6:40–42.

Plumb, J. A. 1975. An 11-year summary of fish disease cases at the Southeastern Cooperative Fish Disease Laboratory. Proc. Annu. Conf. Southeast. Assoc. Game Fish Comm. 29:254–260.

Plumb, J. A. 1978. Epizootiology of channel catfish virus disease. Mar. Fish. Rev. 40:26–29.

Plumb, J. A. 1984. Immunization of warm water fish against five important fish pathogens. Pages 199–222 in P. de Kinkelin and C. Michel, eds. Symposium on fish vaccination. Theoretical background and practical results on immunization against infectious diseases, Paris, France, February 20–22, 1984. Off. Int. Epizoot., Paris, France.

Plumb, J. A., and J. Chappell. 1978. Susceptibility of blue catfish to channel catfish virus. Proc. Annu. Conf. Southeast. Assoc. Fish Wildl. Agencies 32:680–685.

Plumb, J. A., and J. L. Gaines, Jr. 1975. Channel catfish virus disease. Pages 287–302 in W. E. Ribelin and G. Migaki, eds. The pathology of fishes. Univ. Wisconsin Press, Madison.

Plumb, J. A., and D. A. Jezek. 1983. Channel catfish virus disease. Pages 33–49 in D. P. Anderson, M. Dorson, and P. Dubourget, eds. Antigens of fish pathogens. International Symposium, Talloires, France, May 10–12, 1982. Collection Fondation, Marcel Merieux, Lyon, France.

Plumb, J. A., L. D. Wright, and V. L. Jones. 1973. Survival of channel catfish virus in chilled, frozen, and decomposing channel catfish. Prog. Fish-Cult. 35:170–172.

Plumb, J. A., J. L. Gaines, E. C. Mora, and G. G. Bradley. 1974. Histopathology and electron microscopy of channel catfish virus in infected channel catfish, *Ictalurus punctatus* (Rafinesque). J. Fish Biol. 6:661–664.

Plumb, J. A., O. L. Green, R. O. Smitherman, and G. B. Pardue. 1975. Channel catfish virus experiments with different strains of channel catfish. Trans. Am. Fish. Soc. 104:140–143.

Plumb, J. A., R. L. Thune, and P. H. Klesius. 1981. Detection of channel catfish virus in adult fish. Dev. Biol. Stand. 49:29–34.

Plumb, J. A., V. Hilge, and E. E. Quinlan. 1985. Resistance of the European catfish (*Silurus glanis*) to channel catfish virus. J. Appl. Ichthyol. 1:87–89.

Robin J., and A. Rodrigue. 1978. Purification of channel catfish virus, a fish herpesvirus. Can. J. Microbiol. 24:1335–1338.

Robin, J., and A. Rodrigue. 1980a. Resistance of herpes channel catfish virus (HCCV) to temperature, pH, salinity, and ultraviolet irradiation. Rev. Can. Biol. 39:153–156.

Robin, J., and A. Rodrigue. 1980b. Isolation and preliminary characterization of herpes channel catfish virus DNA. Can. J. Microbiol. 26:130–134.

Walczak, E. M., E. J. Noga, and J. X. Hartmann. 1981. Properties of a vaccine for channel catfish virus disease and a method of administration. Dev. Biol. Stand. 49:419–429.

Wellborn, T. L. 1969. A viral disease of channel catfish. Pages 50–54 in Proceedings of the 1969 fish farming conference, Tex. Agric. Ext. Serv. Dep. Wildl. Sci. College Agric., Texas A&M Univ., College Station.

Wise, J. A., and J. A. Boyle. 1985. Detection of channel catfish virus in channel catfish, *Ictalurus punctatus* (Rafinesque): use of a nucleic acid probe. J. Fish Dis. 8:417–424.

Wise, J. A., P. R. Bowser, and J. A. Boyle. 1985. Detection of channel catfish virus in asymptomatic adult channel catfish, *Ictalurus punctatus* (Rafinesque). J. Fish Dis. 8:485–493.

Wolf, K. 1973. Herpesviruses of lower vertebrates. Pages 495–520 in A. S. Kaplan, ed. The herpesviruses. Academic Press, New York.

Wolf, K., and R. W. Darlington. 1971. Channel catfish virus: a new herpesvirus of ictalurid fish. J. Virol. 8:525–533.

Wolf, K., R. L. Herman, and C. P. Carlson. 1972. Fish viruses: histopathologic changes associated with experimental channel catfish virus disease. J. Fish. Res. Board Can. 29:149–150.

Yasutake, W. T. 1975. Fish viral diseases: clinical, histopathological, and comparative aspects. Pages 247–269 in W. E. Ribelin and G. Migaki, eds. The pathology of fishes. Univ. Wisconsin Press, Madison.

2 Cichlid Virus

Synonym: Ramirez' dwarf cichlid virus

Definition

Ramirez' dwarf cichlid virus is a viral entity that has been visualized in the spleen of the South American tropical fish *Apistogramma ramirezi*, which was a victim of an acute disease with high-morbidity. Other pathogens were not detected, and although isolation was not attempted, the virus is considered to be the etiologic agent of the lethal disease (Leibovitz and Riis 1980a, b).

History

Nine cases, each consisting of three to five young adult specimens, were submitted for examination. The specimens came from five separate shipments. They were necropsied by Leibovitz and Riis (1980a, b) within 1 to 3 days after importation to North America. Additional cases have not been reported.

Signs and Pathologic Changes

Behavior and External Signs

Specimens submitted were inappetent and pale, and showed respiratory distress and uncoordinated swimming interpreted as weakness. Hemorrhages were present in the skin and iris, and, as the disease progressed, degenerative changes in the eye became pronounced and the iris developed an irregular outline. The transient scoliosis that occurred was attributed to muscle spasms. Before death, fish were listless and respiration was slow and shallow. The disease lasted 3 to 4 weeks; morbidity was 100% and mortality 40 to 80%.

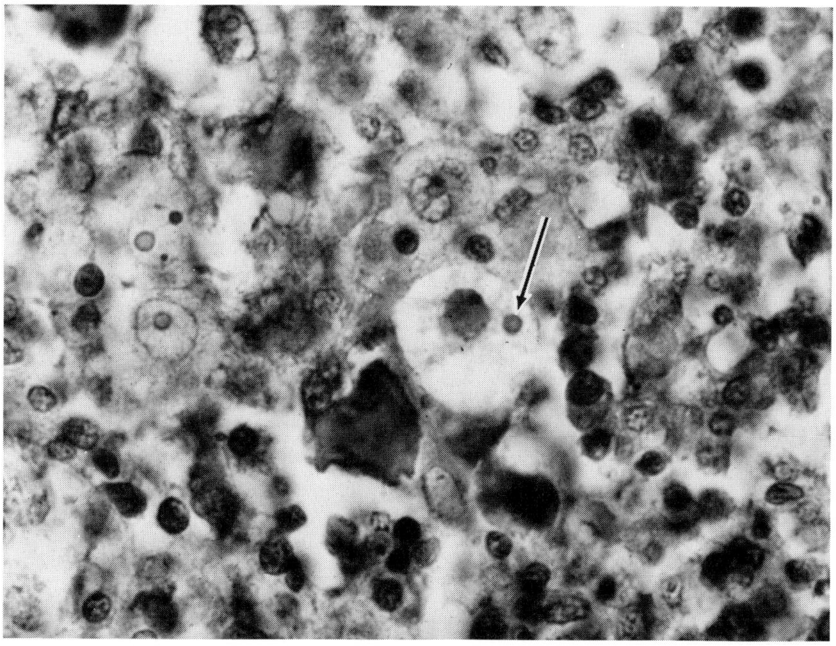

Figure 1. Section through the spleen of an infected specimen. Cellular degeneration and swelling are evident. Late-stage infection shows eosinophilic cytoplasmic inclusion (arrow). Courtesy of L. Leibovitz and R. C. Riis.

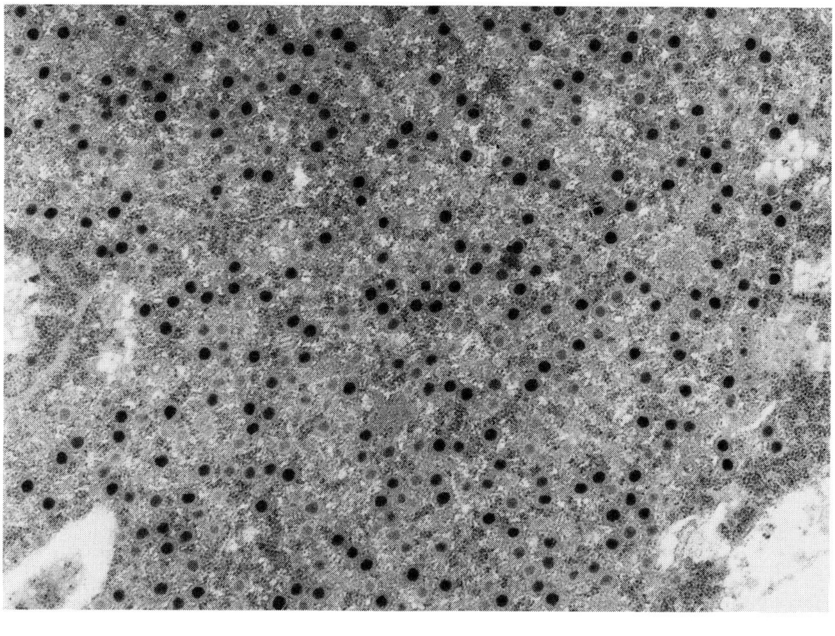

Figure 2. Abundant cytoplasmic virions in spleen cells during an early stage of infection. Particle size is about 120 nm. From Leibovitz and Riis (1980a). Reprinted with the permission of the American Veterinary Medical Association.

Internal Signs

Internally, the viscera were contracted and displaced anteriorly. The liver and kidneys were smaller than normal and pale, but the spleen, though also pale, was 3 to 10 times larger than normal. Organs of the digestive tract were shrunken and devoid of food, but contained a clear mucoid fluid. Bacteriologic examination of kidneys, liver, and spleen showed negative results.

Histopathologic Findings

Degenerative changes consisting of focal necrosis and petechial hemorrhage were found in spleen, liver, intestine, kidneys, pancreas, and eyes. Cytoplasmic inclusions were abundant in spleen cells early in the disease course, but were less common during the late stages. Cells considered to represent early infection were swollen, basophilic, and irregular in outline, and contained blue amorphous material. Cells in the late stage of infection contained a small round eosinophilic inclusion in the cytoplasm and remnants of a nuclear membrane (Fig. 1), but were otherwise devoid of recognizable cellular components. The retrobulbar sinus was filled with mononuclear cells instead of the normal content of erythrocytes, and was altered in architecture. Focal areas of hemorrhage were present and organized in a fibrin network. Viral inclusions were not found.

Etiology

Viral arrays were found in a reticulated matrix in the cytoplasm of spleen cells (Fig. 2). Virus particles ranged in size from 110 to 130 nm. In the absence of other pathogens, they were considered to be the causal agent of the disease. Attempts at isolation were not reported, nor was experimental transmission.

Host and Geographic Range

Thus far, *A. ramirezi* is the only known host, and it is logical to conclude that the infection was contracted at the South American source. Until it is shown otherwise, other cichlids, particularly the young, should be considered susceptible.

References

Leibovitz, L., and R. C. Riis. 1980a. A viral disease of aquarium fish. J. Am. Vet. Med. Assoc. 177:414–416.

Leibovitz, L., and R. C. Riis. 1980b. A new viral disease of aquarium fish. Fish Health News 9(1):iv–vi.

3 Eel Virus European

Definition

Eel virus European (EVE) is a birnavirus that is serologically very closely related to the Ab strain of infectious pancreatic necrosis virus (IPNV). The EVE is treated separately here because it has been so reported in the literature and because the virus has been implicated in Japan in a disease of problem proportions among Japanese eels (*Anguilla japonica*). That disease—variously known as branchionephritis or eel virus kidney disease—is an acute virulent and renotropic to viscerotropic problem that results in mortality during the period of seasonal low temperatures.

Isolations of EVE have also been made in Taiwan during surveys of nonanguillid fishes, and from specimens of *A. japonica* with nephroblastoma; however, a causal role in the nephroblastoma has not been shown.

History

The first report of branchionephritis and attendant mortality of eels was made in 1970 by Egusa in a Japanese language publication; it was later discussed by Egusa (1976) and Sano (1976). The disease appeared during 1969, a year after elvers of *A. anguilla* were imported from Europe. Losses were particularly severe among *A. japonica*, amounting to more than 50% of some stocks. In a single year, 2700 tons of eels were lost in Shizuoka Prefecture alone, and some operations were forced out of business.

In the absence of uniform evidence for other pathogens in branchionephritis, a viral etiology was suspected, and in 1973 virus was isolated from eels showing severe renal pathology. Because the source fish were European specimens of *A. anguilla*, the agent was simply designated EVE (Sano 1976).

Sano, who characterized EVE, showed that it closely resembled the French 21 (Ab) strain of IPNV. He described the signs and histopathologic changes of affected eels and showed that EVE was virulent for young eels. He later briefly mentioned EVE in a discussion of salmonid viruses in Japan (Sano 1977) and noted that it produced no change in trout pancreatic tissues. In a more widely available publication, McAllister et al. (1977) reviewed EVE and the several other eel viruses then known.

Several years passed without reports on EVE, but two papers appeared in the same issue of the *Journal of Fish Diseases* in 1981. Sano et al. (1981), in an amplification of his earlier report (Sano 1976), gave details of his 1973 isolations, his characterization of EVE, and his testing for virulence in experimentally infected eels. From the United Kingdom, Hudson et al. (1981) reported findings of a survey of normal-looking young eels from which IPNV strain Ab virus was isolated. That isolant proved to be avirulent in European eels and in rainbow trout. The British investigators noted that they had determined EVE to be IPNV serotype Ab. They commented on the apparent difference in virulence of the British and Japanese isolants and suggested that comparative tests for pathogenicity be made in *A. anguilla*. Nishimura et al. (1981) described methods for concentrating EVE with polyethylene glycol and purifying it on sucrose and cesium chloride gradients.

A careful antigenic comparison of 11 representative isolants of IPNV showed that three groups (I to III) could be recognized (Okamoto et al. 1983); moreover, EVE was most closely related to strain Ab, which constituted group III. The relatedness of EVE and strain Ab was additionally confirmed by analyses of polypeptides; however, slight differences in genome mobilities were sufficient to induce Hedrick, Okamoto, et al. (1983) to state that each strain was unique.

Isolations of Ab strains of IPNV were next reported from Japanese eels and tilapia (*Tilapia mossambica*) in Taiwan, but again variations were found in mobilities of the genome segments and those of EVE, and each strain was again considered unique (Hedrick, Fryer, et al. 1983). Considering that recognized bacterial pathogens were also present in the eels and tilapia, the investigators questioned the role of the virus isolants in causing the observed clinical signs of disease. They conceded, however, that the viruses might be pathogenic under certain conditions.

An association of virus and specific disease in Taiwanese fish was found when Ueno et al. (1984) isolated EVE-like virus from Japanese eels with nephroblastoma. Virus could be isolated from, but not visualized in, several visceral organs. On injection of virus grown in a culture of eel cells, a low-level mortality developed in eels, common carp (*Cyprinus carpio*), and a hybrid tilapia at 20 to 25°C; at 10 to 16°C, however, tilapia sustained an 80% mortality, whereas control fish injected with noninfective culture medium suffered a 30% loss. No injected eels developed nephroblastoma.

In a brief overview of the Taiwanese virus study, Chen et al. (1985) stated that an outbreak of branchionephritis in 1981 caused nearly total mortality in certain eel ponds. Moreover, virus "similar to EVE" was isolated from 80 to 90% of the randomly sampled eels.

The role of EVE in naturally occurring disease among eels is therefore highly suspect, but confirmation with specific-pathogen–free fish under carefully controlled conditions is needed.

Signs and Pathologic Changes

Eel virus kidney disease, or branchionephritis, is the problem with which EVE is most closely associated, and this disease is first evident as untoward mortality during winter. The relevant signs and pathologic changes were described by Sano (1976) and repeated by Sano et al. (1981). Except for the gill hyperplasia and fusing of lamellae that occur only in pond outbreaks, the major findings among natural victims were duplicated among those that died after experimental injection with EVE.

Moribund specimens show transient muscle spasms or episodes of body rigidity. Congestion is evident in the anal fin and on the abdomen of some. Food is absent from the gut of natural victims, but renal hypertrophy and ascites are common in both natural and experimental cases.

Histopathologic examination of naturally and experimentally infected eels shows proliferative glomerulonephritis. Glomeruli are severely damaged, and focal necrosis occurs in renal interstitial tissue. Cloudy swelling of epithelial cells is seen in some tubules and frank necrosis is evident in others. Necrosis is focal in the liver and extensive in the spleen of some specimens.

Clinical and histologic findings have not been reported in cases of nephroblastoma (Chen et al. 1985; Ueno et al. 1984).

Etiology

Evidence for a causal role for EVE in outbreaks of the kidney disease, or branchionephritis, is strongly suggestive, but additional data are needed. Quantification of virus recovered from victims could indicate whether lethal levels were present or whether the fish were simply carriers. Also, the presence or absence of virus in ponds of unaffected eels should be determined.

Results of laboratory trials favor the consideration that EVE is a pathogen for eels; in most trials, mortality was lower in controls than in eels that had been exposed to or injected with the virus (Sano 1976; Sano et al. 1981).

Elvers that were exposed by immersion sustained 60% mortality, compared with 26% in nonexposed controls (Sano 1976). The number of trials conducted with larger eels injected with virus is not clear; Sano (1976) described a single test with 4- to 26-g eels at 8 to 14°C and noted that 55% died in a 20-day period, during which mortality in controls was only 5%. In further work, in which Sano et al. (1981) held injected eels of about 12 g at 8 to 14°C, the losses were 55% in one replicate and 75% in the second but only 10% in the controls. In a final trial with 9-g elvers held at 14°C, neither the injected eels nor the controls died.

The EVE-like virus isolated by Ueno et al. (1984) from eels with nephroblastoma was not oncogenic experimentally, but was virulent for Japanese eels, common carp, and hybrid tilapia held at 20 to 25°C; mortality was greatest in the tilapia held at 10 to 16°C.

The biophysical characteristics of EVE are those of IPNV.

Diagnosis

Viral kidney disease of eels involving EVE is suspect if, during low temperatures, mortality and findings of the foregoing described signs and pathologic changes are present, if neither significant parasite loads nor microbial pathogens are evident, and if an IPNV-like virus can be demonstrated.

Isolation

Virus can be isolated from homogenates of internal organs. Cell lines susceptible to Ab strain IPNV—BF-2, CHSE-214, or RTG-2—are suggested. Ueno et al. (1984) preferred the EK-1 and EO-2 lines of Japanese eel origin in the belief that virus attenuation would be less likely to occur. Because some lineages are refractory to EVE, the FHM line should not be used. Incubation should be in the range of 10 to 20°C.

Identification

Isolants of EVE are identified with a neutralization test or fluorescent antibody technique in which anti-IPNV strain Ab serum is used.

Tranmission and Incubation

Experimental transmission to elvers was achieved by immersion for 1 hour at 15 to 20°C in water containing $10^{5.8}$ $TCID_{50}$/mL. The disease course also took place in about 20 days in elvers that had been given $10^{6.8}$ $TCID_{50}$/5 g body weight then held at 8 to 14°C. Mortality began at 6 to 8 days and ultimately reached 55 to 75%. The pattern is similar to that of IPNV in young trout.

Source

The source of EVE in Japan is believed to be elvers introduced from Europe. Eyed trout eggs from Japan might have transmitted the virus to Taiwan.

Host and Geographic Range

The association of EVE with problem diseases of eels has thus far been reported only from Japan and Taiwan, and particularly among young *Anguilla japonica*. The impact on fish of market size (200 to 400 g) is not known. Closely related strain Ab virus has been found among normal-looking eels in Great Britain.

Control

Avoidance, if it can be realistically implemented, is undoubtedly the best method of control. Eels raised in proximity to trout that carry IPNV might be at risk.

References

Chen, S. N., G. H. Kou, R. P. Hedrick, and J. L. Fryer. 1985. The occurrence of viral infections of fish in Taiwan. Pages 313–319 in A. E. Ellis, ed. Fish and shellfish pathology. Academic Press, Orlando, Florida.

Egusa, S. 1976. Some bacterial diseases of freshwater fishes in Japan. Fish Pathol. 10:103–114.

Hedrick, R. P., J. L. Fryer, S. N. Chen, and G. H. Kou. 1983. Characteristics of four birnaviruses isolated from fish in Taiwan. Fish Pathol. 18:91–97.

Hedrick, R. P., N. Okamoto, T. Sano, and J. L. Fryer. 1983. Biochemical characterization of eel virus European. J. Gen. Virol. 64:1421–1426.

Hudson, E. B., D. Bucke, and A. Forrest. 1981. Isolation of infectious pancreatic necrosis virus from eels, *Anguilla anguilla* L., in the United Kingdom. J. Fish Dis. 4:429–431.

McAllister, P. E., T. Nagabayashi, and K. Wolf. 1977. Viruses of eels with and without stomatopapillomas. Ann. N.Y. Acad. Sci. 298:233–234.

Nishimura, T., H. Fukuda, H. Yamazaki, and T. Sano. 1981. Concentration and purification of eel virus, EVE. Fish Pathol. 16:75–83. In Japanese.

Okamoto, N., T. Sano, R. P. Hedrick, and J. L. Fryer. 1983. Antigenic relationships of selected strains of infectious pancreatic necrosis and European eel virus. J. Fish Dis. 6:19–25.

Sano, T. 1976. Viral diseases of cultured fishes in Japan. Fish Pathol. 19:221–226.

Sano, T. 1977. Viral diseases of cultured salmonids in Japan. Pages 120–123 in Proceedings from the international symposium and diseases of cultured salmonids, Seattle, Washington, April 4–6, 1977.

Sano, T., N. Okamoto, and T. Nishimura. 1981. A new viral epizootic of *Anguilla japonica* Temminck and Schlegel. J. Fish Dis. 4:127–139.

Ueno, Y., S. Chen, G. Kou, R. P. Hedrick, and J. L. Fryer. 1984. Characterization of a virus isolated from Japanese eels (*Anguilla japonica*) with nephroblastoma. Bull. Inst. Zool. Acad. Sin. (Taipei) 23:47–55.

4 Esocid Lymphosarcoma

Synonyms: leukosarcoma, lymphoma, sarcoma

Definition

Esocid lymphosarcoma or lymphoma is an infectious disease of probable viral etiology. The disease occurs among certain populations of adult northern pike (*Esox lucius*) and muskellunge (*E. masquinongy*). The victims bear unsightly malignant lymphoid tumors that typically occur externally but may also occur internally.

Lymphosarcomas are known from nothern pike in Finland, Ireland, and Sweden and from northern pike and muskellunge in North America. Although some differences have been noted among cases of the disease in the four geographic regions, several major features are common to all. Accordingly, until it is shown that they do not share the same cause, it is considered justified to discuss the tumors of esocids as a single disease.

History

Lymphosarcoma of pike has been recognized for over 80 years and is probably an affliction that has long been part of the biology of its host. The early history of the condition consists of occasional case-type reports or records of occurrence. In time, its infectious nature became evident, and in the absence of microscopic pathogens a viral etiology was proposed. Later, interest in the tumors increased and several investigators began research on the disease. As a result, it has since been shown that some of the lymphomas indeed have an associated virus. Because the disease can be transmitted with cell- free tumor filtrates, it seems that the virus could have a causal role.

Ohlmacher (1898), an American medical pathologist, is credited with an early and possibly the first report of the condition. His paper on comparative pathology of tumors included examples from a frog, chicken, and swine. Ohlmacher caught his specimen pike while "trolling from a rowboat in the ordinary fashion"—a point that bears on the behavior of afflicted fish and reflects further an additional pursuit by the author. Although Ohlmacher's pike was without external lesions, the physician-angler noticed a swelling beneath the dorsal fin and found a large tumor involving the musculature and vertebrae. Multiple smaller, nodular sarcomatous growths were present in peritoneal tissues and on the surface of visceral organs. He therefore concluded that metastasis had occurred.

Ohlmacher described the gross and microscopic pathology as follows. The largest or primary tumor was strikingly similar to "small round cell sarcoma" of humans. Some smaller tumors, which differed appreciably from the sarcoma, were composed of cuboidal cells and therefore judged to be endotheliomata rather than sarcomata. Evidence that the "round cell sarcoma" is the same as today's lymphoma lies in Ohlmacher's careful notation that neither myxosporidans (with which he was "somewhat familiar" as a cause of neoplasms in fish) nor microbial forms could be found in appropriately stained preparations.

In Scotland, Haddow and Blake (1933) reviewed the literature on fish neoplasms and described noninfiltrating tumors of lymphatic origin on fin bases and in the opercles of a decomposed female northern pike. It was not until the 1940s, however, that new information was reported. Nigrelli (1943) compiled a listing of diseases and causes of mortality among fishes of the New York Aquarium, among which was a lethal "lympho-sarcoma" of adult northern pike. He considered the condition to be infectious. The lymphoid tumors occurred in liver and kidneys, and a possibility of metastasis was proposed. Several years later, Nigrelli (1947) published the histopathologic details of lymphosarcoma in *Esox* and gave additional evidence for metastasis. In contrast to later publications describing typical lymphoma, the pike showed no external signs. Nigrelli (1952) included lymphosarcoma of *Esox* in his review of virus and tumors in fishes; he reported the tumor in muskellunge and stated that "the evidence indicates that viruses may be the causative agents." Two years later, lymphosarcoma was included in a review of fish tumors and atypical cell growths, and still another aspect—the seasonal nature of the disease—was brought to light (Nigrelli 1954).

In Ireland, Mulcahy published the first of her reports in 1963 and coauthored a popular article the following year (Mulcahy and O'Rourke 1964). She described and illustrated the gross and microscopic pathology of 13 specimens, and enlisted the help of the renowned virologist, K. M. Smith. In what was the first virologic examination of lymphosarcoma, Smith made electron microscopic examinations of sections of three tumors and of material that had been pelleted by ultracentrifugation. However, none of the preparations showed virus (Mulcahy 1963).

In Sweden sarcoma was reported in pike taken along the Baltic Coast (Ljungberg and Lange 1968). The authors were aware of Nigrelli's and Mulcahy's work and noted that the Baltic specimens differed somewhat from those in North America, in that they were taken in brackish waters and showed no involvement of internal organs.

In 1970, exciting results of research on the disease— the first experimental transmission of lymphosarcoma—came from Ireland. Although the number of test fish was small, serial passage was made with a cell-free filtrate of tumor homogenate (Mulcahy and O'Leary 1970). Mulcahy (1970b) determined that the thymus was not universally affected histologically but that its involution or malfunction could be a prerequisite factor in tumor induction. She also found that tumorous pike had altered hematologic values (Mulcahy 1970a). Mulcahy et al. (1970) reported on light and electron microscopy and concluded that the tumor cells were stem cells—hemocytoblasts. Virus particles, however, were not found.

The first electron micrographs of virus in skin sarcomas came from Sweden: Winqvist et al. (1973) found a few cytoplasmic retrovirus C-type particles in 2 of 17 cases. The authors also mentioned that epidermal proliferations occurred in the same population and that many similar C-type particles were present in the epithelial lesions. Although their opinion was later changed, the authors considered tumors of Irish and Baltic pike to be dissimilar.

In Canada, Sonstegard began studies on lymphosarcoma of the muskellunge. In part, the work led to the development of a suspension culture of cells from lymphosarcomatous tissue (Sonstegard and Sonstegard 1973). The culture was the first fish cell line to be grown in suspension, and although some cells were multinucleated, electron microscopy and other relevant assays failed to show evidence of virus.

At the University of Chicago Medical School, interest developed in fish diseases as indicators of pollution. The first report described a relationship between lymphosarcoma and pollution—about a 5% incidence in a polluted watershed and about 1% in an unpolluted lake (Brown et al. 1973). Two years later, the Chicago group reported an 89% success rate in experimental transmission of lymphosarcoma using a cell-free tumor homogenate (Brown et al. 1975).

The hematology of normal and lymphomatous pike was investigated by Mulcahy (1975), who found marked depression of the various red blood cell values among diseased specimens. Also, many red cells were immature and abnormally fragile. As a part of some extensive studies of lymphosarcoma, Sonstegard (1975) detailed the histopathology and progress of the tumor as it occurred at a 16% prevalence among muskellunge in southern Ontario, Canada, and later (Sonstegard 1976) described the etiology and epizootiology. The two papers are key references on the subject. Sonstegard alluded to the disease in Ireland, but rather curiously made no mention of the work by Ljungberg on specimens from the Baltic Sea.

The year 1976 saw notable progress in lymphosarcoma research and a record number of pertinent publications. Papas et al. (1976) demonstrated C-type particles in North American tumor preparations, determined their buoyant density, and showed that reverse transcriptase was associated with the virus. Brown, Dolowy, et al. (1976) obtained tumor material from Mulcahy and, though they could not achieve serial passage, they were able to transmit the disease on initial inoculation. In Ireland, Casey et al. (1975) found that the ability of lymphocytes to respond to phytohemagglutinin was reduced in pike with advanced lymphosarcoma. Mulcahy (1976) added information on the epizootiology of the tumors in Ireland and discounted the pollution aspect reported by Brown et al. (Brown et al. 1973; Brown, Sinclair, et al. 1976). From Sweden, Ljungberg (1976) reported several significant advances: he achieved a high degree of success in serial transmission by injection of tumor cells and also in contact transmission (by cohabiting fingerlings with a tumorous specimen), initiated two cell lines from experimentally induced tumors, and showed by in vivo tests that one of the cell lines was tumorigenic.

Dawe et al. (1976) described two kinds of intracytoplasmic structures in lymphoma cells from a North American northern pike. One kind was termed a "cylindroid lamella-particle complex" and the other a "nucleoid body," but the significance of the structures was not determined. In a second report on the cylindroid complexes from North American fish Banfield et al. (1976) cited work that reported similar structures had also been found in pike from Ireland. Whang-Peng et al. (1976), who analyzed chromosomes of normal and neoplastic cells from tumorous American pike, found evidence for a consistent and pathognomonic karyotype.

In 1977, the reverse transcriptase found in pike lymphoma was characterized and compared with those associated with avian and mammalian tumors. Several biochemical properties were found to be similar but the temperature of optimal activity was below that of homeotherm polymerase (Papas et al. 1977; Sonstegard and Papas 1977).

The Chicago group added to their earlier data and reported some reduction in esocid neoplasia both in polluted and nonpolluted study areas, but again there was a greater overall prevalence in polluted water (Brown et al. 1977).

Structures comparable to the cytoplasmic "cylindroid lamella-particle complexes" described earlier in tumor cells from North American pike were found in specimens from Ireland; however, the structures in the Irish material were termed a "particle-filament-complex" (Hiraki et al. 1978).

Early in the 1980s the tumors were reported from Finland in brackish waters of the Baltic Sea (Thompson 1982). Although the numbers were few, about 90% of the affected pike were males. Lymphosarcoma of pike was brought to the attention of oncologists and virologists when Gross (1983) reviewed the tumorous condition in his two-volume work on oncogenic viruses.

Signs and Pathologic Changes

The pathology of lymphoma among pike from the four regions where it has been reported is generally similar, but some differences also occur (Table 1).

Behavior

Information on the behavior of wild victim fish is meager; however, specimens with prominent tumors have been taken by angling. Accordingly, the fish seem not to be unduly impaired physically, except during terminal stages.

External Signs

External signs of full-blown tumors consist of soft protruding growths that are several centimeters in diameter and have irregular surfaces (Fig. 1). North American northern pike show involvement of flank, fins, and head, in descending order of frequency; for muskellunge, it is fins, flank, and head (Sonstegard 1976). It should be noted, however, that neither Ohlmacher's single North American pike nor those reported by Nigrelli (1943) showed external lesions; Nigrelli, moreover, believed that the tumors originated in the kidneys. The Irish specimens most often bore tumors on the head, mouth, and buccal cavity (Mulcahy 1963). Pike from Sweden showed

Table 1. A comparison of features of esocid lymphosarcomas from four locations

	Location			
Feature	Finland	Ireland	North America	Sweden
Age of victim	Adult	Adult	Adult	Adult
Environment	Brackish	Freshwater	Freshwater	Brackish
Maximum prevalence	Not reported	12%	21%	10%
Season of occurrence	Not reported	Year-round	Fall and winter	Fall peak
Tumor				
Cell type	Stem cell	Stem cell	Lymphoblast-like	Nondifferentiated sarcoma cell
Growth	Invasive	Invasive	Invasive	Infiltrative
Location	Trunk > fins > head	Head > viscera > trunk	Trunk > fins > head	Trunk only
Metastases	No	Yes	Yes	No
Mitoses	Infrequent	Numerous	Present but few	Frequent
Multinucleated cells	Yes	No	Yes	Yes
Experimental transmission				
With cells	Not tried	Not tried	Yes	Yes
Cell free	Not tried	Yes	Yes	No
Contact	Not tried	Not tried	Not tried	Yes
Virus visualized	Inconclusive	No	Yes	Yes

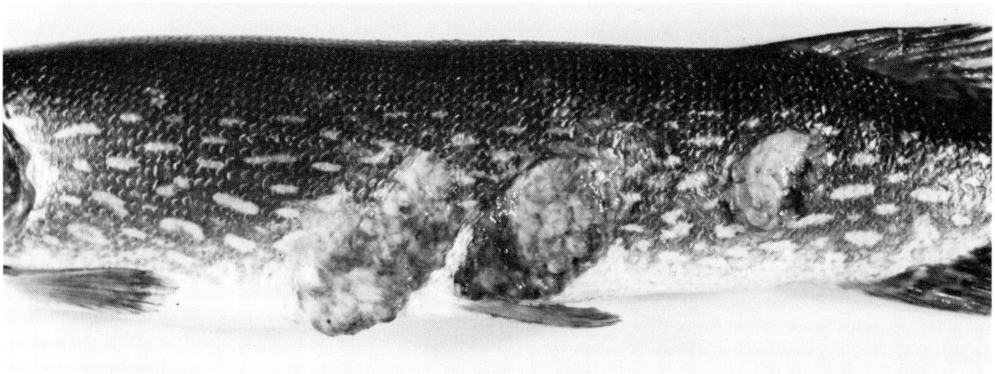

Figure 1. Swedish specimen of a northern pike showing multiple flank tumors of lymphosarcoma. Courtesy of O. Ljungberg.

lesions on the skin, but none on the jaws or in the mouth (Ljungberg and Lange 1968). Specimens from Finland had lesions predominantly on the body and only rarely on the head (Thompson 1982).

The color of the tumor has been variously described as whitish or gray, creamy, yellowish, pink, or red. The range of colors no doubt derives from the lymphocyte-like cells, vascularity, and stage of tumor development. Early tumors can be covered with intact pigmented skin and late tumors can be ulcerated and necrotic. Mulcahy (1963) described most of the growths as being pink to red and some being pale, buff, or cream colored. Ljungberg and Lange (1968) described the color as yellowish gray to yellowish red. The most extensive descriptions of external tumors are those of the muskellunge reported by Sonstegard (1975). The common flank tumors begin in connective tissue spaces and scale pockets as cloudy pink accumulations of fluid. These accumulations appear as single or multiple purple blisters several millimeters in diameter and height. In time, the blisters break, affected scales slough, and the underlying connective tissue is exposed to the environment and becomes pale. About 1% of the muskellunge recover from the early stage of the tumor. More commonly, the tumor margins advance as blistered areas and the central underlying muscle becomes pinkish as it grows outward and increases in volume. The external muscle eventually fragments and the tumor ulcerates. The center is then soft and moist, and thick creamy fluid can be scraped from it.

Investigators in North America, Finland, Ireland, and Sweden uniformly agree that no capsule is formed around the tumor; instead, the underlying connective tissue and musculature are infiltrated (Table 1). In time, muscle bundles are surrounded by the neoplastic cells (Fig. 2).

Internal Signs

Lymphomatous tissue occurs in and about internal organs of pike from North America and Ireland, but not in the Baltic pike from Sweden and Finland. This seemingly basic difference must be considered in light of the fact that Ljungberg (1976) was able to transmit the disease to fingerlings by intraperitoneal injection of

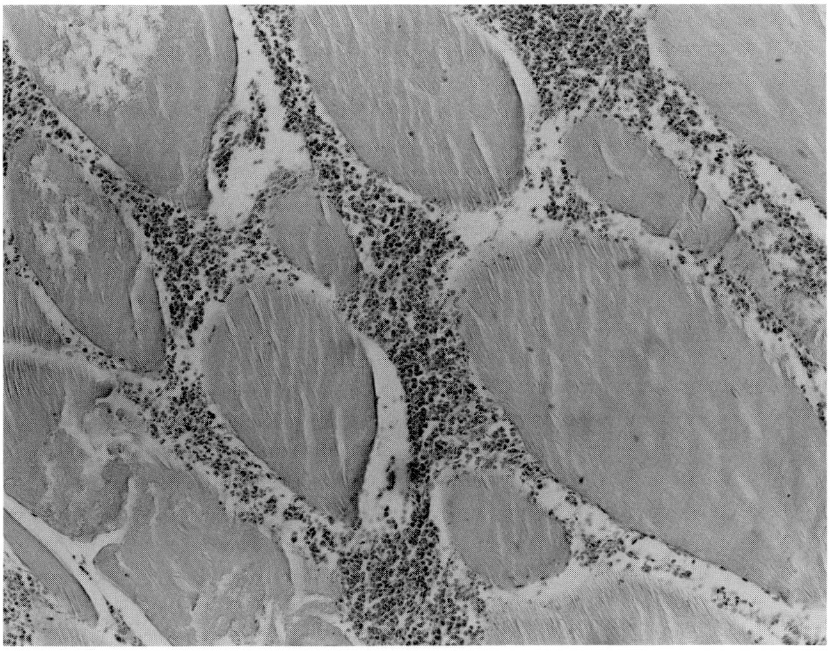

Figure 2. Lymphosarcoma cells infiltrating skeletal muscle bundles.

lymphoma cells and that massive tumor growth and death resulted. According to Sonstegard (1975), all tissues and organs of the American muskellunge may be infiltrated late in the disease. The kidneys and spleen are commonly affected, the liver less so, and the gonads only rarely. The kidneys and spleen sometimes become four times larger than normal, but the involvement is diffuse rather than nodular. Mulcahy (1963) noted that several of her specimens showed discrete internal nodules in several organs but that the thymus was also infiltrated.

Histopathologic Findings

Tumor cells resemble lymphocytes but are usually twice as large. They have a round to oval nucleus that sometimes shows a marginal indentation and clumped or marginated chromatin. Nuclei are surrounded by relatively little cytoplasm, which is usually eosinophilic. Mulcahy et al. (1970), who made both microscopic and ultrastructural comparisons of the tumor cells, found many features to be similar under both methods of observation. Overall, signs of marked nuclear and cytoplasmic immaturity were evident; accordingly, the cells were judged to be hemocytoblasts or stem cells. Features peculiar to the neoplastic cells and not found in renal hematopoietic tissues were juxtanuclear regions that were weakly periodic acid–Schiff (PAS) positive and strongly positive for acid phosphatase. Tumor cells also regularly showed intracytoplasmic vacuoles.

The tumor cells have been variously identified as lymphoreticular stem cells, blastoid cells, hemocytoblasts, lymphoblasts, or reticulum sarcoma cells. Histologically, the tumors consist of masses of infiltrating neoplastic cells that cause atrophy of the stroma without stimulating fibroblastic growth or an inflammatory response. Mulcahy (1963) noted a well-defined organizational pattern: the neoplastic cells were arranged in vertical rows or cords and were interspersed with discontinuous collagen fibers and a few reticulin fibers. She described the vascular network as being relatively sparse and noted that some capillaries were lined with tumor cells.

Sonstegard (1975) stated that the peripheral blood showed large numbers of lymphoblasts, which he equated with the infiltrating neoplastic cells. Mulcahy (1963) also found a great increase in the number of circulating lymphocytes, and a later hematologic study (Mulcahy 1975) found that pike with lymphoma had below normal values for lymphocyte and red blood cell numbers, hemoglobin, hematocrit, and total serum protein. Red blood cells showed a marked increase in fragility and an increase in circulating immature cells. She noted the similarity of her findings with those that occur in Rauscher leukemia virus infections.

There have been two karyologic studies of pike lymphoma cells—one a rather extensive investigation of preparations taken directly from naturally tumorous American pike (Whang-Peng et al. 1976), and the other a smaller study of experimentally transmitted neoplasms from Baltic pike (Ljungberg 1976). A comparison shows some agreement in the respective results but many of the data do not agree.

There was a consistent pattern among 19 American pike tumors: metaphase mitoses showed one submetacentric chromosome, three to five pairs of smaller than normal chromosomes, and a modal diploid number of 50 (range 46 to 54). The normal pike karyotype shows 50 acrocentric chromosomes. Whang-Peng et al. considered the neoplasm to have a distinctive karyotypic "signature." Some of the experimentally transmitted tumors from six Baltic pike showed one submetacentric chromosome, but more often there were two or three. There were also 10 to 15 small or minute chromosomes. Even allowing for differences in terminology, there is no agreement with the American data. Also, the modal number from the Swedish fish was 58 (range 45 to 60). Both the American and the Swedish specimens showed some cells with a tetraploid karyotype.

Mulcahy (1970b) sought to determine the role of the thymus in lymphosarcoma because the thymus was involved in certain homeotherm lymphoid leukemias and 14 of 66 lymphomatous pike had tumors in the thymus. She found no histologic abnormalities in the thymus of normal fish, but among 11 fish with spontaneous tumors, 4 had gross lesions in the thymus, 1 had neoplastic changes, another showed only involution, and 1 was unaltered. Two pike that had been injected with tumor homogenate eventually showed structural changes: one was judged precancerous and the other neoplastic. Mulcahy concluded that the thymus was not always the site of primary tumors but that alteration of thymus structure or function might be a prerequisite for tumor development.

Etiology

The etiology of lymphoma remains to be formally demonstrated, but there is significant evidence for a viral cause. Epizootiologic findings and experimental transmission (particularly trials in which cell-free preparations were used) support the concept of an infectious process. Absolutely no evidence has been found for other microbial forms. Electron microscopy has shown abundant C-type virus particles from North American specimens and a few from Swedish pike. In 2 of 17 specimens examined, Winqvist et al. (1973) found a few cells with one or more C-type particles measuring 100 nm and having a dense core of about 50 nm.

In America, Papas et al. (1976), who processed normal tissue and tumor homogenate from pike, demonstrated the presence of reverse transcriptase in a particulate cytoplasmic fraction from lymphoma. The enzyme activity sedimented in sucrose gradients at 1.16 g/mL, and was additionally processed for electron microscopy: "Numerous type C virus-like particles were found." Although the size was not given, calculations made from the micrographs give a mean diameter of about 75 nm. Reverse transcriptase is found in all known RNA tumor viruses; the enzyme from pike lymphoma showed peak activity at 20°C, and 82% of that activity at 5°C. The low-temperature activity in vitro was discussed as being related to the temperature at which tumors develop in nature. Papas et al. (1977) subsequently reported that the properties of the reverse transcriptase from pike lymphoma were like those of avian myeloblastosis virus, except for a difference in optimum temperature.

Until more is learned, the C-type virus particles known from esocid epithelial hyperplasia should not be excluded as a possible etiologic agent for lymphoma. The epithelial growths may be another expression of host response to a pathogen—a more benign and more common response.

Diagnosis

Esocid lymphoma may be presumptively diagnosed by finding cutaneous lesions of the type described. More accurate determinations require histopathologic examination and demonstration of abundant lymphoblastic cells composing the neoplasm. The presence of C-type particles or experimental transmission or both would provide the best evidence.

Isolation

Thus far, all attempts to isolate the virus in cell culture have been unsuccessful. As inocula, Sonstegard (1976) used "cell-free preparations," presumably of lymphoma homongenates, and living cultures of both pike and muskellunge neoplasms. As indicator systems, he tried BB, BF-2, CAR, FHM, and RTG-2 cell lines and primary cultures of fin, swim bladder, kidneys, and peripheral blood leukocytes; however, he found no evidence of transformation. He took the precaution of using six serial blind passages. Thompson (1982) cultured cells from several tumors and in time found multinucleated giant cells and abnormal mitoses; however, treatment with bromodeoxyuridine failed to induce production of virus.

Transmission and Incubation

Sonstegard and Hnath (1978) believed that natural transmission occurred percutaneously during spawning, but the disease does not seem to be transmitted vertically.

Lymphomas from three of the four geographic regions of occurrence have been experimentally transmitted by transplantation of tumor material or by injection of the neoplastic cells. Cell-free transmission has been accomplished only with pike tumors from North America and Ireland. In no case has there been 100% success. Mulcahy and O'Leary (1970) transmitted the tumor to two adult pike by injecting, intraperitoneally, a 0.22-μm membrane filtrate of tumor homogenate. One pike died at 8 weeks and showed possible thymus involvement. The other developed histologically confirmed tumors at about 6 months. Homogenates of jaw and kidney neoplasms were prepared from the second fish and used to inject four additional pike. One of the four died at 7 weeks and another at 8 weeks; the remaining two were moribund at 8 weeks and were killed. None of the four showed macroscopic growths, but all showed neoplasia and typical infiltration in the several tissues examined. In separate tests, four control fish were injected with normal tissue filtrates but developed no neoplasms. Water temperatures prevailing during the work were not given.

Brown, Dolowy, et al. (1976), who had obtained frozen tumor tissue from pike inoculated by Mulcahy with a cell-free filtrate or unfiltered tumor extract, held the fish at 4 to 7°C and 16 to 20°C. Tumors occurred only at the lower temperature, only on first inoculation, and at most in only 25% of the injected fish. Incubation time was 2.5 to 5.5 weeks after subcutaneous inoculation in the medial side of the jaw, and the tumors arose in the gills. In another trial, when the injection was made intramuscularly on the medial side of the jaw, the tumors appeared at the base of a fin in 15% of the fish. The incubation time was 4 to 6 weeks. In the opinion of the authors, optimal results were obtained with fish 20 to 25 cm long and less than 1 year old that received an inoculum of about 4×10^5 cells. Brown's group specifically stated, however, that material passed through a 90-μm mean porosity filter failed to transmit the tumor.

In Sweden, serial transmission of lymphoma has been clearly achieved by inoculating healthy fingerlings 10 to 18 cm long (2 to 13 months old) with living cells, or by holding fingerlings with tumorous fish (Ljungberg 1976). Cell-free preparations of tumors, serum from a naturally infected specimen, and medium from a tumor cell culture were also tried, but yielded no tumors. Uninoculated, nonexposed control fingerlings remained tumor free. Successful tumor transmission occurred in water temperatures of 8 to 18°C, and rapid growth of the neoplasm was favored at the higher temperature.

Tumor cells obtained by abrading the tissue between screens were usually counted and then inoculated (0.2 to 0.3 mL per fish) intraperitoneally, intracutaneously, or subcutaneously. Cells from a natural tumor produced tumors in 8 of 9 fish after an incubation time of 33 weeks. Cells from the experimentally induced neoplasm were used in a second group of fish and a tumor from that group was, in turn, used to inoculate a third and fourth group. A tumor from the third transfer was used on a fifth lot, and a sixth lot received an intraperitoneal inoculation of 2 million cells from a suspension culture (PS 13) of tumor origin. In all, sarcoma was transmitted to 44 of 51 inoculated fingerlings. Latency shortened with passage: the fourth and fifth groups showed tumors in 10 and 12 weeks, but water temperature during June and July was also more favorable for rapid growth.

Fingerlings that had been inoculated intraperitoneally lost normal pigmentation and became distorted due to volume increase of internal neoplastic cells. The lymphoma cells grew in spaces between organs of the body cavity and also penetrated the "peritoneal lining of the organs" (*sic*). The mass increased in size until the fish were either killed for examination or died. Fish that were inoculated intraperitoneally also developed small tumors of the skin at the site of inoculation. Fish that had been inoculated intracutaneously and subcutaneously developed tumors where injected, but Ljungberg (1976) did not state whether they spread.

Five fish were exposed by contact, and transmission occurred in three, but the elapsed time was 11 to 12 months. Tumors were first seen as tiny yellowish gray coatings of the skin but eventually developed into masses similar to those found on wild specimens.

Although he did not specify the temperature used for his work nor the percentage of success, Sonstegard (1976) provided what might prove to be the key to critical answers in lymphoma transmission. He found the tumor to be transplantable to homologous hosts and from pike to muskellunge. Five to seven months after intramuscular transplant, tumors appeared at the inoculation site and sometimes internally. Intraperitoneal grafts were also successful, producing fish with acute leukemia and massive kidneys and spleen, but no skin tumors. Cell-free transmission was partly effective with both species, but the incubation time was prolonged to 7 to 18 months. What appears to be the most significant finding of all is that the successful trials involved the use of tumors taken during spring and the trials that failed involved material taken in late summer.

Sonstegard expressed the possibility that esocid lymphomas might in some respects be similar to the Lucke' tumor of frogs. The frog adenocarcinoma story has advanced considerably from what was known at the time of the references cited by Sonstegard. There is now conclusive proof that the causal Lucke' herpesvirus is replicated, and that host cells show Cowdry type A intranuclear inclusions only when temperatures are 12°C or less; also, the agent is abundantly present in the urine of frogs at 4°C. Conversely, neither virus particles nor inclusions are present in frog adenocarcinoma cells, in vivo or in vitro, at temperatures above 12°C. However, "at the higher temperatures, virus-specific RNA and membrane antigens are evident in such cells" (Granoff and Naegele 1978). If this condition held true for pike lymphoma, it could account for some of the failures to find virus with electron microscopy and to transmit the disease. A careful experimental approach using both cool and warm temperatures is in order.

Sonstegard and Ljungberg each initiated suspension-type cell lines from lymphoma. The American muskellunge cells grow in static suspension both singly and as macroscopic aggregates (Sonstegard and Sonstegard 1973). The cultured muskellunge cells resemble those of the original lymphoma. Sonstegard (1976) also initiated a lymphoma cell line from northern pike, but electron microscopy has not shown the presence of virus in either line, and no tests for oncogenicity have been reported.

Ljungberg (1976) initiated a cell line from intraperitoneal sarcoma that he had experimentally transmitted. The line, which was designated PS 13 (presumably for pike sarcoma, number 13), grew predominantly as a suspension culture. It too showed giant and irregular cells and proved oncogenic for five fish that received sarcoma cells by intraperitoneal injection. The PS 13 cultures showed some spindle cells that were eventually grown as a monolayer and designated PS 12; their oncogenic potential was not determined.

Host and Geographic Range

Lymphoma has been found in Canada, the United States, Ireland, presumptively in Scotland (Haddow and Blake 1933), and in Europe from Finnish and Swedish coastal Baltic waters. Sonstegard and Hnath (1978) concluded, from responses they received to questionnaires circulated throughout Europe and Asia, that the disease was probably widespread among pike in Europe.

Nothing is known about the susceptibility or resistance of other esocids to lymphoma. The three species of pickerel in North America inhabit some of the same waters as northern pike and muskellunge, and natural hybridization is known to occur. The smaller adult size of pickerel would facilitate their use in laboratory investigations.

Of the two susceptible species, the muskellunge is the more severely affected. It would be interesting to learn how the pike-muskellunge hybrid, the "tiger muskellunge," responds to the disease.

Control

It is unrealistic to consider control of lymphosarcoma in natural waters. Sonstegard and Hnath (1978) strongly advised that adult esocids should not be stocked; instead, stocking should be with eggs, fry, and fingerlings that have been reared in facilities to which adults do not have access.

References

Banfield, W. G., C. J. Dawe, C. E. Lee, and R. Sonstegard. 1976. Cylindroid lamella-particle complexes in lymphoma cells of northern pike (*Esox lucius*). J. Natl. Cancer Inst. 57:415–419.

Brown, E. R., J. J. Hazdra, L. Keith, I. Greenspan, J. B. G. Kwapinski, and P. Beamer. 1973. Frequency of fish tumors found in a polluted watershed as compared to nonpolluted Canadian water. Cancer Res. 33:189–198.

Brown, E. R., L. Keith, J. J. Hazdra, and T. Arndt. 1975. Tumors in fish caught in polluted waters: possible explanations. Pages 47–57 in Y. Ito and R. M. Dutcher, eds. Leukemogenesis. Univ. Tokyo Press, Tokyo.

Brown, E. R., W. C. Dolowy, T. Sinclair, L. Keith, S. Greenberg, J. J. Hazdra, P. Beamer, and O. Callaghan. 1976. Enhancement of lymphosarcoma transmission in *Esox lucius* and its epidemiologic relationship to pollution. Bibl. Haematologia 43:245–251.

Brown, E. R., T. F. Sinclair, L. Keith, J. J. Hazdra, O. H. Callaghan, and W. R. Inch. 1976. Lymphoma in *Esox lucius* (northern pike): viral and environmental interactions. Proc. Am. Assoc. Cancer Res. Abstr. 17:2.

Brown, E. R., T. Sinclair, L. Keith, P. Beamer, J. Hazdra, V. Nair, and O. Callaghan. 1977. Chemical pollutants in relation to disease in fish. Ann. N.Y. Acad. Sci. 298:535–546.

Casey, N., M. Mulcahy, and W. O'Connell. 1975. Leucocyte characteristics and phytohemagglutinin-stimulated response in healthy and lymphoma-bearing pike *Esox lucius* L. Bibl. Haematologica 43:456.

Dawe, C. J., W. G. Banfield, R. Sonstegard, C. W. Lee, and H. J. Michelitch. 1976. Cylindroid lamella-particle complexes and nucleoid intracytoplasmic bodies in lymphoma cells of northern pike (*Esox lucius*). Prog. Exp. Tumor Res., 20:166–180.

Granoff, A., and R. F. Naegele. 1978. The Lucke' tumor: a model for persistent virus infection and oncogenesis. Pages 15–25 in J. Stevens, G. J. Todaro, and C. F. Fox, eds. Persistent viruses. Academic Press, New York.

Gross, L. 1983. Tumors, leukemia and lymphosarcoma in fish. Pages 103–116 in L. Gross, ed. Oncogenic viruses, 3rd ed. Vol. 1, Pergamon Press, Oxford.

Haddow, A., and I. Blake. 1933. Neoplasms in fish: a report of six cases with a summary of the literature. J. Pathol. Bacteriol. 36:41–47.

Hiraki, S., M. F. Mulcahy, and L. Dmochowski. 1978. "Particle-filament-complex" in tumor cells of northern pike, Esox lucius L. Tex. Rep. Biol. Med. 36:111–120.

Ljungberg, O. 1976. Epizootiological and experimental studies of skin tumors in northern pike (*Esox lucius* L.) in the Baltic Sea. Prog. Exp. Tumor Res., 20:156–165.

Ljungberg, O., and J. Lange. 1968. Skin tumors of northern pike (*Esox lucius* L.). I. Sarcoma in a Baltic pike population. Bull. Off. Int. Epizoot. 69:1007–1022.

Mulcahy, M. F. 1963. Lymphosarcoma in the pike, *Esox lucius* L., (Pisces; Esocidae) in Ireland. Proc. R. Ir. Acad. Sect. B Biol. Geol. Chem. Sci. 63:103–129.

Mulcahy, M. 1970a. Hemic neoplasms in cold blooded animals: lymphosarcoma in the pike *Esox lucius*. Bibl. Haematologica 36:644–645.

Mulcahy, M. F. 1970b. The thymus glands and lymphosarcoma in the pike, *Esox lucius* L. (Pisces; Esocidae) in Ireland. Bibl. Haematologica 36:600–609.

Mulcahy, M. F. 1975. Fish blood changes associated with disease: a hematological study of pike lymphoma and salmon ulcerative dermal necrosis. Pages 925–944 in W. E. Ribelin and A. G. Migaki, eds. The pathology of fishes. Univ. Wisconsin Press, Madison.

Mulcahy, M. F. 1976. Epizootiological studies of lymphomas in northern pike in Ireland. Prog. Exp. Tumor Res. 20:129–140.

Mulcahy, M. F., and A. O'Leary. 1970. Cell-free transmission of lymphosarcoma in the northern pike *Esox lucius* L. (Pisces; Esocidae). Experientia 26:891.

Mulcahy, M. F., and F. J. O'Rourke. 1964. Cancerous pike in Ireland. Ir. Nat. J. 14:312–315.

Mulcahy, M. F., G. Winqvist, and C. J. Dawe. 1970. The neoplastic cell type in lymphoreticular neoplasms of the northern pike, *Esox lucius* L. Cancer Res. 30:2712–2717.

Nigrelli, R. F. 1943. Causes of diseases and death of fishes in captivity. Zoologica 28:203–216.

Nigrelli, R. F. 1947. Spontaneous neoplasms in fishes. III. Lymphosarcoma in *Astyanax* and *Esox*. Zoologica 32:101–108.

Nigrelli, R. F. 1952. Virus and tumors in fishes. Ann. N.Y. Acad. Sci. 54:1076–1092.

Nigrelli, R. F. 1954. Tumors and other atypical cell growths in temperate freshwater fishes of North America. Trans. Am. Fish. Soc. 83:262–296.

Ohlmacher, A. P. 1898. Several examples illustrating the comparative pathology of tumors. Bull. Ohio Hosp. Epileptics 1:223–236.

Papas, T. S., J. E. Dahlberg, and R. A. Sonstegard. 1976. Type C virus in lymphosarcoma in northern pike (*Esox lucius*). Nature 261:506–508.

Papas. T. S., T. W. Pry, M. P. Shafer, and R. A. Sonstegard. 1977. Presence of DNA polymerase in lymphosarcoma in northern pike (*Esox lucius*). Cancer Res. 37:3214–3217.

Sonstegard, R. 1975. Lymphosarcoma in muskellunge (*Esox masquinongy*). Pages 907–924 in W. E. Ribelin and G. Migaki, eds. The pathology of fishes. Univ. Wisconsin Press, Madison.

Sonstegard, R. A. 1976. Studies of the etiology and epizootiology of lymphosarcoma in *Esox* (*Esox lucius* L. and *Esox masquinongy*). Prog. Exp. Tumor Res. 20:141–155.

Sonstegard, R. A., and J. G. Hnath. 1978. Lymphosarcoma in muskellunge and northern pike: guidelines for disease control. Am. Fish. Soc. Spec. Publ. 11:235–237.

Sonstegard, R. A., and T. S. Papas. 1977. Descriptive and comparative studies of type C virus DNA polymerase of fish (northern pike, *Esox lucius*). Page 202 in Proceedings of the 68th annual meeting of the American Association for Cancer Research, Denver, Colorado, May 18–21, 1977. Abstract.

Sonstegard, R. A., and K. S. Sonstegard. 1973. Establishment of a cell line from lymphosarcoma of the muskellunge (*Esox masquinongy*). In Vitro 8:410. Abstract.

Thompson, J. S. 1982. An epizootic of lymphoma in northern pike, *Esox lucius* L., from the Åland Islands of Finland. J. Fish Dis. 5:1–11.

Whang-Peng, J., R. A. Sonstegard, and C. Dawe. 1976. Chromosomal characteristics of malignant lymphoma in northern pike (*Esox lucius*) from the United States. Cancer Res. 36:3554–3560.

Winqvist, G., O. Ljungberg, and B. Ivarsson. 1973. Electron microscopy of sarcoma of the northern pike (*Esox lucius* L.). Pages 26–30 in R. M. Dutcher and L. Chieco-Bianchi, eds. Unifying concepts of leukemia. S. Karger, Basel.

5 Grass Carp Reovirus

Definition

The grass carp reovirus (GCRV) is a newly described pathogen that causes an acute hemorrhagic disease leading to mortality in fingerling and yearling specimens of *Ctenopharyngodon idella*.

History

In the south of the Peoples Republic of China (PRC), hemorrhagic disease and mortality were recognized as a serious problem in husbandry of the grass carp. In time, a viral etiology was proposed, and an agent was isolated. At the Sixth International Congress of Virology (held in Sendai, Japan), Chen and Jiang (1984), from the Laboratory of Fish Disease, Wuhan, PRC, reported on the reovirus characteristics of the agent and noted that purified virus produced typical hemorrhagic disease in fingerling grass carp.

At the 1985 Asian Symposium on Freshwater Fish Culture held at Beijing, PRC, 6 of the 30 papers presented concerned relevant aspects of GCRV: cell lines that had been developed from the host species and that replicated the virus (Deng et al. 1985; Zhang and Yang 1985; Zuo et al. 1985); characterization of the virus, including features of its RNA (Hong et al. 1985); gross pathology of the disease and development of an inactivated vaccine, which when injected provided a high level of protection (Chen et al. 1985); and hematologic studies of experimentally and naturally infected specimens (Zhu et al. 1985). Abstracts of the papers are in English, and it is hoped that the full papers will be translated, for there is evidently a high level of sophistication in fish virology in the PRC.

Signs and Pathologic Changes

In summer, at temperatures of 24 to 30°C, infections run an acute course and result in significant mortality, especially in fingerlings and to a lesser extent in yearlings (Chen et al. 1985). Behavior changes have not been described. Victims are exophthalmic and have hemorrhagic or pale gills and hemorrhagic fin bases or gill covers. Hemorrhages occur throughout the musculature, and in the oral cavity, intestinal tract, liver, spleen, and kidneys. No histopathologic findings have been reported.

Hematologic changes were noted by Zhu et al. (1985). In experimental infections, erythrocytes, plasma protein, and urea nitrogen were all significantly reduced by day 4 or 5. By day 8, erythrocytes, hemoglobin, and plasma protein were depressed even further, but blood glucose was not altered. Cellulose acetate electrophoresis revealed changes in serum lactic acid dehydrogenase isozyme activity, and an additional band was found in some specimens.

Among naturally infected specimens showing obvious signs, the changes in blood characteristics were similar to those found in experimental infections. In addition, serum potassium was increased and calcium was significantly reduced.

Etiology

Abstracts of reports from different laboratories, although lacking details, stated that hemorrhagic disease was reproduced with purified virus; however, reisolation was not mentioned (Chen and Jiang 1984; Chen et al. 1985; Hong et al. 1985).

Size and Shape

The virus is a nonenveloped, doubly encapsidated icosahedron with 5:3:2 symmetry, 92 capsomeres, an overall diameter of 60 to 70 nm, and an inner capsid that measures about 40 nm.

Biophysical Properties

The outer capsid is removable with alpha chymotrypsin treatment, but infectivity is not degraded by pH 3, chloroform, or heat—nor is replication inhibited by iododeoxyuridine. The molecular weight of the genome is about 14.5×10^6, comprising 10 segments of dsRNA that range in weight from 0.76 to 3.02×10^6. Sedimentation coefficient of the virus is about 550S.

The GCRV is replicated by several fish cell lines, but notably to highest titer by those of host fish origin. Titers of 10^8–10^9 $TCID_{50}$/mL were obtained from clones of the GCK-84 of kidney origin (Deng et al. 1985). Titers of about 10^7 $TCID_{50}$/mL were obtained by Chen et al. (1985) from the gonad line GCG and the fin line GCF. Susceptibility was reported in the CIK kidney line (Zuo et al. 1985) and in the ZC-7901 snout line (Zhang and Yang 1985), but titers were not given.

Certain cell lines other than those of grass carp replicated GCRV, but virus yield was not appreciable (Deng et al. 1985).

According to Chen et al. (1985), in vitro replication occurred at 20 to 35°C, and optimally at 28 to 30°C. At 28°C, CPE appeared 3 to 4 days after inoculation. Infected cells showed cytoplasmic aggregations of virus, occasionally in crystalline array.

Diagnosis

Young grass carp that show hemorrhagic disease and mortality, in the absence of microbial pathogens and parasites, are suspect.

Isolation

Primary cultures or lines of grass carp origin should be inoculated with decontaminated visceral organ homogenates. Other lines of cyprinid origin, such as EPC and FHM, should be tried.

Identification

Identification is based on the fish of origin and reovirus characteristics of the isolant. Until shown otherwise, a possibility exists that GCRV is the golden shiner virus, a reovirus of the North American bait minnow. Serologic methods have not been reported.

Transmission and Incubation

Fish-to-fish (i.e., waterborne) transmission is assumed and, judging from experimental results, signs of disease and mortality could be expected within 1 or 2 weeks after exposure of fish in water at temperatures of 25°C or higher.

Source

The source is the normal host species, but other cyprinids are not excluded.

Host and Geographic Range

The disease and the virus are now known only from grass carp in the PRC, but the species has been widely introduced to other parts of the world. Chen et al. (1985) stated that the disease can also occur in two other cyprinids, the black carp (*Mylopharyngodon piceus*) and the so-called chebachek (*Pseudorasbora parva*).

Immunity

Experimentally, immunity has been induced to a high but not absolute level by the injection of inactivated vaccine prepared from infected liver, spleen, and kidneys. An 80% level of immunity was obtained and lasted as long as 14 months; protection was evident by the fourth day at temperatures above 20°C, but 20 days were required at 15°C, and 30 days at 10°C (Chen et al. 1985).

Control

Immunization holds promise, but oral or immersion methods are needed for practical application.

References

Chen, Y., and Y. Jiang. 1984. Morphological and physico-chemical characterization of the hemorrhagic virus of grass carp. Page 373 in Sixth International Congress of Virology, Sendai, Japan, September 1–7, 1984. Abstract.

Chen, Y., H. Gao, Y. Jiang, and Y. Wang. 1985. Studies on causative and inactivated vaccine of hemorrhage of grass carp (*Ctenopharyngodon idellus*). Page 47 in Asian symposium on freshwater fish culture, Beijing, Peoples Republic of China, October 10–15, 1985. Abstract.

Deng, C., X. Yang, and H. Chen. 1985. The susceptibility to grass carp reovirus of grass carp kidney cell line and its clones. Page 55 in Asian symposium on freshwater fish culture, Beijing, Peoples Republic of China, October 10–15, 1985. Abstract.

Hong, S., L. Yuan, L. Yu, S. Miu, W. Zuo, H. Qian, and Y. Cai. 1985. Isolation and identification of the etiological factor causing hemorrhage of grass carp-fish reovirus (FRV). Page 44 in Asian symposium on freshwater fish culture, Beijing, Peoples Republic of China, October 10–15, 1985. Abstract.

Zhang, N., and G. Yang. 1985. The establishment and characteristics of cell line ZC-7901 from the snout tissue cells of grass carp. Page 45 in Asian symposium on freshwater fish culture, Beijing, Peoples Republic of China, October 10–15, 1985. Abstract.

Zhu, X., L. Jia, and M. Zang. 1985. Haemapathologic study of hemorrhagic disease on the grass carp fingerlings in the latency and development period. Pages 56–57 in Asian symposium on freshwater fish culture, Beijing, Peoples Republic of China, October 10–15, 1985. Abstract.

Zuo, W., H. Qian, Y. Xu, S. Du, and X. Yang. 1985. A cell line derived from grass carp (*Ctenopharyngodon idellus*) kidney and its biological characteristics. Page 67 in Asian symposium on freshwater fish culture, Beijing, Peoples Republic of China, October 10–15, 1985. Abstract.

6 *Herpesvirus salmonis* Disease

Definition

Herpesvirus salmonis is a serologically distinct agent first isolated from normal-appearing adult stocks of captive and anadromous steelhead trout (*Salmo gairdneri*). Experimentally, the virus causes subacute to chronic viscerotropic disease in rainbow trout fry held at temperatures below 10°C. Fingerling chinook salmon are also susceptible.

History

During the early 1970s, as much as 50% postspawning mortality occurred among rainbow trout at the Winthrop (Washington) National Fish Hatchery. W. G. Taylor, who investigated the problem, found no evidence of a bacterial or a parasitic cause; instead, cell cultures inoculated with ovarian fluids showed CPE selectively for the RTG-2 line and not for FHM. Taylor suspected virus and sent affected cultures to a reference laboratory where the presence of virus was promptly confirmed. The Winthrop agent was characterized as a previously undescribed syncytium-inducing virus and provisionally named *Herpesvirus salmonis* (Wolf 1976; Wolf and Taylor 1975; Wolf et al. 1978).

The virus was then tested and found to be virulent for young rainbow trout (Wolf, Nagabayashi, and Quimby 1975), and a preliminary trial was carried out to determine histopathologic changes in victim fry (Wolf, Herman, et al. 1975). A definitive illustrated report of the pathology was published by Wolf and Smith (1981).

The role of *H. salmonis* in the source brood stock could not be determined, nor were there clues to indicate an original source of virus. Although hatcheries that had received eggs from the Winthrop stock were canvased, none were found to have had suspect viral mortality in progeny from the eggs.

A 2-year program of surveillance for the virus was conducted by the U.S. Fish and Wildlife Service. Hatchery biologists used RTG-2 cells and low incubation temperature in inspections of hatchery salmonids, but found no evidence of *H. salmonis*.

Recently, serologically closely related herpesvirus was isolated from normal-appearing adult steelhead trout in California (Hedrick et al. 1986).

Signs and Pathologic Changes

H. salmonis causes disease and high mortality only in young rainbow trout that have been injected and then held at 6 to 9°C (Wolf and Smith 1981).

Behavior

At about 2 weeks postinfection, affected fry refused food and were lethargic. Some lay on their back or side but responded to alarm or physical stimulation with a brief burst of erratic swimming. Terminal behavior appeared as prostration and weak respiration.

External Signs

A few victims were abnormally dark, but the coloration of most was normal. On the other hand, most victims were exophthalmic, some extremely so (in these, orbital hemorrhages were also frequent). Gills were typically pale, and in most the abdomen was noticeably distended. External hemorrhages occurred predominantly in the fins, but on fewer than half of the infected animals. Beginning about 2 weeks postinfection, some of the infected fry trailed unusually thick fecal pseudocasts.

Internal Signs

Internal organs were pale and flaccid, and any food present was localized in the posterior digestive tract. In some, the liver was mottled or hemorrhaged and had a friable texture. Kidneys were pale to grayish but not swollen. Slight to abundant ascitic fluid, present in some fish, was bloody or gelatinous.

Limited work with blood films from terminal specimens showed that as many as 10% of the circulating erythrocytes were immature, and blast cells were unusually abundant. In contrast to blood films from victims of infectious hematopoietic necrosis, which typically show obvious cellular debris, blood films from the herpesvirus specimens were generally free of cell debris.

Histopathologic Findings

The pathology of the virus was determined from two separate lots of experimentally infected fry (Wolf and Smith 1981). Major changes occurred in the heart and in the respiratory and visceral organs, and dramatic syncytia in the pancreas were judged to be pathognomonic.

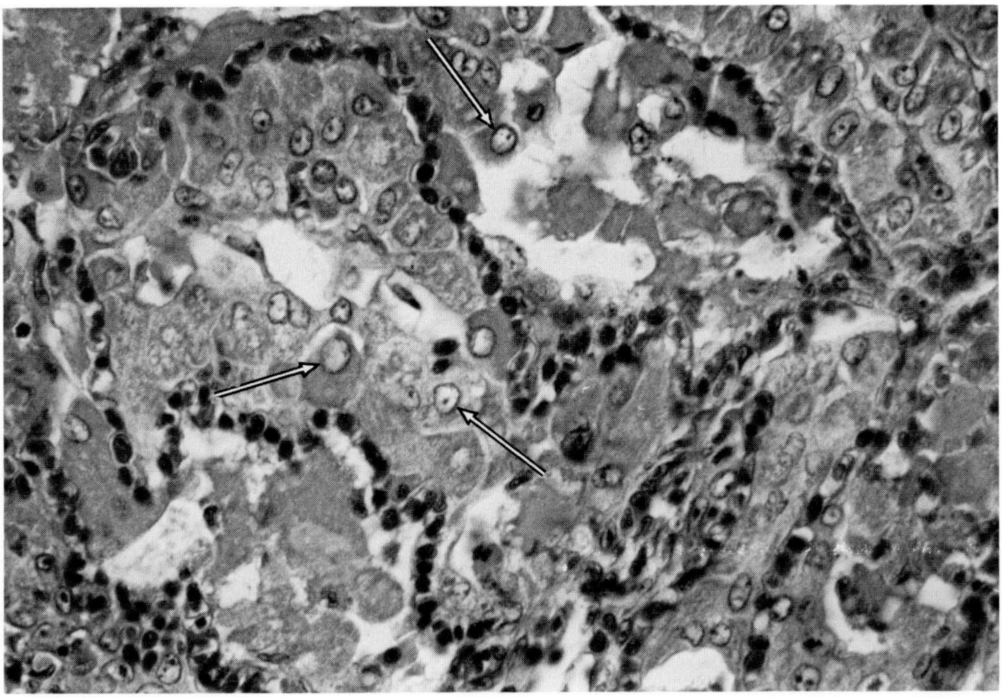

Figure 1. Edema and necrosis of pseudobranch epithelium. Some cells are hypertrophied, and the nuclei show margination of the chromatin (arrows). From Wolf and Smith (1981). Reprinted with the permission of Blackwell Scientific Publications.

The heart of most specimens was edematous and necrotic. Striation was lost in muscle fibers and there was occasional hyaline necrosis. Pronounced leukocytic infiltration was evident and, although lymphocytes predominated, polymorphonuclear cells, macrophages, and blast cells were also involved. Localized hemorrhages were found in some areas of cortical tissue.

Gill epithelium was edematous and hypertrophied. Areas of lamellar epithelium were separated from underlying connective tissue and some cells had been sloughed. Gill hemorrhage was found in several specimens. Pseudobranchs showed slight to extensive edema and in some there were areas of obvious necrosis. Some pseudobranch cells were hypertrophied and had marginated chromatin (Fig. 1).

To various degrees, kidneys showed edema, hyperplasia, congestion, and necrosis. The severity of change progressed from anterior to posterior. Edema was slight to moderate in the anterior kidneys of half the specimens but was severe in the rest and present in the posterior kidneys of all. Anterior hematopoietic tissue was hyperplastic in about half the specimens. Besides showing edema, the posterior kidneys of all but one fish had necrosis of both hematopoietic tissue and tubules. Lumens of tubules of about half the fish contained serous deposits or debris and tubule cells showed some hydropic degeneration. The suspicion that kidneys were prime targets for the virus effects was confirmed by the finding that these organs yielded maximum infectivity.

Liver tissue was prominently altered by edema, necrosis, and hemorrhage or

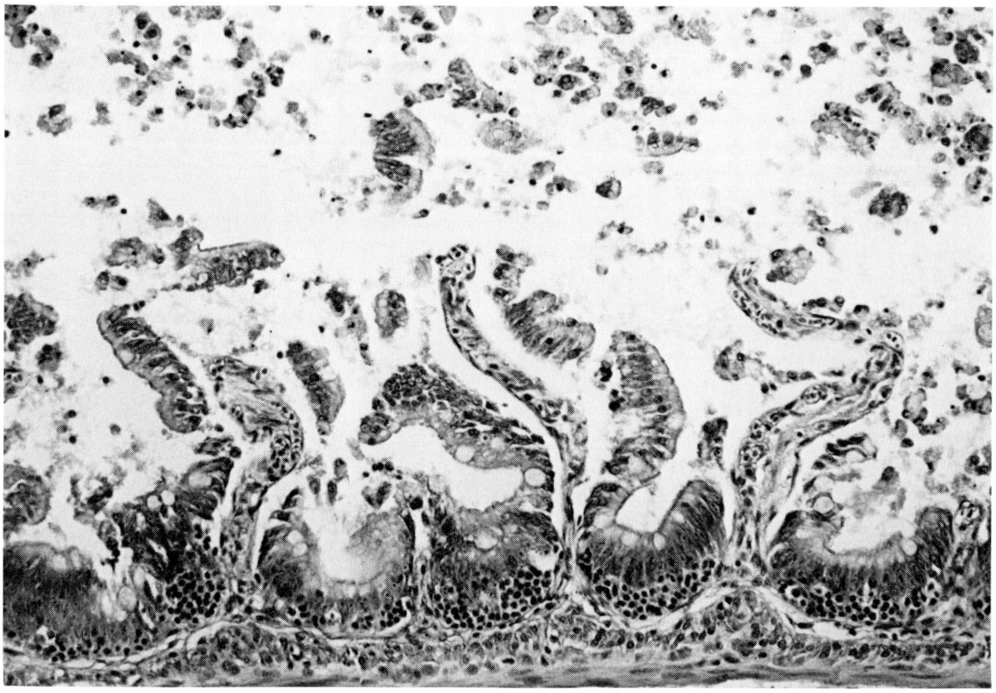

Figure 2. Rectal tissue showing sloughing of mucosal cells and tissues. From Wolf and Smith (1981). Reprinted with the permission of Blackwell Scientific Publications.

congestion. Normal-looking hepatocytes surrounded central veins, but elsewhere there were signs of focal vacuolation and necrosis. Hemorrhage and sinusoidal dilation were present at the periphery of the liver, and a few specimens showed marked coagulation necrosis or phagocytosis of erythrocytes. Intranuclear eosinophilic inclusions were present in a cluster of liver cells in a single specimen and, as could be expected, livers yielded appreciable amounts of virus.

Stomach tissues contained abundant virus but pathologic changes were not proportional. Half the specimens in one trial showed edema around cardiac glands and tunica propria and focal necrosis of the muscularis.

Anteriorly, the intestine with appended pyloric caeca was normal, but marked pathologic changes were consistently present posteriorly; the yield of virus ranked next to that of the kidneys. Rectal tissue was necrotic, the mucosa sloughed into the lumen, and leukocytic infiltration was evident in the submucosa (Fig. 2). Sloughed mucosal tissue undoubtedly forms much of the cast material trailing from the vent of moribund specimens.

Pancreatic syncytia involving up to 30 acinar cells were found in all the specimens of one series, but in only a single specimen of another series (Fig. 3). Neither edema nor necrosis was found in pancreatic tissue.

Spleens showed moderate to advanced edema, and commonly extensive congestion, a decrease in lymphoid tissue, and an increase in connective tissue.

Reproductive tissues were among the least affected, and testes showed no

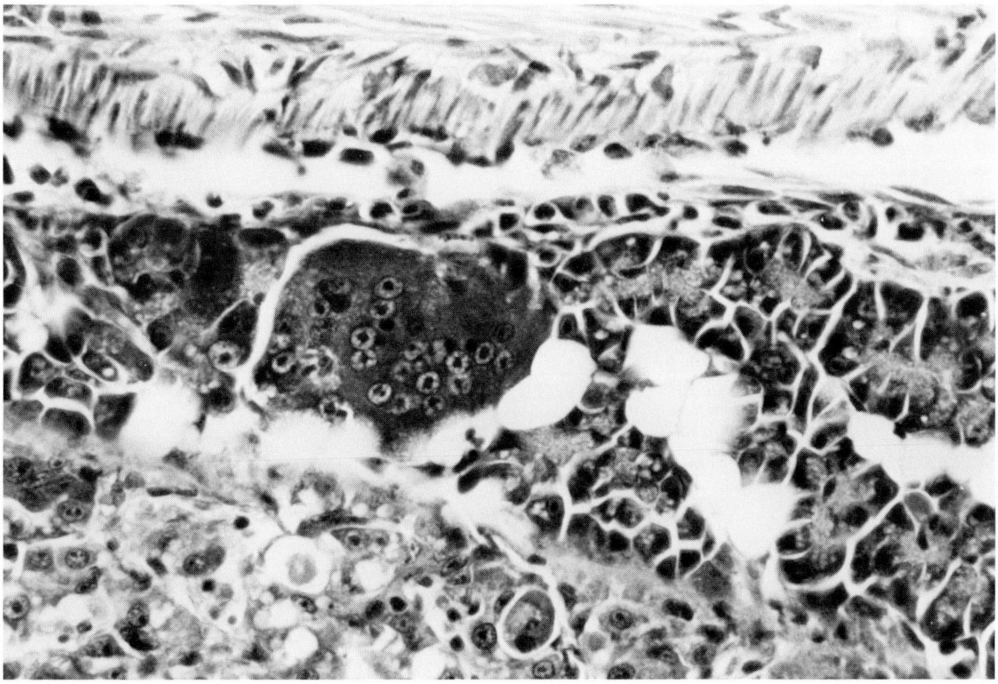

Figure 3. Syncytia in the pancreatic acinar cells of an experimentally infected rainbow trout. From Wolf and Smith (1981). Reprinted with the permission of Blackwell Scientific Publications.

change. Ovaries of two specimens showed edema, inflammation, and necrosis of interstitial tissue and atresia of ova.

Skeletal muscle of some specimens showed hemorrhage and necrosis. The eyes of some specimens showed periorbital edema and retinal distortion.

The brain was slightly to moderately edematous and leukocytic infiltration was evident in subependymal cells of the cerebellum. Some subependymal cells were separated from their attachment, and their nuclei showed vesiculation and margination of chromatin. Meninges or the spaces above the meninges were locally congested or inflamed.

Clinical Findings

Infected fry weighed less than half as much as those that were injected with heat-inactivated virus.

Etiology

The virus has been deposited with the American Type Culture Collection as VR-868.

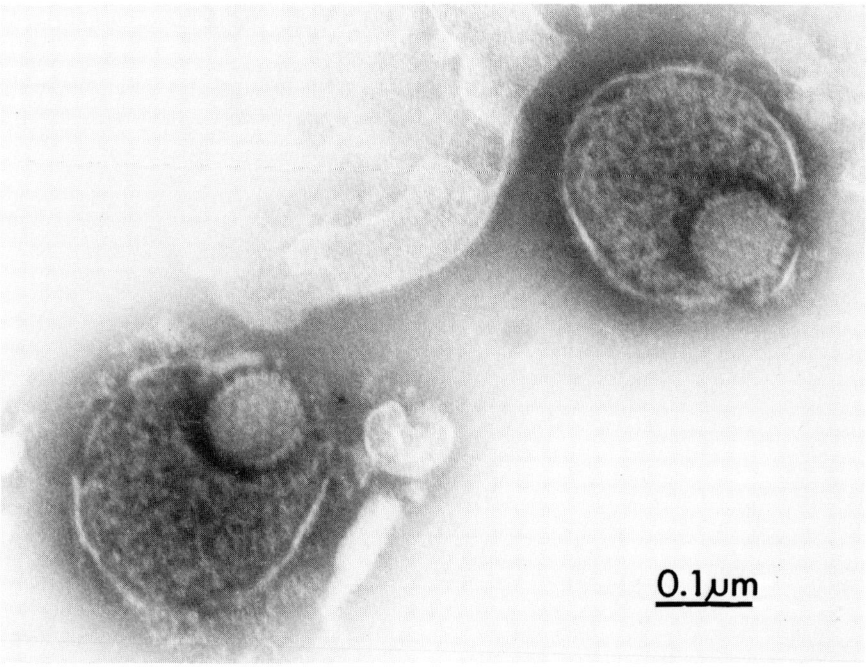

Figure 4. Phosphotungstic acid negative stain of *Herpesvirus salmonis.* Note the distinct membrane-bound envelope and the eccentric placement of the nucleocapsid. ×130,000. From Wolf et al. (1978). Reprinted with the permission of the American Society for Microbiology.

Size and Shape

In size and shape, *H. salmonis* is typical of the family: capsids are icosahedral and measure 90 to 95 nm. Enveloped virions are about 150 nm (Fig. 4).

Replication begins in the nucleus, and capsids with nucleoids are present by 36 hours. The process continues, and postinoculation virus is evident in many cells by 96 hours. As in channel catfish herpesvirus, electron-dense coarse filaments of undetermined nature occur in infected nuclei. Envelopment occurs at cellular membranes, and complete virus is found in the cytoplasm and extracellular spaces.

Biophysical Properties

The virus is heat, ether, and acid labile, and does not hemagglutinate. In cesium chloride the buoyant density of its DNA is 1.709 g/mL and the guanine-cytosine value is 50%. Kimura et al. (1984), who used sodium dodecyl sulfate polyacrylamide gel electrophoresis (SDS-PAGE) to separate virus-induced polypeptides, found that *H. salmonis* had 25 such polypeptides that ranged in molecular weight from 19,500 to 250,000. Moreover, its profile was distinct from that of OMV, the salmonid herpesvirus found in Japan. Infectivity is generally maintained during three cycles of

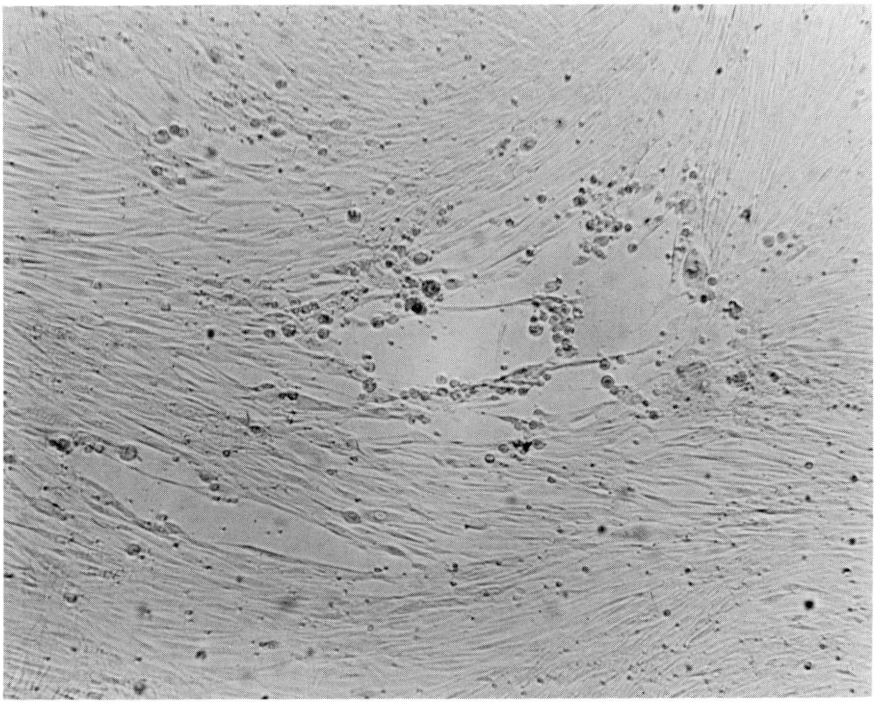

Figure 5. Pyknosis and syncytium formation in a living sheet of RTG-2 cells by *Herpesvirus salmonis*. From Wolf (1983). Reprinted with the permission of the Plenum Publishing Corp.

freezing and thawing and during sonication if the preparation is kept cold. In storage at −20°C or lower, the medium should contain 10% serum and have a pH of about 7.4; nevertheless, after 1.5 years at −80°C, material showed a 90% loss of the initial infectivity. The effect of lyophilization is not known.

Temperatures for replication are critical. Replication at 15°C—a temperature favorable for other salmonid viruses—is inconsistent for *H. salmonis*. It is replicated best at 10°C or less; although it causes cell fusion at 0°C, no new infectivity is produced.

A one-step growth curve in RTG-2 cells at 10°C requires about 4 days. During the first 16 hours, released virus (RV) and cell-associated virus (CAV) show little change. At about 24 hours the amount of virus increases measurably and syncytia appear (Fig. 5). From hours 24 to about 60, exponential growth occurs and CAV exceeds RV by a factor of 10. Peak titers are about 3×10^5 PFU/mL for CAV and about 5×10^4 PFU/mL for RV. In true herpesvirus fashion, *H. salmonis* is replicated only by salmonid fish cell lines, and the highest titers are achieved most readily in RTG-2 or CHSE-214 cells (Table 1).

The CPE of susceptible cultures begins with focal areas of refractile cells that become pyknotic and eventually fuse (Fig. 5). As infection proceeds, pyknosis and fusion produce holes in the cell sheet and the cycle concludes with lysis. Lysis, however, is seldom complete and some masses of fused cells persist for weeks. In plaque assay, *H. salmonis* produces a characteristic plaque that includes syncytia.

Table 1. Comparative yield of *Herpesvirus salmonis* by seven fish cell lines incubated at 10°C for 2 weeks

Cell line	Common name of source fish	Virus yield[a]
BB	Brown bullhead	None
BF-2	Bluegill	None
CHSE-214	Chinook salmon	3×10^5
FHM	Fathead minnow	None
KF-1	Kokanee	3×10^4
RTF-1	Rainbow trout	3×10^5
RTG-2	Rainbow trout	3×10^5

[a]Expressed as $\log_{10}$ of the plaque-forming units per milliliter.

Infected cells stained with May-Grünwald-Giemsa show Cowdry type A intranuclear inclusions.

Diagnosis

The disease can be suspected if rainbow trout fry and small fingerlings sustain a chronic mortality at 10°C or lower. Histologic sections should show little or no necrosis of pancreatic acinar tissues—thus providing presumptive elimination of infectious pancreatic necrosis (IPN), infectious hematopoietic necrosis (IHN), and viral hemorrhagic septicemia (VHS), all of which cause pancreatic necrosis. Herpesvirus disease is accompanied by marked edema of most tissues and hyperplasia of renal hematopoietic tissue. Pancreatic syncytia are pathognomonic. Stained blood films show abundant immature erythrocytes and blast cells; cellular debris is virtually absent. More reliable diagnosis requires electron microscopy and the finding of an agent with herpesvirus morphology.

Isolation

Cultures of susceptible cells should be sparingly confluent, 2 to 5 days old, have a pH of 7.3 to 7.5, and be incubated at 10°C or lower for at least 1 week. Cell sheets should be drained for inoculation and the inocula should be adsorbed at 10°C for at least 2 hours, with occasional agitation.

Whole victim fry, or the kidneys, stomach, liver, and intestines of fingerlings, should be sampled. Where killing adult fish is precluded, ovarian and seminal fluids can be used. If adults can be sacrificed, kidneys, stomach, liver, and intestines should be sampled. Reference *H. salmonis* should be included as a positive control. If suspect material fails to show evidence of CPE after 7 to 10 days, the material should be blind passaged.

Plaquing may reveal two different viruses, if they are present.

Identification

H. salmonis is identified serologically by a neutralization test, and cross-reactions with steelhead herpesvirus (SHV) show that the two isolants are "remarkably similar"; moreover, they are distinct from the salmonid herpesvirus from Japan (Hedrick et al. 1987).

Accordingly, the identification of *H. salmonis* is presumptive and relies on the low temperature requirement and specific pathologic changes.

Transmission and Incubation

Data on natural transmission are lacking; fish-to-fish infection is considered probable, although it could not be demonstrated. The presence of virus in adults indicates a possibility of vertical transmission.

Work with experimental infections, in which several thousand PFU per fish and temperatures of 8 to 10°C were used, revealed the incubation time to first mortality in trout fry and small fingerlings was 25 or more days. Mortality continued until about 50 days postinoculation.

Source

Other than that it was found in normal-appearing rainbow trout and steelhead, the source of *H. salmonis* is an enigma.

According to W. G. Taylor (personal communication, 1976), eggs were shipped repeatedly during 1953–1975 from Winthrop hatchery to many hatcheries in the United States and to several hatcheries in Costa Rica, Hawaii, Japan, and Venezuela—in all, to 35 facilities. One can ask if the Winthrop stock was a source of virus. For now, the answer to that question is "no" or "there are no supporting data."

Until *H. salmonis* was recognized, virologic investigations of salmonids were concerned only with IPN, IHN, and VHS, and cultures were routinely incubated at 15°C. Accordingly, there was no reason to use a lower temperature that would favor the appearance of *H. salmonis*. From mid-1976 to mid-1978, fish health workers of the U.S. Fish and Wildlife Service used 10°C incubation in virologic examinations of salmonids, and paid particular attention to stations that received eggs from Winthrop, but the virus was not found at other locations.

Delving into the history of the Winthrop brood stock, W. G. Taylor (personal communication, 1976) also collected information on incoming materials. During the 1950s, the Cape Cod strain of rainbow trout was brought in from a State of Washington hatchery and crossed with stock from Bishop, California. In 1965, the so-called Donaldson strain males were bred with Winthrop females. In 1971, semen from Oregon fish was brought to Winthrop for use with local eggs, "but there was evidence of a problem prior to the introduction." In essence, the epizootiologic study of *H. salmonis* must include stocks of fish from several locations in North America's Pacific Northwest.

Host and Geographic Range

In addition to the suceptibility of rainbow trout, T. Kimura (personal communication, 1979) found that chum salmon (*Oncorhynchus keta*) fry were susceptible by injection. Cell culture results have also suggested that chinook salmon (*O. tshawytscha*) might be susceptible; indeed, Hedrick et al. (1987) confirmed the susceptibility.

Herpesvirus salmonis was replicated by a line of kokanee cells. Surprisingly, kokanee fry from the population of tissue donors were neither visibly affected nor did they yield virus after an injection of an amount of virus that killed most rainbow trout fry.

Fingerling brook trout (*Salvelinus fontinalis*), brown trout (*Salmo trutta*), and Atlantic salmon (*Salmo salar*) have been tested and found to be resistant (Wolf, Herman, et al. 1975).

Immunity

Nothing is known about the immune response, but it is anticipated that it will be found to be weak. Herpesviruses are generally not regarded as potent antigens, and the response of rabbits to *H. salmonis* is weak.

Control

A discussion of control measures is theoretical because no experience reports exist. Accordingly, one can only advise avoidance of the virus.

Because of its temperature sensitivity, the virus could be inactivated by holding eggs or fry at 16 to 20°C, but the length of time necessary for such an effect would have to be determined.

Kimura et al. (1983a), who compared the in vitro effect of four antiviral compounds on three herpesviruses from fish, found that 9-(2-hydroxyethoxymethyl) guanine (Acyclovir) was highly efficacious; inhibitory concentrations against *H. salmonis* and OMV were as low as 3.2 μg/mL. However, its in vivo efficacy against OMV, the only virus tested, was only partial; treatment consisted of daily immersion of fry for 30 minutes in a solution of 25 μg/mL (Kimura et al. 1983b).

Disinfection procedures should be effective; hypochlorite solutions, alkaline detergents, drying, or gentle warming will undoubtedly inactivate infectivity.

References

Hedrick, R. P., T. McDowell, W. D. Eaton, L. Chan, and W. Wingfield. 1986. *Herpesvirus salmonis* (HPV): first occurrence in anadromous salmonids. Bull. Eur. Assoc. Fish Pathol. 6:66–68.

Hedrick, R. P., T. McDowell, W. D. Eaton, T. Kimura, and T. Sano. 1987. Serological relationships of five herpesviruses isolated from salmonid fishes. J. Appl. Ichthyol. 3:87–92.

Kimura, T., M. Yoshimizu, M. Tanaka, and H. Sannohe. 1981. Studies on a new virus (OMV) from *Oncorhynchus masou*. I. Characteristics and pathogenicity. Fish Pathol. 15:143–147.

Kimura, T., S. Suzuki, and M. Yoshimizu. 1983a. In vitro antiviral effect of 9-(2-hydroxyethoxymethyl) guanine on the fish herpesvirus, *Oncorhynchus masou* virus (OMV). Antiviral Res. 3:93–101.

Kimura, T., S. Suzuki, and M. Yoshimizu. 1983b. In vivo antiviral effect of 9-(2-hydroxyethoxymethyl) guanine on experimental infection of chum salmon (*Oncorhynchus keta*) fry with *Oncorhynchus masou* virus (OMV). Antiviral Res. 3:103–108.

Kimura, T., M. Yoshimizu, and Y. Yano. 1984. Comparison of virus-induced polypeptides from fish herpesvirus—OMV, *H. salmonis*, and CCV. In Sixth International Congress of Virology, Sendai, Japan, September 1–7, 1984. Abstract W37–8.

Wolf, K. 1976. Fish viral diseases in North America, 1971–75, and recent research at the Eastern Fish Disease Laboratory, U.S.A. Fish Pathol. 10:135–154.

Wolf, K. 1983. Biology and properties of fish and reptilian herpesviruses. Pages 319–366 in B. Roizman, ed. The herpesviruses, Vol. 2. Plenum Press, New York.

Wolf, K., and C. E. Smith. 1981. *Herpesvirus salmonis*: pathological changes in parenterally-infected rainbow trout, *Salmo gairdneri* Richardson, fry. J. Fish Dis. 4:445–457.

Wolf, K., and W. G. Taylor. 1975. Salmonid viruses: a syncytium-forming agent from rainbow trout. Fish Health News 4:3.

Wolf, K., R. L. Herman, R. W. Darlington, and W. G. Taylor. 1975. Salmonid viruses: effects of *Herpesvirus salmonis* in rainbow trout fry. Fish Health News 4:8.

Wolf, K., T. Nagabayashi, and M. C. Quimby. 1975. *Herpesvirus salmonis*: tests of species susceptibility, suggestions for isolation, geographic distribution, and control measures. Fish Health News 4:6–7.

Wolf, K., R. W. Darlington, W. G. Taylor, M. C. Quimby, and T. Nagabayashi. 1978. *Herpesvirus salmonis*: characterization of a new pathogen of rainbow trout. J. Virol. 27:659–666.

7 Hirame Rhabdovirus

Synonym: *Rhabdovirus olivaceus*

Definition

Hirame rhabdovirus (HRV), is a bullet-shaped agent (Fig. 1) that was isolated from *Paralichthys olivaceus* and ayu (*Plecoglossus altivelis*) suffering from a severe hemorrhagic disease under husbandry.

History

The HRV was first isolated at facilities on Honshu Island, Japan, in 1984, and then on Hokkaido Island in 1985 (Kimura et al. 1986).

Signs and Pathologic Changes

Diseased hirame with natural infection show gonad congestion, abdominal distention and ascites. Overall, hemorrhages are prominent in fins, skeletal muscle, and internal organs; as a consequence, the gills are pale. The gall bladder is pale yellow. Histologic changes include pyknosis and necrosis of hematopoietic tissue. Ayu show exophthalmia and opercular petechia.

Etiology

The virus is acid and ether labile and replicates in vitro at temperatures of 5 to 20°C, but not at 25°C. Titers of more than 10^9 $TCID_{50}$/mL are produced by FHM and EPC cells; slightly less virus is produced by the YNK, RTG-2, and STE-137 cell lines. A one-step growth curve at the optimal temperature of 15°C requires about 4 days for completion.

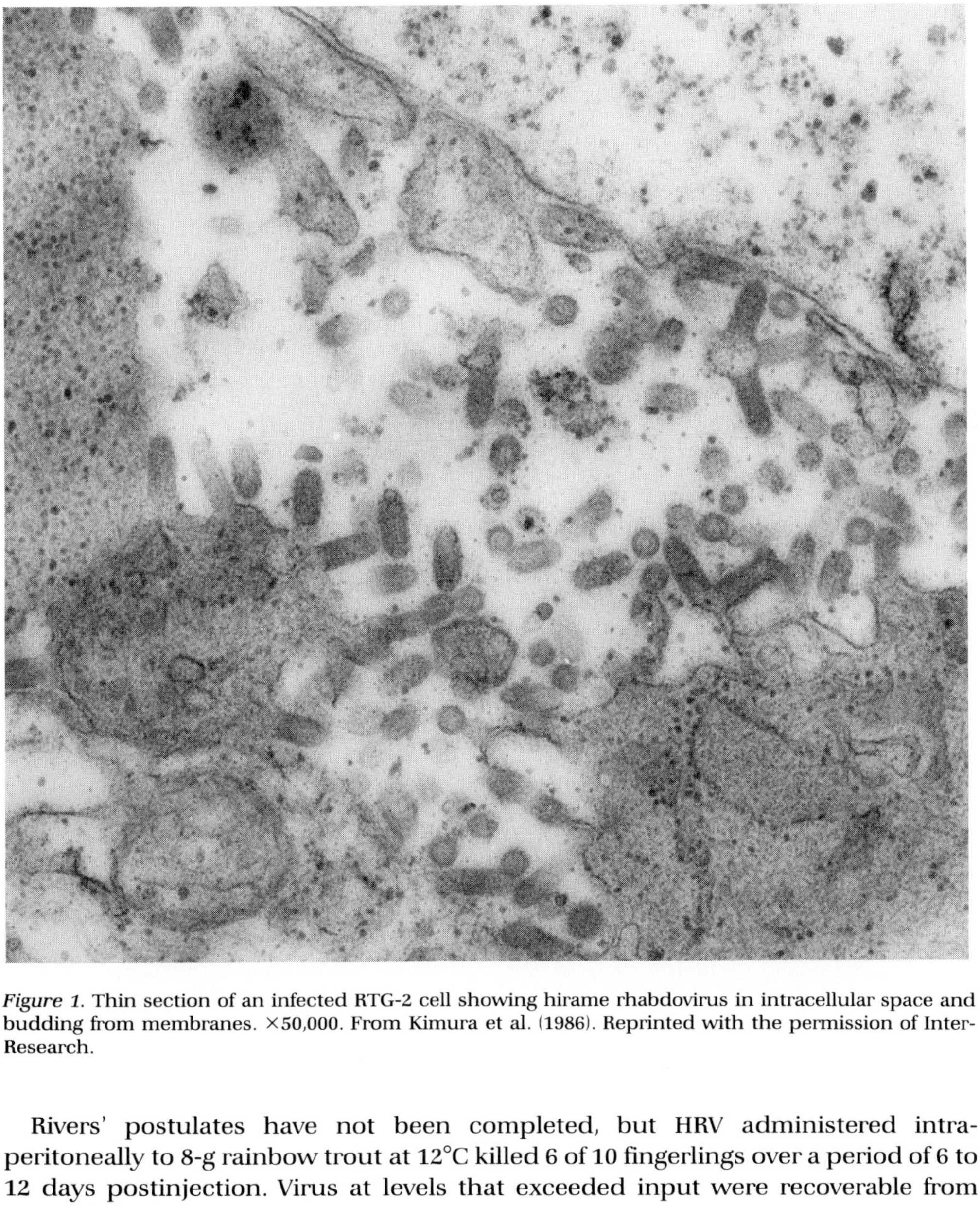

Figure 1. Thin section of an infected RTG-2 cell showing hirame rhabdovirus in intracellular space and budding from membranes. ×50,000. From Kimura et al. (1986). Reprinted with the permission of Inter-Research.

Rivers' postulates have not been completed, but HRV administered intraperitoneally to 8-g rainbow trout at 12°C killed 6 of 10 fingerlings over a period of 6 to 12 days postinjection. Virus at levels that exceeded input were recoverable from internal organs, but titers were highest in the kidneys and spleen. Chum, coho, and masu salmon and ayu were refractory to HRV.

Diagnosis

The disease is presumptively diagnosed on the basis of hemorrhagic disease in the host species, the absence of other pathogens, and the presence of HRV.

Isolation

Pooled visceral homogenates should be inoculated on FHM or EPC cells, with incubation at 15°C.

Identification

Neutralization tests carried out on HRV with antiserum against IHNV, VHSV, spring viremia of carp virus (SVCV), eel virus American (EVA), pike fry rhabdovirus (PFRV), and eel virus European unknown (EVEX) showed that it was a serologically different rhabdovirus. HRV is serologically identified by a neutralization test with HRV-specific antiserum. Possible relations to the sheatfish and Rio Grande perch (cichlid) viruses do not seem likely, but HRV should be compared with the perch virus.

Transmission and Incubation

Natural infection is assumed to be initiated by waterborne virus; however, immersion methods have thus far failed to produce disease experimentally. When HRV was given by injection, minimal incubation time to death of rainbow trout fry was 6 days at 12°C and among 190-g hirame, 14 to 19 days.

Host and Geographic Range

Thus far, HRV is known to occur only among hirame and ayu in Japan. Rainbow trout fry are susceptible when injected with the virus and held at 12°C, but chum salmon fry are not. Fry of ayu and of chum, coho, and masu salmon are resistant to exposure by immersion (Kimura et al. 1986).

Immunity

Nothing has been reported.

Control

Other than avoidance, practical control measures are not known.

Reference

Kimura, T., M. Yoshimizu, and S. Gorie. 1986. A new rhabdovirus isolated in Japan from cultured hirame (Japanese flounder) *Paralichthys olivaceus* and ayu *Plecoglossus altivelis*. Dis. Aquat. Org. 1:209–217.

8 Infectious Hematopoietic Necrosis

Synonyms: chinook salmon virus disease, Coleman disease, Columbia River sockeye disease, Cultus lake virus disease, Oregon sockeye virus, Sacramento River chinook disease, sockeye salmon virus disease

Definition

Infectious hematopoietic necrosis (IHN) is an acute, systemic, and usually virulent rhabdoviral disease that occurs in the wild, but is more typically seen in epizootic proportion among young trout and certain Pacific salmon under husbandry in coastal North America from California to Alaska, and in Japan.

History

During the late 1940s and 1950s, serious epizootics were reported among young salmon in certain facilities in North America's Pacific Northwest (Guenther et al. 1959; Rucker et al. 1953; Watson et al. 1954). It is generally considered that those reports are early accounts of what is now regarded as the viral disease IHN. With the benefit of hindsight, one can recognize still earlier reports of mortality of young salmon that was probably due to virus. For example, Burrows et al. (1951) discussed mortalities in 1944 and 1948 among young sockeye salmon (*Oncorhynchus nerka*) that were being fed rations containing raw salmon viscera; they noted hemorrhagic areas beneath the skin, but believed the cause to be a vitamin deficiency. The occurrence of the 1948 mortality during a cold period also suggests IHN; the authors showed that losses peaked when the water temperature was below 7°C, declined as temperature approached 10°C, and remained low when the temperature was near 13°C. Although proof is lacking, three essentials for occurrence of IHN were present: low temperature, susceptible young fish, and a probable source of virus—the viscera of spawned salmon.

In other examples, Rucker et al. (1953) gave some details of similar epizootics as early as 1948; Parisot et al. (1965) reported an occurrence of sockeye disease in 1946; and Amend et al. (1969) speculated that virus might have been the cause of annual epizootics at a sockeye salmon hatchery on the Fraser River, British Columbia, that claimed an average loss of 96% and dated from 1925 to abandonment of the hatchery in 1936.

During 1950 and 1951, sockeye salmon production in America's Pacific Northwest was plagued with unprecedented high mortality. Stanislas F. Snieszko was called in as a consultant and pointed out the probable cause: the increased mortality followed the introduction of the nutritionally excellent, but hygienically execrable, raw viscera of adult salmon as a component of the rations.

In a telling criticism of the feeding of adult viscera to fry and fingerlings, Snieszko likened the practice to the use of viscera from adult dogs to rear puppies. Guenther et al. (1959) later noted that fish rearing was again successful after marine fish were substituted for the salmon viscera. In time, salmon viscera were reintroduced, but only after the product had been pasteurized. Such processing is part of today's preparation of the widely used Oregon Moist Pellet.

The story of the sockeye salmon virus disease soon began to unfold, and with it the documentation of an agent having what seemed to be high host specificity. Although the presumed virus eluded visualization, a filterable agent present in tissues of returning adults was demonstrated to be virulent for young sockeye salmon (Rucker et al. 1953; Watson 1954; Watson et al. 1954). Because other salmonids were present but unaffected in facilities with natural outbreaks, sockeye salmon appeared to be the only natural host.

Attempted transmission of the sockeye agent to fingerlings of chinook salmon (*O. tshawytscha*), coho salmon (*O. kisutch*), rainbow trout (*Salmo gairdneri*), and cutthroat trout (*S. clarki*) either failed or resulted in mortality so slight as to be considered insignificant (Rucker et al. 1953). In all, five reports covered various aspects of the earliest investigations of epizootics among sockeye salmon fry: Guenther et al. (1959), Rucker et al. (1953), M. E. Watson et al. (1956), S. W. Watson et al. (1954), and Wood and Yasutake (1956). The era predated by several years the first applications of fish cell and tissue culture. Isolants from those early epizootics were not taken, and frozen stocks of victim fish eventually lost infectivity.

As a consequence, the virology of the early sockeye salmon epizootics is conjectural. Nevertheless, the bath or immersion method of infection described by Watson et al. (1954) is widely used today, and the discovery of the sparing effect of warm water has been sustained. The application of elevated temperature to reduce losses provides a measure of control that has value under some circumstances.

New epizootics occurred among young sockeye salmon in 1958, and Parisot et al. (1965) noted that virus was isolated by J. L. Fryer from affected fry at an Oregon hatchery.

In their work in vivo, Parisot et al. (1965) showed that Oregon sockeye virus was capable of infecting young albino rainbow trout. Later, Wingfield et al. (1970) reported sockeye salmon to be highly susceptible, whereas only low mortality—comparable to that in controls—occurred in experimentally challenged chinook salmon and rainbow trout. However, the number of fish used for the work was small.

In the meantime, Ross et al. (1960) described a viruslike disease causing high mortality in chinook salmon fry at the Coleman National Fish Hatchery (NFH) in northern California. Again, a concept of high host specificity was presented. The authors noted that, from the time the Coleman NFH began operation in 1941, fingerling stocks had sustained occasional high losses when they were moved from spring water in the hatchery building to outside raceways supplied with river water. Parisot and Pelnar (1962) reported that the problem occurred annually. They also found that sockeye salmon, cutthroat trout, and steelhead (migratory rainbow trout) were susceptible to the injection of infective material. Curiously, nonmigrating strains of rainbow trout were said to be refractory. Apparently referring to natural outbreaks, they stated, "The infectious agent seems to be specific for chinook salmon." Later work clearly showed that sockeye salmon, rainbow trout, and cutthroat trout could be experimentally infected (Parisot et al. 1965).

First comparisons of the sockeye salmon and the chinook salmon viruses showed similarities between histopathologic changes in experimentally infected fry. Some biophysical properties of the viruses were also similar, but others differed and the concept of high but not absolute host specificity was perpetuated (Klontz et al. 1965; Parisot et al. 1965; Yasutake et al. 1965).

Reviews of the fish virus disease literature (Malsberger and Wolf 1966; Parisot 1963; Wellings 1970; Ross and Rucker 1960; Wolf 1958) and of fish viruses (Wolf 1964, 1966) perpetuated the idea that two host-specific salmon viruses were present in North America's Pacific Northwest.

Cell culture studies of the agent isolated by J. L. Fryer and his associates were first described in a doctoral dissertation by Wingfield (1968), in which he characterized the virus. Portions of the work were soon published in collaboration with his associates (Wingfield et al. 1969). Wingfield was also the first to apply electron microscopy, but he mistakenly concluded that the agent had cubic symmetry and might be an arbovirus.

Outbreaks of acute disease and high mortality occurred in 1967 among rainbow trout and sockeye salmon at one locality on the Fraser River watershed in southern British Columbia. Virus was isolated from both species and was shown to produce a particular nuclear degeneration in cell cultures (Amend et al. 1969); similar in vitro CPE had been noted earlier for sockeye salmon virus (Wingfield 1968). The significance of the work lay in the comparison by Amend et al. (1969) of their agent with the previously isolated sockeye and chinook salmon viruses and the finding that all three produced similar CPE and that the diseases also had certain features in common. Because the renal blood-forming tissues showed evidence of being a specific target, the name infectious hematopoietic necrosis was applied.

Although it was not to be reported in western literature for 9 years, the IHNV was introduced into Hokkaido Island, Japan, in 1968 with a shipment of sockeye salmon eggs from Alaska (Sano et al. 1977). The virus was later found also on Honshu Island, but neither the origin nor source was determined.

McCain (1970), who carried out a biophysical and biochemical characterization of what was then termed Oregon sockeye virus, included serologic comparisons with IHNV and the virus from chinook salmon. He established that the Oregon sockeye virus was bullet shaped and hence had helical, not cubic, symmetry. Antigenic cross-neutralization tests showed that the three isolants were all related and that sockeye virus and IHNV were serologically indistinguishable. Details were published later (McCain et al. 1971, 1974).

Amend and Chambers (1970a) were the first to compare the morphology of the viruses of IHN, Oregon sockeye disease, and chinook salmon virus disease. They concentrated virus by ultracentrifugation and found by negative staining that all three were bullet shaped and had similar diameters. Ultrastructural details were not resolved because the fragile rhabodvirions had been damaged during ultracentrifugation. Later investigators of fine structure took precautions against the notorious fragility of the rhabdoviruses and showed that the three North American agents had the same diameter, length, and ultrastructural detail and that they were almost identical with particles of vesicular stomatitis virus (Darlington et al. 1972). Also noted were truncated forms of defective interfering particles. Cohen and Lenoir (1974), who compared the morphology of the four fish rhabdoviruses then known, found that IHNV was most like Egtved virus, the causal agent of viral hemorrhagic septicemia. They took particular notice of the truncated or cone-shaped particles found among the regular bullet-shaped virions of IHNV. Salmonid virus histopathology was reviewed by Yasutake (1970), who reaffirmed the similar effects produced by the three agents.

Although reports by Wingfield and Chan (1970) on chinook salmon virus disease and by Wingfield et al. (1970) on sockeye salmon virus disease appeared back-to-back, neither acknowledged the existence of the other's work nor of the unifying studies of Amend and his co-workers (Amend 1970a, b; Amend and Chambers 1970a, b; Amend et al. 1969).

As so frequently happens, new information brought change: the viruses of Oregon sockeye disease, chinook salmon virus disease, and IHN became recognized as one virus.

In a symposium on the major communicable fish diseases in Europe, held at Amsterdam in 1972, one of the published panel reviews concerned IHN in salmonids (Amend et al. 1973). In his doctoral research on the pathophysiology of fish with IHN, Amend (1973), used microprocedures and sought characteristics that might have diagnostic value. He later formally published the work with his mentor (Amend and Smith 1974, 1975). Although the approach and the results were scientifically interesting, little application has been made of the work. Also, during the early 1970s, a progression of IHN outbreaks occurred from west to east across the United States. All involved shipments of eggs or fish from the Pacific Northwest; the sequence progressed from South Dakota and Montana to Minnesota, Virginia, and West Virginia (Holway and Smith 1973; Plumb 1972; Wolf et al. 1973). In time, IHN was also reported from New York (Carlisle et al. 1979). Surprisingly, the disease did not recur anywhere.

The year 1974 was significant in the history of IHN because the first molecular and biochemical studies of the virus were reported. Doctoral research on the biophysical and biochemical properties of IHNV was completed (McCain 1970), and the partial characterization of the virus nucleic acid was reported (McCain et al. 1974).

McAllister, who completed his doctoral research on the characterization of IHNV in 1973, studied early virus replication by quantifying infectivity and using an indirect fluorescent antibody technique. He used tritiated uridine to show that the rate of RNA synthesis by infected cells was similar to that of infectious virus. The work was soon published (McAllister and Pilcher 1974; McAllister et al. 1974 a, b). The IHNV was used by de Kinkelin and Le Berre (1974) to induce interferon production in rainbow trout. McAllister and Wagner (1975) described five structural proteins of Egtved virus and IHNV and related the composition of IHNV to that of rabies virus. In a comparative study of four fish rhabdoviruses, Lenoir and de Kinkelin (1975) found similar values for IHNV and rabies virus and agreed that the salmonid viruses resembled rabies virus, except that molecular weights for the L and G proteins were lower in IHNV. Hill et al. (1975), who also reported on the composition of structural proteins of fish rhabdoviruses, found that the sedimentation coefficients for all were in the range of 38 to 40S and confirmed the relatedness of the IHNV and the Egtved virus to rabies virus. Their values for polypeptide molecular weights were greater than those found by earlier workers and they showed some evidence of a sixth entity. Moore et al. (1976), who compared the membrane microviscosity of IHNV with that of vesicular stomatitis virus, found that at their respective optimal temperatures the values were essentially the same. McAllister and Wagner (1977) found RNA-dependent RNA polymerases to be associated with both Egtved virus and IHNV.

Among the noteworthy events in the history of IHNV in the mid to late 1970s, an international symposium on comparative pathology of fishes was held in Washington, D.C., in 1975. The proceedings were published under the editorship of Ribelin and Migaki. As part of that volume, Yasutake gave an updated and comprehensive review of the clinical and histopathologic aspects of selected major viral diseases of fishes, including IHN. Amend (1975), in an account of the results of his epizootiologic studies of IHN transmission and hatchery control measures, noted that adults of both sexes carried the virus and that it could be readily detected in sexually mature animals. In contrast, virus was rarely detected in prespawning carriers and never in embryos or progeny of carriers unless the fry were clinically diseased.

The highly productive fish disease research efforts of Fryer's group at Oregon State University scored another major advance in 1976, when they reported on a 5-year program of attenuation of IHNV and demonstrated a high degree of protection among small lots of sockeye salmon that had been immunized either by injection or by bath (Fryer et al. 1976). The occurrence of IHN was noted in British Columbia by Hoskins et al. (1976), and IHNV was found among 16 stocks of Alaskan sockeye salmon by Grischkowsky and Amend (1976). The report on the Alaskan stocks tended to substantiate a 1960 personal communication to me from W. Hublou recounting an observation of IHN among young sockeye salmon in Oregon that had been fed viscera of adult salmon, presumably taken in Alaska.

The introduction of IHNV into Japan with sockeye salmon eggs from Alaska in the early 1970s was duly described in English language reports by Kimura and Awakura (1977), Sano (1976), and Sano et al. (1977).

The first virologically documented epizootic of IHN under natural conditions was reported by Williams and Amend (1976), and the disease was included in a review by Amend (1976) of the prevention and control of viral diseases of salmonids.

Amend and Nelson (1977) showed a nearly twofold differential survival among various pairings of two stocks of sockeye salmon that were challenged with virulent IHNV; they proposed a genetic basis for the resistance. At about the same time, Pietsch et al. (1977) investigated the survival of IHNV under a range of natural conditions and determined optimal conditions for laboratory preservation of infectivity.

In tests of chlorine and ozone for inactivation of the viruses of IHN and IPN in waters of different chemical composition, Wedemeyer et al. (1978, 1979) found that both chemical agents were more virucidal in distilled than in soft water, and that IHNV was the more sensitive to ozone. They suggested that ozone be used as a fish disease control agent.

The histopathology of IHN was known in fry, but Yasutake (1978) was the first to report findings in experimentally infected sockeye salmon up to 14 months old. Surprisingly, fecal casts and necrotic granular cells of the intestinal wall, key signs in fry, were lacking in the older victims. As a possible long-range approach to control, McIntyre and Amend (1978) analyzed heritability of fish survival and proposed a scheme for artificially selecting brood stock for resistance to IHNV. Properties of IHNV and four other fish rhabdoviruses then known were reviewed (in German) by Ahne (1978).

On the basis of collaborative studies at Fryer's laboratory in Oregon, Hetrick, Fryer, and Knittel (1979) concluded that the elevation of temperature in hatcheries was too costly to be a practical method of controlling loss to IHN. In addition, other research showed that chronic exposure to sublethal levels of copper could increase mortality twofold (Hetrick, Knittel, and Fryer 1979).

A highly sophisticated and productive investigation of the molecular aspects of IHNV was undertaken at Oregon State University. A brief early report by Leong and Turner (1979) gave virus levels in effluents from two epizootics, but noted that virus was not isolated from effluent of a third lot and that methods with greater sensitivity were needed.

Although the literature that appeared in 1980–1985 was diversified, more than one-third of it dealt with aspects of the disease as a serious problem affecting salmonid husbandry. D. Mulcahy, of the U.S. Fish and Wildlife Service laboratory in Seattle, Washington, was a major contributor to new knowledge on the disease. Several papers concerned the prevalence of IHNV in sex products and adult tissue (Mulcahy and Pascho 1984; Mulcahy et al. 1982, 1983a, b; Mulcahy, Jenes, and Pascho 1984) and others addressed the presence and impact of the virus on populations of incubating eggs and fry (Mulcahy and Bauersfeld 1983; Mulcahy and Pascho 1985). Reports dealing with virus detection or identification were published by Dixon and Hill (1984), Grischkowsky and Mulcahy (1982), Hill et al. (1981), Hsu and Leong (1985), and Schultz et al. (1985).

About one-third of the 1980–1985 reports dealt with properties or characteristics of the virus itself and studies of its molecular biology (Hsu et al. 1984; Kurath and Leong 1985; Kurath et al. 1985; Leong et al. 1983). Other investigations dealt with the ability of the virus to induce interferon or to persistently infect cell lines (Engelking and Leong 1981; Okamoto et al. 1983; Sano and Nagakura 1982; Watanabe 1983). Methods for plaquing IHNV were described by Burke and Mulcahy (1980), Leong, Fendrick, et al. (1981), and Okamoto et al. (1985).

Still other authors investigated the stability or lability of infectivity as influenced by temperature, salinity, and the nature of materials stored (Barja et al. 1983; Burke and Mulcahy 1983; Gosting and Gould 1981; Toranzo and Hetrick 1982). A few papers appeared on miscellaneous aspects such as chemotherapy or replication by an invertebrate cell line, but there were none on histopathology, nor were there current reviews.

Signs and Pathologic Changes

The first sign of IHN is usually greater-than-normal mortality—typically among fry or fingerlings, where cumulative losses can approach 100%. Although IHN sometimes occurs among smolts (fish of near-yearling or yearling age), the attendant mortality is not as severe as in the young. Most IHN occurs at temperatures of 12°C or less, although some outbreaks have occurred at 15°C.

Pathologic findings in victims have largely been reported from fry and fingerlings, but significant differences have been noted where IHN has killed older fish.

Behavior

Victim fry are generally lethargic, try to avoid current, and move to the edges of ponds or raceways. Swimming is usually feeble, but immobility may be interspersed with frenzied abnormal action—rolling, swimming vertically or in circles, and flashing. In chinook salmon, affected fry at raceway edges sometimes appear to be totally exhausted, but when stimulated (as by attempted netting) they briefly show surprising vigor as they attempt to escape. Such behavior is usually terminal, and the affected fry eventually drift downstream, lodge against outlet screens, and die.

Behavioral changes in yearlings with IHN have not been reported.

External Signs

The usual young victims are exophthalmic and many are darker than normal. In a given lot, the largest tend to be affected sooner and more severely than those of average or smaller size. Abdomens are swollen, and a fecal pseudocast of opaque white or tan material may trail from the vent (Fig. 1). Gills are usually pale, and fin bases may be noticeably hemorrhagic. Petechiation sometimes occurs laterally along the body and also in the mouth. Among chinook salmon fry, an abnormally dark reddish area may be present immediately behind the head; this discoloration is due to subcutaneous pooling of extravasated blood from the anterior kidneys.

Burke and Grischkowsky (1984) described an outbreak in sockeye salmon smolts in which most affected fish had large cutaneous lesions on the caudal peduncle and (to a lesser extent) around the pelvic and pectoral fins. Hemorrhages were evident in the gills and eyes of some smolts. The cutaneous lesions were judged nonbacterial.

Spinal deformities are common among surviving sockeye salmon. Amend et al. (1969) cited 5% scoliosis. Rucker et al. (1953) reported deformities in half of the survivors: of these 60% showed a retracted head, 30% scoliosis, and 10% lordosis. Vertebral deformity has not been reported in chinook salmon or rainbow trout.

According to Yasutake (1978), affected yearling sockeye salmon did not trail fecal casts nor have distended abdomens, and only a few showed slight exophthalmia, scoliosis, or darkening.

Figure 1. Thick castlike excretion trails from the vent of a moribund rainbow trout fry with IHN.

Internal Signs

As would be expected from petechiation and other blood loss, the overall internal appearance of fry is anemic. Food is characteristically absent from the digestive tract; instead, the stomach and intestine contain mucuslike fluid that is clear to yellowish and slightly cloudy. The body cavity may contain dilute serum-colored ascitic fluid. Petechiation is sometimes seen in visceral fat, mesenteries, peritoneum, swim bladder, meninges, and pericardium. Liver, kidneys, and spleen are usually pale. The hind gut may be hemorrhagic. Large fingerlings may show subdued pathologic changes and be somewhat compressed laterally, probably due to a combination of debilitation and lack of food in the digestive tract. In chinook salmon, the dorsal discoloration is clearly recognizable as accumulated blood if the skin over the area is laid back.

Yearling sockeye salmon showed no ascites.

Histopathologic Findings

In affected fry, major degenerative and necrotic changes occur in kidneys, hematopoietic tissues, pancreas, the gastrointestinal tract, and adrenal cortex. The liver is affected in some specimens. Necrosis of the granular cells of the digestive tract, when present, is considered to be pathognomonic for IHN because this change does not occur in other viral diseases of salmonids.

Renal and splenic hematopoietic tissues are generally affected first and most severely (Amend et al. 1969; Yasutake 1970, 1975; Yasutake and Amend 1972; Yasutake et al. 1965). In the anterior kidneys, an increase in macrophages and a

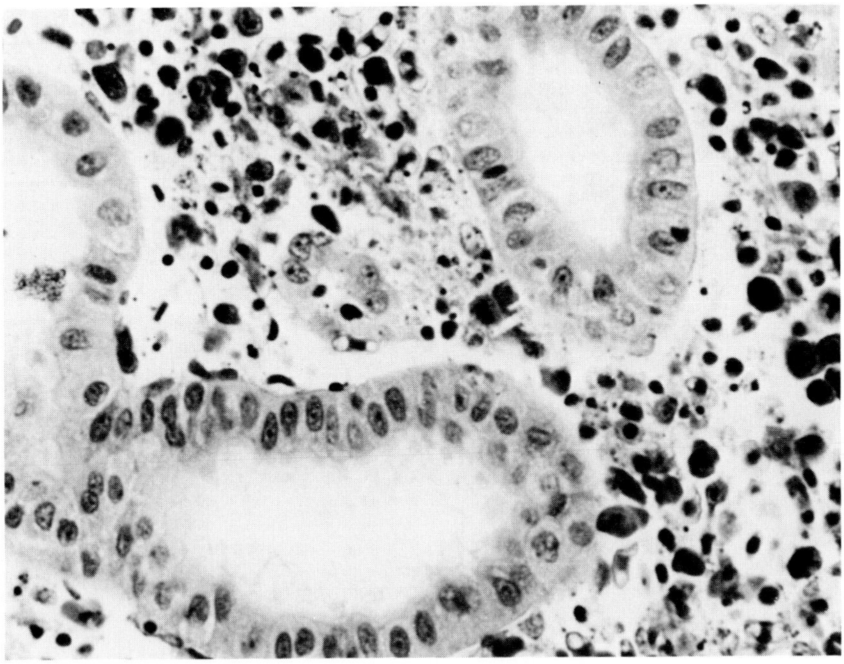

Figure 2. Section through a sockeye salmon kidney showing degenerative and necrotic hematopoietic tissue. ×1000. H & E. From Yasutake (1975). Reprinted with the permission of the University of Wisconsin Press.

decrease in nondifferentiated blast cells are accompanied by degenerative changes. Focal areas of cells showing pyknosis, nuclear polymorphism, and margination of chromatin become evident as the disease progresses. The frank necrosis of renal tissues that follows (Fig. 2) may also involve the interrenal tissue (adrenal cortex, between the anterior lobes of the kidneys). Similar necrosis is then found in pancreatic tissue. Eosinophilic granular cells of the gastrointestinal stratum compactum and stratum granulosum first show degeneration and then necrosis (Fig. 3). This necrosis serves to distinguish the histopathologic changes of IHN from those of viral hemorrhagic septicemia, in which it does not occur. Livers may show focal areas of degeneration and necrosis. Chinook salmon fry from natural epizootics may show hepatic deposits of ceroid. In the final stages of the disease, necrosis may also be found in glomeruli and kidney tubules, as well as in renal hematopoietic tissue (Fig. 4).

In chinook salmon fry from natural epizootics, an examination of victims having dorsal subcutaneous effusion of blood shows that this blood probably leaks from the anterior kidneys. Little or no histopathologic change takes place in the muscle through which it leaks. Yasutake (1970) noted that this clinical finding in the disease in chinook salmon is the principal feature differentiating it from the disease in sockeye salmon.

It has been postulated that the mucosal sloughing of the gastrointestinal tract found in some specimens gives rise to the fecal casts often seen in IHN (Amend et al. 1969; Yasutake 1970, 1975).

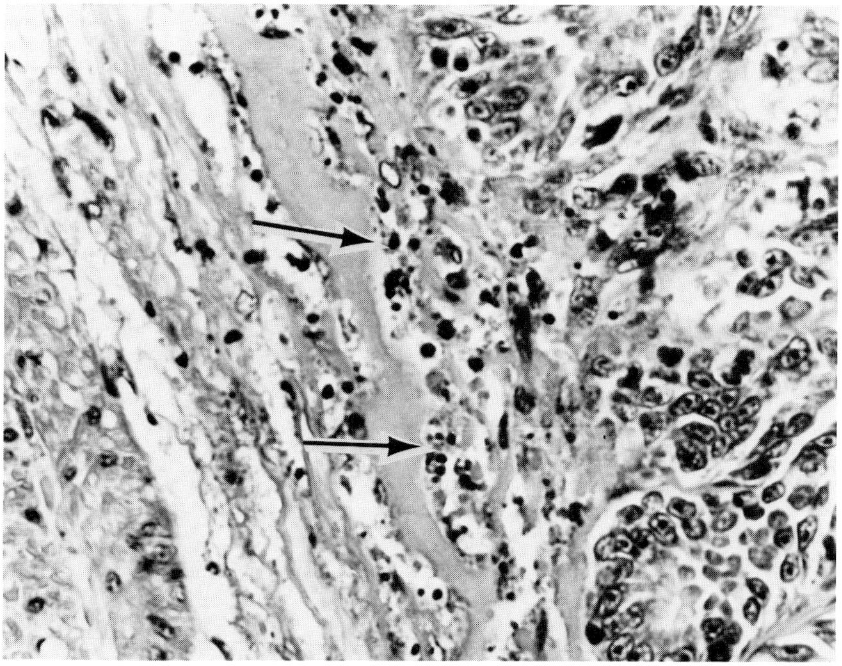

Figure 3. Section through a rainbow trout stomach showing extensive necrosis of granular cells (arrows). ×700. H & E. From Yasutake (1975). Reprinted with the permission of the University of Wisconsin Press.

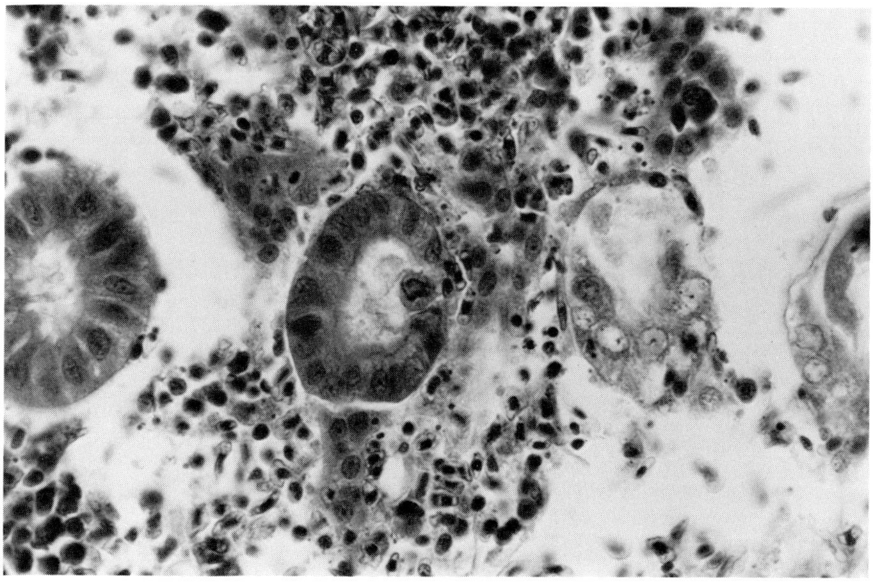

Figure 4. Section through chinook salmon kidneys showing a gradation of necrosis of tubules (left to right) and necrosis of hematopoietic tissue. Specimen courtesy of T. J. Parisot and W. T. Yasutake.

Wood and Yasutake (1956), who made a detailed study of the histopathologic changes in sockeye salmon fry from the late stages of an epizootic (80% mortality had occurred) found necrosis in about 90% of the livers, accompanied by deposits of ceroid. Spleens were normal except for a 90% occurrence of ceroid deposits. The kidneys were without necrosis; this observation was emphasized by the authors because infectivity has been shown to be concentrated in the kidneys in other species. The intestinal tract of about 30% of the victim sockeye salmon fry of an earlier period repeatedly showed small necrotic foci. It should be kept in mind that it was not possible to determine the virology of early epizootics; more than one disease might have been present. Nevertheless, Wood and Yasutake (1956) recognized that hematopoietic tissue was a prime target of the virus.

Compared with the effects of IHN in fry, histopathologic changes in yearlings may be less severe. In yearling sockeye salmon, Yasutake (1978) found subtle cellular degeneration and necrosis in the anterior kidneys and spleen. Renal hematopoietic tissue was affected and hyaline droplet degeneration was evident in tubular epithelium. Moderate sloughing was seen in intestinal mucosa, but fecal casts were not evident. In sockeye salmon smolts from an Alaskan epizootic, renal hematopoietic tissue showed significant necrosis, some spleen and liver cells were pyknotic, and the pancreatic tissue of some fish was hemorrhagic and degenerated. Gills were notably clubbed, epithelium showed hyperplasia, and lamellae were fused (Burke and Grischkowsky 1984). The gill pathology might be discounted as not due to IHN, were it not for the fact that the authors found gills to harbor the highest titers of IHNV.

The significance of gills in the diagnosis and study of IHN is growing: Mulcahy, Jenes, and Pascho (1984) found IHNV most frequently in the gills of adult sockeye salmon just before the fish spawned.

Clinical Findings

Marked hematologic and blood chemical changes occur in IHN. The studies carried out by Watson et al. (1956), and nearly two decades later by Amend and Smith (1974, 1975), are in general agreement, even though only presumptive virologic data were available for the earlier work. In a comparison of normal sockeye salmon fry with fry having natural or experimental infections, Watson et al. (1956) introduced not only a new technique for microhematocrit determinations, but also a bath or immersion method for experimentally infecting fish. They held fry for 15 minutes in a 1% filtrate of homogenized victim fish.

Blood of affected fry shows leukopenia, degenerating leukocytes and thrombocytes, and relatively large quantities of cellular debris; stained imprints of the anterior kidneys show even more debris and the presence of macrophages with vacuolated cytoplasm (Amend and Smith 1974, 1975; Holway and Smith 1973; Watson et al. 1956). Immature erythrocytes are increased significantly and some are obviously bilobed.

Compared with an average red blood cell count of $1.33 \times 10^6/\mu L$ in normal fry, the average was only 0.97×10^6 in moribund fry with IHN (Amend and Smith 1975).

For their normal fish—sockeye salmon for Watson et al. (1956) and rainbow trout for Amend and Smith (1975)—the mean hematocrit value was 47. In rainbow trout with IHN, the mean was 35. A time course was followed with the young sockeye salmon: the mean value rose to 55 at 4 days postinfection, but had dropped to only 16 on day 8; thereafter, it rose during recovery and was again normal on day 27.

Although their observations were not confirmed by others, Watson et al. (1956) believed that a true basophilia occurred and that there was a dissemination of extracellular basophilic granules, usually enclosed in cytoplasm. Parisot et al. (1965), who examined specimens taken early in an epizootic, concluded that the basophilic granules were really fragments of immature erythrocytes—portions of nuclear material surrounded by thin wrapping of cytoplasm. They noted this finding in both sockeye and chinook salmon.

Cellular debris (also termed necrobiotic bodies) occurs in imprints of anterior kidneys and is considered to have presumptive diagnostic value.

Amend and Smith (1975) described the blood picture of IHN victims as a normocytic aplastic anemia. The hemoglobin level, normally 9.4 g/dL, was 7.1 g/dL in IHN. Mean corpuscular volume, corpuscular hemoglobin, and corpuscular hemoglobin concentration were not significantly different from normal. Lymphocytes increased slightly from about 91% to 96% and neutrophils decreased from 4% to less than 1%; monocyte numbers did not change.

Watson et al. (1956) reported that the clotting time of blood was as long as 3 minutes in infected fry, compared with 15 to 20 seconds in normal fry.

Klontz et al. (1965) recorded the immunopathologic changes in chinook salmon and rainbow trout after they were infected with chinook and sockeye salmon strains of virus. Inoculation sites soon were infiltrated with macrophages; in turn these cells were cleared primarily in the anterior kidneys and secondarily in the spleen. At 10 to 12°C, infiltration was followed by an increase in circulating macrophages on days 3 and 4 and a return to normal by about day 7. The numbers of small lymphocytes and neutrophils increased significantly at the inoculation sites by days 3 and 4. When infections were lethal, erythrocytes were severely degenerated. Ratios of spleen weight to total body weight did not change in young chinook salmon, but increased transiently in rainbow trout.

Amend and Smith (1974, 1975) found severe changes in some of the biochemical constituents of blood. Bicarbonate, bilirubin, calcium, chlorides, osmolality, and phosphorus were reduced (Table 1), but glucose and kidney ascorbate remained

Table 1. Comparison of constituent levels in plasma from normal and IHNV-infected rainbow trout

Constituent and unit	Condition of fish	
	Healthy	Diseased
Bicarbonate (mEq/L)	10.15	5.95
Bilirubin (mg/dL)	2.1	1.2
Calcium (mg/dL)	10.8	8.07
Chloride (mEq/L)	117	110
Osmolality (mosmol)	315	277
Phosphorus (mg/dL)	15.38	12.11

Source: Adapted from Amend and Smith (1975).

unchanged. Total plasma protein did not change, but disk electrophoresis showed significant changes in globulins: alpha-2 increased (from 11.7% to 18.9%) and alpha-3 decreased (from 27.8% to 17.6%).

Plasma esterase, glutamic oxylacetic transaminase, and peptidase isozymes did not change in infected fish. Lactic acid dehydrogenase (LDH) assay on plasma and several tissues showed no changes in the tissues, but a decrease (of unknown significance) in the $B_4^{2\prime}$ LDH isozyme in the plasma.

Amend and Smith (1975) concluded that renal failure resulted in electrolyte imbalance, hemodilution, and death. Their conclusion was borne out by the observation that starvation for 19 days resulted in a loss of 6.5% of body weight in healthy fish but a gain of 20% in fish with IHN.

Etiology

The etiology of IHN has been firmly established: the causal agent is the rhabdovirus known as IHNV. An isolant of IHNV is available from the American Type Culture Collection as VR-714.

Size and Shape

As measured in thin section by Darlington et al. (1972), the mean length of three IHNV isolants was 170 nm (range, 150 to 190 nm) and the diameter was 70 nm (65 to 75 nm). An outer fringe measured about 10 nm (Fig. 5). In differential filtration, IHNV passes membranes of 220-nm mean pore diameter, but is retained by membranes of 100-nm porosity (Wingfield et al. 1969).

Because IHNV is fragile, fixation in glutaraldehyde precedes negative staining. Cohen and Lenoir (1974) found a bimodal distribution of negatively stained particles:

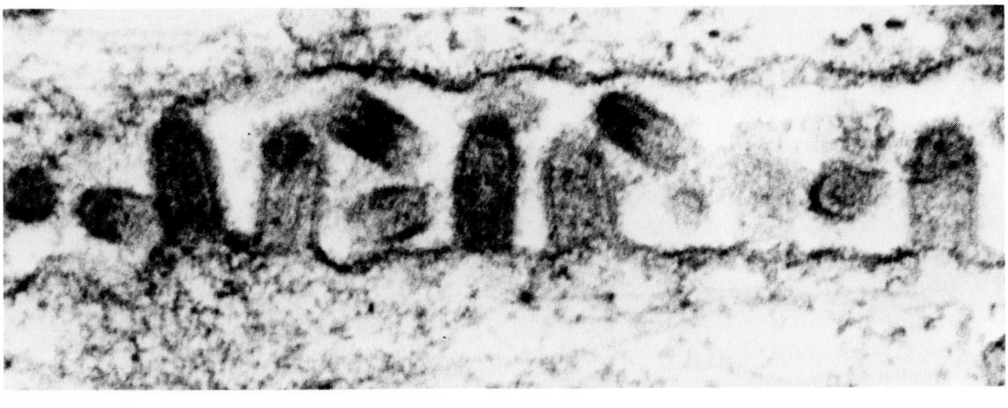

Figure 5. IHNV (chinook salmon isolant) budding from an FHM cell membrane. The plasma membrane is continuous with the outer membrane of virus. An irregular fringe extends beyond the outer membrane. ×103,400. From Darlington et al. (1972). Reprinted with the permission of Springer-Verlag.

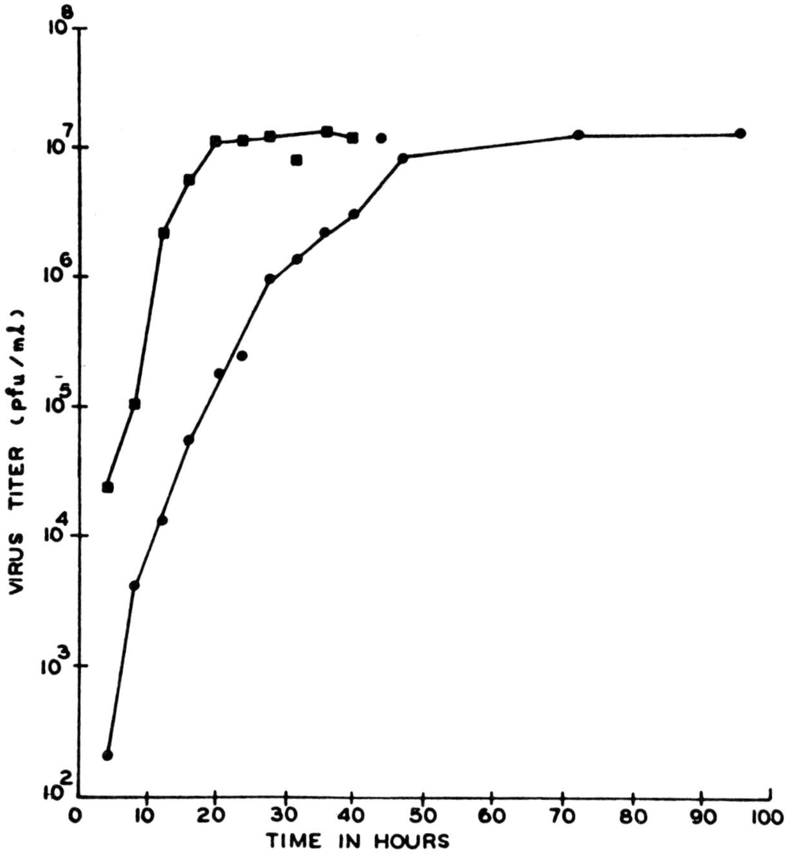

Figure 6. One-step growth curve of IHNV in CHSE-214 cells at 18°C. The squares represent total virus and the circles represent released virus. From McAllister et al. (1974a). Reprinted with the permission of Springer-Verlag.

the larger ones were 188 × 62 nm, whereas shorter and slightly conical forms were 120 × 95 nm. As in Egtved virus, the outer spikes of IHNV were difficult to visualize.

Biophysical Properties

Infectious hematopoietic necrosis virus is heat, acid, and ether labile. It is readily replicated by commonly used fish cell lines such as CHSE-214, EPC, FHM, RTG-2, and STE-137. Preferences for specific cell lines differ among investigators (Kelly et al. 1978; Kurath et al. 1985; Mulcahy, Pascho, and Jenes 1984).

Temperature of replication ranges from 4 to 20°C; the optimum is about 15°C—as could be expected for a coldwater fish pathogen—and 23°C is nonpermissive. A one-step growth curve in CHSE-214 cells at 18°C produces new virus by hour 4; exponential replication follows during the next 16 hours; a plateau is then reached at which maximum titer is about 10^7 PFU/mL (Fig. 6).

Infectious hematopoietic necrosis virus has been found capable of agglutinating erythrocytes of specific domestic geese, but only under additionally restrictive conditions. Consequently, that property has not been widely exploited (McAllister 1979). Similarly, little application has been made of the fact that IHNV has been replicated (with CPE) by certain mammalian and reptilian cell lines (Clark and Soriano 1974). Scott et al. (1980) grew the virus in mosquito cells at 16°C. Although a maximum titer of 10^8 $TCID_{50}$ was attained, CPE was not evident; further use of the insect cells for growing IHNV has not been reported.

Providing that preparations contain 10% serum or other source of protein, laboratory stocks of IHNV are probably best maintained by lyophilization. Pietsch et al. (1977) found that when serum was present survival was good at −70°C and infectivity could be expected to persist for at least several years at −20°C. Moreover, freezing and thawing were not inimical when serum was present. The preferred pH range was 6 to 8. At 4°C, a loss of 99.9% infectivity required more than 20 weeks; accordingly, Pietsch et al. (1977) concluded that IHNV could be easily stored and handled at room temperature for several hours without special precaution.

Loss of IHNV infectivity is hastened by electrolytes or salinity—brackish water or seawater (McAllister et al. 1974a; Pietsch et al. 1977). Virus added to freshwater at 15°C survives for 25 days, or about twice as long as in brackish water or seawater (Barja et al. 1983; Toranzo and Hetrick 1982).

The kinetics of inactivation at temperatures of 8 to 38°C were reported by Gosting and Gould (1981), who found a two-component pattern. The initial rate was rapid until 99.9% of the infectivity was inactivated; therefore, the rate was slower. However, no significant inactivation occurred during 5 hours at 22°C—an observation that supports claims that IHNV can be handled in the laboratory without special precautions to control temperature.

Burke and Mulcahy (1983), in addressing the problem of maintaining infectivity of field-collected specimens, added IHNV to ovarian fluids and homogenates of whole fry or organ tissues and later assayed them. At 4°C, infectivity persisted for several weeks in the most commonly collected materials—ovarian fluids and homogenates. They recommended short-term storage at 4°C for ovarian fluids and homogenates of fry, eggs, spleen, and brain; kidney and liver homogenates could be stored at −20°C for a month, but infectivity was not found after 1 year.

In cesium sulfate the buoyant density of the RNA is 1.59 g/mL; the base composition is 25.4% cytosine, 22.5% adenine, 27.2% uridine, and 24.2% guanine (McCain et al. 1974). Hill et al. (1975) found the sedimentation of the RNA in 5 to 25% sucrose to be 38 to 40S, and Leong et al. (1983) determined that RNA constituted 4.8% of the molecular weight of the virion.

Kurath and Leong (1985) isolated genome RNA and six messenger RNA (mRNA) species, determined their respective molecular weights, and identified the protein encoded for each (Table 2). Kurath et al. (1985) determined coding assignments for the mRNA species, and by R-loop analyses obtained a physical map of the IHNV genome and constituent genes.

However dissimilar serologically, the structual proteins of IHNV have been determined to resemble those of Egtved virus. Moreover, both of these viruses of coldwater fishes resemble rabies virus. Most investigators identified five polypeptides, but Leong et al. (1983) and Hsu et al. (1984) identified two forms of glycoprotein (Table 3).

Table 2. Characteristics of the RNAs of infectious hematopoietic necrosis virus

RNA	Molecular weight ($\times 10^6$)	Protein encoded
Genome	3.700	—
mRNA1	2.260	Polymerase (L)
mRNA2	0.563	Glycoprotein (G)
mRNA3	0.484	Nucleocapsid (N)
mRNA4	0.300	Matrix (M_1 and M_2)
mRNA5	0.195	Nonvirion (NV)

Source: Adapted from Kurath and Leong (1985).

In order of their appearance during replication, nucleocapsid (N) protein appears 2 to 3 hours after infection, matrix proteins (M_1 and M_2) at 6 to 7 hours, and glycoproteins (G_1 and G_2) at 9 to 10 hours. Polymerase (L) can not be distinguished from host cell proteins until after host cell synthesis is completely inhibited. Virus production begins 12 to 14 hours after infection (Leong et al. 1983).

Diagnosis

The clinical picture of fish with IHN offers clues that, coupled with epizootiologic information, can provide a basis for reasonably accurate diagnosis. The typical victims are advanced fry or small fingerlings that are abnormally dark and have exophthalmia and distended abdomens. Fecal pseudocasts are comparatively larger, and longer, and more coarsely textured than those encountered in IPN (Fig. 1). Diagnostic accuracy is increased when sick fish originate in husbandry facilities with low water temperatures and with stock from known carriers or from areas where IHN is enzootic. Clinically based diagnoses are additionally reinforced by the histopathologic findings of renal and hepatic necrosis, and especially by degeneration and necrosis of the granular cells of the stratum compactum and stratum granulosum.

Blood smears and imprints of anterior kidneys stained with Leishman-Giemsa have been advocated as being useful in the clinical diagnosis of IHN. Such preparations, particularly kidney imprints (Fig. 7), show cellular debris (necrobiotic bodies).

Table 3. Molecular weights ($\times 10^3$) of structural proteins of IHNV

Polymerase L	Glycoprotein G	Nucleocapsid N	Matrix M_1	Matrix M_2	Reference
>150	88	55	40	35	Hill et al. (1975)
157	72	40	25	20	McAllister and Wagner (1975)
190	80	38	25	19	Lenoir and de Kinkelin (1975)
153–145	71–63	47–41	26–22	21–17	Leong, Hsu, et al. (1981)[a]
150	67 (G_1) 65 (G_2)	40.5	22.5	17.5	Leong et al. (1983)
150	67 (G_1) 65 (G_2)	40.5	22.5	17.0	Hsu et al. (1984)

[a]Value ranges for 11 isolants.

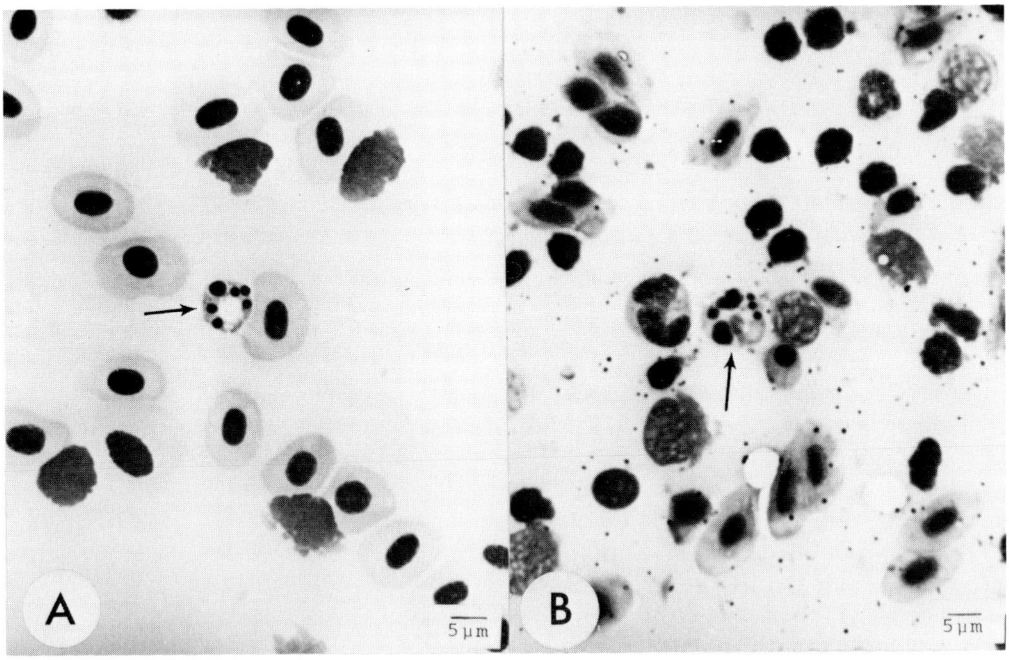

Figure 7. Necrobiotic bodies of IHN. *A.* Peripheral blood smear from a fingerling rainbow trout showing necrobiotic body (arrow). Leishman-Giemsa stain. Courtesy of W. T. Yasutake. *B.* Kidney imprint from a 7- to 8-month-old sockeye salmon showing necrobiotic body (arrow). Leishman-Giemsa stain. From Yasutake (1978). Reprinted with the permission of the Japanese Society of Fish Pathology.

It has recently been found that similar but perhaps lesser amounts of debris accompany IPN and viral hemorrhagic septicemia (W. T. Yasutake, personal communication). Erythrocytes in circulating blood show several abnormalities: many are immature, some are pleomorphic, cytoplasm may be vacuolated in some, and bilobed or dividing cells are occasionally found. The cytoplasm of the large numbers of circulating macrophages may show both ingested cellular debris and vacuolation (Amend and Smith 1975; Holway and Smith 1973).

As in all virologic infections of fishes (except lymphocystis), clinical diagnoses are fallible, and the critical cases should be resolved by virologic examination consisting of isolation and serologic identification.

Dual infections are known, although their relative frequency is low. Mulcahy and Fryer (1976) showed that rainbow trout fry could sustain a superinfection of IHN while maintaining the carrier state for IPN. In that case, IHNV was about 20 times more abundant than IPNV and was therefore considered to be the cause of mortality.

Isolation

Infectious hematopoietic necrosis virus is readily isolated in BF-2, CHSE-214, EPC, FHM, RTG-2, and STE-137 cell lines. Cell lines from warm-water fishes—BF-2, EPC, and FHM—offer the advantage of fast growth; they can be grown at 30°C and used for isolation at 12 to 15°C. BF-2 and FHM cells should be monitored for viral susceptibility; some lineages have lost susceptibility to one or more viruses.

The RTG-2 cells offer the distinct advantage of enabling one to distinguish among Egtved virus, *Herpesvirus salmonis,* IHNV, and IPNV by the characteristics of their plaques (see Chapter 18, Fig. 7). The IHNV plaques are distinctive: the infected cell sheet tends to retract or pile up at the inner margins of the opening, and the centers are generally open and show coarsely granular debris in discontinuous distribution (Wolf and Quimby 1973).

The CPE of IHNV in susceptible cells consists of a general rounding in grapelike fashion. Careful examination of nuclei along the edges of focal lesions shows margination of chromatin.

As in all salmonid viruses, sensitivity is highest when actively growing cultures are inoculated. Cultures for inoculation should be somewhat less than confluent; 80 to 90% confluency is suggested. For IHNV, pH is important and should be about 7.6 at the time inoculation; however, excellent results have been reported for cultures with pH as high as 8.0.

Temperature of incubation should not exceed 18°C, but more consistent results can be obtained at 15°C, and Amend (1970a) recommended 12°C.

For greatest sensitivity, cultures used for isolation should be drained of medium and the inoculum added directly to the cell sheet. Adsorption should proceed for 60 minutes at 20°C or less before the cells are again covered with medium—liquid for routine isolation and appropriate overlays for plaquing.

Under optimal conditions, CPE first becomes evident at 2 days at 15°C. Lower temperatures delay CPE, as do poor culture conditions and liquid media. When IHNV is to be plaqued, maximum numbers of plaques will be present after 4 to 7 days, depending on the method used and the temperature of incubation.

Several methods have been described for plaquing IHNV. Wolf and Quimby (1973) used RTG-2 cells and a two-phase agarose-liquid overlay that yielded plaques in 4 days. Okamoto et al. (1985), who used an agar overlay on RTG-2 cells, recommended 5 to 7 days of incubation at 18°C. Burke and Mulcahy (1980) preferred EPC cells and gum tragacanth gel and incubation at 16°C for 7 days. Leong, Fendrick, et al. (1981) effected a fivefold increase in infectivity by pretreating CHSE-214 cells for 30 minutes with 5 μg/mL polybrene and incubating under gum tragacath at 16°C for 7 days. The advantages and disadvantages of the different methods vary with the convictions and preferences of the investigator.

Amend (1970a) recommended a minimum of 14 days incubation at 12°C when tube or other cultures under liquid are inoculated directly; however, that period may be excessively long—a week is usually adequate. When microcultures are used and seeded with both virus and cell suspension, incubation for 4 to 5 days at 15°C is usually sufficient to show all infectivity. Cultures that have been inoculated and show no CPE should be blind passaged.

Moribund or dead fry from epizootics or from experimental infections pose no problem in isolation, for virus is usually abundant. When fry are small, several specimens should be pooled, homogenized (as a 5 to 10% preparation), and decontaminated.

Filtration, commonly through 0.45- or 0.22-μm membranes, is widely used for decontaminating extracts of homogenates. Alternative methods require clarifying centrifugation plus antibiotics. Amend and Pietsch (1972a) used a 2-hour treatment with final concentrations of 800 IU penicillin, 800 μg streptomycin, and 400 IU nystatin. Some workers simply add gentamicin to clarified inocula; the bactericidal antibiotic is used at concentrations of ≥100 μg/mL.

Whereas IHNV is abundant in and readily isolated from epizootic specimens, the virus is occult or latent in carriers until the fish become sexually mature and ready to spawn (Amend 1975; Mulcahy, Jenes, and Pascho 1984). From 2 or 3 weeks before spawning to actual spawning, the population prevalence of IHNV sometimes changes from zero to 100%. Mulcahy, Jenes, and Pascho (1984) attributed this sudden appearance of virus among sockeye salmon to senility and an impaired immune response. They noted that virus was most frequently found in gills just before the fish spawned; during spawning, it was present in all organs and fluids except brain and serum, and was most abundant in the pyloric caeca and lower gut.

Nonspawning carriers could conceivably be examined by the fluorescent antibody technique or other methods to reveal IHNV antigen. Renewed efforts at isolation could then be directed to specific tissues.

Identification

Several methods have been applied to the identification of IHNV, and strains or geographic isolants of the virus have been recognized. Although all strains cross-react serologically, tests clearly distinguish IHNV from other fish rhabdoviruses and, what is more important, from the Egtved virus.

Neutralization tests are the most commonly used definitive method of identifying IHNV, but enzyme-linked immunosorbent assay (ELISA) is gaining in popularity. Fluorescent antibody techniques have been developed, but have been applied primarily in studies of replication and only indirectly for pathogen identification.

Infectious hematopoietic necrosis virus has not been a particularly potent antigen, and the antibody response in rabbits has varied from low to moderate. McCain et al. (1971) prepared antiserum against isolants from Oregon (sockeye salmon), California (chinook salmon), and British Columbia (rainbow trout and sockeye salmon) using pelleted virus and Freund's complete adjuvant. The three isolants cross-reacted in plaque-neutralization tests and homologous titers ranged from 1:158 to 1:3650 (Table 4). The IHN and Oregon sockeye salmon isolants could not be distinguished. McAllister et al. (1974a) used the protocol of McCain et al. (1971) to prepare antiserum that was used for an indirect fluorescent antibody study of IHNV replication. Neutralization titer of the serum was not given.

Hill et al. (1975) pelleted IHNV and four other fish rhabdoviruses and gave initial intravenous inoculations and injections of virus with Freund's complete adjuvant in both hind legs of rabbits. They gave intravenous boosters at the third and fourth weeks and harvested the serum in the fifth week. Homologous IHNV plaque reduction titer was 1:1700, but the antiserum had a low level of activity (≤1:50) against the pike fry and spring viremia of carp viruses. Homologous titers against the pike and carp viruses ranged from 1:200 to 1:6500.

In a review of the preparation of antisera against various fish viruses, Hill et al.

Table 4. Results of cross-neutralization tests in which IHN, chinook, and sockeye salmon strains of virus were used with homologous and with heterologous antisera

	Serum		
Virus strain	Anti-IHN	Anti-chinook	Anti-sockeye
IHN	158	1465	410
Chinook	450	3650	660
Sockeye	142	1680	340

Source: Adapted from McCain et al. (1971).
Note: Titers shown are reciprocals of mean dilutions giving a 50% plaque reduction against 100 PFU.

(1981) showed that titers were generally higher for sera taken after two booster inoculations than for those taken after a single booster. Dixon and Hill (1984), using the basic protocol, produced a rabbit anti-IHNV serum with a plaque reduction titer of 1 : 50,000. The use of gamma globulin from that serum in ELISA revealed that IHNV was distinct from six other fish rhabdoviruses. The ELISA offers a significant saving of time: results were obtained in 2 hours. Also, ELISA enabled direct detection of IHNV in homogenates of infected fry.

Monoclonal antibodies have been produced against IHNV, and the immunoglobulin G (IgG) has been found to bind antigen but to have little neutralizing activity (Schultz et al. 1985). Details of virus identification were not given.

Leong, Hsu, et al. (1981) collected 11 isolants of IHNV, representing various locations in coastal North America, from Alaska to California, and five host species. Immunologic methods of strain identification were tried but proved to be unsatisfactory. When structural proteins were examined by SDS-PAGE, the investigators were able to distinguish four groups on the basis of differences in the molecular weights of the N and G proteins (Table 5). In addition, two isolants from California showed inhibition of growth at 18°C, and the isolant from the Coleman (California) NFH produced a plaque that was half the diameter of the plaque of most other strains.

As measured by neutralizing activity, immunization of rabbits with IHNV usually produces antiserum of low to moderate titer. Hsu and Leong (1985) found that alternative immunologic assays of low-titered antisera showed significantly greater levels of activity. As examples, sera with 50% plaque reduction titers of 1 : 32 and 1 : 250 measured 1 : 4096 and 1 : 32,768 respectively, when assayed by solid-phase direct binding with ^{125}I-iodinated protein A from *Staphylococcus aureus* and with peroxidase-labeled anti-rabbit IgG. The authors stated that the reactions were specific and sensitive to less than 10 ng of virus protein.

In a method developed by Hsu and Leong (1985), infected cultures were lysed and subjected to SDS-PAGE, and the proteins were transferred to nitrocellulose membranes. The resulting protein blots were then developed with peroxidase-labeled anti–rabbit IgG or with ^{125}I-iodinated protein A, and read by color development in the tests with the peroxidase reagent or by autoradiography of x-ray film exposed to the isotope-labeled protein A.

Hsu and Leong (1985) stated that the immunoperoxidase method is less sensitive

Table 5. Identification of IHNV strains based on structural proteins

	Protein		Source of isolants	
Group	Identity and relative size	Molecular weight ($\times 10^3$)	Location	Host species
1	N, small	39.5–40.5	Suttle Lake, Oregon	Brown trout
			Round Butte Hatchery, Oregon	Steelhead trout
			Tamagas Creek, Alaska	Pink salmon
2	N, large	42.7–43.5	Cedar Creek, Washington	Sockeye salmon
			Nan Scott Lake, Oregon	Rainbow trout
			Elk River, Oregon	Chinook salmon
3	G, large	71	Coleman Hatchery, California[a,b]	Chinook salmon
4	No distinctive differences		Karluk River, Alaska	Sockeye salmon
			Lewis River, Washington	Chinook salmon
			Feather River, California	Chinook salmon
			Trinity Hatchery, California[a]	Chinook salmon

Source: Adapted from Leong, Hsu, et al. (1981).
[a]Growth inhibition at 18°C.
[b]Small plaque size.

than neutralization, but is faster, requiring less than 55 hours. Assay of the transferred blots with radioactive protein A requires about 66 hours. The authors also suggested that the methods can be used to type virus strains because the G, N, and M_1 proteins are reactive with the binding activity of antiserum. Because the L and M_2 proteins are present in only small quantity—they are denatured in transfer—they are unrecognized in the two assays.

Transmission and Incubation

Horizontal transmission between fry or fingerlings is a major component of spread during epizootics, both in nature and in husbandry. Also, contact transmission from carriers to noninfected adults at spawning time is considered to be a distinct possibility. The gills are strongly implicated as a major portal of entry. Experimentally, IHN has been transmitted by immersion, by injection, and by mouth (but it must be noted that feeding does not exclude virus contact with gills).

There is strong circumstantial evidence for vertical transmission, but virus has yet to be found inside the eggs. Virus may be abundant in the environment during spawning time. As an example, 32–1600 PFU/mL were measured downstream of adult sockeye salmon showing 100% virus prevalence and being held at high density (Mulcahy et al. 1983b). Mulcahy and Pascho (1984) found that sperm from steelhead trout and chinook salmon strongly adsorbed IHNV: 99% of the available virus was adsorbed within 1 minute. Clearly, the sperm could provide a vehicle for entrance into eggs.

Mulcahy and Pascho (1985) reported two cases of apparent vertical transmission involving eggs or fry of sockeye salmon. In one case, eggs were collected from several redds, brought to a laboratory, and incubated in virus-free water. After hatching, several fry died and were found to have low levels of IHNV. In the other case, eggs

Table 6. Approximate incubation time for several different methods of infecting sockeye salmon fingerlings with infectious hematopoietic necrosis virus of the sockeye salmon strain

Method of infection	Approximate elapsed time in days	
	First death	50% mortality
Intraperitoneal injection	3–6	7–9
Bath (15 to 60 min) in infective homogenate	3	17
Contact with victims	9	14
Feeding	8	18
Water flow from victims	10	24

Source: Adapted from graphic data of Watson et al. (1954).
Note: Water temperature varied from 9 to 16.5°C.

taken from six carrier females were brought to the laboratory, fertilized with virus-free sperm, and incubated in virus-free water. Assay showed virus present with the eggs at 3 hours postfertilization, but not at 24 hours. On day 13 and again on day 21, dead eggs from one of the fish showed levels of one million or more PFU per gram. Although ovarian fluid from all six salmon contained virus, only the eggs from one retained virus during incubation.

Iodine compounds have been used to disinfect eggs from suspect or known source stocks; although the treatment has generally been successful, transmission has sometimes still occurred—once even after the eggs had been disinfected twice. Again, the inference is that vertical transmission was involved.

Incubation time for IHN is variable and depends on such factors as temperature, route of infection, strain of fish, amount of virus, and age of the fish. The usual incubation time (the interval between initial infection and the first deaths) ranges from 5 days to 2 weeks. Under exceptional conditions, death has occurred within 1 to 3 days after massive virus injection, or not until several weeks after a contact infection.

Watson et al. (1954), who compared several different means of infecting fingerling sockeye salmon, showed that the time to first mortality fell within the range of 3 to 10 days (Table 6).

Using chinook salmon fingerlings and chinook salmon strain virus, Ross et al. (1960) found that the shortest incubation time for injected virus was 3 days and that most of the test fish had died by 13 days. The temperature was not given. Ross and Rucker (1960) gave the incubation period as 3 days for inoculated fingerling chinook salmon and stated that the first death after exposure of healthy fish to infected fish came after 12 days. Amend et al. (1969), who used immersion to infect sockeye salmon and rainbow trout fingerlings, gave the shortest incubation time for IHN as 4 to 5 days to first mortality. By day 14, 60 to 90% of the fish thus infected were dead. Wingfield and Chan (1970) reported contact transmission by holding healthy young chinook salmon downstream from fish suffering a natural epizootic. The test fish "began to show symptoms within 7–10 days."

Hetrick, Fryer, and Knittel (1979) infected rainbow trout fingerlings by immersion in 10^5 PFU IHNV/mL for 24 hours and then moved the fish to six units with temperatures that ranged from 3 to 18°C, in increments of 3°C. They found a near-linear relationship between holding temperature and mean day to death: the mean was only 7 days at 18°C, compared with 17 days at 3°C. Each 3°C drop in temperature thus extended the incubation time by about 2 days.

Source

In unknown numbers, some survivors of IHN epizootics become carriers, apparently for life, but the virus is occult and has not yet been demonstrated until the fish have reached full sexual maturity. Other than during epizootics, IHNV has been isolated only from adults during the time immediately before, during, and after spawning. Brood fish, therefore, are the reservoir or source of IHNV and ensure safe passage of the pathogen from generation to generation.

Among Pacific salmon, more females than males are carriers. In a 3-year study of chinook salmon by Wingfield and Chan (1970), the examination of ovarian and seminal fluids revealed virus in 34% of the females but in only 5% of the males. Moreover, in a hatchery spawning run that lasted 3 months, the first carriers were detected after one-third of the population had been spawned. In surveys of sockeye salmon populations, Grischkowsky and Amend (1976) found an average carrier percentage of 44% (range 7 to 94%) in females and 13% (0 to 48%) in males.

Among Pacific salmon taken at sea, T. P. T. Evelyn (personal communication) was unable to isolate the virus. In contrast, 2% showed virus when they were in freshwater about a month before spawning; the rate increased to 50 to 80% during spawning and to 100% in postspawning salmon.

In adult rainbow trout, Amend (1974, 1975) found no difference between males and females in the prevalence of IHNV.

Host and Geographic Range

In North America the only species in which natural IHN outbreaks have occurred are brown trout, chinook salmon, pink salmon, sockeye salmon, and rainbow (steelhead) trout. In Japan, Sano et al. (1977) reported sustained epizootics in three species of salmon: chum (*Oncorhynchus keta*), amago (*O. rhodurus*), and yamame (*O. masou*). When injected intraperitoneally with IHNV, Atlantic salmon proved to be susceptible (K. Wolf, unpublished data). Parisot et al. (1965) reported that cutthroat trout were refractory to virus of sockeye salmon from two sources but that they were susceptible up to 4 weeks of age to the virus from chinook salmon. Wingfield and Chan (1970) listed the coho salmon as being completely resistant to isolants originally made from sockeye and chinook salmon.

The original range of IHNV was seemingly confined to America's Pacific Northwest—coastal streams from northern California to Alaska. As an example of distribution, surveys of 38 stocks of salmon in Washington by Amend and Wood (1972) yielded virus in 2 of 6 stocks of sockeye salmon, but in none of 32 stocks (collectively) of coho, chum, and chinook salmon. In Alaska, IHNV was isolated from all 16 stocks of sockeye salmon examined (Grischkowsky and Amend 1976).

The virus has been inadvertently introduced elsewhere, and in notable places it has become established. The present range includes Japan's Honshu and Hokkaido Islands, and America's Snake River Valley in Idaho.

The IHNV has been shipped with eggs or fry to some locations in the United States where outbreaks occurred in the original stock but where the virus apparently failed to become established. Holway and Smith (1973) found IHN in Montana, and outbreaks have been reported orally from Colorado and South Dakota. Plumb (1972) reported a case in Minnesota. In 1971, isolation of presumptively identified IHNV was made from fish in Virginia, and in 1972 serologically identified virus was reported from West Virginia (Wolf et al. 1973). In 1979, Carlisle et al. reported IHN among fish in New York. The sequential appearance of IHN from west to east could be interpreted as spread; however, the epizootiology implicated an origin from a known source of IHNV in the Pacific Northwest.

Parisot et al. (1965) and Ashburner (1970) cited clinical signs of a suspect viral epizootic among introduced chinook salmon fry in Australia. Although histopathologic examination was not conclusive, and virologic examination was not made, virus was clearly suspect. As judged by water temperature and the source of eggs—Coleman (California) NFH—the tentative conclusion was that "chinook salmon virus" (IHNV) was involved. Five years later, Ashburner (1970) reported that there had been no recurrence of the problem. Langdon et al. (1986), reporting on findings of virologic surveys made during 1981–1983 on 27 separate populations or year classes of salmonids in Australia, wrote that no evidence of virus was found, and cautiously suggested that the Australian populations are virus free.

Immunity

The immune response of young salmonids to IHNV involves production of antibody and interferon. Circulating interferon was demonstrated in subadult rainbow trout 48 hours after they were injected with the virus; however, its role in the host is uncertain (de Kinkelin and Le Berre 1974). Interferon is also produced in vitro (Sano and Nagakura 1982); Okamoto et al. (1983) reported that less was produced in young and sparsely confluent cell sheets than in older and denser cultures.

Some victims of IHN survive and grow to sexual maturity, but infection reappears at spawning time and massive amounts of virus are produced by multiple tissues. The renewed replication of virus could be a result of weakening or complete loss of immunity during this final life stage of salmon. Alternatively, the huge amount of virus produced might simply overwhelm even fully functional immune systems.

Early workers developed evidence of an immune response by showing that sockeye salmon survivors of IHN lived after challenge with levels of crude infectivity that killed previously nonexposed fish. The results were interpreted as evidence for acquired immunity (Guenther et al. 1959; Watson et al. 1954).

Amend and Smith (1974) provided other evidence that supported the concept of acquired immunity. Adult rainbow trout inoculated with IHNV produced virus-neutralizing activity that was specific for homologous virus but nonfunctional against IPNV. Furthermore, when fingerlings were injected with serum from immunized adults, passive immunization was effected. The protected fingerlings survived challenge with IHNV that killed more than half the controls that received virus alone or virus plus normal serum.

Immunogenicity of IHNV is retained during attentuation by repeated cell culture passage. Fryer et al. (1976) used virus that had undergone a 100-fold reduction in virulence to immunize sockeye salmon fry, either by intraperitoneal injection or by immersion for 48 hours. The immunized fry had survival rates of 72 to 100%, whereas the rates in nonimmunized controls were only 3 to 6%. Immunity was maintained for at least 110 days.

The response of salmonids to subunits of IHNV—especially to the G protein, but also to the M_1 and N proteins—is unknown. Potentially, one or more of the subunits could prove to be protectively immunogenic. Mass production by genetically engineered organisms could then lead to practical large-scale vaccine application at a low unit cost.

Control

Avoidance is the surest and most effective control measure for IHN, but realistic implementation requires that known sources of the virus—either infected fish or contaminated water supplies—be detected and identified. Water from protected springs or wells generally poses no risk of virus, and a critical assessment is easily made by exposing sentinel fish. In contrast, other sources—streams or lakes, particularly those in which carrier salmonids live or to which anadromous forms have access—present a continuing threat to the culture of young susceptibles. Unfortunately, in some places such water is all that is available for husbandry.

Avoidance of IHN is not always possible; other measures must then be applied. Carriers of the virus are common in many populations of America's Pacific salmon; even though the virus cannot be avoided, its impact can be greatly reduced. Mulcahy (1983) reviewed the topic of control and described a method of detecting carriers, termed brood stock culling, which has resulted in significantly lowered losses: Salmon that carry IHNV are detected during the spawning run. Eggs and ovarian fluid are taken from pools of three to five fish, but ideally from individual females. Samples of ovarian fluid are examined by plaque assay, and each source lot of eggs is incubated in a separate container. Virologic assay is carried out for 7 to 10 days under 15°C incubation. When samples are found to contain virus, the source lot of eggs is removed and destroyed. The approach is labor intensive and requires discipline to prevent contamination during spawning and incubation. The method does not lead to the detection of all carriers, but does reveal those that shed the most virus. Accordingly, the principal sources of virus are eliminated, and when the individual lots of eggs or fry are combined, the amount of residual virus is greatly reduced. At one hatchery, this application of brood stock culling reduced the usual losses of 60 to 97% to only 4 to 14%.

Although the prevalence of carriers is lower in males than in females, virologic examination of seminal fluid is suggested.

Amend and Pietsch (1972b) reported that treatments with iodine products disinfected eggs. Indeed, under laboratory conditions at near-neutral pH, a 5-minute exposure to iodophores giving a 25-ppm level of iodine destroyed the infectivity of IHNV, IPNV, and Egtved viruses. As a margin of safety, Amend (1974) recommended a 15-minute treatment with 100 ppm at an optimal pH of 6.0 or slightly higher. Where known carriers are present, disinfection is recommended. Even though vertical transmission might occur, treatment reduces the amount of virus on egg surfaces.

Temperature manipulation has also been recommended for practical control of IHN (Amend 1970b); in practice this procedure consists of raising the temperature to a point tolerated by eggs or fry but inimical to the virus. Elevation of temperature does not eradicate the virus and the warming must be started early in the incubation of the infection to forestall mortality.

Early investigators recognized that IHN mortality was appreciable at low temperatures (those below 10°C) and that survival increased at higher temperatures (Rucker et al. 1953; Watson et al. 1954). The effects of temperature were confirmed by others, and Parisot and Pelnar (1962) stated that temperatures above 13°C could be used to control epizootics.

The work of Amend (1970b) reinforced the recommendation for temperature manipulation. He found that mortality could be prevented by holding newly infected fish at 18°C for 4 to 6 days, but that treatment had to begin within 24 hours after infection. Under hatchery conditions, mortality was less than 10% when water was warmed to 20°C immediately after IHN was diagnosed; however, mortality exceeded 50% if warming was delayed for 1 week and was total if warming was withheld.

The concept of temperature manipulation is valid today, but subject to limitations imposed by energy costs for heating water, plus the difference in temperature sensitivity of different isolants of IHNV. Isolants from chinook salmon from northern California are sensitive, and the Coleman National Fish Hatchery has minimized losses by heating to 14°C. In contrast, IHNV in the Snake River Valley is a persistent problem even at the nearly constant 15°C of those waters.

Several antiviral compounds have shown promise in reducing mortality when administered to experimentally infected rainbow trout fry at 13°C. Hasobe and Saneyoshi (1985) tested 24 compounds, principally nucleoside analogues, and identified 14 that were effective in vitro at levels of 10 μg/mL or less. Five of the compounds were administered daily or on alternate days by flush treatment of unspecified duration. With daily administration, the most effective drugs were 6-thioinosine (0.1 mg/mL), virazole (0.1 μg/mL), and 5-hydroxyuridine (10 μg/mL). Survival at 28 days ranged from 20 to 34% in treated lots, compared with only 8% in nontreated virus controls. Treatment on alternate days was generally less beneficial.

Vaccination is a logical consideration for control of IHN; indeed, vaccines that offer protection (both killed and attenuated live virus preparations) have been developed and tested experimentally. Fryer et al. (1976) showed that IHNV could be attentuated by repeated passage in cell cultures and that significant protection could be provided by waterborne vaccine. Nishimura et al. (1985) demonstrated that a formalin-killed product was protective when administered intraperitoneally or by hyperosmotic infiltration. That preparation was intended for vaccination of brood stock, with the hope of clearing the virus from carrier adults.

Notwithstanding potential benefits, killed and attenuated live IHNV vaccines each have limitations that have effectively forestalled practical or hatchery-scale applications. The killed IHNV would be costly to produce and difficult to administer to large populations of young. Attenuated live virus has neither of those limitations; it could be produced economically and administered easily. However, there are two serious objections to its practical application: the stability of IHNV avirulence is not known, and no simple and rapid method of identifying attentuated IHNV is known. In the absence of a so-called marker, the use of live virus would interfere with existing virologic inspection programs because wild virulent virus could not be distinguished from a vaccine strain. The same kind of conflict would confront brood stock culling operations. As Mulcahy (1983) stated, "The use of attenuated vaccines, at least for the near future, should be anathema to fish health professionals."

A hope for the future is that a subunit IHNV vaccine can be developed that will provide protection, be producible at low cost, and be noninfective—and thus in no way conflict with programs of virologic detection and inspection.

References

Ahne, W. 1978. Eigenschaften fischpathogener Rhabdoviren. Pages 4–17 in H. H. Reichenbach-Klinke, ed. Fisch und Umwelt 5. Gustav Fischer Verlag, Stuttgart.

Amend, D. F. 1970a. Approved procedure for determining absence of infectious hematopoietic necrosis (IHN) in salmonid fishes. U.S. Fish Wildl. Serv. Fish Dis. Leaflet 31.

Amend, D. F. 1970b. Control of infectious hematopoietic necrosis virus disease by elevating the water temperature. J. Fish. Res. Board Can. 27:265–270.

Amend, D.F. 1973. Pathophysiology of IHN virus disease in rainbow trout. Ph.D. dissertation. Univ. Washington, Seattle.

Amend, D. F. 1974. Infectious hematopoietic necrosis (IHN) virus disease. U.S. Fish Wildl. Serv. Fish Dis. Leaflet 39.

Amend, D. F. 1975. Detection and transmission of infectious hematopoietic necrosis virus in rainbow trout. J. Wildl. Dis. 11:471–478.

Amend, D. F. 1976. Prevention and control of viral diseases of salmonids. J. Fish. Res. Board Can. 33:1059–1066.

Amend, D. F., and V. C. Chambers. 1970a. Morphology of certain viruses of salmonid fishes. I. In vitro studies of some viruses causing hematopoietic necrosis. J. Fish. Res. Board Can. 27:1285–1293.

Amend, D. F., and V. C. Chambers. 1970b. Morphology of certain viruses of salmonid fishes. II. In vivo studies of infectious hematopoietic necrosis virus. J. Fish. Res. Board Can. 27:1385–1388.

Amend, D. F., and J. R. Nelson. 1977. Variation in the susceptibility of sockeye salmon *Oncorhynchus nerka* to infectious hematopoietic necrosis virus. J. Fish Biol. 11:567–573.

Amend, D. F., and J. P. Pietsch. 1972a. An improved method for isolating viruses from asymptomatic carrier fish. Trans. Am. Fish. Soc. 101:267–269.

Amend, D. F., and J. P. Pietsch. 1972b. Virucidal activity of two iodophores to salmonid viruses. J. Fish. Res. Board Can. 29:61–65.

Amend, D. F., and L. Smith. 1974. Pathophysiology of infectious hematopoietic necrosis virus disease in rainbow trout (*Salmo gairdneri*): early changes in blood and aspects of the immune response after injection of IHN virus. J. Fish. Res. Board Can. 31:1371–1378.

Amend, D. F., and L. Smith. 1975. Pathophysiology of infectious hematopoietic necrosis virus disease in rainbow trout: hematological and blood chemical changes in moribund fish. Infect. Immun. 11:171–179.

Amend, D. F., and J. W. Wood. 1972. Survey for infectious hematopoietic necrosis (IHN) virus in Washington salmon. Prog. Fish-Cult. 34:143–147.

Amend, D. F., W. T. Yasutake, and R. W. Mead. 1969. A hematopoietic virus disease of rainbow trout and sockeye salmon. Trans. Am. Fish. Soc. 98:796–804.

Amend, D. F., W. T. Yasutake, J. L. Fryer, K. S. Pilcher, and W. H. Wingfield. 1973. Infectious hematopoietic necrosis (IHN). Pages 80–98 In W. A. Dill, ed. Symposium on the major communicable fish diseases in Europe and their control. EIFAC (Eur. Inland Fish. Advis. Comm.) Tech. Paper 17, Suppl. 2.

Ashburner, L. D. 1970. Some aspects of fish diseases. Aust. Soc. Limnol. Bull. 2:21–22.

Barja, J. L., A. E. Toranzo, M. L. Lemos, and F. M. Hetrick. 1983. Influence of water temperature and salinity on the survival of IPN and IHN viruses. Bull. Eur. Assoc. Fish Pathol. 3:47–50.

Burke, J., and R. Grischkowsky. 1984. An epizootic caused by infectious hematopoietic necrosis virus in an enhanced population of sockeye salmon, *Oncorhynchus nerka* (Walbaum), smolts at Hidden Creek, Alaska. J. Fish Dis. 7:421–429.

Burke, J. A., and D. Mulcahy. 1980. Plaquing procedure for infectious hematopoietic necrosis virus. Appl. Environ. Microbiol. 39:872–876.

Burke, J., and D. Mulcahy. 1983. Retention of infectious hematopoietic necrosis virus infectivity in fish tissue homogenates and fluids stored at three temperatures. J. Fish Dis. 6:543–547.

Burrows, R. E., L. A. Robinson, and D. D. Palmer. 1951. Tests of hatchery foods for blueback salmon (*Oncorhynchus nerka*) 1944–1948. U.S. Fish Wildl. Serv. Spec. Sci. Rep. Fish. 59.

Carlisle, J. C., K. A. Schat, and R. Elston. 1979. Infectious hematopoietic necrosis in rainbow trout *Salmo gairdneri* Richardson in a semi-closed system. J. Fish Dis. 2:511–517.

Clark, H. F., and E. Z. Soriano. 1974. Fish rhabdovirus replication in non-piscine cell culture: new system for the study of rhabdovirus-cell interaction in which the virus and cell have different temperature optima. Infect. Immun. 10:180–188.

Cohen, J., and G. Lenoir. 1974. Ultrastructure et morphologie de quatre rhabdovirus de poissons. Ann. Rech. Vet. 5:443–450.

Darlington, R. W., R. Trafford, and K. Wolf. 1972. Fish rhabdoviruses: morphology and ultrastructure of North American salmonid isolates. Arch. Gesamte Virusforsch. 39:257–264.

de Kinkelin, P., and M. Le Berre. 1974. Nécrose hématopoïétique infectieuse dans Salmonidés: production d'interferon circulant induite après l'infection expérimentale de la truite arc-en-ciel (*Salmo gairdneri*). C. R. Acad. Sci. [D] (Paris) 279:445–448.

Dixon, P. F., and B. J. Hill. 1984. Rapid detection of fish rhabdoviruses by the enzyme-linked immunosorbent assay (ELISA). Aquaculture 42:1–12.

Engelking, H. M., and J. C. Leong. 1981. IHNV persistently infects chinook salmon embryo cells. Virology 109:47–58.

Fryer, J. L., J. S. Rohovec, G. L. Tebbit, J. S. McMichael, and K. S. Pilcher. 1976. Vaccination for control of infectious diseases in Pacific salmon. Fish Pathol. 10:155–164.

Gosting, L. H., and R. W. Gould. 1981. Thermal inactivation of infectious hematopoietic necrosis and infectious pancreatic necrosis viruses. Appl. Environ. Microbiol. 41:1081–1082.

Grischkowsky, R. S., and D. F. Amend. 1976. Infectious hematopoietic necrosis virus: prevalence in certain Alaskan sockeye salmon, *Oncorhynchus nerka*. J. Fish. Res. Board Can. 33:186–188.

Grischkowsky, R. S., and D. Mulcahy. 1982. Effects of injection of hormones on the expression of infectious hematopoietic necrosis virus in spawning sockeye salmon (*Oncorhynchus nerka*). Pages 341–344 in B. R. Melteff, and R.. A. Neve, eds. Proceedings of the North Pacific aquaculture symposium, Newport, Oregon, August 25–27, 1980.

Guenther, R. W., S. W. Watson, and R. R. Rucker. 1959. Etiology of sockeye salmon "virus" disease. U.S. Fish Wildl. Serv. Spec. Sci. Rep. Fish. 296.

Hasobe, M., and M. Saneyoshi. 1985. On the approach to the viral chemotherapy against infectious hematopoietic necrosis virus (IHNV) *in vitro* and *in vivo* on salmonid fishes. Fish Pathol. 20:343–351.

Hetrick, F. M., J. L. Fryer, and M. D. Knittel. 1979. Effect of water temperature on the infection of rainbow trout *Salmo gairdneri* Richardson with infectious hematopoietic necrosis virus. J. Fish Dis. 2:253–257.

Hetrick, F. M., M. D. Knittel, and J. L. Fryer. 1979. Increased susceptibility of rainbow trout to infectious hematopoietic necrosis virus after exposure to copper. Appl. Environ. Microbiol. 37:198–201.

Hill, B. J., B. O. Underwood, C. J. Smale, and F. Brown. 1975. Physico-chemical and serological characterization of five rhabdoviruses infecting fish. J. Gen. Virol. 27:369–378.

Hill, B. J., R. F. Williams, and J. Finlay. 1981. Preparation of antisera against fish virus disease agents. Dev. Biol. Stand. 49:209–218.

Holway, J. E., and C. E. Smith. 1973. Infectious hematopoietic necrosis of rainbow trout in Montana: a case report. J. Wildl. Dis. 9:287–290.

Hoskins, G. E., G. R. Bell, and T. P. T. Evelyn. 1976. The occurrence, distribution and significance of infectious diseases and of neoplasms observed in fish in the Pacific region up to the end of 1974. Dep. Environ. Fish. Mar. Serv. Res. Dev. Dir. Tech. Rep. 609.

Hsu, Y. L., and J. C. Leong. 1985. A comparison of detection methods for infectious hematopoietic necrosis virus. J. Fish Dis. 8:1–12.

Hsu, Y., M. H. Engelking, and J. Leong. 1984. Analysis of quantity and synthesis of the virion proteins of infectious hematopoietic necrosis virus. Pages 45–46 in International seminar on fish pathology for the 20th anniversary of the Japanese Society of Fish Pathology, Tokyo, September 8–10, 1984. Abstract.

Kelly, R. K., B. W. Souter, and H. R. Miller. 1978. Fish cell lines: comparisons of CHSE-214, FHM, and RTG-2 in assaying IHN and IPN viruses. J. Fish. Res. Board Can. 35:1009–1011.

Kimura, T., and T. Awakura. 1977. Current status of disease of cultured salmonids in Hokkaido, Japan. Pages 124–160 in Proceedings from the international symposium on diseases of cultured salmonids, Seattle, Washington, April 4–6, 1977.

Klontz, G. W., W. T. Yasutake, and T. J. Parisot. 1965. Virus diseases of the Salmonidae in western United States. III. Immunopathological aspects. Ann. N.Y. Acad. Sci. 126:531–542.

Kurath, G., and J. C. Leong. 1985. Characterization of infectious hematopoietic necrosis virus mRNA species reveals a nonvirion rhabdovirus protein. J. Virol. 53:462–468.

Kurath, G., K. G. Ahern, G. D. Pearson, and J. C. Leong. 1985. Molecular cloning of the six mRNA species of infectious hematopoietic necrosis virus, a fish rhabdovirus and gene order determination by R-loop mapping. J. Virol. 53:469–476.

Langdon, J. S., J. D. Humphrey, J. Copland, R. Carolane, N. Gudkovs, and C. Lancaster. 1986. The disease status of Australian salmonids: viruses and viral diseases. J. Fish Dis. 9:129–135.

Lenoir, G., and P. de Kinkelin. 1975. Fish rhabdoviruses: comparative study of protein structure. J. Virol. 16:259–262.

Leong, J., and S. Turner. 1979. Isolation of water-borne infectious hematopoietic necrosis virus. Fish Health News 8(2):vi–viii.

Leong, J. C., J. L. Fendrick, S. Youngman, and A. Lee. 1981. Effect of polybrene on the infectivity of infectious hematopoietic necrosis virus in tissue culture. J. Fish Dis. 4:335–344.

Leong, J. C., Y. L. Hsu, H. M. Engelking, and D. Mulcahy. 1981. Strains of infectious hematopoietic necrosis (IHN) virus may be identified by structural protein differences. Dev. Biol. Stand. 49:43–55.

Leong, J. C., Y. Hsu, and H. M. Engelking. 1983. Synthesis of the structural proteins of infectious hematopoietic necrosis virus. Pages 61–71 in J. H. Crosa, ed. Bacterial and viral diseases of fish. Molecular studies. Wash. Sea Grant Publ., Univ. Washington, Seattle.

McAllister, P. E. 1973. The Oregon sockeye salmon virus (IHN): A. Replication and autointerference, B. The effect of temperature on infection in kokanee salmon (*Oncorhynchus nerka*) and on virus stability. Ph.D. dissertation. Oregon State Univ., Corvallis.

McAllister, P. E. 1979. Fish viruses and viral infections. Pages 401–470 in H. Fraenkel-Conrat and R. R. Wagner, eds. Comprehensive virology, Vol. 14. Plenum, New York.

McAllister, P. E., and K. S. Pilcher. 1974. Autointerference in infectious hematopoietic necrosis virus of salmonid fish. Proc. Soc. Exp. Biol. Med. 145:840–844.

McAllister, P. E., and R. R. Wagner. 1975. Structural proteins of two salmonid rhabdoviruses. J. Virol. 15:733–738.

McAllister, P. E., and R. R. Wagner. 1977. Virion RNA polymerases of two salmonid rhabdoviruses. J. Virol. 22:839–843.

McAllister, P. E., J. L. Fryer, and K. S. Pilcher. 1974a. Further characterization of infectious hematopoietic necrosis virus of salmonid fish (Oregon strain). Arch. Gesamte Virusforsch. 44:270–279.

McAllister, P. E., J. L. Fryer, and K. S. Pilcher. 1974b. An antigenic comparison between infectious hematopoietic necrosis virus (OSU strain) and the virus of hemorrhagic septicemia of rainbow trout (*Salmo gairdneri*) (Denmark strain) by cross-neutralization. J. Wildl. Dis. 10:101–103.

McCain, B. B. 1970. The Oregon sockeye virus: A. Biophysical and biochemical characteristics. B. Antigenic relationship to two other salmonid viruses. Ph.D. dissertation. Oregon State Univ., Corvallis.

McCain, B. B., J. L. Fryer, and K. S. Pilcher. 1971. antigenic relationships in a group of three viruses of salmonid fish by cross neutralization. Proc. Soc. Exp. Biol. Med. 137:1042–1046.

McCain, B. B., J. L. Fryer, and K. S. Pilcher. 1974. Physicochemical properties of RNA of salmonid hematopoietic necrosis virus (Oregon strain). Proc. Soc. Exp. Biol Med. 146:630–634.

McIntyre, J. D., and D. J. Amend. 1978. Heritability of tolerance for infectious hematopoietic necrosis in sockeye salmon (*Oncorhynchus nerka*). Trans. Am. Fish. Soc. 107:305–308.

Malsberger, R. G., and K. Wolf. 1966. Virus diseases of fishes. Pages 677–684 in J. E. Prier, ed. Basic medical virology. The Williams and Wilkins Co., Baltimore.

Moore, N. F., Y. Barenholz, P. E. McAllister, and R. R. Wagner. 1976. Comparative membrane microviscosity of fish and mammalian rhabdoviruses studied by fluorescence depolarization. J. Virol. 19:275–278.

Mulcahy, D. 1983. Control of mortality caused by infectious hematopoietic necrosis virus. Pages 51–71 in J. C. Leong and T. Y. Barila, eds. Proceedings of a workshop on viral diseases of salmonid fishes in the Columbia River Basin, Portland, Oregon, October 7–8, 1982.

Mulcahy, D., and K. Bauersfeld. 1983. Effect of loading density of sockeye salmon, *Oncorhynchus nerka* (Walbaum), eggs in incubation boxes on mortality caused by infectious hematopoietic necrosis. J. Fish Dis. 6:189–193.

Mulcahy, D. M., and J. L. Fryer. 1976. Double infection of rainbow trout fry with IHN and IPN viruses. Fish Health News 5:5–6.

Mulcahy, D., and R. J. Pascho. 1984. Adsorption to fish sperm of vertically transmitted fish viruses. Science 225:333–335.

Mulcahy, D., and R. J. Pascho. 1985. Vertical transmission of infectious hematopoietic necrosis virus in sockeye slamon, *Oncorhynchus nerka* (Walbaum): isolation of virus from dead eggs and fry. J. Fish Dis. 8:393–396.

Mulcahy, D., J. Burke, R. Pascho, and C. K. Jenes. 1982. Pathogenesis of infectious hematopoietic necrosis virus in adult sockeye salmon (*Oncorhynchus nerka*). Can. J. Fish. Aquat. Sci. 39:1144–1149.

Mulcahy, D., R. J. Pascho, and C. K. Jenes. 1983a. Titer distribution patterns of infectious hematopoietic necrosis virus in ovarian fluid of hatchery and feral salmon populations. J. Fish Dis. 6:183–188.

Mulcahy, D., R. J. Pascho, and C. K. Jenes. 1983b. Detection of infectious hematopoietic necrosis virus in river water and demonstration of waterborne transmission. J. Fish Dis. 6:321–330.

Mulcahy, D., C. K. Jenes, and R. Pascho. 1984. Appearance and quantification of infectious hematopoietic necrosis virus in female sockeye salmon (*Oncorhynchus nerka*) during their spawning migration. Arch. Virol. 80:171–181.

Mulcahy, D., R. Pascho, and C. K. Jenes. 1984. Comparison of in vitro growth characteristics of ten isolants of infecious hematopoietic necrosis virus. J. Gen. Virol. 65:2199–2207.

Nishimura, T., H. Sasaki, M. Ushiyama, K. Inoue, Y. Suzuki, F. Ikeya, M. Tanaka, H. Suzuki, M. Kohara, M. Arai, N. Shima, and T. Sano. 1985. A trial of vaccination against rainbow trout fry with formalin killed IHN virus. Fish Pathol. 20:435–443.

Okamoto, N., T. Shirakura, Y. Nagakura, and T. Sano. 1983. The mechanism of interference with fish viral infection in the RTG-2 cell line. Fish Pathol. 18:7–12.

Okamoto, N., T. Shirakura, and T. Sano. 1985. Precision of a plaque assay: eel virus European—and infectious hematopoietic necrosis virus—RTG-2 cell systems. Fish Pathol. 19:225–230.

Parisot, T. J. 1963. Sacramento river chinook disease (SRCD). U.S. Fish Wildl. Serv. Fish. Leaflet 562.

Parisot, T. J., and J. Pelnar. 1962. An interim report on Sacramento river chinook disease: a viruslike disease of chinook salmon. Prog. Fish-Cult. 24:51–55.

Parisot, T. J., W. T. Yasutake, and G. W. Klontz. 1965. Virus diseases of the Salmonidae in western United States. I. Etiology and epizootiology. Ann. N.Y. Acad. Sci. 126:502–519.

Pietsch, J. P., D. F. Amend, and C. M. Miller. 1977. Survival of infectious hematopoietic necrosis virus held under various conditions. J. Fish. Res. Board Can. 34:1360–1364.

Plumb, J. A. 1972. A virus-caused epizootic of rainbow trout (*Salmo gairdneri*) in Minnesota. Trans. Am. Fish. Soc. 101:121–123.

Ribelin, W. E., and G. Migaki, editors. 1975. The pathology of fishes. Univ. Wisconsin Press, Madison.

Ross, A. J., and R. R. Rucker. 1960. A "virus" disease of chinook salmon. U.S. Fish Wildl. Serv. Fish. Leaflet 497.

Ross, A. J., J. Pelnar, and R. R. Rucker. 1960. A virus-like disease of chinook salmon. Trans. Am. Fish. Soc. 89:160–163.

Rucker, R. R., W. J. Whipple, J. R. Parvin, and C. A. Evans. 1953. A contagious disease of salmon, possibly of virus origin. U.S. Fish Wild. Serv. Fish. Bull. 54:35–46.

Sano, T. 1976. Viral diseases of cultured fishes in Japan. Fish Pathol. 10:221–226.

Sano, T., and Y. Nagakura. 1982. Studies on viral diseases of Japanese fishes—VIII. Interferon induced by RTG-2 cell infected with IHN virus. Fish Pathol. 17:179–185. In Japanese, English abstract.

Sano, T., T. Nishimura, N. Okamoto, T. Yamazaki, H. Hanada, and Y. Watanabe. 1977. Studies on viral diseases of Japanese fishes. VI. Infectious hematopoietic necrosis (IHN) of salmonids in the mainland of Japan. J. Tokyo Univ. Fish. 63:81–85.

Schultz, C. L., B. C. Lidgerding, P. E. McAllister, and F. M. Hetrick. 1985. Production and characterization of monoclonal antibody against infectious hematopoietic necrosis virus. Fish Pathol. 20:339–341.

Scott, J. L., J. L. Fendrick, and J. C. Leong. 1980. Growth of infectious hematopoietic necrosis virus in mosquito and fish cell lines. Wassman J. Biol 38:21–29.

Toranzo, A. E., and F. M. Hetrick. 1982. Comparative stability of two salmonid viruses and poliovirus in fresh, estuarine and marine waters. J. Fish Dis. 5:223–231.

Watanabe, T. 1983. Persistent infection in YNK cells with infectious hematopoietic necrosis virus (IHNV). Bull. Jpn. Soc. Sci. Fish. 49:157–163.

Watson, M. E., R. W. Guenther, and R. D. Royce. 1956. Hematology of healthy and virus-diseased sockeye salmon *Oncorhynchus nerka*. Zoologica 41:27–38.

Watson, S. W. 1954. Virus diseases of fish. Trans. Am. Fish. Soc. 83:331–341.

Watson, S. W., R. W. Guenther, and R. R. Rucker. 1954. A virus disease of sockeye salmon: interim report. U.S. Fish Wildl. Serv. Spec. Sci. Rep. Fish. 138.

Wedemeyer, G. A., N. C. Nelson, and C. A. Smith. 1978. Survival of the salmonid viruses infectious hematopoietic necrosis (IHNV) and infectious pancreatic necrosis (IPNV) in ozonated, chlorinated, and untreated water. J. Fish. Res. Board Can. 35:875–879.

Wedemeyer, G. A., N. C. Nelson, and W. T. Yasutake. 1979. Potentials and limits for the use of ozone as a fish disease control agent. Ozone Sci. Eng. 1:295–318.

Wellings, S. R. 1970. Biology of some virus diseases of marine fish. Pages 296–306 in S. F. Snieszko, ed. A symposium on diseases of fishes and shellfishes. Am. Fish. Soc. Spec. Publ. 5.

Williams, I. V., and D. F. Amend. 1976. A natural epizootic of infectious hematopoietic necrosis in fry of sockeye salmon (*Oncorhynchus nerka*) at Chilko Lake, British Columbia. J. Fish. Res. Board Can. 33:1564–1567.

Wingfield, W. H. 1968. Characterization of the Oregon sockeye salmon virus. Ph.D. dissertation. Oregon State Univ., Corvallis.

Wingfield, W. H., and L. D. Chan. 1970. Studies on the Sacramento river chinook disease and its causative agent. Pages 307–318 in S. F. Snieszko, ed. A symposium on diseases of fishes and shellfishes. Am. Fish. Soc. Spec. Publ. 5.

Wingfield, W. H., J. L. Fryer, and K. S. Pilcher. 1969. Properties of the sockeye salmon virus (Oregon strain). Proc. Soc. Exp. Biol. Med. 130:1055–1059.

Wingfield, W. H., L. Nims, J. L. Fryer, and K. S. Pilcher. 1970. Species specificity of the sockeye salmon virus (Oregon strain) and its cytopathic effects in salmonid cell lines. Pages 319–326 in S. F. Snieszko, ed. A symposium on diseases of fishes and shellfishes. Am. Fish. Soc. Spec. Publ. 5.

Wolf, K., 1958. Virus disease of sockeye salmon. U.S. Fish Wildl. Serv. Fish. Leaflet 454.

Wolf, K. 1964. Characteristics of viruses found in fishes. Pages 140–148 in C. F. Koda, ed. Developments in industrial microbiology, Vol. 5. American Institute of Biological Sciences, Washington, D.C.

Wolf, K. 1966. The fish viruses. Pages 35–101 in K. M. Smith and M. A. Lauffer, eds. Advances in virus research, Vol. 12. Academic Press, New York.

Wolf, K., and M. C. Quimby. 1973. Fish viruses: buffers and methods for plaquing eight agents under normal atmospheres. Appl. Microbiol. 25:659–664.

Wolf, K., M. C. Quimby, L. L. Pettijohn, and M. L. Landolt. 1973. Fish viruses: isolation and identification of infectious hematopoietic necrosis in eastern North America. J. Fish. Res. Board Can. 30:1625–1627.

Wood, E. M., and W. T. Yasutake. 1956. Histopathologic changes of a virus-like disease of sockeye salmon. Trans. Am. Microsc. Soc. 75:85–90.

Yasutake, W. T. 1970. Comparative histopathology of epizootic salmonid virus diseases. Pages 341–350 in S. F. Snieszko, ed. A symposium on diseases of fishes and shellfishes. Am. Fish. Soc. Spec. Publ. 5.

Yasutake, W. T. 1975. Fish viral diseases: clinical, histopathological, and comparative aspects. pages 247–271 in W. E. Ribelin and G. Migaki, eds. The pathology of fishes. Univ. Wisconsin Press, Madison.

Yasutake, W. T. 1978. Histopathology of yearling sockeye salmon (*Oncorhynchus nerka*) infected with infectious hematopoietic necrosis (IHN). Fish Pathol. 14:59–64.

Yasutake, W. T., and D. F. Amend. 1972. Some aspects of pathogenesis of infectious hematopoietic necrosis (IHN). J. Fish Biol. 4:261–264.

Yasutake, W. T., T. J. Parisot, and G. W. Klontz. 1965. Virus diseases of the Salmonidae in western United States. II. Aspects of pathogenesis. Ann. N.Y. Acad. Sci. 126:520–530.

9 Infectious Pancreatic Necrosis

Synonym: acute catarrhal enteritis

Definition

Infectious pancreatic necrosis (IPN) is an acute contagious systemic birnavirus disease of fry and fingerling trout. The disease is widespread and commonly results in mortality that is inversely proportional to the age of the fish, being typically highest in the youngest fish and relatively rare in older fish (in which infections are often inapparent).

The IPN virus (IPNV) occurs as multiple strains that differ in virulence and serologic response. Although it was originally considered to be a virus of only salmonids, strains of IPNV have been found in a wide range of nonsalmonid fishes. Such infections are usually subclinical, but a few are associated with disease and mortality. In addition, IPNV (or IPNV-like viruses) have been isolated from mollusks (Hill 1976) and crustaceans, but the true role of such isolants (passenger or pathogen) remains to be determined.

Worldwide, IPNV is the most widely distributed of the aquatic viruses and, without doubt, has the most varied and complex role. In trout, it is likely that IPN is the same disease that early investigators termed acute catarrhal enteritis or, more simply, enteritis—a long recognized problem in North American trout husbandry. As an example, mention of "whirling sickness" of trout fry at a New York State fish hatchery could well have been an early report of IPN (State of New York Conservation Commission 1923).

History

In earlier fish husbandry in North America, acute "enteritis" among some lots of newly feeding hatchery trout fry was associated with a rise in mortality. Bacteriologic examinations revealed nothing, but a flagellate protozoan was abundant in the gut of such fish. *Hexamita*, or *Octomitus* as it was then named, was considered to be the etiologic agent. M'Gonigle (1941), a Canadian physician, showed that "enteritis" accompanied by whirling behavior and death occurred in the absence of the protozoan. He noted that if anything, fish that had the protozoan "seemed even more healthy than those not affected by the parasite." M'Gonigle considered acute catarrhal enteritis (his term) to be due to nutritional factors or to faulty feeding practices. In view of findings 30 years later that low temperature had a sparing effect on IPN mortality (Frantsi and Savan 1971b), tribute is owed to M'Gonigle's observation: he reported that during cold seasons enteritis "does not develop at all, or only slightly." It is noteworthy that at the conclusion of his presentation to the American Fisheries Society, there is a brief transcription of an oral discussion by Emmeline Moore, America's first fish pathologist. She voiced her skepticism of the pathogenicity of the protozoan.

Fourteen years later in the eastern United States, Wood et al. (1955) described a decidely infectious disease having the features of acute catarrhal enteritis. A histopathologic examination was carried out and, on the basis of the principal damage, the condition was named infectious pancreatic necrosis. The authors postulated a viral cause. During the same year K. Wolf (unpublished data) found histologic evidence of IPN in specimens of trout fry from Alberta, Canada. In the next studies, by Snieszko et al. (1957, 1959), infectivity was again demonstrated and susceptibility was found to decline with age.

In the meantime, methods for culturing fish cells and tissues were being developed in Snieszko's laboratory. Wolf and Dunbar (1957) published procedures for culturing trout tissues and in late 1957 made the first fish virus isolation. At the 1958 annual meeting of the Tissue Culture Association, Wolf et al. reported the isolation of IPNV—the first fish virus to be grown in vitro. The abstract was published the following year (Wolf et al. 1959) and full reports of the work appeared later (Wolf, Dunbar, and Snieszko 1960; Wolf, Snieszko, et al. 1960). Next, Wolf and Quimby (1962) initiated the RTG-2 cell line and noted its susceptibility to IPNV. In 1960 an epizootic of IPN erupted at the Leetown (West Virginia) National Fish Hatchery. Half of the infected population of rainbow trout fingerlings died, and a large stock of homogenized viscera was frozen. Virus was isolated, later cloned, and in 1963 deposited in the American Type Culture Collection as the prototype IPNV strain VR-299.

During the early 1960s IPNV began to attract other investigators. At the University of Miami, Moewus (1962) found that IPNV survived in *Tetrahymena*, a protozoan ciliate; the work was predicated on the possibility that the organism played a vector role. At Lehigh University, Malsberger began work in fish virology and over the years directed a series of doctoral research studies on fish viruses. The earliest of those efforts provided the initial characterization of IPNV (Malsberger and Cerini 1963). Cerini (1964) based his dissertation on the multiplication and morphology of the virus and, although his micrographs showed particles 18 nm in diameter that he mistook for IPNV, he was the first to employ electron microscopy of the virus. Results of the morphologic study were published (Cerini and Malsberger 1965), and six other reports on IPN were given at a conference of the New York Academy of Sciences.

Malsberger and Cerini (1965) published their expanded growth studies and reported that metabolic inhibitors indicated that IPNV was an RNA virus. Moewus-

Kobb (1965) found that IPNV survived, but did not multiply, in two of five species of marine fish for as long as a month, and that survival was comparable in seawater and in cultures of a pathogenic marine ciliate. Three related papers appeared from the U.S. Fish and Wildlife Service Western Fish Disease Laboratory (Seattle, Washington): Parisot et al. (1965) reviewed the literature on diseases of salmonids in America, with particular reference to etiology and epizootiology, compared several agents, and documented the occurrence of IPN in 8 of 10 western states; Yasutake et al. (1965) compared the pathogenesis of the viral diseases then known in North American salmonids; and Klontz et al. (1965), in describing the immunopathologic responses of three species of fingerling salmonids to experimental infections with the western viruses, reported that macrophage and leukocyte changes that accompanied IPNV were qualitatively similar to those in isolants from chinook and sockeye salmon. The sixth of the IPN papers in the symposium of the New York Academy of Sciences was by Gravell and Malsberger (1965), who described development of the fathead minnow (FHM) cell line, its susceptibility to IPNV, and the production of interferon.

Although others had tried therapy with iodine and found no benefit, Economon (1963) reported some reduction in IPN mortality among brook trout that were fed polyvinylpyrrolidone-iodine. He later published further observations on this topic (Economon 1973).

Tracing back and carrying out a virologic examination of brood stock brook trout that were circumstantially implicated as a source of IPNV, Wolf et al. (1963) found the virus in ovarian and seminal fluids. They also showed virus-neutralizing activity in the serum from virus carriers. At about the same time, Moewus and Sigel (1963) reported the in vitro growth of IPNV without CPE; theretofore all culture systems susceptible to the virus had shown marked lysis.

Beasley et al. (1965) described the development of a subline of grunt fin (GF) cells that were persistently infected with IPNV. Their work also implied the production of an interferon-like material by the cells. Infectious pancreatic necrosis was reported from France in 1965—the first occurrence of the disease outside North America (Besse and de Kinkelin 1965a, b; de Kinkelin and Besse 1966; Wolf 1966).

During 1966 two reviews of fish viruses brought the existing information on IPN up to date (Malsberger and Wolf 1966; Wolf 1966).

Continuing toward the objective of a practical method of control, Wolf et al. (1968) showed that a sequence of virologic examinations could be carried out on a population of IPNV carrier brood stock and that virus-free individuals could be selected for propagation. Also, Moss and Gravell (1968) gave the first correct size and capsomere number for IPNV. A later full report of that work (Moss and Gravell 1969) included a preliminary finding that the RNA of IPNV was double stranded. The size of IPNV was confirmed by Cohen et al. (1971) and Lightner and Post (1969).

Work intensified as the 1960s closed: Lyophilization and stability of IPNV were reported by Wolf, Bullock, et al. (1969). The clinical and immune responses of adult brook trout and rainbow trout were reported by Wolf and Quimby (1969). Yu et al. (1969, 1970) described their work on the immune response of normal and splenectomized blue gourami (*Trichogaster trichopterus*) following injection of IPNV. Billi and Wolf (1969), who compared peritoneal washes and feces for nondestructive detection of IPNV in adults, found the feces to yield significantly more accurate results. They also found that periodic fluctuation occurred in the amount of virus shed by individual carriers. Wolf, Quimby, and Carlson (1969), in tests for virulence of 15 IPNV isolants, showed that mortality ranged from less than 10% to more than 90%. They also measured neutralization indices of rabbit anti–VR-299 serum against 28 isolants of IPNV; again, there was a large spread of values. The results of their work provided the rationale for the formulation of a polyvalent antiserum comprising five component strains of the virus. Wolf and Vestergard Jorgensen (1970) showed cultures of the RTG-2 cell lines to be capable of simultaneous replication of Egtved virus and IPNV, and Yasutake (1970) described and illustrated the comparative pathology of the salmonid viruses then known.

The 5-year period of 1971–1975 was one of unusual productivity in research on IPN. Several doctoral dissertations and other contributions considered such diverse subjects as the pathogenesis and morphology of IPNV (Lightner 1971), the synergistic effect of zinc on viral replication in vitro (Hiller and Perlmutter 1971), and the epizootiology of IPN and esocid lymphosarcoma (Sonstegard 1970). A description of the almost complete inhibition of IPNV replication by actinomycin D (Nicholson 1971a) was followed by an autoradiographic study of protein and nucleic acid synthesis in infected cells (Nicholson 1971b, c). A preliminary determination (later supported but ultimately refuted) that the virus RNA was single stranded evolved from this work. Frantsi and Savan (1971b) demonstrated that, in routine sampling for virus, assays of kidney revealed more carriers than did those of feces; however, they noted that stress increased the number of carriers that shed virus in feces.

During 1971 more information appeared on the diversity of strains of IPNV. After a year's stay at the National Fish Health Research Laboratory, Sano returned to Japan and there found that an "unknown disease" of rainbow trout was really IPN, and that his isolants were serologically like a North American and a French strain (Sano 1970, 1971a, b). Wolf and Quimby (1971), who isolated several IPN viruses from French specimens, showed that they differed serologically from VR-299, that they were markedly labile at pH values much above 7.0, and that they were especially vulnerable to freezing and thawing. In Denmark, Vestergard Jorgensen and Grauballe (1971) found two IPNV strains—one designated Sp and the other Ab—of which the Ab strain was the less virulent and serologically distinct. In the same year Vestergard Jorgensen and Kehlet (1971), in surveys of Danish trout farms, found that the Sp strain accounted for more than 80% of their isolations and that the low-virulence Ab strain evoked CPE only in RTG-2 cells, not in FHM cells.

In 1972, Grauballe completed his doctoral dissertation on the etiology and diagnosis of IPN. Kelly and Loh (1972a, b) added significantly to the characterization of the virus, determining buoyant density and molecular weight of the virion, and the sedimentation value of the RNA. On the basis of several biochemical analyses they concluded that the genome was single stranded—a conclusion that agreed with that of Nicholson (1971c); however, they noted that in sucrose gradients the RNA of IPNV behaved as if it were double stranded. Theirs was the first suggestion that IPNV should be placed in a new classification group and that it was distinct from reovirus. Work by Argot (1969) on the intracellular replication of IPNV and the results of acridine orange staining by Argot and Malsberger (1972) led them to conclude that the nucleic acid of IPNV was double stranded. Cohen and Scherrer (1972) supported the size determinations of the virus and contributed details on its ultrastructure.

Continuing his serologic studies, Vestergard Jorgensen (1972b) compared incomplete and complete Freund's adjuvant for immunizing rabbits with IPNV and showed the superiority of the complete form. In the same year, Vestergard Jorgensen and Meyling (1972) briefly mentioned the first successful application of a direct fluorescent antibody technique on IPNV-infected cell cultures.

Sonstegard et al. (1972) called attention to previously unknown aspects of the virus; they found the agent in a nonsalmonid fish and showed that IPNV retained infectivity after passage through a mammal and two species of birds. Similar findings were reported for gulls by Eskildsen and Vestergard Jorgensen (1973). Sano (1972) isolated IPNV from fish in 24 of 27 hatcheries surveyed in Japan.

Although decontamination of materials for virus isolation had previously been routinely by filtration, Amend and Pietsch (1972) introduced a procedure employing antibiotics that they claimed was more economical and sensitive. Wolf and Quimby (1973a, b) described a two-phase method of plaquing whereby IPNV and the other fish viruses then isolated could be assayed. They noted that strains of IPNV plaqued differently: VR-299 showed easily discerned plaques in 48 hours, whereas others—notably the French isolants and certain of the less pathogenic North American isolants—required 72 hours to develop plaques of similar size.

The year 1973 was another banner year in IPN research. Scherrer (1973) reviewed IPN as then understood. Vestergard Jorgensen (1973b) reported on the effects of drying and chemical treatment for inactivation of the virus. New reports appeared from Japan, where Sano (1973a, b, c) confirmed a sparing effect of low temperature on fish with IPN, and found that two Japanese species of *Oncorhynchus* were susceptible to the disease. Piper et al. (1973), who applied a direct fluorescent antibody technique to a kinetic study of IPNV replication in vitro, showed that antigen could be detected within a mere several hours after infection.

Vestergard Jorgensen (1973a), in a scholarly report on the nature and activity of IPNV-neutralizing antibodies in normal and immunized rainbow trout, showed that immunoglobulin M had the neutralizing capacity but that serum from IPNV-free stock also had neutralizing activity in lower—but still appreciable—amounts. In an application of Wolf and Quimby's 5-strain polyvalent antiserum, Lientz and Springer (1973) tested 42 cloned isolants of IPNV and found that 2 were not neutralized; those 2 were later used as antigen to produce a 7-strain polyvalent antiserum. Perlmutter et al. (1973) described an unusual study of the immune response of a tropical fish to IPNV, reporting a pheromone-like immunosuppressant that was extractable with methylchloroform. Cohen et al. (1973) contributed significantly to characterization of the polypeptides and nucleic acid of IPNV and, as a result, clearly established the double strandedness of the virus genome.

In 1974, an immunofluorescent cell assay for IPNV was found to be about three times more sensitive than the plaque assay used for comparison (Tu et al. 1974). Homologous interference of IPNV replication by defective particles was reported (Nicholson and Dunn 1974), as well as the inhibition of cellular DNA synthesis by the virus (Lothrop and Nicholson 1974). Dorson and de Kinkelin (1974) found that serum from most salmonids tested had an IgM component that neutralized IPNV—especially the Sp and the Ab strains—and that it had a sedimentation coefficient of 6S and occasionally 16S. Scherrer and Cohen (1974) reported on the kinetics of replication of IPNV and its induction of interferon.

The replication of IPNV in a cell line from a tropical fish was described and the sensitivity of the virus to actinomycin D, at least early in the replication cycle, was confirmed by Kelly and Loh (1975). Their acridine orange staining led them to conclude again that the genome was RNA. Other work in 1975 was centered on immunologic and serologic studies. The mystery of neutraliztion of Sp virus by the 6S antibody was solved when it was found that the susceptibility was acquired by cell culture passage of the virus in as few as seven transfers; newly isolated virus was not neutralized, and it was lethal for trout fry (Dorson et al. 1975).

When the complement fixation test was used to type seven IPNV strains, strains Sp, Ab, and VR-299 were found to be distinct, and a polyvalent serum derived from only these three antigens neutralized all seven of the strains tested (Finlay and Hill 1975). Among the seven isolants were the Buhl (Idaho), Powder Mill (New Hampshire), and Bonnamy and d'Honnincthum (France). In Oregon, McMichael et al. (1975) carried out antigenic comparisons of three isolants of IPNV and found that differences in cross-plaque reduction tests were probably due to the binding of antibody by noninfective virus that resulted from lability in freezing and thawing. Invoking a related phenomenon—defective interfering particles—Nicholson and Dexter (1975) showed that infective tissue homogenates did not produce CPE in susceptible cell cultures unless the material was sufficiently diluted. Agniel (1975) quantified passive immunization of rainbow trout fry, using serum from immunized adult brook trout, and showed a dose response in the protection that was provided.

Alayse et al. (1975) described the synthesis of the RNA of INPV, and two reports appeared on the stability and inactivation of the virus (Desautels and MacKelvie 1975; MacKelvie and Desautels 1975).

From Germany, Schlotfeldt and Frost (1975) reported the first naturally occurring double infection of a fish with two viruses: rainbow trout were found to be infected with both Egtved virus and IPNV. During the same year, De Sena and Rio (1975) characterized the interferon produced in RTG-2 cells by IPNV and noted that it protected against heterologous IHNV, as well as against homologous IPNV. Within months after the Schlotfeldt and Frost (1975) report, Yamamoto (1975a) found IPNV and bacterial kidney disease in the same fish, and Mulcahy and Fryer (1976) described a dual infection of rainbow trout with IHNV and IPNV.

Comparatively speaking, the year 1976 saw few reports on IPN. Dobos (1976a) presented a paper on virus protein synthesis, and reported on the size and structure of the virus RNA (Dobos 1976b). McKnight and Roberts (1976) provided new information on the pathology of the infection. Hill (1976) isolated a virus from a marine mollusk with morphology like that of IPNV; however, it was antigenically different. Munro et al. (1976) reported finding IPNV in wild nonsalmonids during an epizootiologic study. Elazhary et al. (1976) reported the occurrence of low mortality associated with IPNV in yearling brook trout.

In 1977, Dobos and his collaborators published a sequence of sophisticated molecular biological studies of IPNV, and contributed new and complex information on the polypeptides synthesized by the two double-stranded segments of the virus genome (Dobos 1977; Dobos and Rowe 1977; Dobos et al. 1977). Macdonald and Yamamoto (1977) showed by a modified Kleinschmidt technique that the RNA was double stranded and that its length was about 0.92 μm. The single-stranded hypothesis of earlier years was laid to rest. Macdonald et al. (1977) then gave further support for IPNV having two species of RNA, each with a guanine-cytosine content of 54%.

Dorson (1977) gave a brief account of immunization trials and reported that protection was provided by inoculation with formalin-killed virus, even though neutralizing antibody was not detected. Hill and Dixon (1977) also reported on immunization and developed a standard procedure for determining pathogenicity. Baudouy and Castric (1977) added to the information on long-term stability of IPNV in water. Underwood et al. (1977) characterized the *Tellina* virus of Hill (1976), showing how the molluskan agent differed from IPNV but also pointing out features that were shared by the two. The authors suggested that the two agents belong to the same virus group—a new genus. Hill (1977) highlighted the then-current status of IPNV, and Munro and Duncan (1977) reviewed the problems of IPN in hatchery management. Ahne (1977a, b) announced the isolation of IPNV from northern pike; the isolant was avirulent for trout but sensitive to the 6S antibody described by Dorson et al. (1975). Ahne used the new isolant to develop a line of CHSE-214 cells that were persistently infected. Somewhat off the mainstream of IPN research, Seeley et al. (1977) showed that vertical transmission of the virus could be effected in a tropical fish by injecting the female, but that virus-injected males did not transfer infectivity with semen. Interestingly, the F_1 progeny of the injected females harbored the virus for at least 5 months, but apparently showed no sign of clinical disease.

The emphasis of reports appearing during the late 1970s was decidedly on the virus rather than on the disease. Reno et al. (1978) confirmed that adult trout could be infected with IPNV and that they could mount an immune response and clear the infection. In contrast, when brook trout fry were experimentally infected, disease followed and most fish became carriers. Whereas Reno (1976) had used an indirect immunoperoxidase test for IPNV identification, Nicholson and Henchal (1978) found a direct immunoperoxidase test to be the more specific. Nicholson et al. (1979) noted the differential replication of eight isolants by RTG-2 and FHM cell lines and postulated a variant IPNV. Dorson et al. (1978) showed that IPNV strains were highly changeable in culture, depending on conditions that were imposed. Ahne (1978) characterized his isolate of IPNV from northern pike, and Hedrick et al. (1978a, b) described cell lines persistently infected with IPNV.

Macdonald (1978) discovered a new phenomenon of IPNV in cell culture: after somewhat extended incubation in CHSE-214 cells, six isolants produced the traditional clear plaque, whereas four produced a ringed plaque that was correlated with a stable heritable trait of production of defective interfering particles. Macdonald and Yamamoto (1978) quantified the production of defective interfering particles and showed that they did not increase as a result of undiluted passage of virus and that they were incompletely separable from infective virus. Some features of apparent heterogeneity were clarified by the work of Chang et al. (1978), who compared 10 isolants of the virus and showed that they all were 71 nm in size, all had the same buoyant density of 1.33 g/mL, and all had a trio of polypeptides with molecular weights of 50,000, 30,000, and 27,000.

Two brief popular-type reviews on various aspects of IPN appeared (Hill 1978; McAllister 1978), and Busch (1978) reported the unintentional transmission of IPN by a batch of immunizing bacterin that had been contaminated with virus-carrying fish. Busch (1978) also noted the triggering of IPN losses among asymptomatic young fish that resulted from stress during the immunization procedures. Addressing their investigations to matters of control, Wedemeyer et al. (1978) determined ozone and chlorine inactivation curves (time and concentration) for IHNV and IPNV in waters of different chemical composition; and Elliott and Amend (1978) reported on the disinfecting efficacy of six different compounds routinely used in fish health.

The closing year of the decade saw IPNV and infectious bursal disease virus of poultry compared; the viruses were found to have similar size, shape, and RNA genome, but to be otherwise distinct (Todd and McNulty 1979). Work of broader scope included the drosophila X virus and molluskan viruses and the proposal of a new group—the birnaviruses—for the agents that had a bisegmented double-stranded RNA genome and that were distinct from reoviruses (Dobos et al. 1979).

In further developments in 1979, the persistent infection of cell lines with IPNV was shown to be due to defective interfering particles and not to interferon (Macdonald and Kennedy 1979); the inability of some IPNV isolants to grow in FHM cells was found to be due to a variant virus (Nicholson et al. 1979); and a methodology was developed for recovering as little as 1 PFU IPNV/L of water (Grinnell and Leong 1979). Two review papers on IPN were published for veterinarians (Schlotfeldt 1979a, b), and a complement fixation test for IPNV and other fish pathogens was described by Dorson et al. (1979). The replication of IPNV and IHNV was found to be enhanced in fish cell cultures infected with mycoplasma (Emerson et al. 1979). Pathogenesis of subclinical IPNV infection in Atlantic salmon was described (Swanson and Gillespie 1979). Yamamoto and Kilistoff (1979) reported the long-term persistence of virus in carrier brook trout stocked in an isolated lake but its disappearance within 6 years after annual plantings of virus-free fish were begun.

The IPN literature of the 1980s is voluminous—more than 100 relevant references appeared in 1980–1985. Accordingly, historical aspects are dealt with here only in broad generalization and only selected representative examples and key references are cited. Three reviews merit special attention: Dorson (1983) gave general and comprehensive coverage of the topic and current problems; Hill (1982) discussed in detail the virus and its virulence; and Dobos and Roberts (1983) reviewed the molecular biology of the virus.

About 20% of the 1980–1985 references were case reports that dealt with new hosts or outbreaks and isolations from new geographic areas, e.g., Chile, Korea, and Taiwan (Table 1). The reviews of Ahne (1985) and Hill (1982) included listings of species from which the virus had been isolated, but also included additional new hosts (Tables 2 and 3). Moreover, the new hosts included several species of aquacultural importance that showed signs of disease and sustained mortality: the yellowtail, *Seriola quinqueradiata* (Sorimachi and Hara 1985); the sea bass, *Dicentrarchus labrax* (Bonami et al. 1983); and the loach, *Misgurnus anguillacoubulatus* (Kou and Chen 1984).

Another 20% of the reports dealt with virus isolation and its detection or identification. Much attention was given to serologic techniques; those used in fish virology were reviewed by Ahne in 1981. Methods for rapid identification of IPNV include fluorescent antibody (Swanson and Gillespie 1981), ELISA (Dixon and Hill 1983a), and coagglutination (Kimura et al. 1984). Identification of virus strains was important, and two schools of thought remained. One group used genome and polypeptide

Table 1. Countries in which IPNV has been found

Country	Reference
Canada	MacKelvie and Artsob (1969)
Chile	McAllister and Reyes (1984)
Denmark	Vestergard Jorgensen and Bregnballe (1969)
England	Hill (1982)
France	Besse and de Kinkelin (1965a)
Germany	Schlotfeldt et al. (1975)
Greece	C. Carlson (personal communication)
Italy	Ghittino (1972)
Japan	Sano (1971a)
Korea	Hah et al. (1984)
Norway	Hastein and Krogsrud (1976)
Scotland	Ball et al. (1971)
Sweden	Ljungberg and Vestergard Jorgensen (1973)
Taiwan	Hedrick et al. (1983)
United States	Wood et al. (1955)
Yugoslavia	Fijan (1974)

profiles and considered most isolants to be unique. The other group used serologic methods; they recognized that cross-reactivity generally occurred among all isolants of IPNV, but considered that three major divisions were represented by prototype strains VR-299, Ab, and Sp (Macdonald and Gower 1981; Okamoto, Sano, et al. 1983).

The dichotomy of interests in the virus continued. One branch dealt with traditional aspects of host-virus relationships. The other branch—about one-tenth of the period's papers—dealt strictly with molecular biology (Dobos and Mertens 1983; Macdonald and Dobos 1981; Persson and Macdonald 1983; Stephens and Hetrick 1983). The first international symposium on double-stranded RNA viruses was held in the U.S. Virgin Islands in 1982. The proceedings, published in the following year,

Table 2. Salmonoid fishes from which IPNV has been isolated

Common name	Scientific name	Reference
Amago	*Oncorhynchus rhodurus*	Sano (1973b)
Arctic char	*Salvelinus alpinus*	Ljungberg and Vestergard Jorgensen (1973)
Atlantic salmon	*Salmo salar*	MacKelvie and Artsob (1969)
Brook trout	*Salvelinus fontinalis*	Wolf, Dunbar, and Snieszko (1960)
Brown trout	*Salmo trutta*	Wolf, Dunbar, and Snieszko (1960)
Chinook salmon	*Oncorhynchus tshawytscha*	Hill (1977)
Chum salmon	*Oncorhynchus keta*	Ahne (1985)
Coho salmon	*Oncorhynchus kisutch*	Wolf and Pettijohn (1970)
Cutthroat trout	*Salmo clarkii*	Parisot et al. (1963)
Danube salmon	*Hucho hucho*	Ahne (1985)
Grayling	*Thymallus thymallus*	Ahne (1978)
Lake trout	*Salvelinus namaycush*	D. Locke (personal communication)
Mountain whitefish	*Prosopium williamsoni*	Yamamoto and Kilistoff (1979)
Rainbow trout	*Salmo gairdneri*	Parisot et al. (1963)
Sockeye salmon	*Oncorhynchus nerka*	Sano (1973b)
Splake	*Salvelinus fontinalis* × *Salvelinus namaycush*	Yamamoto and Kilistoff (1979)

Table 3. Families of fish in which one or more species has been found infected with IPNV

Family number[a]	Family[a]	Common name
6	Petromyzonidae	Lampreys
147	Clupeidae	Herrings
163	Salmonidae	Trouts and salmon
165	Coregonidae	Whitefishes
166	Thymallidae	Grayling
202	Anguillidae	True eels
236	Catostomidae	Suckers
237	Cyprinidae	Minnows
239	Cobitidae[b]	Loaches
288	Esocidae	Pikes
294	Poecilidae	Top minnows
326	Bothidae	Lefteye flounders
327	Paralichthyidae	Bastard halibuts
333	Soleidae	Soles
373	Atherinidae[c]	Silversides
401	Carangidae[d]	Cavallas
410	Percidae	Perches
423	Moronidae[e]	White bass
461	Sciaenidae[c]	Drums
534	Cichlidae	Cichlids

Source: Adapted from Ahne (1985) and Hill (1982).
[a]Classification according to Jordan (1963).
[b]*Misgurnus anguillicaubulatus* (Kou and Chen 1984).
[c]*Menidia menidia* and *Leiostomus xanthurus* (McAllister et al. 1984).
[d]*Seriola quinqueradiata* (Sorimachi and Hara 1985).
[e]*Morone saxatilis* (Schutz et al. 1984).

included three papers on IPNV: Dobos and Mertens (1983), Hedrick and Okamoto (1983), and Persson and Macdonald (1983).

Other subjects of interest were addressed in the relevant publications of the period. Immunization and vaccines were discussed by Dixon and Hill (1983b) and Hill et al. (1980). Findings on transmission of IPNV, particularly on aspects of vertical transmission, were presented by Ahne and Negele (1985a, b) and Dorson and Torchy (1985). The reports of Okamoto et al. (1984) and Swanson et al. (1982) dealt with pathogenesis, and the effects of antiviral drugs and disinfectants were reported in works such as those by Ahne and Held (1980) and Migus and Dobos (1980). Smaller numbers of reports concerned minor and miscellaneous aspects of the virus and the infection.

As 1985 ended, promising research on monoclonal antibodies was under way under the leadership of J. C. Leong at Oregon State University, B. L. Nicholson at the University of Maine, and F. M. Hetrick at the University of Maryland. New research will undoubtedly lead in time to effective immunization and to resolution of the relationships among strains. Nevertheless, recent progress tended to be offset by the appearance of IPNV among more and more nonsalmonid fishes and in more and more geographic areas.

Signs and Pathologic Changes

IPN is best known in several species of salmonids, notably brook trout and rainbow trout. However, IPNV has also been implicated in disease among several non-salmonid fishes. Factors such as age and physiologic condition of the host, water temperature, and strain of virus affect the outcome. In hatchery trout, infection varies from inapparent or subclinical, in which little or no loss occurs, to acutely virulent, in which mortality is total or nearly so. Total mortality is the exception, however; it is more usual to have survivors. In trout, however, the survivors serve as virus carriers, and one study has shown that they additionally have a reduced condition factor (Rosenlund 1977).

In hatchery populations of trout, the first indication of an outbreak is usually a sudden increase in mortality among fry or fingerlings, and commonly a preferential kill of the larger and more robust individuals.

Behavior

Whirling that alternates with prostration is typical. Victim trout rotate about their long axis in a motion that may be slow and feeble or rapid and frantic. Whirling is commonly a terminal sign, but it may also be a transient change among fish that recover. Whirling also occurs with other diseases. After a siege of agonal swimming, the victims may sink to the bottom or lodge against the outlet screen, respiring rapidly and shallowly. Swimming may resume or the fish may die. When whirling is not readily apparent, it can be elicited by startling the fish. Whirling behavior may be lacking in fish of marginal quality and among the very young.

External Signs

Generally, in trout an overall darkening occurs. Mild to moderate exophthalmia and abdominal distention are common and hemorrhages are sometimes present in ventral areas, including the ventral fins. Gills are typically pale. Many victims trail long, thin, whitish, castlike excretions from the vent. The casts are thinner and more fragile than those of IHN and salmonid herpesvirus disease.

Internal Signs

In the older fingerling trout victims, IPN sometimes produces abundant petechial hemorrhages throughout the visceral mass. In contrast, the opposite condition is usual among the more typical victims—the fry. Spleen, heart, liver, and kidneys of fry are abnormally pale and the digestive tract is almost always devoid of food. If, on rare occasions, food residue remains in the gut, the quantity is small and confined to the far distal or rectal portion. In some fry, the pyloric caecae and anterior adipose tissue are flecked with petechiae, and the body cavity may contain ascitic fluid. The stomach and anterior intestine, although devoid of food, characteristically contain a pathognomonic clear to milky cohesive mucus. Since the mucoid material does not coagulate in the usual 5 to 10% formalin fixative, its presence may have diagnostic value in preserved specimens.

Histopathologic Findings

As the name indicates, the infection among trout produces marked pancreatic necrosis, but histopathologic changes sometimes also occur in adjacent adipose tissue, in renal hematopoietic tissue, in the gut, and in the liver.

Pancreatic necrosis is typically evident in acinar cells (Fig. 1), and extensive areas are sometimes involved. Some areas show a gradation of change, from necrotic cellular debris through intact but obviously abnormal acini to acini that appear to be normal. Some affected acinar cells show pyknotic nuclei and basophilic cytoplasmic inclusions; the inclusions, however, are products of cellular breakdown and not true viral inclusions. In the usual hatchery trout, adipose tissue is prominent about the pyloric caeca, and adipose cells may also show necrosis—whether by action of released zymogen or by direct viral attack.

Histopathologic changes can also occur in renal excretory and hematopoietic tissues, as first reported by Yasutake et al. (1965). Although renal damage was consistent with the high titer of virus typically found in the kidneys, at least of carrier fish, confirmation was first reported by Wolf and Quimby (1971). Sano (1971b, 1973c) also described and additionally illustrated renal damage from IPN in rainbow trout and in amago (*Oncorhynchus rhodurus*). There was congestion or hemorrhage in glomeruli, edema, and destruction or desquamation of tubule epithelium. Wolf and Quimby (1971) and Sano (1971b) reported still another change, the necrosis and sloughing of the intestinal mucosa (Fig. 2). McKnight and Roberts (1976) also found mucosal damage; they described it as acute enteritis that contributed to the castlike exudate and speculated that this damage might be a more lethal change than the necrosis of the pancreas.

Sano (1971b) found that some rainbow trout with IPN had congestion and necrosis of liver tissue, and later reported the same changes in *O. rhodurus* (Sano 1973c). Swanson and Gillespie (1979) similarly noted focal degeneration of liver parenchymal cells in yearling Atlantic salmon that had been inoculated with IPNV. Liver damage is consistent with virologic findings, and Kudo et al. (1973) showed that virions were present in hepatocytes.

Hyaline degeneration of skeletal muscle was described by Snieszko et al. (1957) as one of the pathologic changes of IPN, but such degeneration has since been established as being common in the absence of the disease.

At the ultrastructural level, electron microscopy shows virus in pancreatic tissue (Lightner and Post 1969) and in kidney tissue (Yamamoto 1974). Cytoplasmic aggregates, usually membrane bound and sometimes in crystalline array, occur in cells showing only slight pathologic change, but more virus becomes evident as necrosis approaches. At the ultrastructural level, inclusion bodies were described by Lightner and Post (1969); they consist of degenerating cytoplasm and virus, and are much smaller than those seen by light microscopy.

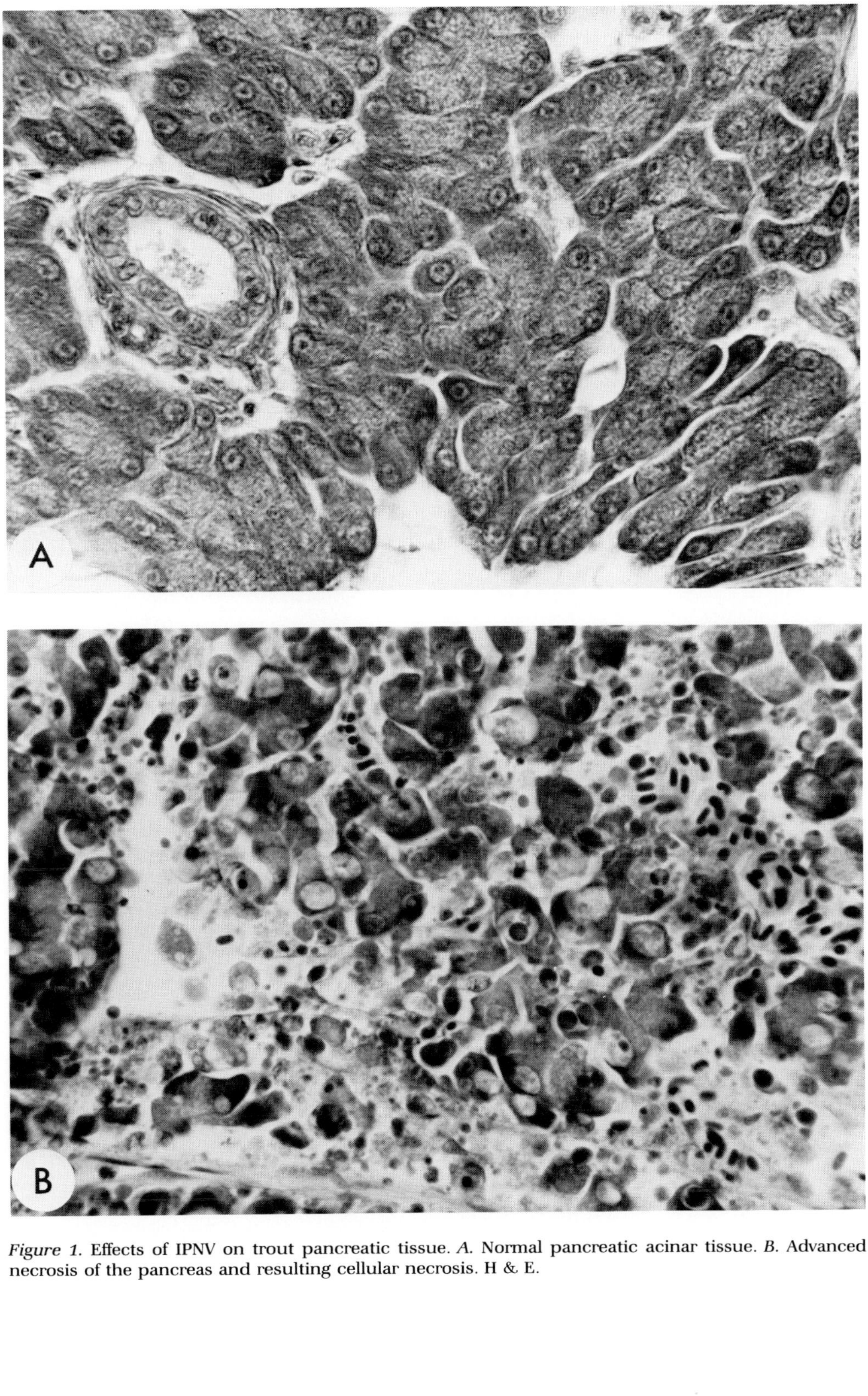

Figure 1. Effects of IPNV on trout pancreatic tissue. *A.* Normal pancreatic acinar tissue. *B.* Advanced necrosis of the pancreas and resulting cellular necrosis. H & E.

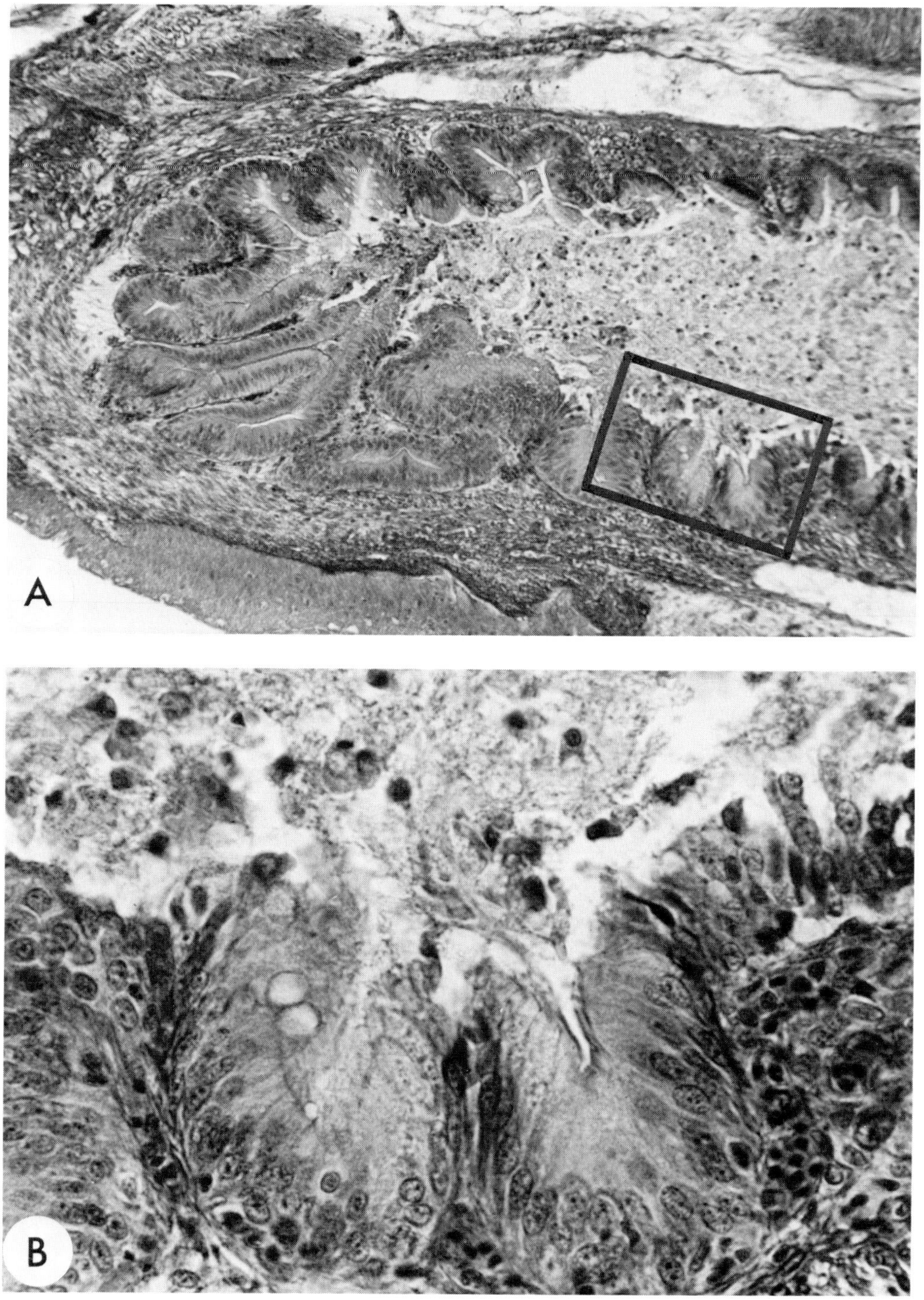

Figure 2. Sloughing of intestinal mucosa in IPN. *A.* Cellular debris fills the lumen of the intestine and constitutes a major portion of the fecal pseudocast. *B.* Enlarged inset of part A showing details of necrosis. H & E.

Clinical Findings

The usually small size of trout fry with IPN and the usual peracute course of the disease have discouraged anything but superficial study of clinical features. As expected, hematocrit values are depressed. From normal levels of 35 to 40, a small group of victims showed values as low as 20.

Strains of IPNV have been isolated during outbreaks of disease among nonsalmonid fishes, but few reports include details of accompanying signs and pathologic changes.

As reported by Sano, Okamoto, and Nishimura (1981), affected young Japanese eels (*Anguilla japonica*) had muscular spasms, a retracted abdomen, congestion of the anal fin, and (in some fish) congestion of the abdomen and gills. Food was absent from the gut, and victims had ascites and slight hypertrophy of the kidneys. Histology showed exudative glomerulonephritis, congestion of renal interstitium, nephrosis with hyaline droplet degeneration, and sloughing of tubule cells into lumens. Some livers and spleens showed focal necrosis. Except for gill changes, similar pathologic changes occurred among experimentally infected eels.

Sorimachi and Hara (1985) reported that fingerling yellowtails sustained acute disease with ascites. Experimental infection also resulted in severe ascites and additionally in liver hemorrhage.

Bonami et al. (1983) noted that larval sea bass showed a spiral swimming behavior, distended swim bladder, fecal casts, exophthalmia, and sloughing of gut epithelium. They withheld judgment about the role of IPNV in the disease.

Other investigators have attributed mortality among certain nonsalmonids to IPNV; although losses from experimental infections were significant, features of the disease were generally omitted. Moreover, virus titers on initial isolation—critically significant data—have usually been omitted.

Etiology

The viral etiology of IPN was established by Wolf, Snieszko, et al. (1960) and later confirmed by other investigators. The prototype strain of virus was isolated in 1960 from an outbreak among fingerling rainbow trout at the Leetown (West Virginia) National Fish Hatchery. Cloned virus from that epizootic was deposited with the American Type Culture Collection (ATCC) and designated VR-299. K. Wolf and M. C. Quimby later deposited six additional cloned strains of IPNV in the ATCC (VR-869, -876, -877, -881, -883, and -890). Stocks of the European reference strains Ab and Sp have not been deposited with the ATCC, but are available from major European laboratories.

Size and Shape

The IPNV is a nonenveloped icosahedron with a single capsid consisting of 92 capsomeres (Fig. 3). The diameter is variously given as 55 to 75 nm. Chang et al. (1978) reported the mean diameter of 10 isolants to be 71 nm; Nicholson et al. (1979) gave a

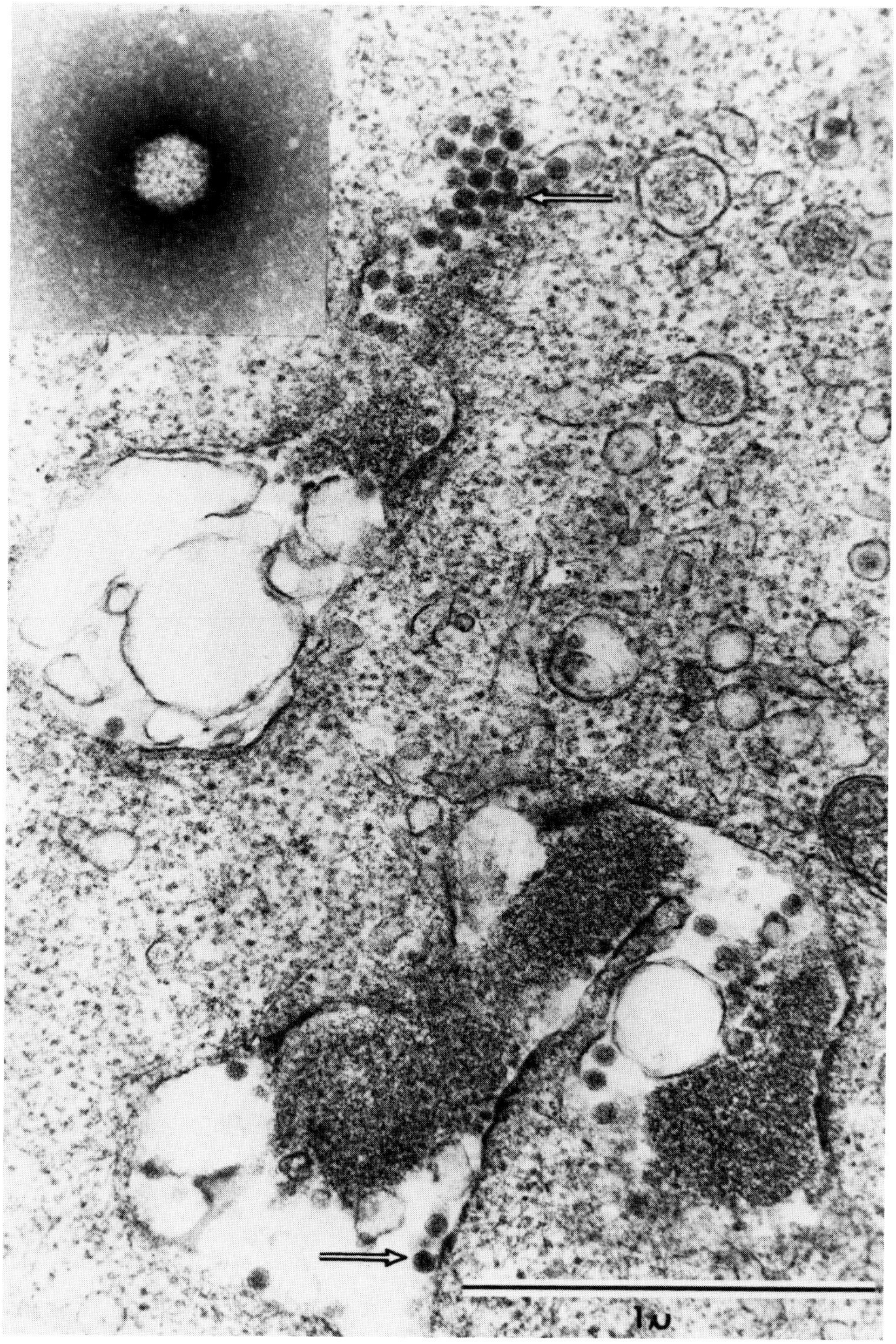

Figure 3. Infected RTG-2 cell showing IPNV in cytoplasmic vesicles (arrows). *Inset:* single negatively stained IPN virion showing a typical hexagonal profile. From Wolf and Quimby (1971). Reprinted with the permission of Springer-Verlag.

value of 69 nm for two isolants; Nishimura et al. (1981) found the mean of two isolants to be about 66 nm; and values of about 60 nm were measured for single isolants by Dobos et al. (1979) and Todd and McNulty (1979). This measurement—60 nm—is most commonly cited as the size of IPNV.

Biophysical Properties

Infectious pancreatic necrosis virus is the most studied of the fish viruses. It has a bisegmented, double-stranded RNA genome; accordingly, it belongs to the birnavirus group. It is acid, ether, and glycerol stable, and relatively heat stable. The Ab, Sp, and VR-299 strains hemagglutinate, but only in cells from particular strains of mice and only at pH 5.75 to 6.2 (Cleator and Burney 1980).

Providing that the suspending medium contains serum or other protein and that the pH is in the range of 5.0 to 7.0, IPNV is stable in storage at 4°C for 4 months, but for long-term storage it should be held at −20°C or lower. Thus far, it is the most stable of the fish viruses and is readily lyophilized in the presence of skim milk, lactose, or lactalbumin hydrolysate (Wolf, Quimby, and Carlson 1969).

Infectious pancreatic necrosis virus is routinely replicated in such commonly used fish cell lines as AS, BF-2, CHSE-214, EPC, and RTG-2. Isolates of strain Ab do not grow in the FHM cell line.

Typically, IPNV results in a rapidly lytic CPE in susceptible cell cultures, and in yields of infectivity of 10^6 to 10^9 PFU/mL. However, cell lines can also be persistently infected and not show CPE. That response was first reported by Moewus and Sigel (1963). More recent findings have shown that defective interfering particles, not interferon, are responsible for the development of persistently infected cells (Hedrick and Fryer 1981; Macdonald and Kennedy 1979; Macdonald and Yamamoto 1978; Nicholson and Dunn 1974).

The normal temperature range for growth of IPNV is 4 to 27.5°C. Scherrer et al. (1974) found that the upper limit of temperature for replication was 27.5°C in FHM cells. By increasing the temperature of incubation by a small increment with each passage, the virus can be adapted to grow at temperatures as high as 30°C.

Malsberger and Cerini (1963, 1965) were the first to determine growth rates for IPNV, and their results are representative of the kinetics of production of titratable infectivity (Figs. 4 and 5). New virus appears by the fifth hour at 24°C; 8 hours are required at 15°C, but the ultimate yield is greater. Cultures of RTG-2 cells have shown CPE after 9 hours of incubation at 26°C and after 18 hours at 20°C, but not until after several days at 4°C.

The buoyant density of IPNV in cesium chloride was determined to be 1.33 g/mL by Chang et al. (1978), Cohen et al. (1973), and Kelly and Loh (1972a, b). Dobos et al. (1977), who made some of the most complex determinations of biophysical properties of IPNV, found the molecular weight of the particle to be 55×10^6 and estimated the total weight of capsid protein to be 50.2×10^6. The difference of 4.8×10^6 was due to the RNA component, which constitutes 8.7% of the particle weight. They gave the sedimentation coefficient of the particle as 435S.

The double-strandedness of the RNA was correctly inferred by Moss and Gravell (1969); although several later investigators concluded that the genome was single

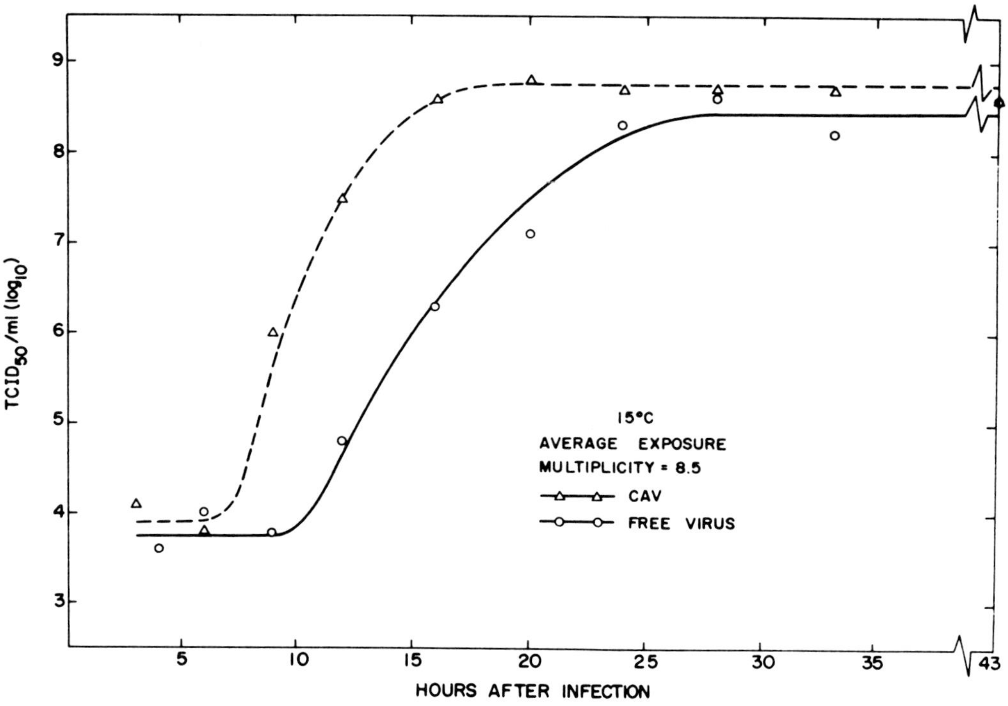

Figure 4. One-step growth curve of IPNV in RTG-2 cells at 15°C. CAV, cell-associated virus. From Malsberger and Cerini (1965). Reprinted with the permission of the New York Academy of Sciences.

stranded, double strandedness was firmly established by the convincing work of Alayse et al. (1975), Cohen (1975), Cohen et al. (1973), Dobos (1976b, 1977), Macdonald and Yamamoto (1977), and Macdonald et al. (1977). In cesium sulfate the RNA has a buoyant density of 1.60–1.615 g/mL and on sucrose gradients the sedimentation value is 14S. Using the Kleinschmidt protein film technique, Macdonald and Yamamoto (1977) showed the RNA to be a linear and double-stranded molecule with an average length of 0.92 ± 0.07 μm. The length corresponded to a molecular weight of $2.4 \pm 0.2 \times 10^6$.

Infectious pancreatic necrosis virus has two species of RNA; Dobos (1976b) found them to weigh 2.3 and 2.5×10^6 and Macdonald and Yamamoto (1977) stated that the molecules could not differ by more than 4×10^5. Macdonald et al. (1977) confirmed the presence of only two RNA molecules in IPNV by oligonucleotide fingerprinting. They also determined that the total guanine-cytosine content was about 54%.

Infectious pancreatic necrosis virus has an RNA-dependent polymerase that catalyzes synthesis of single-stranded RNA from the double-stranded genome and cosediments with viral infectivity (Cohen 1975; Mertens et al. 1982). The RNA of IPNV is unusual in that the polymerase occurs both in free form and probably bound covalently to the 5′ end of each genome strand (Persson and Macdonald 1982, 1983).

In contrast to the general uniformity of biophysical values reported for virus RNA, the results of analyses of structural proteins vary considerably. Understandably, the differences found are due to the use of different methods and reagents. Some

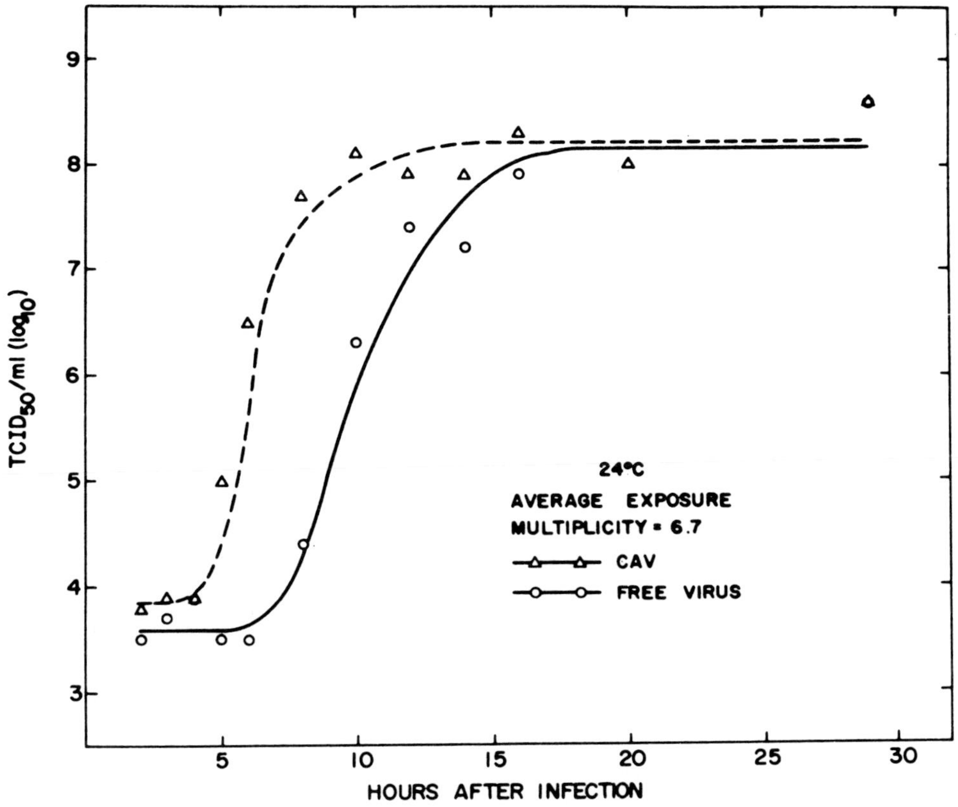

Figure 5. One-step growth curve of IPNV in RTG-2 cells at 24°C. CAV, cell-associated virus. From Malsberger and Cerini (1965). Reprinted with the permission of the New York Academy of Sciences.

agreement is to be found in the results obtained by Chang et al. (1978), Cohen et al. (1973), Dobos (1977), Loh et al. (1974), and Macdonald and Dobos (1981). According to the review by Dobos and Roberts (1983), the structural proteins fall into three size classes. One large polypeptide has a weight of 105,000, constitutes 4% of the total molecular weight, and is the putative RNA-linked polymerase. A medium-sized polypeptide of 54,000 bulks large at 62% of the total weight, is the major capsid protein, and is responsible for stimulating production of neutralizing antibodies. Two small internal proteins have weights of 31,000 (28%) and 29,000 (6%).

The infectivity of IPNV under laboratory storage and test conditions is unusually stable. In 1960, I and my associates at the National Fish Health Research Laboratory homogenized equal volumes of infective viscera and Hanks' balanced salt solution and stored multiple samples at −20°C. The titer was 10^8 $TCID_{50}$/mL after 3 years and 5.7 × 10^6 PFU/mL after 19 years, in spite of two power failures during which the material thawed.

The infectivity of IPNV is unusually persistent—the most persistent of any fish virus. In filter-decontaminated water held at 4°C, infectivity persisted for at least 5 to 6 months (Baudouy and Castric 1977; Desautels and MacKelvie 1975; MacKelvie and Desautels 1975). In municipal tap water held at 10°C, infectivity similarly persisted for more than 7 months, but at 10°C in nontreated canal water with its natural flora, virus was not detectable after only 14 days (Ahne 1982). Comparative survival studies at 15°C in nontreated natural waters showed that infectivity was longest in estuarine water—27 days was required to effect a 3-log loss. Comparable loss occurred in freshwater and seawater in only 17 days (Barja et al. 1983; Toranzo and Hetrick 1982). Infectivity persisted four times longer in filter-decontaminated or autoclaved estuarine water than in nontreated water (Toranzo et al. 1983).

The most comprehensive study of IPNV stability was reported by Ahne (1982), whose results confirmed persistence under various environmental conditions and vulnerability to chemical and physical disinfection measures (Table 4).

Most fish viruses are decidely heat labile, but IPNV is an exception; whereas low pH favors cold preservation, acid conditions enhance heat inactivation. In the range of pH 3.0 to 9.0 the rate of thermal inactivation at 60°C is biphasic, the more rapid activity occurring at pH 3.0 during the first 30 minutes. In the physiologic range and at pH 9.0, significant infectivity was sustained for several hours (MacKelvie and Desautels 1975).

Infectivity is inactivated by the several disinfecting agents commonly used in fish culture or research facilities. Given adequate exposure time and concentration, chlorine, formalin, iodine, ozone, and pH 12.5 are all virucidal. As might be expected, inactivation by pH 2.5 is incomplete (Table 5).

Although the study of interferon induction by IPNV is one of the less exploited properties of the virus, that ability was noted by Gravell and Malsberger (1965) when they described the development of the FHM cell line. Scherrer et al. (1974) confirmed the observation and described the kinetics of replication and interferon production by strain VR-299 in FHM cells at several temperatures. Interferon was produced at 15, 20, 26, and 30°C. At 26°C the virus and interferon followed a comparable high-yield course; at lower temperatures, lesser amounts of interferon were produced. The yield of interferon at 30°C was also reduced but the virus did not replicate. Scherrer et al. (1974) stated that RTG-2 cells were refractory to interferon induction by VR-299 virus, but De Sena and Rio (1975), who also used VR-299, found the opposite. They not only showed that interferon was induced, but characterized it and determined that it had the major attributes of mammalian interferon, in that it was heat and pH stable, nondialyzable, and not sedimented by high-speed centrifugation. The activity was vulnerable to trypsin and 2-mercaptoethanol but

Table 4. Stability of IPNV under various test conditions

Condition	Result
Municipal tap water at 10°C	<1 $TCID_{50}$/mL at 231 days
Nontreated canal water at 10°C	Not detectable after 14 days
Pond mud at 10°C	<1 $TCID_{50}$/mL at 70 days
Drying at 10°C	Persisted more than 4 weeks
10^6 rads gamma irradiation	10% infectivity survived
2540 Å UV irradiation at 5 cm	100% inactivation at 60 min
2% NaOH solution (pH 11.9)	Total inactivation in 5 min
3% formalin solution	Total inactivation in 5 min
520 ppm chlorine	99% inactivation in <2 min

Source: Adapted from Ahne (1982).

Table 5. Response of IPNV to selected disinfecting agents

Agent	Concentration (ppm)	Time (min)	Effect	Reference
Chlorine	40	30	Virucidal	Elliott and Amend (1978)
Chlorine	16	30	Virucidal	Desautels and MacKelvie (1975)
Chlorine	0.7	2	Virucidal	Wedemeyer et al. (1978)
Formalin	20,000	5	Virucidal	Vestergard Jorgensen (1973b)
Formalin	2,000	60	Partly virucidal	Elliott and Amend (1978)
Iodine	35	5	Virucidal	Desautels and MacKelvie (1975)
Iodine	16	5	Virucidal	Elliott and Amend (1978)
Ozone	90	0.5–10	Virucidal	Wedemeyer et al. (1978)
pH 2.5 (HCl)	—	60	Partly virucidal	Vestergard Jorgensen (1973b)
pH 12.5 (NaOH)	—	10	Virucidal	Vestergard Jorgensen (1973b)

resistant to DNAse and RNAse. The interferon that was produced protected RTG-2 cells against homologous virus and against IHNV, but details of that work were not given. The estimated molecular weight of interferon produced by RTG-2 cells was 94,000—heavier by a factor of 2.5 to 4 than interferon from rainbow trout and representative of that in higher vertebrates.

Interferon was found to be induced in RTG-2 cells by IPNV in cultures that were 6 to 14 days old and tightly confluent, but not in 2-day-old cultures that were in active growth and loosely confluent. Conversely, the yield of virus was highest in the youngest cultures (Okamoto, Shirakura, et al. 1983).

Diagnosis

Diagnosis of IPN can be made in several ways, but, because of differences in methods, accuracy ranges from presumptive to absolute. It is sometimes necessary to determine whether IPNV is the sole cause or merely an associated factor in mortality, for deaths can be the result of other infections. In addition, it may be necessary to quantify virus and decide whether it is present in subclinical levels—as in carriers or during early incubation—or (as has happened) is an associated influence in chronic pesticide or pollution mortality.

In trout, a clinical diagnosis can be made on the basis of susceptible species that show mortality, behavior changes, and the internal and external signs that have been described. Overall pallor of visceral organs and especially the absence of food and the presence of mucoid material in the alimentary tract provide a high degree of diagnostic reliability. The experienced diagnostician can expect virologic confirmation of clinical diagnoses in 9 of 10 cases. Epizootiologic data on the source of the affected lot, water supply, and facility history should of course also be considered.

Pathologic diagnosis may be the only practical approach where virologic capability is lacking or where time or distance precludes firsthand examination of fresh specimens. Epizootiologic data are especially useful in such cases, and recently fixed specimens retain many of the features present in fresh victims. In histologic examination, findings of pancreatic necrosis are highly indicative of IPN, but the situation can be complicated if there are accompanying pathologic changes in renal, hepatic, and intestinal tissues, for these also occur in IHN and viral hemorrhagic septicemia (VHS). Necrosis of the granular cells of the digestive tract—a distinct feature of IHN—does not occur in IPN. Histologic examination cannot detect a second virus, if one is present, nor can it be used to quantify virus and hence to determine whether IPNV is merely present or is causing mortality.

Worthwhile adjuncts to pathologic diagnoses include the use of fluorescent antibody techniques on frozen or paraffin sections of suspect specimens. Coagglutination, ELISA, or immunoperoxidase techniques also can be used on suspect tissue homogenates to reveal IPNV antigen. The various methods are rapid and convenient, but their applications—whether on sections or homogenates—are best considered as approaches in IPNV detection. Of themselves the methods have limitations, for they neither provide quantification of virus nor reveal the presence of other infectious agents.

Isolation

Whether the need is to determine the cause of an epizootic or to detect carriers, IPNV can generally be easily isolated with a high degree of reliability. The CHSE-214 and RTG-2 cell lines are recommended for isolation because of their long history of demonstrable susceptibility and lytic response. The RTG-2 cells also give presumptive identification by virtue of their ability to yield distinctive plaques. Although the AS and PG cell lines have a short history, they have also worked well. Some workers prefer FHM or BF-2 cells, but some lineages of these two lines have lost their ability to show CPE. Reasons for their resistance have not been determined, but when lower passages were tried or when clones were selected the lines were again susceptible. It should be again noted that strain Ab does not produce CPE in FHM cells.

The temperature of incubation for isolation of IPNV is not critical and may be anywhere in the range of 4 to 26°C. In RTG-2 cells, CPE becomes apparent only after several days of incubation at 4°C, but can be found within about 9 hours if the cultures are incubated at 26°C (the highest temperature tolerated by the line). If one is certain that only IPNV is involved, the higher temperatures produce results more quickly, but other salmonid viruses may be inhibited. An incubation temperature of 10°C is certain to reveal any of the salmonid viruses. At the nominal temperature of 20°C that is commonly used for isolation of IPNV, CPE usually appears within 48 hours. Under favorable culture conditions, CPE seldom appears after the fifth day. In its absence, however, blind passage should be made.

Depending on whether the need is to detect virus—possibly at low levels—or to determine the cause of mortality, there are precautions to be taken. For detection work, where sensitivity is a requirement, seminal or ovarian fluids should not be diluted because they are physiologic and generally free of toxicity for cells. Conversely, materials such as fecal extracts should be diluted (despite the accompanying reduction in virus number) to reduce toxicity to cells. As a generalization, feces should be diluted at least 1:20 or even 1:50 to reduce the effects of this toxicity. Tissue homogenates, which also may contain inhibitory materials, should be diluted to 1:10 or greater.

Ideally, two or more dilutions of inocula should be used for detection work and

Table 6. Average IPNV infectivity in selected tissues of five adult carrier brook trout

Tissue	Infectivity[a]
Kidneys	6.7
Stomach	4.6
Ovary	4.5
Liver	4.4
Spleen	3.3
Foregut	3.3
Hindgut	1.7
Muscle	0.3

Source: K. Wolf and M. C. Quimby (unpublished data).
[a]Infectivity expressed as exponent of $\log_{10}$ $TCID_{50}/g$.

for diagnostic cases. One reason for this suggestion is the finding by Nicholson and Dexter (1975) that CPE was not evoked at 1:10 dilutions of infected tissue homogenates but was clearly evident at dilutions of 1:1000 or greater. They postulated a possible homologous interference. A second reason for multiple dilutions is that it is often necessary to quantify an isolation and thus be able to judge whether virus is causing mortality, or is merely present. Even in less than optimally sensitive culture systems, one finds anywhere from 10^6 to 10^9 infection units per gram of tissue when IPNV is the cause of death. Titration at the time of isolation is simple and efficient if microculture plasticware is used.

In a population of salmonids that die as a consequence of IPN the virus is present and can be isolated during incubation, not only within the period when disease is clinically evident, but for several years thereafter in some survivors. The same is not true for the other lethal salmonid viruses. On an individual basis, not all survivors of IPN become carriers, and even some of the carriers (particularly those tending to shed small amounts) may shed virus only intermittently (Billi and Wolf 1969).

Virus can be found in many different tissues and body products. In epizootics, homogenates of pooled whole fry or visceral organs of fingerlings are suggested for study, but one should include kidneys and pyloric caeca of the fingerlings. Isolations from valuable nonepizootic specimens—e.g., valuable brood stock—may require the use of nondestructive methods. Seminal and ovarian fluids or feces of carriers commonly harbor virus, and are much more reliable than peritoneal washes. Virus has also been found in leukocytes by co-cultivation with an appropriately sensitive cell line such as RTG-2 or CHSE-214 (Swanson and Gillespie 1982; Yu et al. 1982). Nevertheless, the most certain approach to detection of a carrier population is to kill a sample number and assay the internal organs. The kidneys offer the highest probability of containing virus if IPNV is present (Table 6).

The kidneys were similarly found to be the organs of choice by Frantsi and Savan (1971a), Reno et al. (1978), Yamamoto (1975a, b), and Yamamoto and Kilistoff (1979). Caeca, spleen, intestine, and sometimes the liver yield smaller but nevertheless significant amounts of virus.

In detection of carrier trout, Agius et al. (1982) found that co-cultivation of trypsinized kidney cells on indicator RTG-2 cell line was about twice as sensitive as assay by traditional culture of filtrates of kidney homogenates. In a later comparison they found that filtrates of homogenized pyloric caeca and pancreas were comparable to co-cultivation of kidney cells, and recommended that, for the most sensitive work, both organs be assayed (Agius et al. 1983).

Identification

Whether for detection or diagnostics, serologic methods provide the best and most widely used means of identifying IPNV; in molecular biology, however, phenotypic features and analyses of genome and polypeptide or protein profiles are used. In some cases, the need is simply to ensure that the virus is IPNV and not another of the fish viruses. In contrast, epizootiologic or forensic needs may require identification to major group or even to specific strain of IPNV. Accordingly, serologic methods differ in sensitivity, specificity, equipment needs, and time required for completion.

The most commonly used serologic methods are neutralization tests, fluorescent antibody techniques, and ELISA. Other tests, used less frequently, include immunoperoxidase, complement fixation, immunodiffusion, counterimmunoelectrophoresis, and hemagglutination.

Plaque characteristics of IPNV—but only in RTG-2 cells—can be used for presumptive identification and for distinguishing IPNV from the salmonid viruses IHNV, *Oncorhynchus masou* virus (OMV), Egtved virus, and *Herpesvirus salmonis*. The identifying features of the IPNV plaque are a reticulum of elongated, shrunken dead cells, and coarsely stellate margins from which living cells radiate inward (Wolf and Quimby 1973b).

Infectious pancreatic necrosis virus is an impressive immunogen—the most antigenic of the known fish viruses—and high-titered antisera are readily prepared by simple methods in rabbits, goats, trout, or common carp. Hill et al. (1981) regularly produced rabbit antiserum with 50% plaque neutralization titers in the range of 1 to 10 million. Their rabbits were given initial intramuscular injections of purified and concentrated virus in Freund's complete adjuvant, and intravenous booster injections of virus alone on the following fourth and fifth weeks. Animals were exsanguinated 7 days after the second booster. Comparably high-titered antiserum has also been prepared by simply infecting previously nonexposed adult brook trout (Wolf and Quimby 1969).

Serologically, IPNV is a heterogeneous virus, all isolants of which show some degree of cross-reactivity. Accordingly, various schemes have been proposed for arranging isolants in related groupings. After the isolation of strain VR-299 in North America, two different strains were recognized in Spjarup (Sp) and Abild (Ab), villages in Denmark (Vestergard Jorgensen and Grauballe 1971; Vestergard Jorgensen and Kehlet 1971). Hill (1977) recognized the three serogroups, and Nicholson and Pochebit (1981) used neutralization kinetics to arrive at a similar conclusion. Macdonald and Gower (1981) used neutralization kinetics, phenotypic markers, and analyses of genome and proteins for their scheme of "VR-299 (serotype 1), Ab (serotype 2), and Sp (serotype 3)." Still other methods gave major groupings represented by VR-299, Ab, and Sp: ELISA (Dixon and Hill 1983a; Nicholson and Caswell 1982), cross-neutralization and calculation of r values (Hedrick and Okamoto 1983; Okamoto, Sano, et al. 1983), and coagglutination (Kimura et al. 1984).

Although there is uniform agreement about reference strains of the major groups (Ahne 1981), there is some disagreement about placement of certain isolants; moreover, little effort has been made to reconcile differences between the schemes.

The heterogeneity of IPNV requires that polyvalent antiserum be used to merely identify it; however, group-specific or even strain-specific antisera are needed for more definitive identification.

After the discovery that anti–VR-299 serum incompletely neutralized some field isolants, a five-strain antiserum was prepared with VR-299, Powder Mill (New Hampshire), Buhl (Idaho) and Bonnamy and d'Honnincthun (France). Lientz and Springer (1973) found that this preparation neutralized all but 2 of 42 isolants tested—Reno (Nevada) and West Buxton (Maine). Accordingly, antisera were prepared against the two refractory strains and all indications are that the seven-strain polyvalent antiserum will neutralize any IPNV of fish origin.

Finlay and Hill (1975), who mixed equal parts of anti–VR-299, Ab, and Sp sera, reported that in complement fixation tests this polyvalent antiserum was capable of reacting with the seven strains that they tested.

Neutralization tests are the most widely used means of identifying IPNV. Relationships between several strains have been investigated by cross-neutralization (Ishiguro et al. 1984; Okamoto, Sano, et al. 1983) and by neutralization kinetics (Macdonald and Gower 1981; Nicholson and Pochebit 1981). Kinetic studies showed that binding of virus by antibody is a rapid process, completed in only a few minutes.

In neutralization tests, virus that has been isolated is usually used; however, time and effort can be saved if diagnostic cases include titration of tissue homogenate and concurrent use of polyvalent antiserum.

Freeze-thaw lability of some IPNV isolants results in a predominance of noninfectious particles that bind antibody and distort true serologic relationships (McMichael et al. 1975). Adding to the complications of serologic typing of IPNV isolants are the findings of Dorson et al. (1978), who showed that as few as several passages of strain Sp virus in RTG-2 cell cultures were enough to reduce virulence, increase plaque size, increase sensitivity to the 6S component of normal trout serum, and slightly reduce sensitivity to neutralization.

Fluorescent antibody techniques, direct or indirect, are the second most widely used method of identifying IPNV serologically. For infected cell cultures, direct methods have largely been used by Argot (1969), Grauballe (1972), Piper et al. (1973), Tu et al. (1974), Vestergard Jorgensen (1972a), and Vestergard Jorgensen and Meyling (1972); and indirect methods have been used by Ishiguro et al. (1984), Nishimura (1980), Swanson (1981), Swanson and Gillespie (1981), and Vestergard Jorgensen (1974). If carried out on frozen sections, the test can be completed within 90 minutes (Swanson and Gillespie 1981). The most simple application of fluorescent antibody techniques is carried out on imprint or smear preparations of suspect tissues (Nishimura 1980).

The use of ELISA is gaining popularity for identification because tests can be completed within 1.5 to 7 hours, and specialized equipment is not essential (Dixon and Hill 1983a; Hattori et al. 1984; Nicholson and Caswell 1982). Levels of sensitivity range from $10^{3.5}$ to $10^{5.5}$ $TCID_{50}$; therefore, ELISA can be used on epizootic specimens, and it is specific for IPNV (Dixon and Hill 1983a).

Immunoperoxidase techniques have occasionally been used to identify IPNV in infected cell cultures (Faisal and Ahne 1980; Nicholson and Henchal 1978; Reno 1976). The procedures can be completed in several hours and read with ordinary light microscopy.

Kimura et al. (1984) described a promising serologic test in which cells of *Staphylococcus aureus* sensitized with anti-IPNV serum are used to identify the virus by agglutination. The test requires less than 2 hours, is somewhat more sensitive than isolation, and requires no specialized equipment.

Complement fixation was reported by Finlay and Hill (1975) and Dorson et al. (1979). The method is less sensitive than the neutralization test and requires the use of trout complement and hemolysin.

Still other tests have been reported, but their specialized requirements or limitations have thus far discouraged use by others. Cleator and Burney (1980) described a hemagglutination test that required erythrocytes from specific strains of mice and a pH range of 5.75 to 6.0. Dea and Elazhary (1983) described counterimmunoelectrophoresis; it required less than 2 hours in an electrophoresis chamber, but was most sensitive on virus that had been passaged. Immunodiffusion reactions of IPNV and antiserum were used by Ishiguro et al. (1984), but the test, although simple, lacks sensitivity.

Identification of a pathogen is often sought indirectly or presumptively by showing the presence of specific antibody. The small size of most victims of IPN and the usually acute course of the disease are but two of the reasons that mitigate against practical applications of assaying serum for presumptive evidence of IPN. Among carriers, specific neutralizing antibody may be present in fish that have survived infection and have cleared the virus. Furthermore, high levels of the 6S natural antibody might be present in fish that have never been exposed to IPNV. Finally, laboratory passage of IPNV can result in such extensive alteration that the virus becomes susceptible to neutralization by normal trout serum at dilutions as high as 1:5000 (Dorson et al. 1978). In effect, the serology of the fish, at least for the present, does not lend itself to reliable indications of the presence or absence of IPNV.

Transmission and Incubation

Virus is shed by carriers with their sex products and in their feces. During an active outbreak, victims release large amounts of virus in the feces and presumably in the urine. In one hatchery outbreak, a level of about 25,000 infective particles per liter was found in a tank supplied with water at the rate of 88 L/min. Desautels and MacKelvie (1975) found more than 100,000 $TCID_{50}$/L of water during an epizootic.

Transmission occurs laterally by contact and ingestion, and vertically during spawning; Mulcahy and Pascho (1984) found that IPNV was adsorbed onto rainbow trout sperm. In lateral transmission, it is likely that the digestive tract is the prime site of infection but the gills may be equally important because fry and fingerlings are easily infected by bath or immersion. Large fish are easily infected by injection. Among susceptible fish, the disease is highly contagious.

One of the early reports on IPN noted the association of the disease with eggs from a particular source (Snieszko et al. 1957), and the virus was later found in the ovarian fluid that typically accompanies salmonid eggs (Wolf et al. 1963). Repeated instances of apparent egg transmission have built a strong but nevertheless circumstantial case. The best evidence for egg transmission was obtained by Bullock et al. (1976), who reported that in two different laboratories the disease broke out among a single lot of progeny of carriers, even though the eggs had been disinfected with a virucidal iodophore. It is not known if the virus is truly within the egg or merely an adherent passenger. No one has been able to demonstrate egg-borne virus during egg incubation, but there is reason to suspect that the virus may contribute to embryo mortality. The disease typically appears after hatching and after the fry have begun to feed. Electron microscopy of dead embryos from carrier stock would be one way to resolve the uncertainty.

The incubation time of IPN is typically short—a matter of 3 to 5 days before the first signs and deaths occur—but factors such as host age, low temperature, and fish strain can lengthen the incubation time.

Under experimental conditions with young brook trout fry at about 12°C, the feeding of IPNV results in clinical signs and mortality that begin on the 4th or 5th day. In one such trial with 10 isolants, mortality peaked during the 12th to 18th day and then subsided (K. Wolf and M. C. Quimby, unpublished data). By the time fish reach fingerling size, an incubation period of 8 to 10 days is usually the rule. Incubation time is increased by temperatures lower than 12°C and decreased by higher temperatures, but the comparative trials required to determine incubation time more accurately have not been carried out. If virus is injected in massive quantity, incubation at 12°C can be as short as 3 days before first mortality.

Source

Among salmonids, the principal source of IPNV is infected brood stock and, secondarily, victims of active disease and survivors that become carriers and shed virus. Subclinically infected fishes of any species can be a source of IPNV for other species, but whether IPN disease results has yet to be shown. Similarly, viable IPNV can be transported and excreted by fish-eating birds and mammals. However, such shedding by homeotherms has not yet been shown to have resulted in natural outbreaks of IPN.

Host and Geographic Range

For several years after it was first discovered, IPNV was known only from North America and only as a pathogen that produced disease and mortality in trout fry. In time, the virus was also found as an inapparent infection—the carrier state—in normal-looking trout brood stock, and IPN disease was soon reported in France.

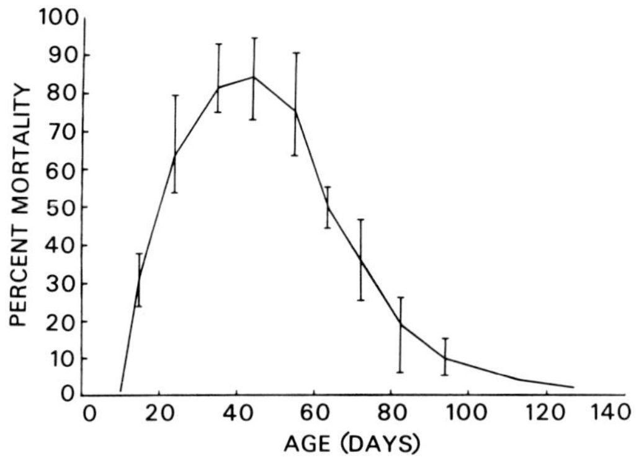

Figure 6.Comparative mortality from IPNV among brook trout exposed to standard challenge at various ages after hatching. From McAllister and Owens (1986). Reprinted with the permission of the American Fisheries Society.

The shipping of contaminated salmonid stock, principally eyed trout eggs, led to the virus being widely introduced. The range now includes much of North America and Europe, Scandanavia, the United Kingdom, Japan, and—in all probability—western areas of the Soviet Union. In recent years, isolations have been made from fish in Chile, Korea, and Taiwan (Table 1). Thus far, vigilance by Australia and New Zealand has seemingly prevented introduction of the virus into those countries.

The lethality of IPNV is age dependent (Fig. 6). That feature was first documented by Snieszko et al. (1959), who recorded losses of 45% among 1 month-old brook trout, 35% among fish exposed 1 month later, and only 7% among brook trout challenged at the age of 4 months. Age dependency was amply demonstrated by later investigators (Dorson and Torchy 1981; Frantsi and Savan 1971b; Lightner 1971; McAllister and Owens 1986), and the general susceptibility of salmonoid species to infection—but not necessarily to mortality—from IPNV has been clearly established (Table 2).

A few cases of mortality among yearling or older trout have been attributed to IPNV, but they are unusual (Elazhary et al. 1976; Parisot et al. 1965). Ivarie (1973) reported that brook trout 25 cm long developed typical clinical signs and began to die about a week after being exposed to infective effluent. The most significant finding was the presence of visceral levels typical of epizootics, 10^7 PFU/g.

For more than a decade after its discovery, the literature on IPN dealt only with salmonids, but in 1972 Sonstegard et al. isolated IPNV from pooled visceral homogenate taken from healthy white suckers (*Catostomus commersoni*). The suckers, however, had been taken below a trout hatchery where IPNV was known to have occurred; therefore, it was possible that the virus had merely been ingested. Since then, however, it has been made clear that nonsalmonid fishes can be infected with IPNV. Indeed, representatives of at least 20 families—from primitive cyclostomes to advanced bony fishes—have been found with IPNV (Table 3). Most isolations were made from fish that appeared to be healthy, but fish of some species were obviously diseased. As examples, Sano (1976) reported the virus to be virulent for anguillid eels, and Kou and Chen (1984) stated that, experimentally, the virus was pathogenic for the cobitid loach (*Misgurnus anguillicaubulatus*), and a cichlid and a cyprinid. Sorimachi and Hara (1985) isolated IPNV from diseased fingerling of the yellowtail (*Seriola quinqueradiata*), a carangid; they also induced mortality experimentally.

The host range of fishes that can be infected with IPNV is undoubtedly much greater than indicated in Table 3. Moreover, IPN-like viruses have been isolated from mollusks and crustaceans (Hill 1982). In addition, Lo and Wang (1984) isolated IPNV from *Clinostomum complanatum*, a trematode parasite of fishes. That isolant was first seen in the form of virus aggregates in mesenchymal cells of the metacercaria.

All indications are that under natural conditions IPNV infects poikilothermic hosts, but that it can at least tolerate the temperatures of homeotherms. Eskildsen and Vestergard Jorgensen (1973) and Sonstegard and McDermott (1972) showed that infectivity was retained during passage through the gut of a bird or mammal. Although the body temperatures of homeotherms exceed the upper limit of IPNV replication, Angiolelli and Rio (1971) found necrosis and other pathologic changes in the pancreas of suckling mice that were inoculated intracerebrally. Moreover, virus was presumptively recovered from the pancreas 10 to 21 days after inoculation.

Immunity

Whether IPNV is alive or gently inactivatd with formalin, it is a potent and nonlethal immunogen when injected into adult trout. The virus is also immunogenic in some of the fry and fingerlings that survive outbreaks of the disease. The indications are, however, that young survivors are protected against homologous strain virus but not necessarily against heterologous virus. Factors involved in protection include virulence of challenge, attainment of threshold immune competence (at about 1 month of age or 1 g in weight), and quantity of antibody that has been produced—hence a time factor.

Features of an IPN vaccine, therefore, should include broad-spectrum effectiveness or protection against the various major strains of the virus. The vaccine should be safe and cause no mortality and be easily administered by mouth or by immersion. Cost effectiveness is critical; therefore, the vaccine should be an attenuated live virus preparation that will be transmitted within a population of fish. Ideally, but perhaps unrealistically, live virus vaccine should be distinguishable from wild virus.

Basic studies of immunity include work by Wolf and Quimby (1969), who injected adult rainbow and brook trout with strain VR-299 live virus and found that neutralization indices of 10^4 to 10^6 were produced after several months. Vestergard Jorgensen (1973a) determined that neutralizing antibody was an IgM-like immunoglobulin with a high molecular weight.

Immunity to IPN can be transferred by intraperitoneal injection of serum from immunized adults (Agniel 1975), but evidence of egg transfer of maternal antibody to fish progeny has not been found.

Although the stability of induced avirulence has not been determined, IPNV can be readily attenuated by passage in cell culture. Sano (1977) found that Buhl strain virus passaged 47 times caused only a 2% mortality in rainbow trout fry; in contrast, a 38% mortality resulted from virus that had been transferred only twice.

Hill et al. (1980) reported that in rainbow trout fry the antigenicity of IPNV after treatment with 1:200 formalin for 24 hours at 4°C was much greater than that of virus treated with beta propiolactone; also, chemical treatment of virus to release polypeptides resulted in loss of antigenicity. The authors found that both live avirulent and attenuated virus showed promise when administered by bath; the mortality in immunized fry was less than half that in nonimmunized controls. The use of hyperosmotic infiltration gave mixed results.

Sano, Tanaka, and Fukuzaki (1981), who immunized adult rainbow trout with concentrated formalin-killed virus, found that the fish later neutralized live virus given intravenously. The apparent intent of the study was to determine if virus-free sex products could be produced by vaccinating brood stock, but application of the approach, particularly to carrier brood stock, was not reported.

Dixon and Hill (1983b), who sought optimal conditions for inactivation of IPNV with minimal loss of antigenicity, confirmed the earlier work of Hill et al. (1980). Formalin at 1:200 at 20°C for 4 or more days at pH 7.2 inactivated virus completely, but the treated preparation was nevertheless antigenic in 6-month-old trout. As determined by in vitro titrations, inactivation with formalin caused a slight loss in antigenicity, whereas similar treatment with beta propiolactone destroyed more than 50% of the antigenicity. The work did not include trials with young trout of susceptible age.

Several immunologic studies involving IPNV and tropical fish have come from A. Perlmutter's laboratory at New York University. Yu et al. (1969) inoculated the blue gourami (*Trichogaster trichopterus*) with strain VR-299, held the fish at about 27°C, and bled them 3 weeks later; they found evidence of neutralizing activity in the serum fraction whose electrophoretic migration resembled that of rat gamma globulin. The authors later reported that splenectomy prevented the production of serum neutralizing activity (Yu et al. 1970). Perlmutter et al. (1973) showed that crowding produced an immunosuppressive effect on the blue gourami, but that a waterborne factor produced by the fish was soluble in methylchloroform.

Control

Avoidance is the most effective and most economical control measure for IPN, and that approach requires that sources of the virus be known and shunned (Wolf and Quimby 1967). Stocks of specific-pathogen–free eggs or fish are available commercially or, alternatively, virus-free brood stock can be selected. Avoidance requires that the hatchery and rearing facilities be virus free and supplied with noncontaminated water.

Wolf et al. (1968) described procedures whereby virus-free brook trout were derived from a population of carrier brood stock. Fish were individually tagged and virologic examinations of feces and peritoneal washings were made well before spawning time. The adults that harbored virus were culled. Eggs and sperm of the remaining fish were examined at spawning, and the eggs of individual pairings were incubated as separate lots until results of the second virologic screening were obtained. A second culling was made and, after hatching, dead fry and samples of living fry from each lot were cultured for virus. Virus-free progeny were pooled and reared to adulthood. With careful virologic screening, strains of trout with desirable heritable characteristics—such as resistance to furunculosis—can be freed of IPNV.

In practice, particularly in commercial operations where IPN cannot be avoided, economic loss is minimized by anticipating the extent of mortality and incubating proportionately more eggs. Alternatively, some hatcheries rear the vulnerable young under virus-free conditions and move them to exposed situations after they have reached a more resistant age.

Significant progress has been made in genetic selection and breeding of brook trout for resistance to IPN, but the work has not been published. In 1968, W. C. Kennedy of the Pennsylvania Fish Commission and J. E. Wright of Pennsylvania State University began a program of developing strains of brook trout for several traits, including resistance to IPN. In the strain with the highest resistance to IPN, survival was 94%. In contrast, IPN mortality in the Colville strain from the White Sulphur Springs (West Virginia) National Fish Hatchery exceeded 95%. Kennedy and Wright suggested that if the Colville strain were converted to an albino form and kept IPN-free, it would be an excellent indicator species. The IPN testing has been with wild homologous virus, probably the ancestral strain of VR-299. The response of these brook trout strains to heterologous strains of IPNV is not known.

Eradication is recommended when outbreaks occur in facilities that can be operated without risk of recontamination. The extent of IPNV should be determined among all lots present, and the premises depopulated as appropriate by marketing, stocking, or destroying. All holding facilities, fish handling equipment, and personal items such as boots and rain gear should be cleaned and decontaminated. Vestergard Jorgensen (1973b) found that a 2% solution of formalin inactivated IPNV rapidly, as did pH 12.2. Field results with 100 ppm chlorine from calcium hypochlorite have been entirely satisfactory, for the compound has the dual virucidal effect of chlorine and high pH. Also, the residual chlorine can either be allowed to dissipate or can be neutralized with sodium thiosulfate.

Practical decontamination of IPNV-contaminated water supplies by ultraviolet irradiation is not likely because the virus is highly resistant to inactivation. Maisse et al. (1980) found that the LD_{90} of Egtved virus was a low 10,000 microwatts per second per square centimeter (mWs/cm^2), whereas a dosage of 330,000 mWs/cm^2 resulted in only an LD_{80} for IPNV.

The first attempt at chemotherapy of IPN was that by Snieszko et al. (1959), who fed iodine compounds to reverse a postulated dietary deficiency that caused the whirling behavior. No beneficial effect was achieved. Economon (1963) fed polyvinlypyrrolidone (PVP) iodine at several levels for 5 days to small numbers of brook trout suffering from a mild IPN outbreak. The loss was 9.5% in the nontreated lot, and 5.5% in a lot treated at the most effective iodine dosage. Economon (1973) used experimental lots numbering 26,000 fry each and fed either PVP or one of five concentrations of PVP-iodine. Losses in the nontreated control group and in fish fed PVP alone were about 34%. All lots receiving PVP-iodine sustained lower mortality, and the fewest deaths occurred at levels of 1:64 g or more of available iodine per kilogram of body weight per day. Economon noted "some impairment of weight gain." J. Hutcheson (personal communication), who carried out a small but carefully controlled trial, found a slight benefit— about a 20% reduction in mortality—from either PVP-iodine or iodine alone, and quantified the growth retardation. During the 30–day trial the surviving controls gained 16% in weight, whereas those receiving either form of iodine gained only 3.5% and showed abnormally light pigmentation. The PVP-iodine has not been registered for legal use in the United States.

A synthetic nucleoside known as ribavarin or virazole (1-D-ribofluoranosyl-1,2,4-

triazole-3-carboxamide) has been tested in vitro and in vivo and been found to have limitations that preclude hatchery applications (Migus and Dobos 1980; Savan and Dobos 1980). When added at 10 μg/mL just before or within a few hours after infection, the drug inhibited IPNV replication in CHSE-214 and RTG-2 cells. However, single-dose additions of levels as high as 400 μg/mL to the water of tanks containing IPNV-infected trout fry did not effect a marked reduction in mortality. Cost and efficacy make hatchery applications impractical.

Incubation at relatively lower temperatures has been found to reduce losses from IPN. This beneficial effect of low temperature was first reported by Frantsi and Savan (1971b), who compared the virulence of strains VR-299 and Pem-P1 fed to brook trout at 15.5, 10, and 4.5°C. Strain VR-299 was the most virulent at 15.5°C (31% mortality) and least virulent at 10 and 4.5°C (10 to 12% mortality). Strain Pem-P1 was most virulent at 10°C (74% mortality), less so at 15.5°C (46% mortality), and least at 4.5°C (10% mortality). After the experiment, the fish held at 4.5°C were moved to 10°C, but the disease did not recur.

Sano (1973a) used comparable temperatures in two separate trials with rainbow trout that experienced natural outbreaks. The mean mortality was 80% at 14°C, 20% at 10.5°C, and about 43% at 6°C. The two reports are somewhat at odds, especially regarding the effects of temperatures of 10.5 and 6°C.

References

Agius, C., H. Mangunwiryo, R. H. Johnson, and D. A. Smail. 1982. A more sensitive technique for isolating infectious pancreatic necrosis virus from asymptomatic carrier rainbow trout, *Salmo gairdneri* Richardson. J. Fish Dis. 5:285–292.

Agius, C., A. Richardson, and W. Walker. 1983. Further observations on the co-cultivation method for isolating infectious pancreatic necrosis virus from asymptomatic carrier rainbow trout, *Salmo gairdneri* Richardson. J. Fish Dis. 6:477–480.

Agniel, L. D., Jr. 1975. An assessment of passive transfer of immunity to infectious pancreatic necrosis virus (IPNV) in trout. M.S. thesis. American University, Washington, D.C.

Ahne, W. 1977a. Persistent infection in CHSE-214 cells with IPN virus isolated from pike (*Esox lucius*). Bull. Off. Int. Epizoot. 87:415–416.

Ahne, W. 1977b. Some properties of an IPN-virus isolated from pike (*Esox lucius*). Bull. Off. Int. Epizoot. 87:417–418.

Ahne, W. 1978. Isolation and characterization of infectious pancreatic necrosis virus from pike (*Esox lucius*). Arch. Virol. 58:65–69.

Ahne, W. 1981. Serological techniques currently used in fish virology. Pages 3–27 in D. P. Anderson and W. Hennessen, eds. Proceedings of the international symposium on fish biologics: serodiagnostics and vaccines, National Fish Health Research Laboratory, Leetown, West Virginia, April 26–30, 1981. S. Karger, Basel.

Ahne, W. 1982. Vergleichende Untersuchungen über die Stabilität von vier fischpathogenen Viren (VHSV, PFR, SVCV, IPNV). Zentralbl. Veterinärmed. [B] 29:457–476.

Ahne, W. 1985. Virusinfektionen bei Fischen: Ätiologie, Diagnose und Bekämpfung. Zentralbl. Veterinärmed. [B] 32:237–264.

Ahne, W., and C. Held. 1980. Untersuchungen über die viruzide Wirkung von Actomar K_{30} auf fischpathogene Viren. Tierärztl. Umsch. 35:308–318.

Ahne, W., and R. D. Negele. 1985a. Studies on transmission of infectious pancreatic necrosis virus via eyed eggs and sexual products of salmonid fish. Pages 261–269 in A. E. Ellis, ed. Fish and shellfish pathology. Academic Press, London.

Ahne, W., and R. D. Negele. 1985b. Transmission of infectious pancreatic necrosis virus with pituitaries of fish. Pages 271–275 in A. E. Ellis, ed. Fish and shellfish pathology. Academic Press, London.

Alayse, A. M., J. Cohen, and R. Scherrer. 1975. Etude de la synthèse des ARN viraux dans les cellules FHM et RTG-2 infectées par le virus de la nécrose pancréatique infectieuse (NPI). Ann. Microbiol. (Paris) 126:471–483.

Amend, D. F., and J. P. Pietsch. 1972. An improved method for isolating viruses from asymptomatic carrier fish. Trans. Am. Fish. Soc. 101:267–269.

Angiolelli, R. F., and G. J. Rio. 1971. The Swiss/ICR (Ha) albino mouse as an experimental host for infectious pancreatic necrosis of trout. J. Fish Biol. 3:139–144.

Argot, J. E. 1969. Infectious pancreatic necrosis virus intracellular replication. Ph.D. dissertation. Lehigh Univ., Bethlehem, Pennsylvania.

Argot, J., and R. G. Malsberger. 1972. Intracellular replication of infectious pancreatic necrosis virus. Can. J. Microbiol. 18:865–867.

Ball, H. J., A. L. S. Munro, A. Ellis, K. G. R. Elson, W. Hodgkiss, and I. S. McFarlane. 1971. Infectious pancreatic necrosis in rainbow trout in Scotland. Nature 234:417–418.

Barja, J. L., A. E. Toranzo, M. L. Lemos, and F. M. Hetrick. 1983. Influence of water temperature and salinity on the survival of IPN and IHN viruses. Bull. Eur. Assoc. Fish Pathol. 3:47–50.

Baudouy, A. M., and J. Castric. 1977. Persistance du pouvoir pathogène du virus de la nécrose pancréatique infectieuse après un séjour prolongé dans l'eau. Bull. Off. Int. Epizoot. 87:409–413.

Beasley, A. R., M. M. Sigel, and L. W. Clem. 1965. Virus-carrier cell cultures from a marine fish. Bacteriol. Proc. 14:99. Abstract.

Besse, P., and P. de Kinkelin. 1965a. Sur l'existence en France de la nécrose pancréatique de la truite arc-en-ciel (*Salmo gairdneri*). Bull. Acad. Vet. 38:185–190.

Besse, P., and P. de Kinkelin. 1965b. La nécrose pancréatique des alevins arc-en-ciel. Piscic. Fr. 2:16–19.

Billi, J. L., and K. Wolf. 1969. Quantitative comparison of peritoneal washes and feces for detecting infectious pancreatic necrosis (IPN) virus in carrier brook trout. J. Fish. Res. Board Can. 26:1459–1465.

Bonami, J. R., F. Cousserans, M. Weppe, and B. J. Hill. 1983. Mortalities in hatchery-reared sea bass fry associated with a birnavirus. Bull. Eur. Assoc. Fish Pathol. 3:41.

Bullock, G. L., R. R. Rucker, D. Amend, K. Wolf, and H. M. Stuckey. 1976. Infectious pancreatic necrosis: transmission with iodine-treated and nontreated eggs of brook trout (*Salvelinus fontinalis*). J. Fish. Res. Board Can. 33:1197–1198.

Busch, R. A. 1978. Protective vaccines for mass immunization of trout. Salmonid 1:10–22.

Cerini, C. P. 1964. Multiplication and morphology of infectious pancreatic necrosis virus. Ph.D. dissertation. Lehigh Univ., Bethlehem, Pennsylvania.

Cerini, C. P., and R. G. Malsberger. 1965. Morphology of infectious pancreatic necrosis virus. Ann. N.Y. Acad. Sci. 126:315–319.

Chang, N., R. D. Macdonald, and T. Yamamoto. 1978. Purification of infectious pancreatic necrosis (IPN) virus and comparison of polypeptide composition of different isolates. Can. J. Microbiol. 24:19–27.

Cleator, G. M., and L. A. Burney. 1980. The hemagglutinating properties of infectious pancreatic necrosis virus. Arch. Virol. 63:81–85.

Cohen, J. 1975. Ribonucleic acid polymerase activity in purified infectious pancreatic necrosis virus of trout. Biochem. Biophys. Res. Commun. 62:689–695.

Cohen, J., and R. Scherrer. 1972. Structure de la capside du virus de la nécrose pancréatique infectieuse (NPI) de la truite. C. R. Acad. Sci. [D] (Paris) 274:1222–1225.

Cohen, J., P. de Kinkelin, and R. Scherrer. 1971. Etude de la structure du virus de la nécrose pancréatique infectieuse de la truite. Colloq. Soc. Fr. Micro. Elect. Caen, 1971. Abstract.

Cohen, J., A. Poinsard, and R. Scherrer. 1973. Physico-chemical and morphological features of infectious pancreatic necrosis virus. J. Gen. Virol. 21:485–498.

Dea, S., and M. A. S. Y. Elazhary. 1983. Counterimmunoelectrophoresis for identification of infectious pancreatic necrosis virus after isolation in cell culture. Can. J. Fish. Aquat. Sci. 40:2200–2203.

de Kinkelin, P., and P. Besse. 1966. Une épizootie de nécrose pancréatique dans les salmonicultures françaises. Bull. Off. Int. Epizoot. 65:999–1010.

Desautels, D., and R. M. MacKelvie. 1975. Practical aspects of survival and destruction of infectious pancreatic necrosis virus. J. Fish. Res. Board Can. 32:523–531.

De Sena, J., and G. J. Rio. 1975. Partial purification and characterization of RTG-2 fish cell interferon. Infect. Immun. 11:815–822.

Dixon, P. F., and B. J. Hill. 1983a. Rapid detection of infectious pancreatic necrosis virus (IPNV) by the enzyme-linked immunosorbent assay (ELISA). J. Gen. Virol. 64:321–330.

Dixon, P. F., and B. J. Hill 1983b. Inactivation of infectious pancreatic necrosis virus for vaccine use. J. Fish Dis. 6:399–409.

Dobos, P. 1976a. Virus-specific protein synthesis in cells infected by infectious pancreatic necrosis virus (IPN). Abstr. Annu. Meet. Am. Soc. Microbiol. S165:232.

Dobos, P. 1976b. Size and structure of the genome of infectious pancreatic necrosis virus. Nucleic Acids Res. 3:1903–1924.

Dobos, P. 1977. Virus-specific protein synthesis in cells infected by infectious pancreatic necrosis virus. J. Virol. 21:242–258.

Dobos, P., and P. P. C. Mertens. 1983. Infectious pancreatic necrosis virus (IPNV)—specific protein synthesis in vivo and in vitro. Pages 237–243 in R. W. Compans and D. H. L. Bishop, eds. Double-stranded RNA viruses. Elsevier, New York.

Dobos, P., and T. E. Roberts. 1983. The molecular biology of infectious pancreatic necrosis virus: a review. Can. J. Microbiol. 29:377–384.

Dobos, P., and D. Rowe. 1977. Peptide map comparison of infectious pancreatic necrosis virus-specific polypeptides. J. Virol. 24:805–820.

Dobos, P., R. Hallett, D. T. C. Kells, O. Sorensen, and D. Rowe. 1977. Biophysical studies of infectious pancreatic necrosis virus. J. Virol. 22:150–159.

Dobos, P., B. J. Hill, R. Hallett, D. T. C. Kells, H. Becht, and D. Teninges. 1979. Biophysical and biochemical characterization of five animal viruses with bisegmented double-stranded RNA genomes. J. Virol. 32:593–605.

Dorson, M. 1977. Vaccination trials of rainbow trout fry against infectious pancreatic necrosis. Bull. Off. Int. Epizoot. 87:405.

Dorson, M. 1983. Infectious pancreatic necrosis of salmonids: overview of current problems. Pages 7–31 in D. P. Anderson, M. M. Dorson, and P. Dubourget, eds. Antigens of fish pathogens. Collection Fondation Marcel Marieux, Lyon, France.

Dorson, M., and P. de Kinkelin. 1974. Nécrose pancréatique infectieuse des salmonides: existence dans le sérum de truites indemnes d'une molécule 6S neutralisant spécifiquement le virus. C. R. Acad. Sci. [D] (Paris) 278:785–788.

Dorson, M., and C. Torchy. 1981. The influence of fish age and water temperature on mortalities of rainbow trout, *Salmo gairdneri* Richardson, caused by a European strain of infectious pancreatic necrosis virus. J. Fish Dis. 4:213–221.

Dorson, M., and C. Torchy. 1985. Experimental transmission of infectious pancreatic necrosis virus via the sexual products. Pages 251–260 *in* A. E. Ellis, ed. Fish and shellfish pathology. Academic Press, London.

Dorson, M., P. de Kinkelin, and C. Torchy. 1975. Virus de la nécrose pancréatique infectieuse:

acquisition de la sensibilité au facteur neutralisant du sérum de truite après passages successifs en culture cellulaire. C. R. Acad. Sci. [D] (Paris) 281:1435–1438.

Dorson, M., J. Castric, and C. Torchy. 1978. Infectious pancreatic necrosis virus of salmonids: biological and antigenic features of a pathogenic strain and of a non-pathogenic variant selected in RTG-2 cells. J. Fish Dis. 1:309–320.

Dorson, M., C. Torchy, and C. Michel. 1979. Rainbow trout complement fixation used for titration of antibodies against several pathogens. Ann. Rech. Vét. 10:529–534.

Economon, P. P. 1963. Experimental treatment of infectious pancreatic necrosis of brook trout with polyvinylpyrrolidone-iodine. Trans. Am. Fish. Soc. 92:180–182.

Economon, P. P. 1973. Polyvinylpyrrolidone-iodine as a control for infectious pancreatic necrosis (IPN) of brook trout. Pages 59–66 in W. A. Dill, ed. Symposium on the major communicable fish diseases in Europe and their control. EIFAC (Eur. Inland Fish. Advis. Comm.) Tech. Paper 17, Suppl. 2.

Elazhary, M. A. S. Y., A. Lagace, G. Cousineau, R. S. Roy, L. Berthiaume, P. Paulhus, and J. L. Frechette. 1976. Outbreak of infectious pancreatic necrosis in yearling brook trout (*Salvelinus fontinalis*). J. Fish. Res. Board Can. 33:2621–2625.

Elliott, D. G., and D. F. Amend. 1978. Efficacy of certain disinfectants against infectious pancreatic necrosis virus. J. Fish Biol. 12:277–286.

Emerson, M., B. L. Nicholson, and R. Bayer. 1979. Effects of *Acholeplasma laidlawii* and an unidentified mycoplasma on selected fish cell cultures and the replication of fish viruses. J. Fish Dis. 2:227–238.

Eskildsen, U. K., and P. E. Vestergard Jorgensen. 1973. On the possible transfer of trout pathogenic viruses by gulls. Riv. Ital. Piscicol. Ittiopatol. 8:104–105.

Faisal, M., and W. Ahne. 1980. Use of immunoperoxidase technique for detection of fish virus antigens. Pages 186–192 in W. Ahne, ed. Fish diseases. Third COPRAQ Session. Springer-Verlag, New York.

Fijan, N. 1974. Zarazna nekroza gušterače pastrva: prvi nalaz virusa u. Zugoslaviji. Vet. Arh. 44:187–192.

Finlay, J., and B. J. Hill. 1975. The use of the complement fixation test for rapid typing of infectious pancreatic necrosis virus. Aquaculture 5:305–310.

Frantsi, C., and M. Savan. 1971a. Infectious pancreatic necrosis virus: comparative frequencies of isolation from feces and organs of brook trout (*Salvelinus fontinalis*). J. Fish. Res. Board Can. 28:1064–1065.

Frantsi, C., and M. Savan. 1971b. Infectious pancreatic necrosis virus—temperature and age factors in mortality. J. Wildl. Dis. 7:249–255.

Ghittino, P. 1972. Malattie esotiche dei pesci che minacciano troticoltura e carpicolutra italiane. Riv. Ital. Piscicol. Ittiopatol. 7:53–62.

Grauballe, P. C. 1972. Infektios pancreasnekrose en ny virussygdom i Danmark hos Laksefisk. Diagnostiske og aetiologiske un der sogelser. Ambulatorisk Klinik den Kongelige Veterinaer-og Landbohojskole, Copenhagen. A Licentiate treatise.

Gravell, M., and R. G. Malsberger. 1965. A permanent cell line from the fathead minnow (*Pimephales promelas*). Ann. N.Y. Acad. Sci. 126:555–565.

Grinnell, B., and J. C. Leong. 1979. Recovery and concentration of infectious pancreatic necrosis (IPN) virus in water. J. Fish. Res. Board Can. 36:1405–1408.

Hah, Y., S. Hong, M. Kim, J. L. Fryer, and J. R. Winton. 1984. Isolation of infectious pancreatic necrosis virus from goldfish (*Carassius auratus*) and chum salmon (*Oncorhynchus keta*) in Korea. Korean J. Microbiol. 22:85–90.

Hastein, T., and J. Krogsrud. 1976. Infectious pancreatic necrosis first isolation of virus from fish in Norway. Acta Vet. Scand. 17:109–111.

Hattori, M., H. Kodama, S. Ishiguro, A. Honda, T. Mikami, and H. Izawa. 1984. In vitro and in vivo detection of infectious pancreatic necrosis virus in fish by enzyme-linked immunosorbent assay. Am. J. Vet. Res. 45:1876–1879.

Hedrick, R. P., and J. L. Fryer. 1981. Persistent infection of three salmonid cell lines with infectious pancreatic necrosis virus (IPNV). Fish Pathol. 15:163–172.

Hedrick, R. P., and N. Okamoto. 1983. Biochemical and antigenic characteristics of eel virus European compared to selected strains of infectious pancreatic necrosis virus. Pages 245–251 in R. W. Compans and D. H. L. Bishop, eds. Double-stranded RNA viruses. Elsevier, New York.

Hedrick, R. P., J. C. Leong, and J. L. Fryer. 1978a. Persistent infection by infectious pancreatic necrosis virus (IPNV) in piscine cells. Abstr. Annu. Meet. Am. Soc. Microbiol. S334:267.

Hedrick, R. P., J. C. Leong, and J. L. Fryer. 1978b. Persistent infections in salmonid fish cells with infectious pancreatic necrosis virus (IPNV). J. Fish Dis. 1:297–308.

Hedrick, R. P., J. L. Fryer, S. N. Chen, and G. H. Koo. 1983. Characteristics of four birnaviruses isolated from fish in Taiwan. Fish Pathol. 18:91–97.

Hill, B. J. 1976. Properties of a virus isolated from the bi-valve mollusc *Tellina tenuis* (DaCosta). Pages 445–452 in L. A. Page, ed. Wildlife diseases. Plenum Press, New York.

Hill, B. J. 1977. Present status of IPN virus. Pages 116–119 in Proceedings of the international symposium on diseases of cultured salmonids. Tavolek, Inc., Seattle, Washington, April 4–6, 1977.

Hill, B. 1978. IPN—a virus at large. Fish Farmer 1:37.

Hill, B. J. 1982. Infectious pancreatic necrosis virus and its virulence. Pages 91–114 in R. J. Roberts, ed. Microbial diseases of fish. Academic Press, London.

Hill, B. J., and P. F. Dixon. 1977. Studies on IPN virulence and immunization. Bull. Off. Int. Epizoot. 87:425–427.

Hill, B. J., M. Dorson, and P. F. Dixon. 1980. Studies on immunization of trout against IPN. Pages 29–36 in W. Ahne, ed. Fish diseases. Third COPRAQ Session. Springer-Verlag, Berlin.

Hill, B. J., R. F. Williams, and J. Finlay. 1981. Preparation of antisera against fish virus disease agents. Pages 209–218 in D. P. Anderson and W. Hennessen, eds. Proceedings of the international symposium on fish biologics: serodiagnostics and vaccines, National Fish Health Research Laboratory, Leetown, West Virginia, April 26–30, 1981. S. Karger, Basel.

Hiller, J., and A. Perlmutter. 1971. Effect of zinc on viral-host interactions in a rainbow trout cell line, RTG-2. Water Res. 5:703–710.

Ishiguro, S., H. Izawa, H. Kodama, M. Onuma, and T. Mikami. 1984. Serological relationships among five strains of infectious pancreatic necrosis. J. Fish Dis. 7:127–135.

Ivarie, R. A. 1973. An unusual outbreak of IPN among 9- to 10-inch brook trout. Fish Health News 2:2.

Jordan, D. S. 1963. The genera of fishes and a classification of fishes. Stanford Univ. Press, Stanford, California.

Kelly, R. K., and P. C. Loh. 1972a. Some properties of infectious pancreatic necrosis and its RNA. Abstr. Annu. Meet. Am. Soc. Microbiol. V139:208.

Kelly, R. K., and P. C. Loh. 1972b. Electron microscopical and biochemical characterization of infectious pancreatic necrosis virus. J. Virol. 10:824–834.

Kelly, R. K., and P. C. Loh. 1975. Replication of IPN virus: a cytochemical and biochemical study in SWT cells. Proc. Soc. Exp. Biol. Med. 148:688–693.

Kimura, T., M. Yoshimizu, and H. Yasuda. 1984. Rapid, simple serological diagnosis of infectious pancreatic necrosis by coagglutination test using antibody-sensitized staphylococci. Fish Pathol. 19:25–33.

Klontz, G. W., W. T. Yasutake, and T. J. Parisot. 1965. Virus diseases of the Salmonidae in western United States. III. Immunopathological aspects. Ann. N.Y. Acad. Sci. 126:531–542.

Kou, G. H., and S. N. Chen. 1984. Pathogenicity of birnavirus isolated from loach *Misgurnus anguillicaubulatus*. Page 50 in International seminar on fish pathology for the 20th anniversary of the Japanese Society of Fish Pathologists, Tokyo, Japan, September 8–10, 1984. Abstract.

Kudo, S., D. Kurosawa, I. Kunimeni, K. Nobusawa, and S. Kobayashi. 1973. Electron microscopic observations of the pancreas and liver in the fingerling rainbow trout with symptoms of IPN. Jpn. J. Ichthyol. 20:163–177.

Lientz, J. C., and J. E. Springer. 1973. Neutralization tests of infectious pancreatic necrosis virus with polyvalent antiserum. J. Wildl. Dis. 9:120–124.

Lightner, D. V. 1971. Infectious pancreatic necrosis (IPN): its pathogenesis and morphology of the virus in rainbow trout. Ph.D. dissertation. Colorado State Univ., Fort Collins.

Lightner, D., and G. Post. 1969. Morphological characteristics of infectious pancreatic necrosis virus in trout pancreatic tissue. J. Fish. Res. Board Can. 26:2247–2250.

Ljungberg, O., and P. E. Vestergard Jorgensen. 1973. Infectious pancreatic necrosis (IPN) of salmonids in Swedish fish farms. Pages 67–70 in W. A. Dill, ed. Symposium on the major communicable fish diseases in Europe and their control. EIFAC (Eur. Inland Fish. Advis. Comm.) Tech. Paper 17, Suppl. 2.

Lo, C., and C. Wang. 1984. The study of *Clinostomum complanatum* (Rud., 1914). V. The influences of metacercaria of *Clinostomum complanatum* on fish. Pages 39–40 in International seminar on fish pathology for the 20th anniversary of the Japanese Society of Fish Pathologists, Tokyo, Japan, September 8–10, 1984. Abstract.

Loh, P. C., M. H. Lee, and R. K. Kelly. 1974. The polypeptides of infectious pancreatic necrosis virus. J. Gen. Virol. 22:421–423.

Lothrop, D., and B. L. Nicholson. 1974. Inhibition of cellular DNA synthesis in cells infected with infectious pancreatic necrosis virus. J. Virol. 14:485–492.

McAllister, P. E. 1978. Viruses and virus diseases of salmonid fishes. Mar. Fish. Rev. 40:21–23.

McAllister, P. E., and W. J. Owens. 1986. Infectious pancreatic necrosis virus: protocol for a standard challenge in brook trout. Trans. Am. Fish. Soc. 115:466–470.

McAllister, P. E., and X. Reyes. 1984. Infectious pancreatic necrosis virus: isolation from rainbow trout, *Salmo gairdneri* Richardson, imported from Chile. J. Fish Dis. 7:319–322.

McAllister, P. E., M. W. Newman, J. H. Sauber, and W. J. Owens. 1984. Isolation of infectious pancreatic necrosis virus (serotype Ab) from diverse species of estuarine fish. Helgol. Meeresunters. 37:317–328.

Macdonald, R. D. 1978. Ringed plaque formation in infectious pancreatic necrosis correlates with defective interfering particle formation. J. Gen. Virol. 41:623–628.

Macdonald, R. D., and P. Dobos. 1981. Identification of the proteins encoded by each genome segment of infectious pancreatic necrosis virus. Virology 114:414–422.

Macdonald, R. D., and D. A. Gower. 1981. Genomic and phenotypic diversion among three serotypes of aquatic birnaviruses (infectious pancreatic necrosis virus). Virology 114:187–195.

Macdonald, R. D., and J. C. Kennedy. 1979. Infectious pancreatic necrosis virus persistently infects chinook salmon embryo cells independent of interferon. Virology 95:260–264.

Macdonald, R. D., and T. Yamamoto. 1977. The structure of infectious pancreatic necrosis virus RNA. J. Gen. Virol. 34:235–247.

Macdonald, R. D., and T. Yamamoto. 1978. Quantitative analysis of defective interfering particles in infectious pancreatic necrosis preparations. Arch. Virol. 57:77–89.

Macdonald, R. D., K. L. Roy, T. Yamamoto, and N. Chang. 1977. Oligonucleotide fingerprints of the RNAs from infectious pancreatic necrosis virus. Arch. Virol. 54:373–377.

M'Gonigle, R. H. 1941. Acute catarrhal enteritis of salmonid fingerlings. Trans. Am. Fish. Soc. 70:297–303.

MacKelvie, R. M., and H. Artsob. 1969. Infectious pancreatic necrosis virus in young salmonids of the Canadian Maritime Provinces. J. Fish. Res. Board Can. 26:3259–3262.

MacKelvie, R. M., and D. Desautels. 1975. Fish viruses—survival and inactivation of infectious pancreatic necrosis virus. J. Fish. Res. Board Can. 32:1267–1273.

McKnight, I. J., and R. J. Roberts. 1976. The pathology of infectious pancreatic necrosis. I. The sequential histopathology of the naturally occurring condition. Br. Vet. J. 132:76–85.

McMichael, J., J. L. Fryer, and K. S. Pilcher. 1975. An antigenic comparison of three strains of infectious pancreatic necrosis of salmonid fishes. Aquaculture 6:203–210.

Maisse, G., M. Dorson, and C. Torchy. 1980. Inactivation de deux virus pathogènes pour les salmonidés (virus de la nécrose pancréatique infectieuse et de la septicémie hémorragique virale) par les rayons ultraviolets. Bull. Fr. Piscic. 278:34–40.

Malsberger, R. G., and C. P. Cerini. 1963. Characteristics of infectious pancreatic necrosis virus. J. Bacteriol. 86:1283–1287.

Malsberger, R. G., and C. P. Cerini. 1965. Multiplication of infectious pancreatic necrosis virus. Ann. N.Y. Acad. Sci. 126:320–327.

Malsberger, R. G., and K. Wolf. 1966. Virus diseases of fish. Pages 677–684 in J. E. Prier, ed. Basic medical virology. Williams and Wilkins, Baltimore.

Mertens, P. P. C., P. B. Jamieson, and P. Dobos. 1982. *In vitro* RNA synthesis of infectious pancreatic necrosis virus-associated RNA polymerase. J. Gen. Virol. 59:47–56.

Migus, D. O., and P. Dobos. 1980. Effect of ribavirin on the replication of infectious pancreatic necrosis virus in fish cell cultures. J. Gen. Virol. 47:47–57.

Moewus, L. 1962. Studies on a marine parasitic *Tetrahymena* species. J. Protozool. 9:34.

Moewus, L., and M. M. Sigel. 1963. In vivo and in vitro cultivation of a fresh water fish virus in marine host systems. Bacteriol. Proc. 26:135. Abstract.

Moewus-Kobb, L. 1965. Studies with IPN virus in marine hosts. Ann. N.Y. Acad. Sci. 126:328–342.

Moss, L. H., and M. Gravell. 1968. Morphology of infectious pancreatic necrosis virus. Bacteriol. Proc. 104:162. Abstract.

Moss, L. H., and M. Gravell. 1969. Ultrastructure and sequential development of infectious pancreatic necrosis virus. J. Virol. 3:52–58.

Mulcahy, D. M., and J. L. Fryer. 1976. Double infection of rainbow trout fry with IHN and IPN viruses. Fish Health News 5:5–6.

Mulcahy, D., and R. J. Pascho. 1984. Adsorption to fish sperm of vertically transmitted fish viruses. Science 225:333–335.

Munro, A. L. S., and I. B. Duncan. 1977. Current problems in the study of the biology of infectious pancreatic necrosis virus and the management of the disease it causes in cultivated salmonid fish. Pages 325–337 in F. A. Skinner and J. M. Shewan, eds. Aquatic microbiology. Academic Press, New York.

Munro, A. L. S., J. Liversidge, and K. G. R. Elson. 1976. The distribution and prevalence of infectious pancreatic necrosis virus in wild fish in Loch Awe. Proc. R. Soc. Edinb. [B] (Nat. Environ.) 75:223–232.

Nicholson, B. L. 1971a. Effect of actinomycin D on the multiplication of the infectious pancreatic necrosis virus of trout. Experientia 27:1362–1363.

Nicholson, B. L. 1971b. Macromolecule synthesis in rainbow trout gonad cells (RTG-2) infected with infectious pancreatic necrosis virus of trout. Bacteriol. Proc. 315:220. Abstract.

Nicholson, B. L. 1971c. Macromolecule synthesis in RTG-2 cells following infection with infectious pancreatic necrosis (IPN) virus. J. Gen. Virol. 13:369–372.

Nicholson, B. L., and P. Caswell. 1982. Enzyme-linked immunosorbent assay for identification of infectious pancreatic necrosis virus. J. Clin. Microbiol. 16:469–472.

Nicholson, B. L., and R. Dexter. 1975. Possible interference in the isolation of IPN virus from carrier fish. J. Fish. Res. Board Can. 32:1437–1439.

Nicholson, B. L., and J. Dunn. 1974. Homologous viral interference in trout and Atlantic salmon cell cultures infected with infectious pancreatic necrosis virus. J. Virol. 14:180–182.

Nicholson, B. L., and E. A. Henchal. 1978. Rapid identification of infectious pancreatic necrosis virus in infected cell cultures of immunoperoxidase techniques. J. Wildl. Dis. 14:465–469.
Nicholson, B. L., and S. Pochebit. 1981. Antigenic analysis of infectious pancreatic necrosis viruses (IPNV) by neutralization kinetics. Pages 35–41 in D. P. Anderson and W. Hennessen, eds. Proceedings of the international symposium on fish biologics: serodiagnostics and vaccines, National Fish Health Research Laboratory, Leetown, West Virginia, April 26–30, 1981. S. Karger, Basel.
Nicholson, B. L., G. W. Thorne, C. Janicki, and A. Hanson. 1979. Studies on a host range variant from different isolates of infectious pancreatic necrosis virus (IPNV). J. Fish Dis. 2:367–379.
Nishimura, T. 1980. Rapid diagnosis of viral diseases of salmonids. Fish Pathol. 14:191–197. In Japanese with English abstract.
Nishimura, T., H. Fukuda, H. Yamazaki, and T. Sano. 1981. Concentration and purification of eel virus, EVE. Fish Pathol. 16:75–83.
Okamoto, N., T. Sano, R. P. Hedrick, and J. L. Fryer. 1983. Antigenic relationships of selected strains of infectious pancreatic necrosis virus and European eel virus. J. Fish Dis. 6:19–25.
Okamoto, N., T. Shirakura, Y. Nagakura, and T. Sano. 1983. The mechanism of interference with fish viral infection in the RTG-2 cell line. Fish Pathol. 18:7–12.
Okamoto, N., N. Taniguchi, Y. Seno, and T. Sano. 1984. The relation between the change of quantities of infectious pancreatic necrosis virus in infected rainbow trout fry and the disease process. Fish Pathol. 19:1–4.
Parisot, T. J., T. Yasutake, and V. Bressler. 1963. A new geographic and host record for infectious pancreatic necrosis. Trans. Am. Fish. Soc. 92:63–66.
Parisot, T. J., W. T. Yasutake, and G. W. Klontz. 1965. Virus diseases of the Salmonidae in western United States. I. Etiology and epizootiology. Ann. N.Y. Acad. Sci. 126:502–519.
Perlmutter, A., D. A. Sarot, M. Y. Yu, R. J. Filazzola, and R. J. Seeley. 1973. The effect of crowding on the immune response of the blue gourami, *Trichogaster trichopterus*, to infectious pancreatic necrosis virus. Life Sci. 13:363–375.
Persson, R. H., and R. D. Macdonald. 1982. Evidence that infectious pancreatic necrosis virus has a genome-linked protein. J. Virol. 44:437–443.
Persson, R. H., and R. D. Macdonald. 1983. A protein is bound (covalently?) to the ends of IPNV genomic RNA. Pages 129–133 in R. W. Compans and D. H. L. Bishop, eds. Double-stranded RNA viruses. Elsevier, New York.
Piper, D., B. L. Nicholson, and J. Dunn. 1973. Immunofluorescent study of the replication of infectious pancreatic necrosis in trout and Atlantic salmon cell cultures. Infect. Immun. 8:249–254.
Reno, P. W. 1976. Qualitative and quantitative aspects of the infectious pancreatic necrosis virus (IPNV) carrier state in trout. Ph.D. dissertation. Univ. Guelph, Ontario, Canada.
Reno, P. W., S. Darley, and M. Savan. 1978. Infectious pancreatic necrosis: experimental induction of a carrier state in trout. J. Fish. Res. Board Can. 35:1451–1456.
Rosenlund, B. D. 1977. Infectious pancreatic necrosis virus at the Willow Beach NFH, Nevada, and in rainbow trout stocked into adjacent Lake Mohave. Fish Health News 6:10–13.
Sano, T. 1970. Etiology and histopathology of several epizootics among fingerling salmonids. J. Tokyo Univ. Fish. 56:13–22.
Sano, T. 1971a. Studies on viral diseases of Japanese fishes—I. Infectious pancreatic necrosis of rainbow trout: first isolation from epizootics in Japan. Bull. Jpn. Soc. Sci. Fish. 37:495–498.
Sano, T. 1971b. Studies on viral diseases of Japanese fishes—II. Infectious pancreatic necrosis of rainbow trout: pathogenicity of the isolants. Bull. Jpn. Soc. Sci. Fish. 37:499–503.
Sano, T. 1972. Studies on viral diseases of Japanese fishes—III. Infectious pancreatic necrosis of rainbow trout: geographical and seasonal distributions in Japan. Bull. Jpn. Soc. Sci. Fish. 38:313–316.
Sano, T. 1973a. Current preventive approach to infectious pancreatic necrosis (IPN) in

Japan. Pages 75–78 in W. A. Dill, ed. Symposium on the major communicable fish diseases in Europe and their control. EIFAC (Eur. Inland Fish. Advis. Comm.) Tech. Paper 17, Suppl. 2.

Sano, T. 1973b. Studies on viral diseases of Japanese fishes—IV. Infectious pancreatic necrosis of rainbow trout: susceptibility of fresh water salmons of genus *Oncorhynchus*. Bull. Jpn. Soc. Sci. Fish. 39:117–120.

Sano, T. 1973c. Studies on viral diseases of Japanese fishes—V. Infectious pancreatic necrosis of amago trout. Bull. Jpn. Soc. Sci. Fish. 39:477–480.

Sano, T. 1976. Viral diseases of cultured fishes in Japan. Fish Pathol. 10:221–226.

Sano, T. 1977. Viral diseases of cultured salmonids in Japan. Pages 120–123 in Proceedings of the international symposium on diseases of cultured salmonids, Tavolek, Inc., Seattle, Washington, April 4–6, 1977.

Sano, T., N. Okamoto, and T. Nishimura. 1981. A new viral epizootic of *Anguilla japonica* Temminck and Schlegel. J. Fish Dis. 4:127–139.

Sano, T., K. Tanaka, and S. Fukuzaki. 1981. Immune response in adult trout against formalin killed concentrated IPNV. Pages 63–70 in D. P. Anderson and W. Hennessen, eds. Proceedings of the international symposium on fish biologics: serodiagnostics and vaccines, National Fish Health Research Laboratory, Leetown, West Virginia, April 26–30, 1981. S. Karger, Basel.

Savan, M., and P. Dobos. 1980. Effect of virazole on rainbow trout *Salmo gairdneri* Richardson fry infected with infectious pancreatic necrosis virus. J. Fish Dis. 3:437–440.

Scherrer, R. 1973. Infectious pancreatic necrosis (IPN) of Salmonidae. Pages 51–58 in W. A. Dill, ed. Symposium on the major communicable fish diseases in Europe and their control. EIFAC (Eur. Inland Fish. Advis. Comm.) Tech. Paper 17, Suppl. 2.

Scherrer, R., E. Bic, and J. Cohen. 1974. Le virus de la nécrose pancréatique infectieuse: étude de la réplication et de l'induction de la synthèse d'interferon en fonction de l'hôte et de la température. Ann. Microbiol. (Paris) 125:455–467.

Schlotfeldt, H. J. 1979a. Die infektiose Pankreasnekrose (IPN) der Salmoniden. Teil I: Ätiologie, Epizootiologie und Bekämpfung. Tierärztl. Umsch. 34:539–546.

Schlotfeldt, H. J. 1979b. Die infektiose Pankreasnekrose (IPN) der Salmoniden. Teil II: Laboratoriumsdiagnostik. Tierärztl. Umsch. 34:599–605.

Schlotfeldt, H. J., and J. W. Frost. 1975. Doppelinfektion von Regenbogenforellen mit dem Virus der viralen hämorrhagischen Septikämie (VHS) und dem Virus der infektiosen Pankreasnekrose (IPN). Berl. Münch. Tierärztl. Wochenschr. 88:455.

Schlotfeldt, H. J., B. Liess, and J. W. Frost. 1975. Erst Isolierung und Identifizierung des Virus der infektiosen Pankreasnekrose (IPN) der Salmoniden in der Bundesrepublik Deutschland. Berl. Münch. Tierärztl. Wochenschr. 88:109–111.

Schutz, M., E. B. May, J. N. Kraeuter, and F. M. Hetrick. 1984. Isolation of infectious pancreatic necrosis virus from an epizootic occurring in cultured striped bass, *Morone saxatilis* (Walbaum). J. Fish Dis. 7:505–507.

Seeley, R. J., A. Perlmutter, and V. E. Seeley. 1977. Inheritance and longevity of infectious pancreatic necrosis virus in the zebra fish, *Brachydanio rerio* (Hamilton-Buchanan). Appl. Environ. Microbiol. 34:50–55.

Snieszko, S. F., E. M. Wood, and W. T. Yasutake. 1957. Infectious pancreatic necrosis in trout. Arch. Pathol. 63:229–233.

Snieszko, S. F., K. Wolf, J. E. Camper, and L. L. Pettijohn. 1959. Infectious nature of pancreatic necrosis. Trans. Am. Fish. Soc. 88:289–293.

Sonstegard, R. A. 1970. Descriptive and epizootiological studies of infectious pancreatic necrosis of salmonids and lymphosarcoma of *Esox masquinongy*. Ph.D. dissertation. Univ. Guelph, Ontario, Canada.

Sonstegard, R. A., and L. A. McDermott. 1972. Epidemiological model for passive transfer of IPNV by homeotherms. Nature 237:104–105.

Sonstegard, R. A., L. A. McDermott, and K. S. Sonstegard. 1972. Isolation of infectious pancreatic necrosis virus from white suckers (*Catostomus commersoni*). Nature 236:174–175.

Sorimachi, M., and T. Hara. 1985. Characteristics and pathogenicity of a virus isolated from yellowtail fingerlings showing ascites. Fish Pathol. 19:231–238. In Japanese with English abstract.

State of New York Conservation Commission. 1923. Page 66 in Twelfth annual report for the year 1922, Legislative Document (1923) No. 29.

Stephens, E. B., and F. M. Hetrick. 1983. Molecular characterization of infectious pancreatic necrosis virus isolated from a marine fish. Pages 72–86 in J. H. Crosa, ed. Bacterial and viral diseases of fish. Washington Sea Grant Publ., Univ. Washington, Seattle.

Swanson, R. N. 1981. Use of the indirect fluorescent antibody test to study the pathogenesis of infectious pancreatic necrosis virus infection in trout. Pages 71–77 in D. P. Anderson and W. Hennessen, eds. Proceedings of the international symposium on fish biologics: serodiagnostics and vaccines, National Fish Health Research Laboratory, Leetown, West Virginia, April 26–30, 1981. S. Karger, Basel.

Swanson, R. N., and J. H. Gillespie. 1979. Pathogenesis of infectious pancreatic necrosis in Atlantic salmon (*Salmo salar*). J. Fish. Res. Board Can. 36:587–591.

Swanson, R. N., and J. H. Gillespie. 1981. An indirect fluorescent antibody test for the rapid detection of infectious pancreatic necrosis virus in tissues. J. Fish Dis. 4:309–315.

Swanson, R. N., and J. H. Gillespie. 1982. Isolation of infectious pancreatic necrosis virus from the blood and blood components of experimentally infected trout. Can. J. Fish. Aquat. Sci. 39:225–228.

Swanson, R. N., J. C. Carlisle, and J. H. Gillespie. 1982. Pathogenesis of infectious pancreatic necrosis virus infection in brook trout, *Salvelinus fontinalis* (Mitchill), following intraperitoneal injection. J. Fish Dis. 5:449–460.

Todd, D., and M. S. McNulty. 1979. Biochemical studies with infectious bursal disease virus: comparison of some of its properties with infectious pancreatic necrosis virus. Arch. Virol. 60:265–277.

Toranzo, A. E., and F. M. Hetrick. 1982. Comparative stability of two salmonid viruses and poliovirus in fresh, estuarine and marine waters. J. Fish Dis. 5:223–231.

Toranzo, A. E., J. L. Barja, M. L. Lemos, and F. M. Hetrick. 1983. Stability of infectious pancreatic necrosis virus (IPNV) in untreated, filtered and autoclaved estuarine water. Bull. Eur. Assoc. Fish Pathol. 3:51–53.

Tu, K., R. S. Spendlove, and R. W. Goede. 1974. Immunofluorescent cell assay of infectious pancreatic necrosis virus. Appl. Microbiol. 27:593–599.

Underwood, B. O., C. J. Smale, F. Brown, and B. J. Hill. 1977. Relationship of a virus from *Tellina tenuis* to infectious pancreatic necrosis virus. J. Gen. Virol. 36:93–109.

Vestergard Jorgensen, P. E. 1972a. Egtved virus: antigenic variation in 76 virus isolates examined in neutralization tests and by means of the fluorescent antibody technique. Symp. Zool. Soc. Lond. 30:333–340.

Vestergard Jorgensen, P. E. 1972b. Freund's adjuvants: their influence on the specificity of viral antisera. Acta Pathol. Microbiol. Scand. [B] 80:931–933.

Vestergard Jorgensen, P. E. 1973a. The nature and biological activity of IPN virus neutralizing antibodies in normal and immunized rainbow trout (*Salmo gairdneri*). Arch. Gesamte Virusforsch. 42:9–20.

Vestergard Jorgensen, P. E. 1973b. Inactivation of IPN and Egtved virus. Riv. Ital. Piscicol. Ittiopatol. 8:107–108.

Vestergard Jorgensen, P. E. 1974. Indirect fluorescent antibody techniques for demonstration of trout viruses and corresponding antibody. Acta Vet. Scand. 15:198–205.

Vestergard Jorgensen, P. E., and F. Bregnballe. 1969. Infectious pancreatic necrosis in rainbow trout (*Salmo gairdneri*) in Denmark. Nord. Vet. Med. 21:142–148.

Vestergard Jorgensen, P. E., and P. C. Grauballe. 1971. Problems in the serological typing of IPN virus. Acta Vet. Scand. 12:145–147.

Vestergard Jorgensen, P. E., and N. P. Kehlet. 1971. Infectious pancreatic necrosis (IPN) viruses in Danish rainbow trout. Nord. Vet. Med. 23:568–575.

Vestergard Jorgensen, P. E., and A. Meyling. 1972. Egtved virus: demonstration of virus antigen by the fluorescent antibody technique in tissues of rainbow trout affected by viral hemorrhagic septicemia and in cell cultures infected with Egtved virus. Arch. Gesamte Virusforsch. 36:115–122.

Wedemeyer, G. A., N. C. Nelson, and C. A. Smith. 1978. Survival of the salmonid viruses infectious hematopoietic necrosis (IHNV) and infectious pancreatic necrosis (IPNV) in ozonated, chlorinated, and untreated waters. J. Fish. Res. Board Can. 35:875–879.

Wolf, K. 1966. The fish viruses. Adv. Virus Res. 12: 35–101.

Wolf, K., and C. E. Dunbar. 1957. Cultivation of adult teleost tissues in vitro. Proc. Soc. Exp. Biol. Med. 95:455–458.

Wolf, K., and L. L. Pettijohn. 1970. Infectious pancreatic necrosis virus isolated from coho salmon fingerlings. Prog. Fish-Cult. 32:17–18.

Wolf, K., and M. C. Quimby. 1962. Established eurythermic line of fish cells in vitro. Science 135:1065–1066.

Wolf, K., and M. C. Quimby. 1967. Infectious pancreatic necrosis (IPN): its diagnosis, identification, detection and control. Riv. Ital. Piscicol. Ittiopatol. 2:76–82.

Wolf, K., and M. C. Quimby. 1969. Infectious pancreatic necrosis: clinical and immune response of adult trouts to inoculation with live virus. J. Fish. Res. Board Can. 26:2511–2516.

Wolf, K., and M. C. Quimby. 1971. Salmonid viruses: infectious pancreatic necrosis virus morphology, pathology, and serology of first European isolations. Arch. Gesamte Virusforsch. 34:144–156.

Wolf, K., and M. C. Quimby. 1973a. Fish viruses: buffers and methods for plaquing eight agents under normal atmosphere. Appl. Microbiol. 25:659–664.

Wolf, K., and M. C. Quimby. 1973b. Fish virology: procedures and preparation of materials for plaquing fish viruses in normal atmosphere. U.S. Fish Wildl. Serv. Fish Dis. Leaflet 35.

Wolf, K., and P. E. Vestergard Jorgensen. 1970. Salmonid viruses: double infection of RTG-2 cells with Egtved and infectious pancreatic necrosis viruses. Arch. Gesamte Virusforsch. 29:337–342.

Wolf, K., S. F. Snieszko, and C. E. Dunbar. 1958. Infectious pancreatic necrosis of fish as studied in cultures of fish tissues. Page 18 in Program of the 9th annual meeting of the Tissue Culture Association, Philadelphia, Pennsylvania, April 1958.

Wolf, K., S. F. Snieszko, and C. E. Dunbar. 1959. Infectious pancreatic necrosis, a virus-caused disease of fish. Excerpta Med. 13:228. Abstract.

Wolf, K., C. E. Dunbar, and S. F. Snieszko. 1960. Infectious pancreatic necrosis of trout. I. A tissue-culture study. Prog. Fish-Cult. 22:64–68.

Wolf, K., S. F. Snieszko, C. E. Dunbar, and E. Pyle. 1960. Virus nature of infectious pancreatic necrosis in trout. Proc. Soc. Exp. Biol. Med. 104:105–108.

Wolf, K., M. C. Quimby, and A. D. Bradford. 1963. Egg-associated transmission of IPN virus of trouts. Virology 21:317–321.

Wolf, K., M. C. Quimby, C. P. Carlson, and G. L. Bullock. 1968. Infectious pancreatic necrosis: selection of virus-free stock from a population of carrier trout. J. Fish. Res. Board Can. 25:383–391.

Wolf, K., G. L. Bullock, C. E. Dunbar, and M. C. Quimby. 1969. Viral diseases of freshwater fishes and other lower vertebrates (general). Pages 136–141 in Progress in sport fishery research 1968. Bureau Sport Fish. Wildl. Resource Publ. 77. Washington, D.C.

Wolf, K., M. C. Quimby, and C. P. Carlson. 1969. Infectious pancreatic necrosis virus: lyophilization and subsequent stability in storage at 4°C. Appl. Microbiol. 17:623–624.

Wood, E. M., S. F. Snieszko, and W. T. Yasutake. 1955. Infectious pancreatic necrosis in brook trout. Arch. Pathol. 60:26–28.

Yamamoto, T. 1974. Infectious pancreatic necrosis virus occurrence at a hatchery in Alberta. J. Fish. Res. Board Can. 31:397–402.

Yamamoto, T. 1975a. Infectious pancreatic necrosis virus and bacterial kidney disease appearing concurrently in populations of *Salmo gairdneri* and *Salvelinus fontinalis*. J. Fish. Res. Board Can. 32:92–95.

Yamamoto, T. 1975b. Infectious pancreatic necrosis (IPN) virus carriers and antibody productions in a population of rainbow trout (*Salmo gairdneri*). Can. J. Microbiol. 21:1343–1347.

Yamamoto, T., and J. Kilistoff. 1979. Infectious pancreatic necrosis virus: quantification of carriers in lake populations during a 6-year period. J. Fish. Res. Board Can. 36:562–567.

Yasutake, W. T. 1970. Comparative histopathology of epizootic salmonid virus diseases. Pages 341–350 in S. F. Snieszko, ed. A symposium on diseases of fishes and shellfishes. Am. Fish. Soc. Spec. Publ. 5.

Yasutake, W. T., T. J. Parisot, and G. W. Klontz. 1965. Virus diseases of the Salmonidae in western United States. II. Aspects of pathogenesis. Ann. N.Y. Acad. Sci. 126:520–530.

Yu, K. K., R. D. Macdonald, and A. R. Moore. 1982. Replication of infectious pancreatic necrosis virus in trout leucocytes and detection of the carrier state. J. Fish Dis. 5:401–410.

Yu, M. L., D. A. Sarot, R. J. Filazzola, and A. Perlmutter. 1969. Immune response of the blue gourami, *Trichogaster trichopterus*, to infectious pancreatic necrosis (IPN) virus. Life Sci. 8:1207–1213.

Yu, M. L., D. A. Sarot, R. Filazzola, and A. Perlmutter. 1970. Effects of splenectomy on the immune response of the blue gourami *Trichogaster trichopterus*, to infectious pancreatic necrosis (IPN) virus. Life Sci. 9:749–755.

10 Japanese Eel Iridovirus

Definition

The Japanese eel iridovirus (EV-102) was isolated from *Anguilla japonica* and in tests proved to be virulent for both leptocephali and elvers of the species (Sorimachi 1984).

History

The virus designated EV-102 was isolated on RTG-2 cells during 1978 as part of a 5-year program of virologic examinations of ponds in which *A. anguilla* and *A. japonica* were cultured. Among 153 ponds of eels examined, the virus was encountered only once, in a single pond of diseased eels in Shizuoka Prefecture (Sorimachi and Egusa 1982). In the same year, Sorimachi (1982) determined the susceptibilities of seven fish cell lines to five eel viruses and showed that EK-1 and EO-2 lines of Japanese eel origin replicated the virus better than did lines of cyprinid or salmonid origin. Virus isolant EV-102 was stored at −80°C and later tested and found to produce disease and mortality in *A. japonica* but not in *A. anguilla* (Sorimachi 1984).

Signs and Pathologic Changes

At temperatures of 15 to 19°C, experimentally infected Japanese eels (1 to 120 g) show signs of disease 3 to 5 days after exposure or inoculation: they lose color; congestion is evident in anal, dorsal, and pectoral fins; and the secretion of mucus increases. Mortality begins soon after the appearance of signs and continues for about 2 weeks. Virulence of the virus diminishes at temperatures of 23°C or higher, as evidenced by fewer eels showing signs of disease and a decrease in mortality (Table 1).

Table 1. Effects of EV-102 infection in *Anguilla japonica* of various body weights and at different temperatures

Body weight (g)	Route of inoculation[a]	Temperature (°C)	Number tested	Mortality (%)	Mean days to death
0.2	Bath	16	50	42	9
1.0	Bath	17	40	53	13
13.0	i.p.	17	30	47	10
15.0	i.p.	23	20	15	13
16.0	i.p.	15	20	75	11
21.0	i.p.	16	20	70	10
50.0	i.p.	24	25	0	—
61.0	i.p.	16	15	60	13
76.0	i.m.	19	10	40	14
123.0	i.p.	17	12	58	12

Source: Adapted from Sorimachi (1984).
[a]i.p., intraperitoneal; i.m., intramuscular.

Etiology

Few characteristics of the virus have been determined. It is heat, acid, and ether labile, and differential filtration has shown the particle size to be 100 to 200 nm. The virus is replicated by EPC and RTG-2 cells with attendant CPE (Sorimachi and Egusa 1982). According to Sorimachi (1984), the maximal titer is about 10^7 $TCID_{50}$/mL.

Isolation

Homogenates should be prepared from any obviously altered tissues and pooled internal organs, inoculated into cultures of EK-1, EO-2, EPC, or RTG-2 cell lines, and incubated at 15 to 20°C. Cytopathic effects are to be expected.

Identification

Serologic methods have not been reported; therefore, only presumptive identification can be made on the basis of host species and features of the virus.

Transmission and Incubation

Transmission has been effected by bath or immersion; waterborne infectivity is therefore assumed to occur in nature. Incubation time to first mortality is about 1 week and mean time to death is about 10 to 14 days (Table 1).

Source

The source is assumed to be eels of the genus *Anguilla*.

Host and Geographic Range

The agent is known only from *A. japonica* on Honshu Island, Japan, and only one case has been reported.

Control

Because the virus seems to lack virulence at 24°C—or because the host is better able to combat infection at that temperature—Sorimachi (1984) suggested that the disease might be controlled by raising the temperature to 24°C or higher. But when one considers the usual pond sizes and unavailability of warm water or economical sources of energy, that approach has considerable limitations.

References

Sorimachi, M. 1982. Susceptibility of fish cell lines to eel viruses. Fish Pathol. 17:119–123. In Japanese with English abstract.

Sorimachi, M. 1984. Pathogenicity of ICD virus isolated from Japanese eel. Bull. Natl. Res. Inst. Aquacult. 6:71–75. In Japanese with English abstract.

Sorimachi, M., and S. Egusa. 1982. Characteristics and distribution of viruses isolated from pond-cultured eels. Bull. Natl. Res. Inst. Aquacult. 3:97–105.

11 *Oncorhynchus masou* Virus

Synonyms: nerka virus Towada Lake, Akita Prefecture (NeVTA); Niigata tumor virus (NTV); yamame tumor virus (YTV)

Definition

Three isolations of herpesvirus have been made from two species of Pacific salmon in Japan. The isolants have been given individual designations, but the viruses are closely related serologically, have similar characteristics in vitro, and show common histopathologic changes in vivo. In addition, two have been shown to be oncogenic. Accordingly, all three are treated as one virus here—the most fully characterized one is known as *Oncorhynchus masou* virus (OMV). Comparative analyses of DNAs and structural proteins will more fully determine if this unification is justified.

Oncorhynchus masou virus is a herpesvirus that infects salmon of the genus *Oncorhynchus* and causes mortality in some of the young victims and epithelial neoplasms in some that survive (Figs. 1 and 2).

History

The pattern of events in the history of OMV is somewhat similar to that seen in America when *Oncorhynchus* salmon were found to be infected with virus. For several years, the salmon infections were named for their assumed exclusive hosts: chinook salmon virus disease and sockeye salmon virus disease. Later work showed that only one agent—a rhabdovirus—was involved, and the clinically descriptive name infectious hematopoietic necrosis (IHN) was applied.

The story of herpesvirus in *Oncorhynchus* in Japan begins on Honshu Island with Sano's 1976 report, in which he noted that as early as 1970 high mortality occurred among progeny of himemasu or kokanee, landlocked *O. nerka*. He isolated herpes-

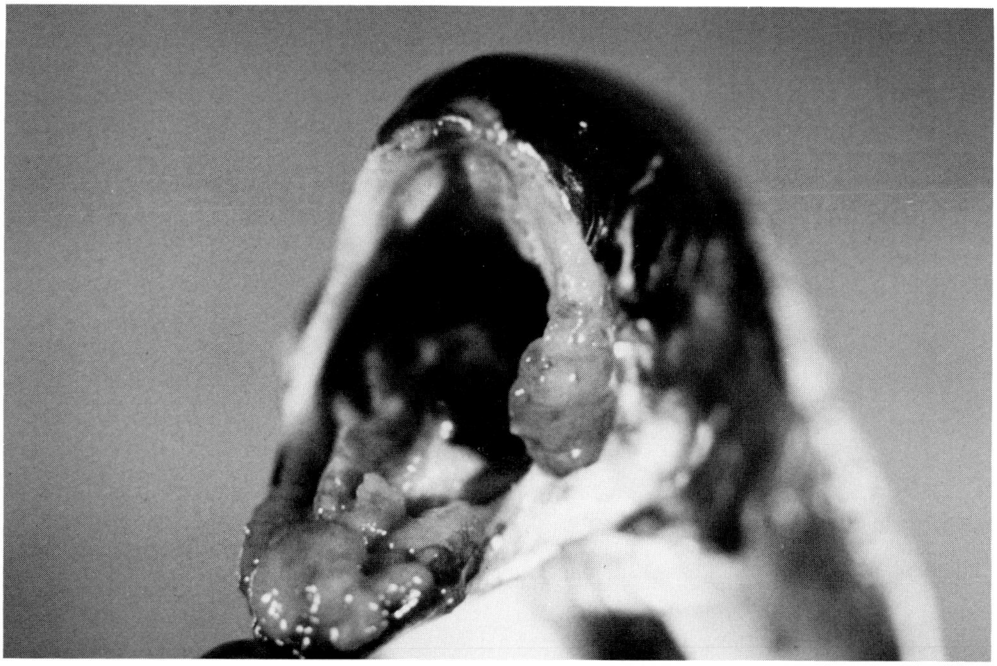

Figure 1. Gross appearance of naturally occurring epithelial neoplasms on mouth parts of *Oncorhynchus masou* from a Japanese hatchery. From Sano et al. (1983). Reprinted with the permission of the Japanese Society of Scientific Fisheries.

virus from moribund fry and coined the name NeVTA—from the source species and location (nerka virus Towada Lake, Akita Prefecture).

Herpesvirus was next found on Hokkaido Island in 1978 among yamame (landlocked *O. masou*). Kimura, Yoshimizu, Tanaka, and Sannohe (1981) isolated virus from ovarian fluids of normal-looking stock that also had a history of poor progeny survival. That agent, named OMV, was shown to be pathogenic for several salmonids; unexpectedly, some survivors developed epithelial tumors. Kimura, Yoshimizu, and Tanaka (1981a, b) noted that OMV was distinct from NeVTA and *Herpesvirus salmonis* that had been found in America.

Next the herpesvirus was isolated directly from naturally occurring tumors on yamame on Honshu Island (Sano et al. 1983). That isolant was named yamame tumor virus (YTV) and shown by cross-neutralization to be closely related to NeVTA. Experimentally, YTV was pathogenic for several salmonids. Moreover, as in OMV infections, some survivors of YTV developed epithelial tumors (Figs. 1 and 2).

Kimura's group published two reports in 1983 on studies of antiherpesvirus compounds, which showed that inhibition of OMV occurred both in vitro and in vivo. Extended daily antiviral treatment of fish effected an increase in survivors of experimental infection and a marked inhibition of tumor development (Kimura, Suzuki, and Yoshimizu 1983a, b). A third report by Kimura, Yoshimizu, and Tanaka (1983), although written in Japanese, has an English summary that reports results of com-

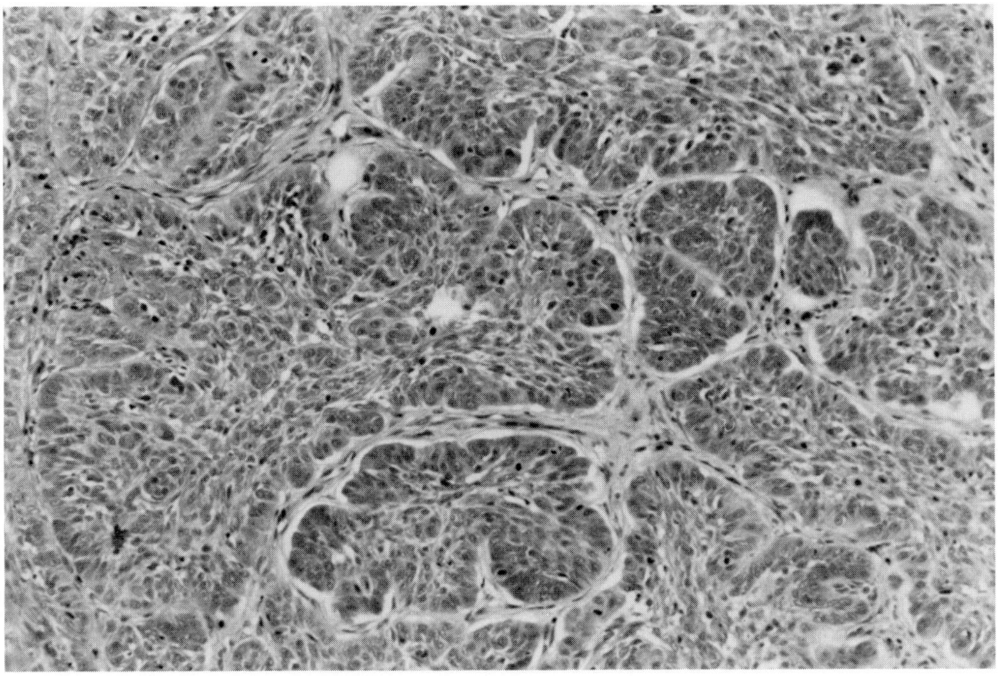

Figure 2. Histologic section of naturally occurring epithelial neoplasm, showing lobular or papillary growth pattern. From Sano et al. (1983). Reprinted with the permission of the Japanese Society of Scientific Fisheries.

parative susceptibilities of fry of five species of salmonids to experimental infection with OMV. The probable relatedness of OMV and YTV (NeVTA) is supported by the finding that the landlocked forms—himemasu and yamame—sustained the highest mortality.

Also from Kimura's laboratory, Tanaka et al. (1984) enlarged on earlier work and described and illustrated the histopathologic effects of OMV. They included comments on differences and similarities of OMV, NeVTA, and *Herpesvirus salmonis*.

Signs and Pathologic Changes

Like most viral infections of fishes, the salmonid herpesvirus from Japan exacts its toll of disease and mortality in the young.

Behavior

Himemasu naturally infected with NeVTA are anorexic, sluggish, and congregated at the water inlet. Chum salmon (*O. keta*)—the preferred experimental host for OMV—experimentally infected by immersion are also anorexic.

External Signs

Naturally infected himemasu are simply described as darkened in coloration. Experimentally infected chum salmon develop marked exophthalmia and external petechiation.

Internal Signs

Chum salmon with experimental OMV have no food in their digestive tract and the kidneys generally appear to be normal. The livers are mottled with white spots and in advanced cases that organ is pearly white. The spleen of some specimens is swollen.

Histopathologic Findings

In naturally infected himemasu, Sano (1976) found syncytia in renal interstitial cells, but other changes were less marked. Renal tubule cells showed necrosis, and proteinaceous fluid surrounded some glomeruli. Granular degeneration was evident in skeletal muscle, pancreatic acinar cells were vacoulated, and gill epithelium was swollen. Further details were not given.

The most extensive histopathologic study was reported by Tanaka et al. (1984), who worked with three species of salmonids infected with OMV by immersion. The kidneys of all month-old yamame showed syncytia and hematopoietic necrosis, but glomeruli and tubules were not affected. Livers were atrophied and necrosis was present in older specimens. That change increased in severity as incubation time was extended. Pancreas and spleen were partly necrotic. Necrosis of epidermal cells was found in and about the mouth and head. That change presaged the sites where tumors appeared on other specimens.

In contrast with younger specimens, yamame that were 3 months old at the time of exposure to OMV showed primary effects in the liver. They exhibited multiple foci of necrosis, accompanied by pyknosis and karyorrhexis. In addition, syncytia were present in some liver foci. Kidneys also showed necrosis and hyperplasia of hematopoietic tissue. Necrosis and hemorrhages were found in the brain.

Kidneys were the apparent key target for the virus in chum salmon fry. Hematopoietic necrosis was evident and tubule cells showed hyaline droplet degeneration and pyknosis. Partial necrosis occurred in spleen, liver, stomach, and pancreas, and the intestinal mucosa was sloughed. No syncytia were noted.

Coho salmon fry showed marked changes in kidneys and liver, much like those in chum salmon. Again, no syncytia were noted.

Etiology

On the basis of results of experimental infections, OMV can be regarded as a significant pathogen and as a probable cause of mortality among young himemasu and yamame.

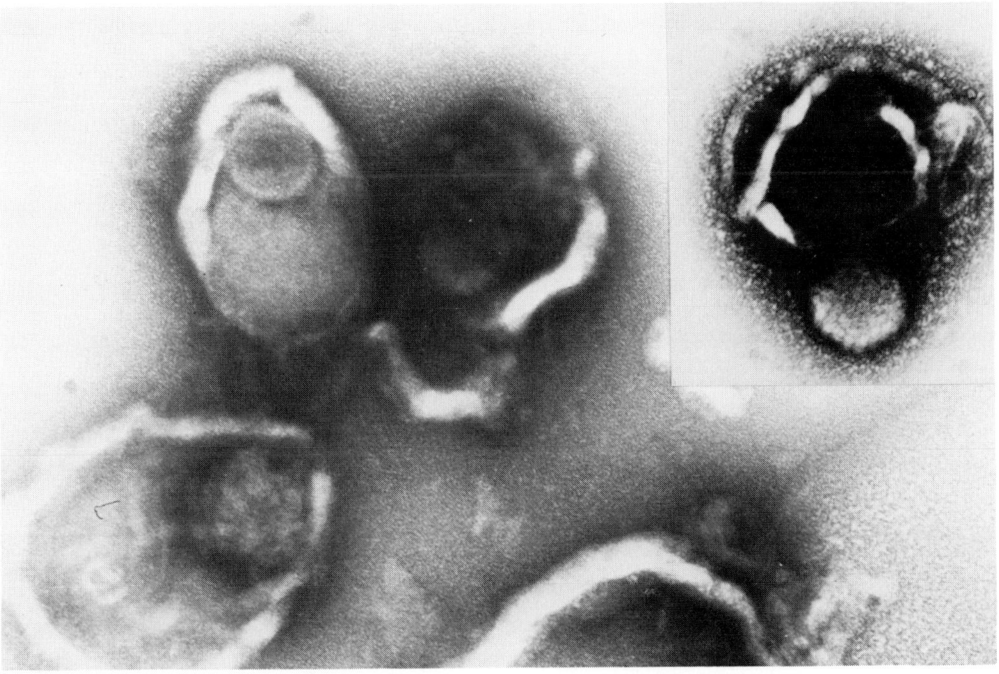

Figure 3. Electron micrograph of negatively stained virions, showing typical herpesvirus features of enveloped capsids. From Sano et al. (1983). Reprinted with the permission of the Japanese Society of Scientific Fisheries.

Size and Shape

Mean dimensions of enveloped particles are 220 to 240 nm and the capsids are about 100 to 115 nm (Fig. 3). Infectivity passes membranes with a mean porosity of 450 nm, but is retained by 220-nm membranes.

Biophysical Properties

As herpesviruses, the agents are heat, ether, and acid labile. They are replicated by salmonid cell lines of fibroblastic or epithelial morphology, but nonsalmonid cells are refractory. The CHH-1, CHSE-214, KO-6, RTG-2, and YNK cell lines produce a maximum yield of about 10^6 $TCID_{50}$/mL at near-optimum temperature. The temperature range for incubation extends from 5 to 18°C, the optimum being 10 to 15°C. One-step growth curves have not been established, but CPE in the form of syncytia appears in 5 to 7 days at near-optimum incubation temperature (Fig. 4). Infected cells show Cowdry type A intranuclear inclusions, and replication terminates with lysis of the cell sheet.

Preservation of infectivity is best maintained at −80°C. In 17 days, OMV lost 99.9% of its infectivity at −20°C and all of its infectivity at temperatures above 15°C. The effects of lyophilization and freeze-thaw cycling are unknown.

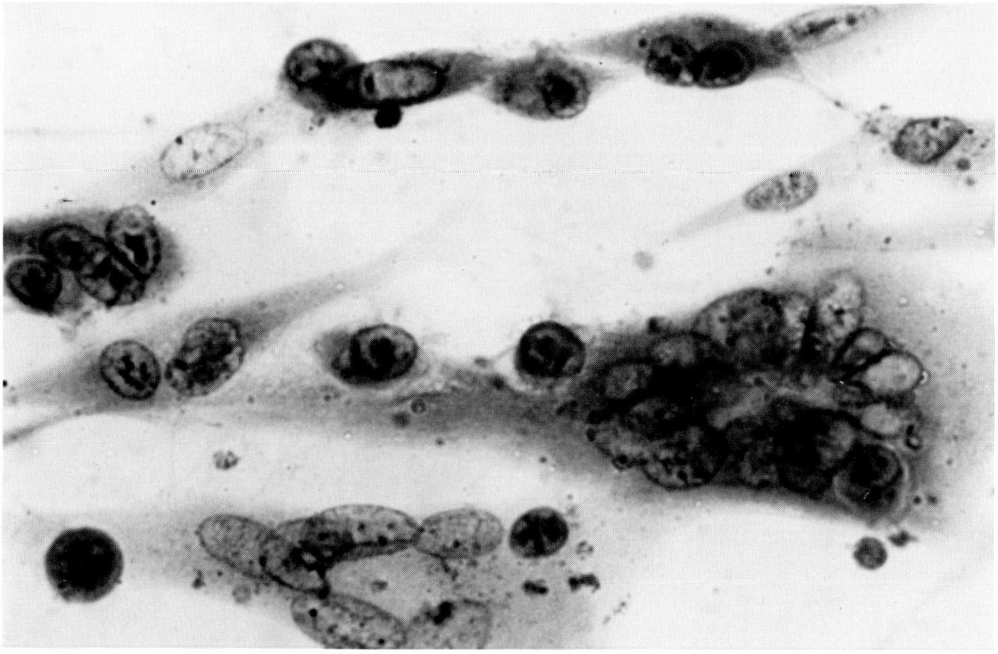

Figure 4. In vitro effect of herpesvirus from *Oncorhynchus masou* on RTG-2 cells includes formation of syncytia, margination of chromatin, and development of Cowdry type A intranuclear inclusions. From Sano et al. (1983). Reprinted with the permission of the Japanese Society of Scientific Fisheries.

Diagnosis

The disease can be suspected when abnormal mortality occurs among young yamame and himemasu and no other causes are evident. However, the himemasu and the kokanee are also susceptible to the rhabdovirus that causes infectious hematopoietic necrosis, which also occurs in Japan. Diagnosis should therefore include virologic examination.

Isolation

Homogenates of whole victim fry should be decontaminated and cultures of salmonid cells inoculated and incubated at 10 to 15°C for at least 2 weeks. The CHH-1, CHSE-214, K0-6, RTG-2, or YNK cell lines are suggested. In the absence of CPE, blind passage is suggested.

Virus has also been isolated from ovarian fluids, and to a lesser extent from tumors. Although Kimura, Yoshimizu, and Tanaka (1981a) isolated virus from tumor tissue of experimentally infected *O. keta*, they were unable to find it in liver, kidneys, heart, or spleen. The presence or absence of virus in tissues of adult carriers merits study.

Identification

Identification is by a serum neutralization test. The degree of cross-reactivity between isolants of OMV, NeVTA, and YTV indicates that the three are serologically tightly clustered, but that they are distinctly different from the North American salmonid herpesvirus isolants *H. salmonis* and SHV (Hedrick et al. 1987). The North American isolants have been designated as salmonid herpesvirus type 1 and the Japanese as type 2 (R. P. Hedrick, personal communication, 1987).

Transmission and Incubation

The presence of virus in ovarian fluids and epizootic disease among progeny indicate that vertical transmission occurs. Horizontal transmission also occurs, since susceptibles can be infected by immersion.

Kimura, Yoshimizu, Tanaka, and Sannohe (1981) demonstrated that incubation time to first mortality increases with age. In chum salmon 80 days old, mortality began 11 days after exposure and 60% of the fish died; in fish 150 days old, incubation time was 20 days and cumulative mortality was 35%. Among fish 240 days old when exposed, neither disease nor mortality occurred, and the fish were additionally refractive to intraperitoneal injection of virus.

Source

The only documented natural sources of virus are ovarian fluids and tumors of yamame. Virus can also be found in homogenates of young victims of experimental infection.

Host and Geographic Range

Natural disease and mortality is known only from Japan and among young himemasu and yamame. Experimentally, young *O. keta*, *O. kistuch*, and *Salmo gairdneri* are also susceptible. Overall, 1-month-old himemasu (kokanee) are most susceptible, followed by 3-month-old *O. keta*. In general, cumulative mortality declines with age and is negligible among fish 6 months old (Kimura, Yoshimizu, and Tanaka 1983).

Immunity

Neutralizing activity against OMV is found in sera of experimentally infected *O. keta* that survived. Against 100 $TCID_{50}$ virus, titers range from 1:160 to 1:320.

Control

Avoidance is suggested. Practical therapy of viral disease in fish has yet to be realized, but progress has been made. Kimura, Suzuki, and Yoshimizu (1983b) showed that the antiherpesvirus compound 9-(2-hydroxyethoxymethyl) guanine (Acyclovir) was efficacious when administered by immersion, but not orally. Mortality due to OMV was reduced and tumor development was inhibited among fingerlings that were given daily immersions for 30 minutes in 25 μg/mL for 2 months. Iododeoxyuridine at an oral dosage of 25 μg per fish per day for 2 months also reduced mortality and completely inhibited tumor development. Benefits of iododeoxyuridine exceeded those of Acyclovir.

References

Hedrick, R. P., T. McDowell, W. D. Eaton, T. Kimura, and T. Sano, 1987. Serological relationships of five herpesviruses isolated from salmonid fishes. J. Appl. Ichthyol. 3:87–92.

Kimura. T., M. Yoshimizu, and M. Tanaka. 1981a. Fish viruses: tumor induction in *Oncorhynchus keta* by the OMV herpesvirus. Pages 59–68 in C. J. Dawe, J. C. Harshbarger, T. Sugimura, S. Takayama, and S. Kodo, eds. Phyletic approaches to cancer. Proceedings of the 11th international symposium of the Princess Takamatsu Research Fund. Jpn. Sci. Soc. Press, Tokyo.

Kimura, T., M. Yoshimizu, and M. Tanaka. 1981b. Studies on a new virus (OMV) from *Oncorhynchus masou*. II. Oncogenic nature. Fish Pathol. 15:149–153.

Kimura, T., M. Yoshimizu, M. Tanaka, and H. Sannohe. 1981. Studies on a new virus (OMV) from *Oncorhynchus masou*. I. Characteristics and pathogenicity. Fish Pathol. 15:143–147.

Kimura, T., S. Suzuki, and M. Yoshimizu. 1983a. In vitro antiviral effect of 9-(2-hydroxyethoxymethyl) guanine on the fish herpesvirus, *Oncorhynchus masou* virus (OMV). Antiviral Res. 3:93–101.

Kimura, T., S. Suzuki, and M. Yoshimizu. 1983b. In vivo antiviral effect of 9-(2-hydroxyethoxymethyl) guanine on experimental infection of chum salmon *Oncorhynchus keta* fry with *Oncorhynchus masou* virus (OMV). Antiviral Res. 3:103–108.

Kimura, T., M. Yoshimizu, and M. Tanaka. 1983. Susceptibility of different fry stages of representative salmonid species to *Oncorhynchus masou* virus (OMV). Fish Pathol. 17:251–258.

Sano, T. 1976. Viral diseases of cultured fishes in Japan. Fish Pathol. 10:221–226.

Sano, T., H. Fukuda, N. Okamoto, and F. Kaneko. 1983. Yamame tumor virus: lethality and oncogenicity. Bull. Jpn. Soc. Sci. Fish. 49:1159–1163.

Tanaka, M., M. Yoshimizu, and T. Kimura. 1984. *Oncorhynchus masou* virus: pathological changes in masu salmon (*Oncorhynchus masou*), chum salmon (*O. keta*) and coho salmon (*O. kisutch*) fry infected with OMV by immersion method. Bull. Jpn. Soc. Sci. Fish. 50:431–437.

12 Perch Iridovirus

Synonyms: epizootic hematopoietic necrosis, Nillahcootie redfin virus

Definition

The redfin perch iridovirus has been circumstantially implicated as the cause of a viscerotropic disease and mortality of epizootic proportion among young-of-the-year and adult *Perca fluviatilis*. Thus far, the virus is known only from Victoria, southeastern Australia (Langdon et al. 1986). Rivers' postulates have yet to be fulfilled.

History

According to Langdon et al. (1986), the Eurasian perch was introduced into Australia in 1861, and there was called the redfin perch. Apparently, only one introduction was made—a shipment of only 10 specimens from England. The species readily adapted to the new location and is now widely distributed throughout Victoria.

During investigations of outbreaks of mortality among juvenile perch in various impoundments in 1984, an iridovirus was isolated and implicated as the etiologic agent (Fig. 1). The agent is the first fish virus to be isolated in Australia. Outbreaks of disease among perch occurred again in 1985 and 1986. However, Langdon and Humphrey (1987) reported that the problem was not newly recognized, but that similar mass mortalities of juvenile perch have been observed in several lakes since 1972. The virus is believed to be widespread—at least in Victoria—as a result of natural spread in water, but also because perch have been indiscriminately moved about by anglers. In addition, stock has been widely distributed from one fish culture facility with a history of serious mortality in juvenile perch.

Clinical signs of the disease, histopathologic findings, and incidence were described by Langdon and Humphrey (1987).

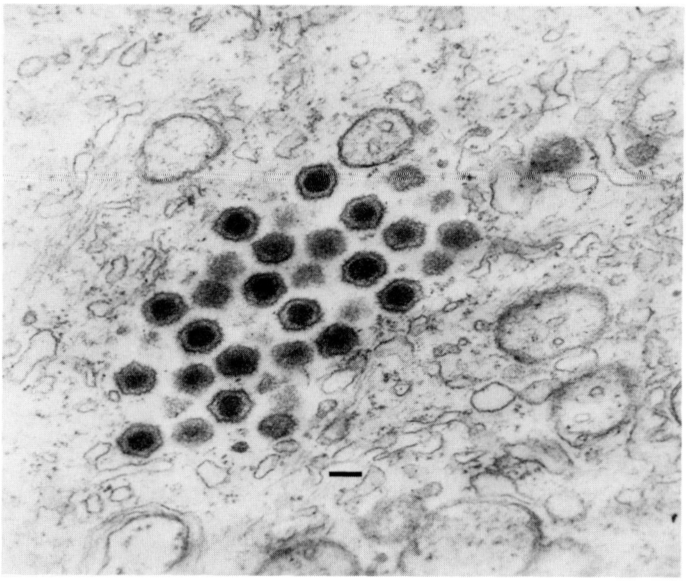

Figure 1. Thin section of an RTG-2 cell with a cytoplasmic array of iridovirus particles. Bar = 100 nm. From Langdon et al. (1986). Reprinted with the permission of Blackwell Scientific Publications Ltd.

Signs and Pathologic Changes

The disease typically occurs among juvenile perch in spring or early summer, and outbreaks run their course in a matter of 2 to 3 weeks. Older fish, including adults, may also be sporadically affected (Langdon and Humphrey 1987).

The initial report noted that many victim perch "displayed focal hepatocellular and hematopoietic necrosis" (Langdon et al. 1986). Detailed findings were later described and illustrated by Langdon and Humphrey (1987).

Behavior

Affected fish are listless and ataxic, respire slowly, and sometimes swim in a slow spiral to the surface.

External Signs

Petechial hemorrhages are common in fin bases—notably the base of the anal fin. Adult victims and some juveniles show reddening about the brain and nares. In juveniles, areas of pale skeletal muscle are common, particularly around the caudal trunk.

Internal Signs

Multiple pale foci 1 to 3 mm in diameter are found in the livers of adults and, less conspicuously, in the paler livers of juveniles. Spleens, particularly of juveniles, are

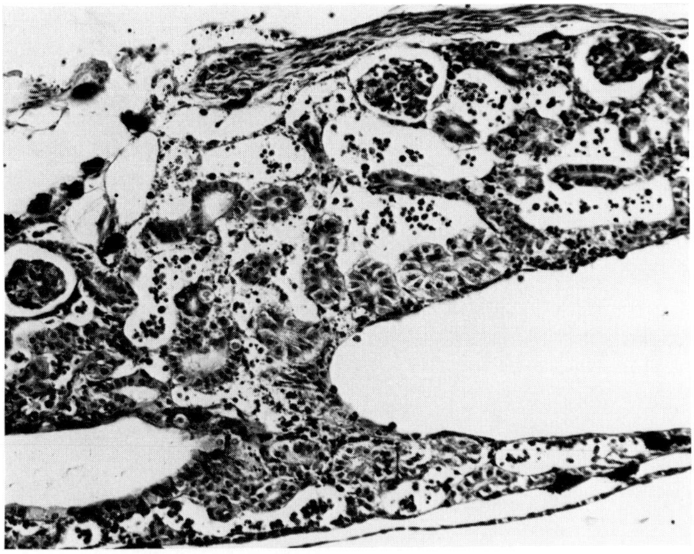

Figure 2. Necrosis of hematopoietic tissue and indications of possible edema in a kidney section from a naturally infected perch. From Langdon and Humphrey (1987). Reprinted with the permission of Blackwell Scientific Publications Ltd.

enlarged, bright red, and gelatinous. Gills are congested. Some adults show subperitoneal hyperemia about the swim bladder and kidneys.

Histopathologic Findings

The kidney, liver, spleen, and (in some specimens) pancreas show necrosis. The changes range from severe and focal to more disseminated, and are most consistent and characteristic in renal hematopoietic tissue and the liver. In addition, necrotic hematopoietic cells are disseminated throughout the vasculature.

Necrosis is evident throughout the hematopoietic tissue of the kidneys and is most severe in the lateral lobes of the anterior kidneys; however, renal tubules show little or no change (Fig. 2). Individual hematopoietic cells in situ and seemingly distributed throughout the vascular system are intensely stained, rounded, and have pyknotic nuclei.

Discrete foci of necrosis, usually centered on an artery or vein, are present in the liver. The walls of such blood vessels appear normal, but the lumens hold the kind of large necrotic hematopoietic cells seen in the kidneys. Hematopoietic cells surrounding blood vessels are affected in what appear to be early lesions, but hepatic cells are eventually involved and result in pale areas of necrosis and cellular debris surrounded by hyperchromatic liver cells with pyknotic nuclei. Many liver cells have clear cytoplasmic vacuoles and some also have basophilic inclusions. Foci of necrosis are surrounded by cords of normal hepatocytes, but sinusoids commonly contain the necrotic hematopoietic cells. Little or no evidence of an inflammatory response is seen.

Spleens show necrosis that ranges from ill-defined foci to diffuse areas. Organization of red and white pulp is lost, but blood vessels are unaffected. The pancreas of some victims shows occasional foci of necrosis, but the digestive tract is normal.

Corresponding to the reddening around the brain, meningeal blood vessels show congestion and occasional thromboses. Suprameningeal hemorrhage is present in some adults.

Histologically, the areas of pale skeletal muscle are edematous.

Experimental infections have not been reported in young Eurasian perch, but Langdon et al. (1986) described them in fingerling Atlantic salmon (*Salmo salar*). They injected the salmon intraperitoneally with about 10^4 $TCID_{50}$ virus and reported signs of neurologic disease and histopathologic changes. However, mortality was not mentioned. Injected salmon were anorexic, behaviorally depressed, markedly dark, and seemingly blind for 4 to 14 days. During the next 5 to 10 days the fish showed hyperexcitability and an increased rate of coughing reflexes; thereafter the depression of behavior and dark pigmentation moderated.

Salmon killed at 4 weeks had severe vacuolating encephalopathy of the optic lobe and lesser focal vacuolation of the optic nerve and fiber tracts of the cerebellum. Small foci of subcapsullar necrosis occurred in the liver. Virus was recovered from the brain and anterior kidney but, surprisingly, not from the liver or spleen; nor was pathology found in kidneys of the salmon.

Etiology

Because the virus was present in victim fish, it was believed to be the causal agent (Langdon et al. 1986). Kidneys, livers, and spleens from three fish were pooled and inoculated onto drained monolayers of RTG-2 cells that were then incubated at 15°C. Ten days later, CPE appeared and proved to be serially transmittable. Electron microscopy showed the presence of virus in infected RTG-2 cells (Fig. 1). The virus proved to be transiently pathogenic for fingerling Atlantic salmon—the only species tested.

Viral particles measure 150 to 170 nm. The capsid of 11 to 15 nm is separated by a 16–nm space from the nucleoid, which is 88 to 110 nm in diameter (Fig. 1). The agent is enveloped and ether sensitive. As judged by available information, the agent is an iridovirus.

Replication occurs in BB, BF-2, FHM, and RTG-2 cells, and wholly in the cytoplasm. Best growth—a titer of $10^{5.8}$ $TCID_{50}$/mL—is obtained in FHM cells at 15°C; CPE first appears in 15 to 17 hours and the cell sheet is destroyed in 2 to 3 days.

Diagnosis

The redfin perch iridovirus can be suspected when outbreaks of mortality occur among young *P. fluviatilis* and affected individuals show the described signs and histopathologic changes.

Isolation

Inocula prepared from visceral homogenates are assayed in FHM or RTG-2 cells and incubated at 15°C.

Identification

The perch iridovirus is identified solely on the basis of the host species and presumed pathologic changes. It certainly is not lymphocystis, nor does it appear to be the agent of viral erythrocytic necrosis (VEN). It must be compared with the iridoviruses from Japanese eels, the cod ulcus syndrome, and goldfish and carp gill necrosis.

Host and Geographic Range

The virus is known only from Australia, and the natural host is *P. fluviatilis*. Other percids should be tested for susceptibility.

References

Langdon, J. S., and J. D. Humphrey. 1987. Epizootic hematopoietic necrosis, a new viral disease in redfin perch *Perca fluviatilis* L. in Australia. J. Fish Dis. 10:289–297.

Langdon, J. S., J. D. Humphrey, L. M. Williams, A. D. Hyatt, and H. A. Westbury. 1986. First virus isolation from Australian fish: an iridovirus-like pathogen from redfin perch, *Perca fluviatilis* L. J. Fish Dis. 9:263–268.

13 Perch Rhabdovirus Disease

Definition

Perch rhabdovirus disease is a slow to develop condition that involves a central nervous disorder of *Perca fluviatilis* and results in a low order of mortality.

History

A single isolation was made from naturally infected yearlings that weighed about 20 g. The source fish were from a lot of about 100 that was received at the Laboratory of Fish Diseases, Thiverval-Grignon, France. The fish soon showed abnormal behavior and more than 30% died. Causal bacteria or parasites could not be found, but virologic examination revealed virus in a homogenate of pooled brain tissue from freshly dead specimens. No evidence of virus was found in inocula prepared from kidneys and spleen. The rhabdovirus was propagated in cell culture with CPE and has proved to be pathogenic for young perch (Dorson et al. 1984).

Signs and Pathologic Changes

The source fish lost equilibrium and swam about erratically before death. Young perch that had been inoculated intracranially with live virus showed similar behavior changes and mortality. Histopathologic changes and other details of the disease have not been reported.

Etiology

Loosely interpreted, Rivers' postulates have been fulfilled. Intracranial inoculation of fingerling perch (~ 8 g) held at 10 to 15°C resulted in disease signs at 4 weeks, and

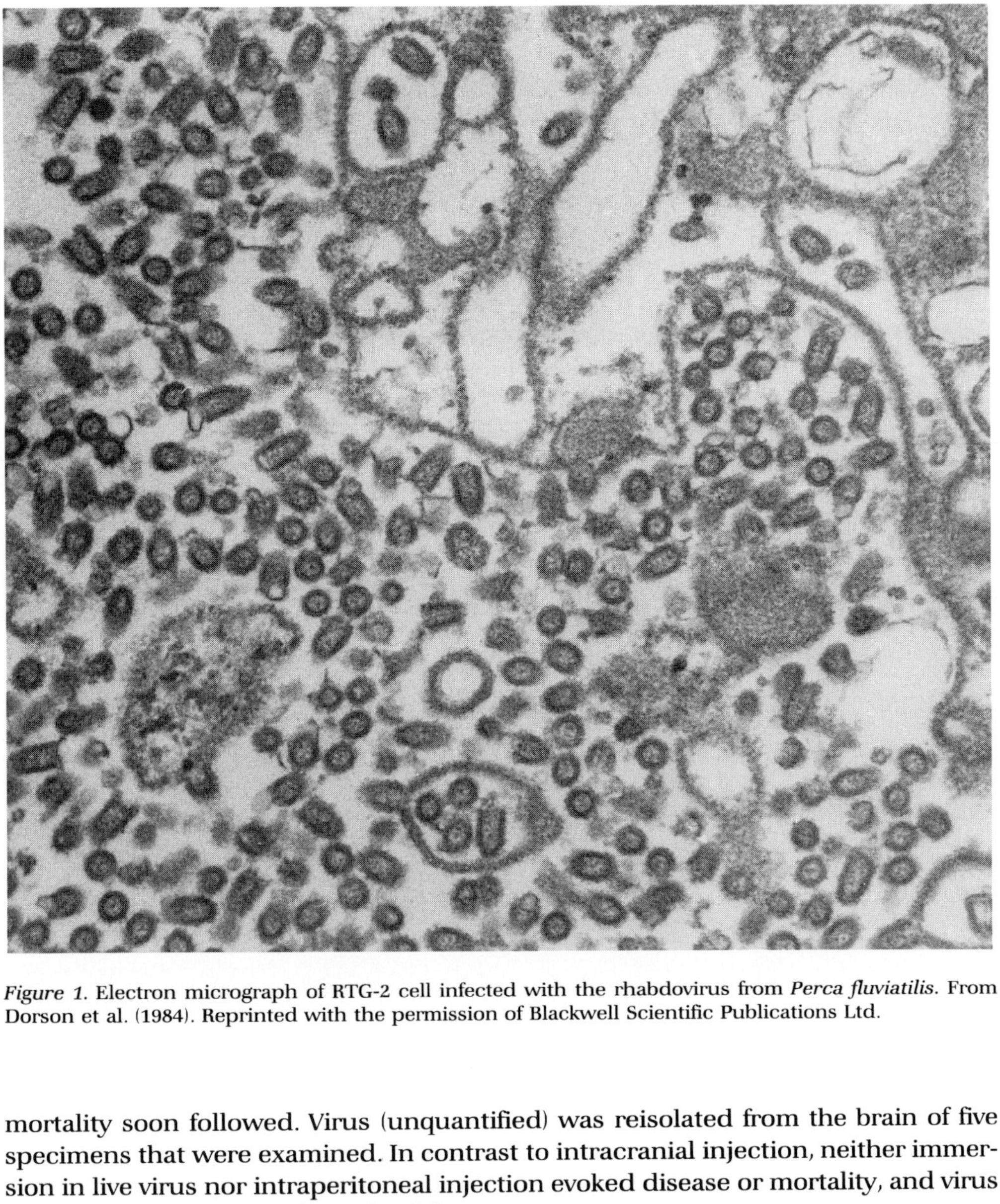

Figure 1. Electron micrograph of RTG-2 cell infected with the rhabdovirus from *Perca fluviatilis*. From Dorson et al. (1984). Reprinted with the permission of Blackwell Scientific Publications Ltd.

mortality soon followed. Virus (unquantified) was reisolated from the brain of five specimens that were examined. In contrast to intracranial injection, neither immersion in live virus nor intraperitoneal injection evoked disease or mortality, and virus could not be reisolated from the fish.

The virus has not been characterized. Its dimensions are about 200 × 100 nm and it is heat labile, losing more than 99% infectivity after 30 minutes at 37°C. The RTG-2 cell line at 14°C replicates the agent (Fig. 1) with CPE to a maximum titer of about 10^5 PFU/mL. The EPC line has been found to be refractory.

Limited serologic comparisons have shown that the perch rhabdovirus is not a known salmonid rhabdovirus, nor is it eel virus American (EVA). Considerations such as cell line susceptibilities suggest that it is a newly discovered virus.

Host and Geographic Range

Thus far, the sole occurrence has been in perch from France. It is expected that the North American *Perca flavescens* and other percid fishes of the northern hemisphere would be susceptible because the host specificity of rhabdoviruses is not tightly restricted.

Reference

Dorson, M., C. Torchy, S. Chilmonczyk, P. de Kinkelin, and C. Michel. 1984. A rhabdovirus pathogenic for perch, *Perca fluviatilis* L.: isolation and preliminary study. J. Fish Dis. 7:241–245.

14 Pike Fry Rhabdovirus Disease

Synonyms: hydrocephalus, red disease of pike

Definition

Pike fry rhabdovirus disease (PFRD) is an acute to subacute hemorrhagic infection causing generalized edema, hematopoietic necrosis, and high mortality in fry of *Esox lucius*.

History

The history of PFRD began in The Netherlands with Bootsma's 1971 report of hydrocephalus and red disease, two pathologic conditions of then-unknown etiology but both with concomitant high mortality. Prophetically, Bootsma noted that, although the clinical signs of hydrocephalus and red disease were widely different, they could be manifestations of the same condition. He stated that what was called "head disease" by fish farmers had first been seen in 1959. He cited a 1969 report that red disease had been observed "as early as 1956" (Bootsma 1976).

Two years after Bootsma's first report, publications appeared showing that the affected pike fry had only one disease, that it was contagious and waterborne, and that the etiologic agent was a rhabdovirus. Furthermore, the disease could be initiated by culture-grown virus (de Kinkelin et al. 1973).

It was then shown that the virus particles were located in renal hematopoietic tissue (Bootsma and van Vorstenbosch 1973) .

The newly recognized pike fry virus was quick to capture the attention of electron microscopists and molecular biologists. In 1974 Cohen and Lenoir published their comparisons of the ultrastructure of the four fish rhabdoviruses then known and showed that the agents could be arranged in two groups—the longer and more fragile particles of salmonid viruses in one group, and the somewhat shorter, structurally stronger PFR and *Rhabdovirus carpio* in the other. Additionally, in agents of the second group, surface spikes were more distinct and prominent. Characterization of PFR was reported by de Kinkelin et al. (1974), who determined the usual biophysical properties, but also made the first separation of the PFR structural proteins by PAGE. Four polypeptides were resolved but molecular weights were not determined. The authors were the first to point out that PFR was the fish rhabdovirus most like vesicular stomatitis virus, the prototype bullet-shaped virus.

The year 1975 saw the publication of eight relevant reports on PFR; numerically, the emphasis was primarily on biophysical properties of the virus and secondarily on the disease. Bootsma et al. (1975) reported transmission of PFRD after the addition of culture-grown virus to fertilized eggs. The resulting mortality was 100%, but infection was prevented when such eggs were treated prophylactically with an iodophore. Elegant physicochemical characterization of PFR was reported by Roy et al. (1975), who gave detailed analyses of the virus structural proteins, including molecular weights, and reported the presence of a protein kinase and an RNA-dependent RNA polymerase. Several months later, in a comparative study of five fish rhabdoviruses, Hill et al. (1975) reported similar protein values for PFR. They also showed by cross-neutralization that PFR was distinct from the other four, and that, like *R. carpio*, PFR had a polypeptide profile similar to that of vesicular stomatitis. Lenoir and de Kinkelin (1975), who also made a comparative study of fish rhabdoviruses and used PAGE to separate structural proteins, grouped the carp rhabdovirus with PFR and also called attention to the similarity of the two virions to the virus of vesicular stomatitis. It had become apparent that data on the structural proteins of PFR differed little . Clerx et al. (1975) confirmed earlier determinations of the molecular weight of the RNA and its sedimentation coefficient, and presented new data on its guanine-cytosine value.

Ahne (1975a, b) isolated a rhabdovirus from a diseased adult grass carp and speculated about it possibly sharing antigens with PFR. [Later work showed that the isolant was indeed indistinguishable from PFR (Clerx and Horzinek 1978; Clerx et al. 1978)].

Although the proposed epizootiologic significance for fish does not stand critical scrutiny, Bussereau et al. (1975) injected fruit flies with PFR and found evidence of virus replication.

The northern pike has been a minor species in aquaculture; accordingly, publications that deal with its pathogens have appeared only irregularly, and partly in reports of comparative studies. Ahne and Held (1980) showed that PFR was like other fish rhabdoviruses in being readily inactivated by an iodophore. The response of PFR to various physical and chemical agents was found to be similar to that of two other fish rhabdoviruses (Ahne 1982).

Reports of PFRD have been infrequent. Five cases were reported from Hungary during the period 1981–1983 (Békési et al. 1984). During the same interval, the virus was isolated in Germany—not from pike fry but from diseased yearlings of two species of cyprinids. The host range for PFR now includes four species (Ahne et al. 1982).

Signs and Pathologic Changes

The PFR produces two forms of disease: in the very young, early swimming fry, the usual clinical sign is one of hydrocephalus (Fig. 1) that has a subacute course, but

Figure 1. Northern pike fry with prominent hydrocephalus. From Bootsma (1976).

when older fry are involved they have acute disease and typically show the hemorrhagic signs from which the name red disease arose (Fig. 2). Bootsma (1976) stated that in exceptional cases both forms of the disease have been observed in a single fish. He also provided most of the available information on pathologic changes resulting from the disease.

Behavior

In either form of the disease, the first sign to appear is a loss of schooling behavior: affected fry scatter throughout a tank. Some swim about slowly at the surface and others lie on the bottom. In severe cases of the hydrocephalus form, fry show nervous system derangement and lose equilibrium or swim in a spiral.

External Signs

In the hydrocephalus form, slight to severe unilateral or bilateral swelling occurs in the central dorsum of the head, and moderate to severe bilateral exophthalmos occurs. The usual sign of the hemorrhagic form is a red and swollen saddlelike area situated over the pelvic fins, plus a conspicuous abdominal distention. Occasionally, there is a prominent red spot between the pectoral fins or at the side of the skull. In the hemorrhagic form the gills are exceptionally pale, and slight bilateral exophthalmos is sometimes evident.

Figure 2. Northern pike fry showing the hemorrhagic sign of rhabdovirus infection. From Ahne and Wolf (1980). Reprinted with the permission of Gustav Fischer Verlag.

Internal Signs

In general, scant information is available on gross internal signs; the emphasis has been on histopathologic changes. The abdominal cavity contains ascitic fluid and in some fish the liver is greenish. When the prominent interpectoral red spot occurs, the internal finding is one of blood accumulation in the pericardial space.

Histopathologic Findings

Overall, the significant pathology of the disease results from edema and hemorrhage. The hydrocephalus results from fluid accumulation in the ventricle of the mesencephalon, and similar accumulations cause the exophthalmos and ascites.

Hemorrhages occur in connective tissue of the muscle, and minor hemorrhages occur in the optic tectum, spinal cord, and renal hematopoietic tissue.

Proximal convoluted tubules of the kidneys commonly show degenerative changes and necrosis. In advanced cases there is both lytic and coagulation necrosis. Nuclei show pyknosis and margination of chromatin, followed by disintegration or lysis. Renal tubules contain hyalin casts or necrotic material. Cellular debris, possibly of tubular origin, occurs in the space of some Bowman's capsules. Distinctive degenerative changes are not to be found in hematopoietic tissue—a somewhat surprising condition, considering that virus is abundant there.

Liver sections show occasional perivascular lytic necrosis, margination of chromatin, and karyolyis in hepatocytes. The spleen may be congested with blood, but the heart, gastrointestinal tract, pancreas, gills, and skin are notably without consistent pathologic change.

Etiology

Pike fry rhabdovirus disease has been produced experimentally and Rivers' postulates have been fulfilled (de Kinkelin et al. 1973). In addition, Bootsma et al. (1975) induced the disease in fry by exposing fertilized eggs to culture-grown virus.

Size and Shape

In negatively stained preparations of culture-grown virus, PFR measures about 125 ± 10 nm × 80 nm (Cohen and Lenoir 1974; de Kinkelin et al. 1973, 1974; Hill et al. 1975). The PFR virions, like those of *R. carpio*, bear a regular array of distinctive surface spicules that measure about 9 nm. Comparisons of the fish rhabdoviruses show that PFR and *R. carpio* are also structurally stronger than the salmonid agents and withstand centrifugation without significant morphologic change.

Biophysical Properties

The virus shares the usual characteristics of the rhabdovirus group: it has the requisite shape, is acid, ether, and heat labile (at 45°C, 15 minutes), and grows well in a variety of cell lines (de Kinkelin et al. 1974). The BB, EPC, RTG-2, and FHM cells support growth, but FHM cells are preferred because their yield is greatest—about 10^9 PFU/mL. The virus also grows in mammalian BHK-21 cells, but with reduced CPE. However, contamination from cellular proteins is presumably reduced—an advantage that led to the choice of BHK-21 cells by Roy et al. (1975) for their work on physicochemical properties of the virus.

Replication of PFR occurs at 10 to 31°C and is optimal at 21 to 28°C. According to de Kinkelin et al. (1974), a one-step growth curve at 21°C shows latency for about 3.75 hours, after which there is exponential growth for 7 hours. Growth plateaus about 12 hours after infection and the yield from the FHM line is 500 to 1500 PFU/cell. In 15 to 45% sucrose gradient, peak infectivity localizes at a buoyant density of 1.16 g/mL. The virus does not induce the production of interferon and in turn is relatively insensitive to interferon produced by FHM cells.

Infectivity of PFR is readily preserved by lyophilization or by freezing at −30°C or lower in medium with 2% serum. In storage at 4°C in medium with 2% serum, infectivity declines to about 1% of the original level at 5 months. The decline is accelerated at higher temperatures and in the absence of serum.

The RNA of PFR is single stranded and its molecular weight is $3.9 \pm 0.1 \times 10^6$

Table 1. Percent base composition of the RNA of pike fry rhabdovirus

Percentage				
Adenine	Cytosine	Guanine	Uridine	Reference
25.1	22.4	21.6	30.9	Clerx et al. (1975)
24.9	22.5	20.2	32.4	Roy et al. (1975)

(Clerx et al. 1975; Roy et al. 1975). Hill et al. (1975), who labeled the RNA with tritiated uridine, found that on a 5 to 25% sucrose gradient, it sedimented as a fairly sharp peak at about 38 to 40S. The base composition of the RNA of PFR is comparable to that of the RNAs of other rhabdoviruses (Table 1).

As for the other rhabdoviruses, PAGE resolves four and sometimes five structural proteins in PFR (Table 2). The first such determination was made by de Kinkelin et al. (1974), who resolved four polypeptides, but whose data show a minor fifth component. Their work, however, did not give molecular weights.

Roy et al. (1975), in the only study of its kind thus far, reported that PFR has a protein kinase and an RNA-dependent RNA polymerase that is capable of transcription in vivo or in vitro.

Diagnosis

Clinical or pathologic diagnoses of PFRD can be made on the basis of signs and pathologic changes described earlier. However, virus isolation and identification should be carried out because northern pike fry with hemorrhagic signs have been found infected with Egtved virus, the agent of viral hemorrhagic septicemia in trout.

Isolation

Cell culture isolation of virus poses no problem. Whole victim fry are homogenized in about 50 volumes of a physiologic solution, clarified by cold centrifugation, and decontaminated by membrane filtration or treatment with antibiotics. The

Table 2. Molecular weights ($\times 10^3$) of the polypeptides of pike fry rhabdovirus

Protein designation					
Large (L)	Glycoprotein (G)	Nucleocapsid (N)	Nucleocapsid (minor, NS)	Membrane (M)	Reference
180	78	47	58	25	Clerx et al. (1978)[a]
>150	80	55	—	28	Hill et al. (1975)
190	80	42	50	21	Lenoir and de Kinkelin (1975)
160	85	45	—	21	Roy et al. (1975)

[a]Essentially the same values were found for the grass carp rhabdovirus and the V75/94 isolant of PFR.

decontaminated material is adsorbed onto FHM or comparably sensitive cell sheets for 0.5 to 1.0 hours at 15 to 20°C; the medium is then replaced and cultures are incubated at 15 to 25°C. Clear-cut CPE can be expected within 48 hours if PFR is present and causing disease. The features of the CPE are granulation and rounding of cells, followed by lysis. Fixed and stained cell sheets show margination of chromatin and granular degeneration of the cells.

Identification

Neutralization of PFR for serologic identification has been reported by several investigators who uniformly agreed that the northern pike agent is separate and distinct from *R. carpio*, infectious hematopoietic necrosis virus, and Egtved viruses (Clerx et al. 1978; de Kinkelin et al. 1974; Hill et al. 1975; Roy et al. 1975). The most definitive work is that of Clerx et al. (1978), who found that the original isolant of PFR, a newer isolant from pike fry designated V75/94 (Clerx and Horzinek 1978), and the grass carp virus (Ahne 1975a, b) were serologically indistinguishable. All indications are that the pike fry rhabdovirus exists as a monotypic agent.

Neutralization tests are generally regarded as being reliable and straightforward, but Clerx et al. (1978) showed that neutralization of PFR with rabbit antiserum is a complement-dependent reaction, whereas neutralization with antiserum prepared in pike is not. Because antisera are commonly prepared in rabbits and because complement is heat labile, precautions should be exercised to ensure the avoidance of errors in neutralization tests. As one such precaution, rabbit anti-PFR serum should be fresh or, if stored, should be lyophilized or held at −65°C or lower. The neutralizing capability of rabbit anti-PFR serum deteriorates at −20°C. Furthermore, rabbit PFR antiserum should not be heat inactivated for use in neutralization tests. Clerx et al. (1978) discovered that rabbit anti-SVCV (spring viremia of carp virus, or *R. carpio*) serum neutralized PFR when guinea pig serum (complement) was added to the reaction. Such a reaction was weak or absent in the homologous SVCV system.

Transmission and Incubation

Pike fry rhabdovirus disease is readily transmitted by contact, i.e., by exposure to waterborne infectivity and by injection. Bootsma et al. (1975) achieved a 100% transmission by adding virus to fertilized eggs. At 13 to 14°C, 14 days elapsed before the first signs appeared. The inference to be drawn from their work is that vertical transmission occurs; however, proof is lacking and the causal virus has yet to be found in adult pike. Incubation time for transmission by contact was not given, but temperature, quantity of virus, and size of test fish all affected time of incubation when the intraperitoneal route was used. At 13 to 14°C, injection of 10^6 PFU/fish (5 to 6 cm long) led to first mortality in 4 days; 10 times more virus (10^7 PFU) killed the first fish within 2 days, and the cumulative mortality was higher. At 21 to 25°C, injection of 10^7 PFU/fish caused the first mortality in 1 to 2 days and cumulative deaths were highest.

Source

The source of PFR is not known, although adult northern pike are logically suspect. Bootsma et al. (1975) examined 18 pike from two locations using positive controls and three blind passages; they found no virus. When PFRD occurs, mortality is usually severe; accordingly, young pike might be a new host. Supporting that speculation is the finding of PFR in three cyprinids—*Ctenopharyngodon idella*, *Tinca tinca*, and *Blicca bjoerkna*. The role of the virus in cyprinids is not known.

Host and Geographic Range

As a naturally occurring disease, PFRD is known only from countries in Europe. *Esox lucius* is the only species known to be susceptible, and vulnerability decreases with age; as an example, Bootsma et al. (1975) found that fish 4 to 5 cm long had considerable resistance to infection by the immersion method. By the time they reach maturity, adults are apparently wholly refractory to the disease.

Immunity

Virtually nothing is known of the immune response to natural infection with PFR. Clerx (1978), who examined the sera of 80 adults, was unable to find what he considered to be significant levels of antibodies. Several sera however, showed neutralizing activity at a 1:100 dilution. Other sera had a component that, at a 1:90 dilution, combined with labeled virus.

Adult pike inoculated with live virus mounted a strong immune response and produced neutralizing antibodies that were effective at endpoints several orders of magnitude greater than the highest levels of natural activity (Clerx et al. 1978). Pike eggs were also found to contain immunoglobulin-like substances that increased soon after pike fry began to swim and feed.

Control

If it were possible to identify sources of the virus, it would be appropriate to recommend avoidance as a most effective measure for control. At the practical level, however, treatment of eggs with an iodophore yielding an iodine level of 50 ppm and maintaining it for 10 to 15 minutes has greatly reduced the frequency of outbreaks at two hatcheries in The Netherlands (Bootsma 1976). Prophylaxis has not been absolute, possibly because egg masses are cohesive and the disinfectant might not always penetrate thoroughly. Bootsma et al. (1975) recommended that at least 20 L of disinfectant be used for each kilogram of eggs and that particular care be taken to disaggregate the egg masses.

International, and especially intercontinental, shipment of northern pike should be avoided. When it is necessary to make such shipments, adequate quarantine should be effected.

References

Ahne, W. 1975a. A rhabdovirus isolated from grass carp (*Ctenopharyngodon idella* Val.). Arch. Virol. 48:181–185.

Ahne, W. 1975b. Biological properties of a virus isolated from grass carp (*Ctenopharyngodon idella* Val.). Pages 135–140 in L. A. Page, ed. Wildlife diseases. Plenum Press, New York.

Ahne, W. 1982. Vergleichende Untersuchungen über die Stabilität von vier fischpathogenen Viren (VHSV, PFR, SVCV, IPNV). Zentralbl. Veterinärmed. [B] 29:457–476.

Ahne, W., and C. Held. 1980. Untersuchungen über die viruzide Wirkung von Actomar K_{30} auf fischpathogene Viren. Tierärztl. Umsch. 35:308–318.

Ahne, W., and K. Wolf. 1980. Viruserkrankungen der Fische. Pages 56–105 in H. H. Reichenbach-Klinke, ed. Krankheiten und Schädigungen der Fische. Gustav Fischer Verlag, Stuttgart.

Ahne, W., H. Mahnel, and P. Steinhagen. 1982. Isolation of pike fry rhabdovirus from tench, *Tinca tinca* L., and white bream, *Blicca bjoerkna* (L.). J. Fish Dis. 5:535–537.

Békési, L., G. Majorus, and E. Szabó. 1984. Csukaivadék (*Esox lucius* L.) elhullását okozó rhabdovirus megjelenése hazánkban. Magy. Allotorv. Lapja 39:231–234.

Bootsma, R. 1971. Hydrocephalus and red-disease in pike fry *Esox lucius* L. J. Fish Biol. 3:417–419.

Bootsma, R. 1976. Studies on two infectious diseases of cultured freshwater fishes. Ph.D. dissertation. Univ. Utrecht, The Netherlands.

Bootsma, R., and C. J. A. H. van Vorstenbosch. 1973. Detection of a bullet-shaped virus in kidney sections of pike fry (*Esox lucius* L.) with red disease. Neth. J. Vet. Sci. 98:86–90.

Bootsma, R., P. de Kinkelin, and M. Le Berre. 1975. Transmission experiments with pike fry (*Esox lucius* L.) rhabdovirus. J. Fish Biol. 7:269–276.

Bussereau, F., P. de Kinkelin, and M. Le Berre. 1975. Infectivity of fish rhabdoviruses for *Drosophila melanogaster*. Ann. Microbiol. (Paris) 126A:389–395.

Clerx, J. P. M. 1978. Studies on pike fry rhabdovirus and the immunoglobulin of pike (*Esox lucius*). Ph.D. dissertation. Univ. Utrecht, The Netherlands.

Clerx, J. P. M., and M. C. Horzinek. 1978. Comparative protein analysis of non-salmonid fish rhabdoviruses. J. Gen. Virol. 40:287–295.

Clerx, J. P. M., B. A. M. van der Zeijst, and M. C. Horzinek. 1975. Some physicochemical properties of pike fry rhabdovirus RNA. J. Gen. Virol. 29:133–136.

Clerx, J. P. M., M. C. Horzinek, and A. D. M. E. Osterhaus. 1978. Neutralization and enhancement of infectivity of non-salmonid fish rhabdoviruses by rabbit and pike immune sera. J. Gen. Virol. 40:297–308.

Cohen, J., and G. Lenoir. 1974. Ultrastructure et morphologie de quatre rhabdovirus de poissons. Ann. Rech. Vét. 5:443–450.

de Kinkelin, P., G. Galimard, and R. Bootsma. 1973. Isolation and identification of the causative agent of "red disease" of pike (*Esox lucius* L. 1766). Nature 241:465–467.

de Kinkelin, P., M. Le Berre, and G. Lenoir. 1974. Rhabdovirus des poissons. I. Propriétés in vitro de virus de la maladie rouge de l'alevin de brochet. Ann. Microbiol. (Paris) 125A:93–111.

Hill, B. J., B. O. Underwood, C. J. Smale, and F. Brown. 1975. Physico-chemical and serological characterization of five rhabdoviruses infecting fish. J. Gen. Virol. 27:369–378.

Lenoir, G., and P. de Kinkelin. 1975. Fish rhabdoviruses: comparative study of protein structure. J. Virol. 16:259–262.

Roy, P., H. F. Clark, H. P. Madore, and D. H. L. Bishop. 1975. RNA polymerase associated with virions of pike fry rhabdovirus. J. Virol. 15:338–347.

15 Rio Grande Cichlid Rhabdovirus

Definition

The Rio Grande cichlid rhabdovirus is a fish pathogen isolated during necropsy of specimens of *Cichlasoma cyanoguttatum* that died of a peracute but otherwise undefined disease after their receipt at a university laboratory (Malsberger and Lautenslager 1980; G. Lautenslager, personal communication, 1982).

History

All fish in a shipment of 50 subadult *C. cyanoguttatum* received in May 1980 showed acute disease signs and died within a week. Virologic examination revealed a rhabdovirus that later proved to be acutely pathogenic for the original species and for two related species.

Signs and Pathologic Changes

In the studies thus far, the emphasis has been on characterization of the virus. Disease signs among specimens in the original shipment were not described, nor were signs or pathologic changes investigated in fish of several species that died after experimental infection.

Etiology

The presumed causal agent of the original disease is indisputably a rhabdovirus. It has a rounded instead of a flat base and, like infectious hematopoietic necrosis virus, it is easily damaged by the forces of high-speed centrifugation. Virus dimensions were not given, but an electron micrograph of an infected FHM cell showed both free and budding particles (Fig. 1). As could be expected for a rhabdovirus, the

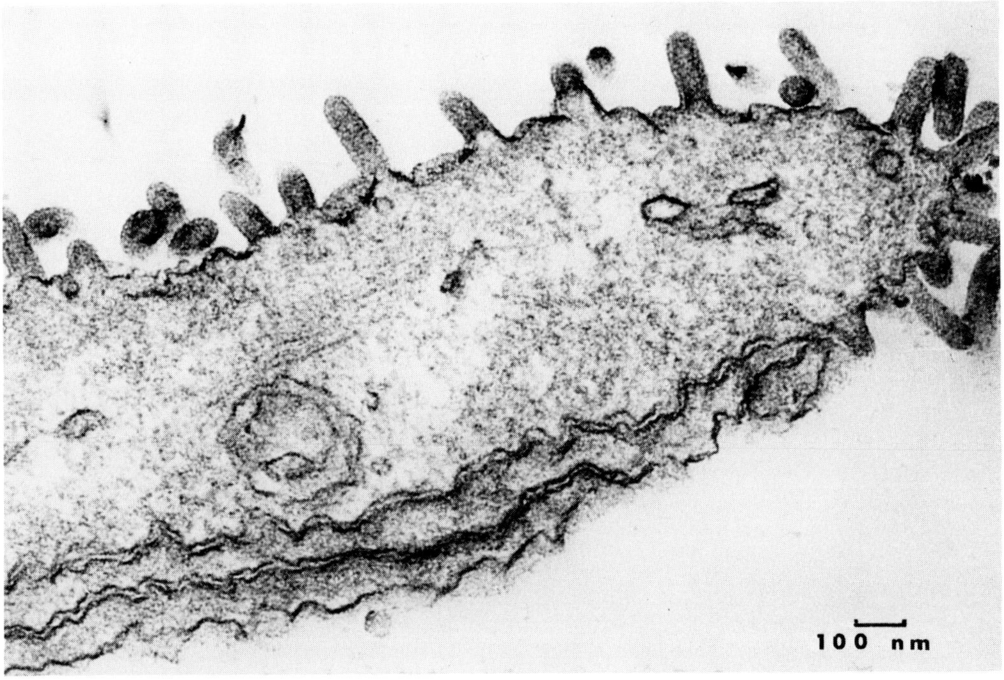

Figure 1. Infected FHM cell showing budding virions of the Rio Grande cichlid agent. From Malsberger and Lautenslager (1980).

agent is ether sensitive and its replication is not inhibited by azaguanine, actinomycin D, or bromodeoxyuridine.

After the original inoculation, the FHM and BF-2 cell lines show CPE within 24 hours at 23°C. Initially, the RTG-2 line seemed to be refractory, but it was later found that CPE and replication occur if cultures are incubated for 1 week. The nominal yield of virus in FHM cells is 3×10^6 PFU/mL at 23°C and 10^7 at 33°C. The RTG-2 line yields 10^6 PFU at 7 days but the BB line yields only 10^2 PFU (Malsberger and Lautenslager 1980).

Infected FHM cell cultures stained with acridine orange show yellow-orange inclusions in the cytoplasm, and inclusions are also evident by electron microscopy.

Tests with antisera against the agents of spring viremia of carp and pike fry rhabdovirus disease show no neutralizing activity against those agents. Other cross-neutralization tests, however, need to be done.

Futher characterization of the rhabdovirus is under way at Lehigh University, Bethlehem, Pennsylvania (G. Lautenslager, personal communication, 1985).

Isolation

Titrations made 3 days after the death of experimentally infected *Tilapia zillii* demonstrated low concentrations of virus in combined skin and muscle, but not in other tissues.

Host and Geographic Range

Duplicate groups of 10 fish each of the original host species, *C. cyanoguttatum*, and of three other species of the family Cichlidae were injected intraperitoneally with about 4×10^6 PFU of virus and observed for 2 weeks. Mean mortality was 95% in *C. cyanoguttatum* and 80% in *T. zillii* and *C. nigrofasciatum*, the convict cichlid. Incubation time, presumably at room temperature, was only 3 to 5 days in the convict cichlid. No specimen of *Sarotherodon aureus* died (Malsberger and Lautenslager 1980; G. Lautenslager, personal communication, 1982).

The Florida supplier of these Rio Grande cichlids had gone out of business and it could not be determined if the fish had been raised in Florida or if they had been taken from the wild and transshipped.

The Rio Grande cichlid is native to Mexico and presumably serves as the source host of the virus. Virulence for related cichlids and the broad pathogenicity characteristic of rhabdoviruses suggests that the virus might also be present in other species of the family.

Reference

Malsberger, R. G., and G. Lautenslager. 1980. Fish viruses: rhabdovirus isolated from a species of the family Cichlidae. Fish Health News 9:i–ii.

16 Snakehead Rhabdovirus

Definition

The snakehead rhabdovirus is a bullet-shaped pathogen isolated from internal organ homogenates of striped snakeheads (*Ophiocephalus striatus*) showing severe ulcerative disease and sustaining attendant mortality.

History

Annual outbreaks of ulcerative disease have been recognized since 1980, but isolation of the virus was first reported by Frerichs et al. in 1986.

Etiology

The virus is associated with—and possibly the cause of—skin ulcerations and mortality, but isolations have been made from filtrates of homogenized liver, kidneys, and spleen. As seen in negatively stained pellets of supernatant fluids of inoculated snakehead cultures, the virus is about 80 nm wide and 120 nm long. Characterization has yet to be reported; however, the principal attributes are assumed to be those of other fish rhabdoviruses. Serologic relations to other fish pathogens are entirely unknown.

Host and Geographic Range

The ulcerative disease, mortality, and virus are known from the snakehead and the so-called ditch or swamp eel (*Fluta alba*)—both eel-like members of the family Flutidae from Thailand and Burma. A similar rhabdovirus-like agent has been encountered in ulcerated barramundi (*Lates calcarifer*) in Australia (J. S. Langdon, Australian Fish Health Reference Laboratory, personal communication, 1986).

Reference

Frerichs, G. N., S. D. Millar, and R. J. Roberts. 1986. Ulcerative rhabdovirus in fish in South-East Asia. Nature 322:216.

17 Spring Viremia of Carp

Synonyms: Contemporary—acute infectious dropsy, infectious dropsy of carp. Historical—*Ascites infectiosus cyprinorum*, hemorrhagic septicemia, infectious ascites, infektiöse Bauchwassersucht, Krasnucha, Rotseuche, rubella, rubeola, spring virus disease

Definition

Spring viremia of carp (SVC) is an acute hemorrhagic and contagious viral infection—typically of cyprinids and more specifically of the common carp, *Cyprinus carpio* (the species of carp referred to here, unless otherwise indicated), in which the disease usually erupts in spring and causes mortality of the adults as well as the young.

History

The name spring viremia of carp was coined by Fijan et al. (1971) when they reported multiple isolations of a rhabdovirus from carp with signs of what had been known as acute infectious dropsy of carp (IDC). The new name was used in a deliberate move to distinguish the viral infection from the confused mixture of disease forms then composing what was known as the IDC complex. Fijan et al. (1971) named the causal virus *Rhabdovirus carpio* and fulfilled Rivers' postulates. From the point of that first report, the history of SVC is brief. From a broader viewpoint, the story of SVC has roots that undoubtedly thread through the tangled skein of conflicting opinion and information about IDC that had accumulated through the years.

From early times in Europe, the most serious cause of carp mortality in pond culture has been attributed to infectious dropsy. Hofer (1904) described and illustrated a hemorrhagic disease termed Rotseuche that was then attributed to a bacterium and that might have been dropsy. According to Bauer and Faktorovich (1969), dropsy was first noted in the USSR in 1915. Otte (1963) and others generally credit Schäperclaus with first describing IDC in 1930. However, in a review of the history of IDC—or infektiöse Bauchwassersucht, as it is called in German—Schäperclaus (1979) conceded that a 1727 description of carp showing ascites and enteritis could be an early account of IDC. Accordingly, it seems likely that the history of SVC goes back to medieval times, and conceivably to Asia, several millenia ago.

Infectious dropsy of carp was initially considered to be a bacterial disease, and that view was held by many investigators, particularly in western Europe. Indeed, systemic gram-negative bacteria could be found in victims of many outbreaks, and antimicrobial prophylaxis or therapy sometimes proved beneficial. However, no bacteria—much less a specific bacterium—could be shown to be present in all outbreaks of IDC; moreover, the disease could be transmitted with bacteria-free filtrates prepared from tissues of some victims. Accordingly, some investigators—notably in eastern Europe and the USSR—believed that IDC had a viral etiology. Still other investigators proposed environmental factors or a mixed etiology of virus and bacteria.

Notwithstanding differences in opinion regarding etiology, two forms of IDC became recognized: an acute or ascitic form and a chronic or ulcerative form. Fijan et al. (1971) showed that *R. carpio* produced the acute form of what had been termed IDC. The chronic or ulcerative form was later named carp erythrodermatitis. In time, the ulcerative disease was shown to be caused by a nonmotile bacterium of the genus *Aeromonas* (Bootsma et al. 1977).

In his 1972 review, Fijan noted that the literature on IDC then consisted of more than 450 references. Understandably, one cannot be certain which reports actually dealt with virus and which concerned bacteria; nevertheless, several publications of the 1950s merit mention: Roegner-Aust et al. (1950), Roegner-Aust and Schleich (1951), and Roegner-Aust (1953). Those reports dealt with the then primitive electron microscopy in which virus preparations were shadowed with vaporized heavy metal. Ascitic fluid taken from carp with IDC, when filtered and then shadowed with palladium, showed abundant round particles 100 to 200 nm in diameter. The particles were considered *Virus der infektiösen Bauchwassersucht der Karpfen*, and although elongated rhabdoviral forms are not to be seen in the illustrations, the ascitic fluid proved infectious for test carp and produced an ulcer-free form of IDC.

Reports of work that appeared during the 1960s presaged events that were to occur in the next decade. Fish tissue cultures were used in three laboratories to study the etiology of IDC. In Leningrad, Tets and Yakovleva (1962) inoculated primary cultures of carp tissues with materials prepared from carp with the ulcerative form of IDC. They found no evidence of a cytopathic agent and electron microscopy of their source materials similarly showed no evidence of virus. Accordingly, they became unequivocally convinced that no virus was present in the ulcerative form of IDC. In Kiev, 2 years later, Osadčaja (1964) also inoculated primary cultures of carp tissues with materials from carp with IDC, but used fish with the acute form of the disease. In cultures incubated at 20°C, CPE appeared in 1 or 2 days and soon destroyed the cell sheet. The CPE was serially transferred seven times and then frozen at −14°C. Infectivity persisted in storage, and after thawing the CPE was serially passaged an additional five times. Electron microscopy was not done nor were fish inoculated with tissue culture harvests.

At Yugoslavia's University of Zagreb, research on fish diseases was an important activity, and methods of fish tissue culture were also used. However, inocula were prepared from the ulcerative form of IDC and, although the authors cautiously considered their results as supporting a viral etiology, the evidence was not conclusive (Tomašec and Fijan 1965; Tomašec et al. 1964). Fijan (1966), who used mostly lesions from skin, reported experimental transmission of ulcerative IDC; he noted that internal organs of such fish were not infective.

Under the sponsorship of the New York Academy of Sciences, an "International Symposium on Viral Diseases of Poikilothermic Vertebrates" was held in 1964 and the contributed papers were published the next year (Whipple 1965). Reports were given on five viral diseases of fishes and six papers discussed the etiology of IDC. Polarization of opinion was evident—three papers favored a viral etiology, two favored bacteria, and one was undecided.

In the USSR, beginning in the 1960s, a condition of carp involving inflammation of the swim bladder was spreading and causing serious losses. It was described by Arshaniza et al. (1968), who inferred that the inflammation was a condition different from IDC. Faktorovich (1968) proposed diagnosing IDC by demonstrating eosinophilic cytoplasmic inclusions in brain ganglia of carp with acute IDC. However, inclusions were also found in some apparently healthy fish from populations with the disease, but not in fish with the swim bladder disease.

A review of the infectious diseases of carp in the USSR by Bauer and Faktorovich (1969) brought into wider circulation the findings described in 50 reports in the Russian language.

The initial isolation of *R. carpio* in Yugoslavia early in the 1970s (Fijan et al. 1971) was followed by confirming isolations elsewhere—including but not limited to Czechoslovakia (Macura et al. 1973; Tesarčík, Macura, and Dedek 1977), Hungary (Békési and Szabó 1977, 1979), and Great Britain (Bucke and Finlay 1979).

Availability of the virus opened avenues of relevant research, and reports on various aspects of virus characterization constituted a significant percentage of the SVC papers published during the decade—e.g., de Kinkelin and Le Berre (1974), and Ahne (1976, 1979). Cohen and Lenoir (1974) compared the ultrastructure of *R. carpio* with that of three other rhabdoviruses of fishes. Other work showed that the virus could be grown not only in various fish cell lines but also in reptilian and mammalian cells (Clark and Soriano 1974), and even in a dipteran insect (Bussereau et al. 1975).

In papers on transmission and incubation, the route by which virus was administered was found to influence incubation time and even the very appearance of clinical signs: intraperitoneal injection of the virus produced typical SVC (Ahne 1977; Varović and Fijan 1973). Several reports concerned transmission of the virus by the carp louse, a copepod of the genus *Argulus* (Ahne 1978; Pfeil-Putzien 1977, 1978).

Some workers concentrated on diagnostics or various aspects of serology. Sulimanović (1973) used an indirect hemagglutination test to determine antibody levels in fish and immunized rabbits. Pfeil (1978) used the indirect fluorescent antibody technique to identify the virus in infected cells. Hill et al. (1975) prepared antiserum in rabbits and showed that *R. carpio* was serologically distinct from other fish rhabdoviruses, with exception of the agent then known as the virus of swim bladder inflammation (SBI); they claimed that the viruses of SBI and SVC were serologically indistinguishable. These authors also used electron microscopy and determined polypeptide profiles of five rhabdoviruses of fishes. Clerx et al. (1978), who compared neutralization of spring viremia and pike fry rhabdoviruses, showed that complement was not involved in neutralization of the carp virus.

Some of the SVC papers of the 1970s addressed various aspects of disease control. From the USSR came word of a program (begun in the early 1960s) for selective breeding of fish for resistance to acute dropsy. Using experimental infections, the investigators achieved a 30% differential occurrence of disease between susceptible and resistant fish (Kirpichnikov and Faktorovich 1972; Kirpichnikov et al. 1972). Certain methods of immunization were found to be promising (Baudouy 1978; Baudouy et al. 1977; Fijan, Petrinec, Stancl, Dorson, and Le Berre 1977; Fijan, Petrinec, Stancl, Kezić, and Teskeredzić 1977; Tesarčík et al. 1978). Hill (1977) discussed virulence and susceptibility as factors in evaluating the efficacy of potential vaccines.

The homeotherm rhabdoviruses of rabies and vesicular stomatitis had long been of interest to virologists; therefore, it was to be expected that *R. carpio* too would be studied—particularly the details of its molecular biology. (The name *R. carpio* was soon to give way to the more popular SVCV—for spring viremia of carp virus.) During the 1970s nearly one-fifth of the SVC literature dealt with virus components and replicative processes. Reports came from the Wistar Institute in Philadelphia (Sokol and Koprowski 1975; Sokol, Clark, et al. 1974; Sokol, Tan, et al. 1974) from the Institute National de la Recherche Agronomique in France (Lenoir 1973; Lenoir and de Kinkelin 1975), and from the Netherlands State University at Yalelaan, Utrecht (Clerx and Horzinek 1978). At the University of Alabama, Roy and her associates began long-term continuing studies of the RNA and proteins of the virus (Gupta et al. 1979; Roy 1976; Roy and Clewley 1978a, b).

The papers from a symposium on the major communicable fish diseases in Europe, held in Amsterdam in 1972, were published in 1973. Three panel reviews concerned carp dropsy. Fijan (1973) entitled his presentation "Spring Viremia of Carp," whereas Havelka (1973) and Tomašec (1973) favored the then more traditional view that infectious dropsy was a complex of two or more diseases.

Three other presentations reflected a growing concern for SBI of carp—a term that in one usage describes a pathologic change caused by *R. carpio*, and in a later usage a specific disease of probable protozoan etiology. Antychowicz (1973) and Liebmann (1973) favored the idea that SBI had a viral etiology, whereas Arshaniza and Bauer (1973) were unsure of the etiology but noted that incubation time was about 2 months. In Munich, Ahne (1973) completed doctoral research on SBI in which he isolated and characterized a virus that later proved to be indistinguishable from *R. carpio*. Publications on the virus appeared under the authorship of Bachmann and Ahne (1973, 1974). Prophetically, Tesarčík (1977) recognized SBI as a relatively new disease, distinct from SVC.

Collectively, about one-fourth of the SVC literature of the 1970s concerned SBI, or consisted of reviews of SVC, or of fish diseases including SVC. In addition to the papers given at the 1972 Amsterdam symposium, Fijan (1972) reviewed the IDC complex at a symposium of the London Zoological Society. Other publications included SVC in more comprehensive coverage of fish diseases (Fijan 1975; McAllister 1979; Schäperclaus 1979).

All told, about two-thirds of the SVCV literature of 1980–1985 dealt with molecular biology, immunology, features of transmission and incubation, or methods relating to serologic identification.

Research on the molecular biology of SVCV at the University of Alabama moved from investigations of virus proteins to studies of genomic and messenger RNA (Gupta and Roy 1980, 1981; Kiuchi and Roy 1984; Roy 1981; Roy et al. 1984). In Germany, the proteins and RNAs of SVCV and Egtved virus were found to have some similarities and some differences (Deuter et al. 1982; Enzmann et al. 1981).

Immunologic reports can be categorized in part as vaccination trials (Fijan and Matašin 1980; Macura et al. 1983; Tesarčík et al. 1984) and serologic surveys for anti-SVCV neutralizing antibody (Wizigmann et al. 1980, 1983). Induction of circulating interferon in carp was reported by de Kinkelin et al. (1982), but its presence was short lived and its role in protection not determined. Information on SVC was included in an overall review of theoretical and practical considerations of fish immunization by de Kinkelin et al. (1984). Fijan (1984) addressed a more specific topic in his discussion of the prospects and limitations of fish immunization in pond culture.

As the name suggests, the seasonal occurrence of spring viremia has long been recognized, but factors in disease transmission were not completely understood. Work in France and Germany showed temperature to be a major factor in that it affected the ability of the host to respond (Ahne 1980; Baudouy et al. 1980a, b, c). Ahne (1985a) also determined that, even though the carp louse and a leech could transmit SVC, the virus was not replicated by the invertebrate; transmission was purely mechanical.

Advances were made in serologic identification. At Ahne's laboratory in Munich, indirect fluorescent antibody and immunoperoxidase techniques were successfully applied (Faisal and Ahne 1980, 1983, 1984). In Weymouth, methods were developed to produce rabbit antiserum against SVCV and other fish viruses (Hill et al. 1981), and the resulting anti-SVCV serum was used in an ELISA (Dixon and Hill 1984).

During the 1980s, uncertainties about the nature and etiology of SBI have been markedly reduced. Little doubt remained—SBI should no longer be considered as an infection by SVCV. Instead, new evidence has been obtained from histologic examinations that SBI involves a protozoan and, more specifically, a myxozoan (Csaba et al. 1984; Körting 1982; Kovács-Gayer 1983).

A succinct review of SVC was presented by Bootsma and Ebregt (1983), and the subject was included in discussions of fish viruses by Ahne (1984), Ahne and Wolf (1980), and Pilcher and Fryer (1980).

Békési and Csontos (1985) reported isolation of SVCV from asymptomatic carp brood stock in Hungary, and Shchelkunov et al. (1984) isolated the virus from silver carp (*Hypophthalmichthys molitrix*).

Signs and Pathologic Changes

The signs and pathologic changes described here are those observed in clear-cut and uncomplicated cases—experimental infections with SVCV under laboratory conditions. The descriptions are taken largely from the works of Bachmann and

Ahne (1973, 1974), Fijan (1972, 1973, 1975), Fijan et al. (1971), and Negele (1977). Under natural conditions of pond husbandry, outbreaks are commonly complicated by the presence of other disease agents, notably bacteria, that are considered secondary invaders. As an example, the presence of open skin lesions—notably absent in uncomplicated SVC—indicates the likelihood of bacterial infection.

Behavior

Concentration of fish at the pond inlet is one of the first signs of an outbreak of SVC. The rate of respiration is reduced, lethargy is common, and some victims rest on the bottom. Loss of balance is evident as fish lean, swim on their side, or otherwise move aimlessly.

External Signs

Victims have a distended abdomen and show petechiation of the gills and skin; hemorrhages may occur in the eyes. Thick, long mucoid fecal casts are shed from an inflamed, protruding, and edematous vent. Slight exophthalmia may be evident and skin pigmentation may be slightly to markedly darker than normal. Gills are pale, and fibrillation of skeletal muscle can be seen in some specimens. When fish are lifted from the water and held in a normal position, bloody ascitic fluid freely drips from (or can be expressed) from the vent.

Internal Signs

Internally, the overall impression is one of a hemorrhagic condition with accompanying ascites and evidence of edema. Multiple focal hemorrhages occur in swim bladder tissue (Fig. 1)—a diagnostic finding in SVC as well as in SBI. The spleen is enlarged, and petechiation is evident in the heart, liver, kidneys, intestine, peritoneum, and skeletal muscle. Catarrhal enteritis evidently gives rise to the fecal casts trailing from the vent.

Histopathologic Findings

Although hematologic and histopathologic changes of various rhabdoviral infections have been described in salmonids, such information on SVC is specious or at best limited. Svobodová and Tesarčík (1973) reported hematologic aspects of several diseases of carp, including spring viremia, and Stankiewicz (1977) described histopathologic changes in brain tissues of carp assumed to be infected with virus; however, in neither study was the virology demonstrated.

Negele (1977) reported histopathologic findings in carp fingerlings experimentally infected with *R. carpio*, but that work was incomplete; it consisted solely of a

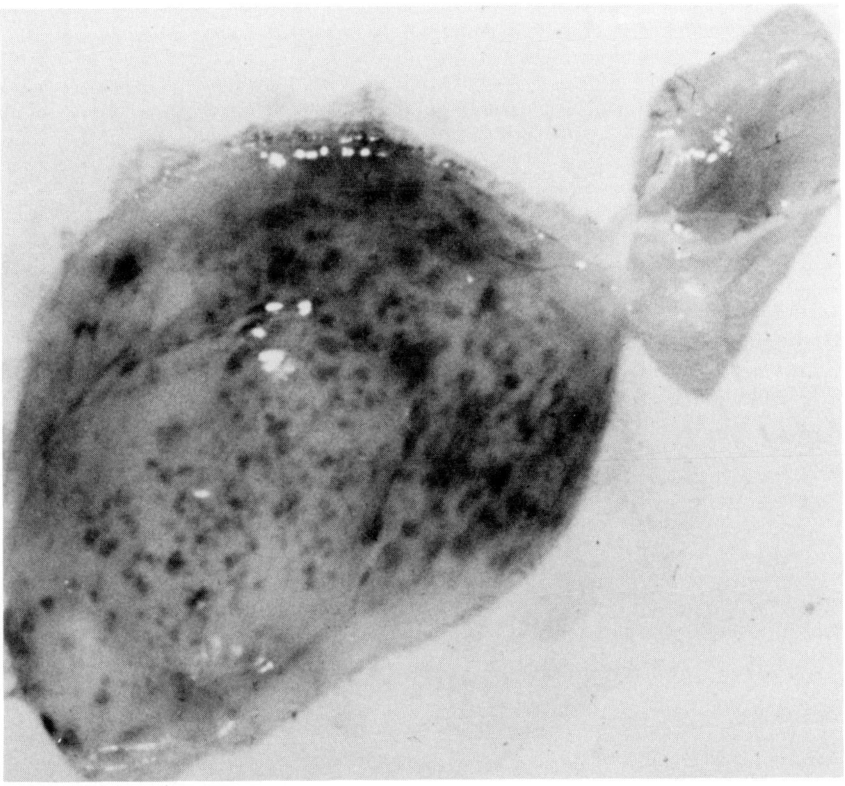

Figure 1. Carp swim bladder showing the widespread petechiation that commonly occurs in spring viremia. From Ahne and Wolf (1977).

summary and was not illustrated. Therefore, the findings are described only briefly here. Perivasculitis and edema occur in blood vessels of the liver and ultimately result in vessel necrosis. Multiple foci of necrosis are present in liver parenchyma, and areas of hyperemia are present throughout. The pancreas shows purulent inflammation and multifocal necrobiosis. Lymph vessels of the peritoneum are dilated and filled with detritus and macrophages, and the serosa shows necrosis. Perivascular inflammation is present in vessels of the intestine; in time, the epithelium sloughs and the villi atrophy. The spleen is hyperemic, and hyperplasia develops in the reticuloendothelium. The melano-macrophage centers increase and enlarge. Excretory and hematopoietic functions of the kidneys are obviously damaged: the renal tubules become progressively clogged with casts and tubule cells show cytoplasmic inclusions and hyaline degeneration. The swim bladder is a specific target; the monolayer of the epithelial lamina becomes a discontinuous multilayer. Vessels of the submucosa are dilated and contain accumulations of extravasated blood, and lymphocytes infiltrate nearby areas. The pericardium becomes inflamed; cardiac muscle is infiltrated; and, in later stages, discontinuous degeneration and necrosis occur.

Clinical Findings

Although the actual value is not given, Fijan (1975) stated that the red cell count of carp with SVC is greatly reduced.

Etiology

As already noted, Fijan et al. (1971) fulfilled Rivers' postulates and thereby established the viral etiology of what had been called acute infectious dropsy. They coined the term spring viremia of carp and named the causal agent *Rhabdovirus carpio*. Later workers also showed that culture-grown virus was able to produce disease and mortality in experimentally infected carp—e.g., Ahne (1980), Hill (1977), and Tesarčík, Macura, Dedek, Valíček, and Šmíd (1977). In addition, some cases of a condition known as SBI or aerocystitis of carp were found to yield a rhabdovirus indistinguishable from SVCV (Bachmann and Ahne 1973, 1974); moreover, the SBI isolant produced inflammation of the swim bladder—a common finding in SVC. However, SBI in carp fry occurs in the absence of virus or bacteria; histologic evidence for a protozoan cause was provided by Csaba et al. (1984), Körting (1982), and Kovács-Gayer (1983).

Size and Shape

R. carpio, or SVCV, is typically bullet shaped, 60 to 90 nm wide, and 90 to 180 nm long (Fig. 2). It passes membranes of 450 and 200 nm porosity, but is retained by membranes of 100 nm porosity.

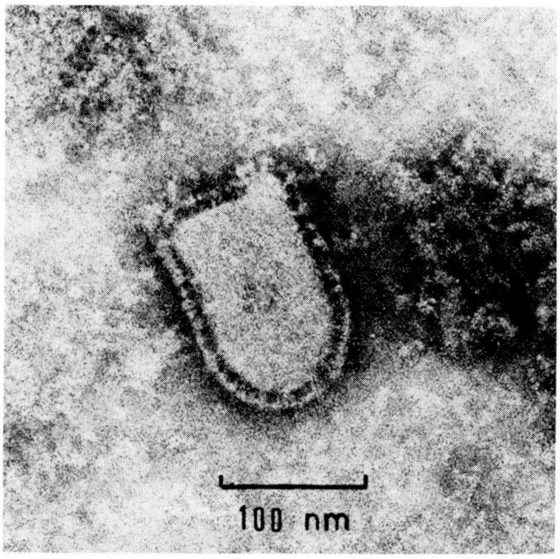

Figure 2. Negatively stained virion of *Rhabdovirus carpio* (or SVCV) showing a prominent array of surface spicules. From de Kinkelin and Le Berre (1974). Reprinted with the permission of the Secritariat des *Annales*, Paris.

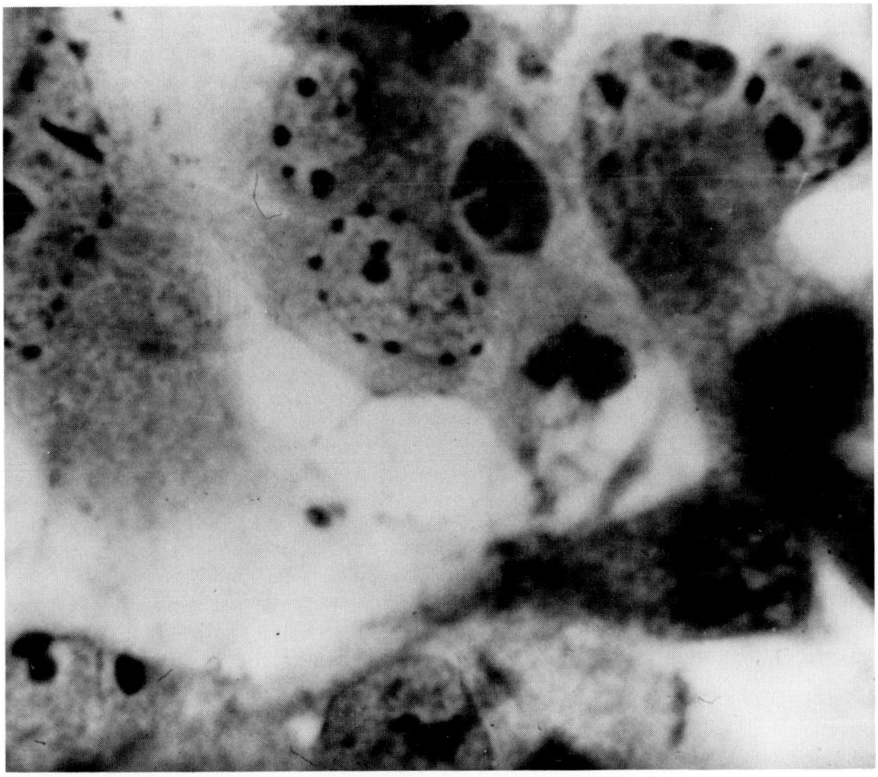

Figure 3. Fathead minnow cells infected with *Rhabdovirus carpio* (or SVCV). Chromatin has condensed along nuclear margins in a discontinuous and punctiform array. H & E. From Ahne and Wolf (1977).

Biophysical Properties

Spring viremia of carp virus is ether, heat, and acid labile, but is stable at pH 7 to 10. As tested with representative mammalian, avian, and fish erythrocytes at 4 to 37°C and pH 5 and 7.6, the virus shows no evidence of an intrinsic hemagglutinating ability (Ahne 1973). The virus is replicated by primary cultures of carp gonads or swim bladders and by such commonly used fish cell lines as BB, BF-2, EPC, FHM, and RTG-2. However, the EPC and FHM lines are preferred and yield maximal infectivity of about 10^9 $TCID_{50}$/mL. The yields of BB and RTG-2 cells are significantly less. The SVCV is also replicated by a wide range of vertebrate cell lines: primary cultures of pig kidney, calf embryo, and chick embryo; the mammalian cell lines BHK-21, VERO, and Hep-2 (Ahne 1973); and the mammalian cell line WI-38 and the reptilian cell lines VSW, VH2, GL1, and TH1 (Clark and Soriano 1974).

The CPE in EPC and FHM cells begins with rounding, followed by detachment and eventually by lysis. Fixed and stained cells show condensations of chromatin along nuclear margins (Fig. 3).

Table 1. Stability of SVCV in Eagle's MEM, with and without serum, at different temperatures

Temperature (°C)	With 5% serum		Without serum	
	Residual infectivity[a]	Holding time	Residual infectivity[a]	Holding time
23	2.2	28 days	4.5	7 days
4	4.2	6 months	2.2	4 months
−20	5.8	6 months	2.8	6 months
−74	7.5	6 months	4.1	6 months

Source: Adapted from Ahne (1976).

[a]Initial infectivity $10^{7.8}$ $TCID_{50}$/mL; values in column are exponents of $\log_{10}$ $TCID_{50}$/mL.

Replication occurs at 4 to 31°C, but the optimal temperature is 20 to 22°C. In cultures incubated at that temperature, a 4- to 6-hour period of latency is followed by exponential production of virus to hours 8 to 10, a plateau at about hour 12, and then a decline in infectivity (Ahne 1973; de Kinkelin and Le Berre 1974).

In the laboratory, SVCV is sufficiently stable to allow handling in routine procedures without undue loss of infectivity, providing that serum is present (Table 1). In its general response to chemical, environmental, and physical factors (Table 2), SVCV is comparable to two other fish rhabdoviruses—Egtved and pike fry rhabdovirus (PFR)—but it is much more drastic than that of the birnavirus infectious pancreatic necrosis virus (IPNV).

However SVCV is handled, the presence of serum has a protective effect on infectivity. During four freeze-thaw cycles of virus the loss of infectivity was only 10% when 2% serum was present, but 95% when the serum was omitted; furthermore, 2% serum protected infectivity when the virus was lyophilized (de Kinkelin and Le Berre 1974).

Table 2. Effects of selected chemicals and environmental and physical factors on infectivity of SVCV

Agent or factor	Effect	
	Inactivation[a] (%)	Period
Formalin 3%	100	5 min
NaOH 2%	100	5–10 min
Chlorine, 500 ppm	99	<2 min
Actomar (I_2) 0.01%	99	5 min
Tap water, 10°C	<100	42 days
Stream water, 10°C	<100	14 days
Mud, pH 7.4 and 4°C	<100	42 days
Drying (4 to 20°C)	<100	21 days
Gamma irradiation at 10^6 rads	100	—
UV irradiation at 254 nm	100	10 min

Source: Adapted from Ahne (1982a, b).

[a]Inactivation <100% indicates that some infectivity (percent unspecified) survived for period shown.

Table 3. Molecular weights ($\times 10^3$) of structural proteins of SVCV

Polymerase L	Glycoprotein G	Nucleocapsid N	Nonstructural NS	Matrix M	Reference
180	76–77	46–47	52–53	25–27	Clerx and Horzinek (1978)
>90	73	50	NS1 NS2 39 26	22	Deuter et al. (1982)
>150	82	52	44	23	Hill et al. (1975)
150	70	40	—	19	Lenoir (1973)
190	80	42	50	21	Lenoir and de Kinkelin (1975)
160	85	45	—	21	Roy and Clewley (1978a)
—	88	52	43	23	Sokol, Clark, et al. (1974)
150	88	52	43	23	Sokol and Koprowski (1975)

The buoyant density of purified SVCV is 1.195 to 1.200 g/mL in cesium chloride (Bachmann and Ahne 1973), and 1.16 g/mL in a linear 15 to 60% sucrose gradient (Lenoir 1973).

The structural proteins of SVCV have been resolved in several independent studies (Table 3). The results uniformly support the conclusion that the polypeptides of SVCV—like those of PFR—are similar to those of prototype vesicular stomatitis virus (VSV) (Clerx and Horzinek 1978; Deuter et al. 1982; Hill et al. 1975; Lenoir 1973; Lenoir and de Kinkelin 1975; Roy and Clewley 1978a; Sokol and Koprowski 1975; Sokol, Clark, et al. 1974).

The RNAs of SVCV have received critical attention. Hill et al. (1975) found that in 5 to 25% sucrose gradients the RNAs of SVCV, Egtved, PFR, and infectious hematopoietic necrosis virus (IHNV) uniformly sedimented at 38 to 40S. Roy and Clewley (1978b) showed that SVCV had an RNA-dependent RNA polymerase and that the 5′ nucleotide of the virus RNA is pppAP. Gupta et al. (1979) determined that the 5′-terminal structure of the RNA species synthesized by the transcriptase resembled that of VSV.

Gupta and Roy (1980), who observed two alternative mechanisms of mRNA capping, found one to be identical to that described for VSV; they considered this mechanism to be a conserved function during rhabdovirus evolution. The alternative capping mechanism indicated that substrate termini are probably independently initiated. They later confirmed this independent mRNA initiation (Gupta and Roy 1981), demonstrating that methylated capped and uncapped mRNA 5′-terminal oligonucleotides were synthesized in the absence of complete mRNA.

Among the structural polypeptides of SVCV, the two principal species of phosphoproteins, NS1 and NS2, are related and have phosphoserine as their major phosphorylated amino acid (Roy 1981). Roy et al. (1984) determined the base sequence at the 3′ end of the RNA of SVCV; a comparison with VSV showed complete homology for the first 20 nucleotides but less thereafter. They also found that the NS mRNA yielded the translation products NS1 and NS2. Kiuchi and Roy (1984) carried out sequence analysis of SVCV mRNA coding for the M protein using a complete DNA copy cloned in a plasmid of *Escherichia coli*. The amino acid sequence of the M protein aligns with that of VSV to the extent of 28% homology. In the predicted amino acid sequence, the M protein of SVCV (at MW 25,599) is slightly smaller than that of VSV (MW 26,064). Kiuchi and Roy (1984) considered the sequence homologies of SVCV and VSV to indicate a shared common ancestor.

Diagnosis

Clinical diagnosis of SVC is possible if serious mortality occurs among carp during spring or early summer and affected specimens show the key features of the behavior changes and signs previously described: aimless swimming; abdominal distention; ascites; an overall hemorrhagic condition of skin, gills, and viscera; and (particularly) petechiation and hemorrhages in the swim bladder. The absence of external lesions and systemic bacteria lends credence to the diagnosis.

Confirmation of the diagnosis requires that SVCV be isolated and identified serologically, or alternatively, that viral antigen be identified serologically in tissue sections.

Isolation

The virus of SVC is readily isolated from typical victims during an epizootic, and from survivors during the several weeks immediately after an outbreak. Preferred source material is internal organ tissues; because kidneys have the highest virus titer, they are suggested during the 7 to 14 days postinfection, but the titer is highest in liver and spleen from days 14 to 24 (Faisal and Ahne 1984). Tissues are homogenized, diluted 1 : 10 with water or physiologic saline, and inoculated into susceptible cell cultures—ideally onto drained sheets of young, active cultures.

Primary cultures of carp ovaries are susceptible to SVCV, but established cell lines are preferred. Most isolations have been made in FHM, EPC, or BF-2 cells, but the BB and RTG-2 lines have also been used. The pH of the medium should be 7.3 to 7.6 and the incubation temperature about 20°C (though virus has also been isolated in cultures incubated at 15°C); cultures should be observed for 7 to 10 days. At 20°C, cytopathic effects of SVCV may appear in less than 2 days, but at least several days are sometimes required. The CPE begins as focal areas of cell rounding and sheet contraction. The effect spreads over the cell sheet, lysis ensues, and the cells slough. The CPE is not overtly distinctive, but in the stained cell sheets individual cells show discontinuous or punctiform condensations of chromatin along the nuclear margins (Fig. 3).

Quantification of virus in various tissues has shown that the levels are generally comparable in natural and experimental infections (Table 4). Virus has also been found in the blood (at least during the viremic stage), in ascitic fluid, and in feces; however, quantifications are lacking.

With little doubt, virus is harbored in some brood stock; however, its location in the host and methods of isolation—particularly nondestructive methods—have yet to be documented. In that regard, the situation with SVCV is much like that of IHNV. Békési and Csontos (1985) found SVCV in 5- to 10-year-old brood stock that was artificially induced to spawn, but detected none in 211 samples of seminal fluids

Table 4. Quantification of SVCV from various tissues

Tissue	Type of infection	
	Natural[a]	Experimental[b]
Liver	6.5	6.8
Spleen	3.8	5.5
Intestine	ND	5.5
Kidney	5.8	5.2
Brain	4.3	4.5
Gills	3.5	ND

Note: Infectivity expressed as exponents of $\log_{10}$ $TCID_{50}$/g; ND, no data.
[a]From Fijan et al. (1971).
[b]From Ahne (1973).

(although they recovered it in 3 of 491 samples of ovarian fluids). Their work thus confirmed earlier unsuccessful efforts to isolate virus from seminal fluid.

Considering that SVCV might be released at or near the time of spawning, isolation attempts could be patterned after nondestructive approaches that have been successful with other viruses. Brood stock could be immunosuppressed and leukocytes co-cultured with susceptible cell lines; graded dilutions of ovarian fluids ought to be tried; the presence of virus in the victims' kidneys suggests that brood stock should be catheterized and isolations attempted from urine samples; low temperature stress should be applied to determine whether it causes virus to be released and demonstrable in culture; and, above all, blind passage should be made of materials that show no CPE on original inoculation.

Identification

SVCV is a distinct agent, serologically separable from the other major bullet-shaped fish pathogens—the several eel viruses, IHNV, PFR, Rio Grande perch rhabdovirus, and Egtved virus. SVCV has yet to be compared with the rhabdoviruses isolated from cod, halibut, and the Eurasian perch. The sheatfish rhabdovirus reported by Fijan et al. (1984) and the rhabdovirus isolated by Ahne (1973) from carp with SBI are isolants of SVCV.

Two strains of SVCV have been recognized (Fijan et al. 1984). The S/30 isolant of Fijan et al. (1971) has been considered serotype I, and the isolant from carp with SBI, the 10/3 of Ahne (1973), has been considered representative of serotype II. In serologic tests, however, Hill et al. (1975) could not distinguish between SVCV and the SBI isolant, and de Kinkelin and Le Berre (1974) considered them to be identical.

Three common serologic tests have been used for identification of SVCV: neutralization (Clerx et al. 1978; de Kinkelin and Le Berre 1974; Hill et al. 1975, 1981; Sulimanović 1973), immunoperoxidase (Faisal and Ahne 1980, 1983, 1984), and fluorescent antibody (Faisal and Ahne 1983, 1984; Pfeil 1978). Sulimanović (1973) also used an indirect hemagglutination test to detect antibodies to SVCV.

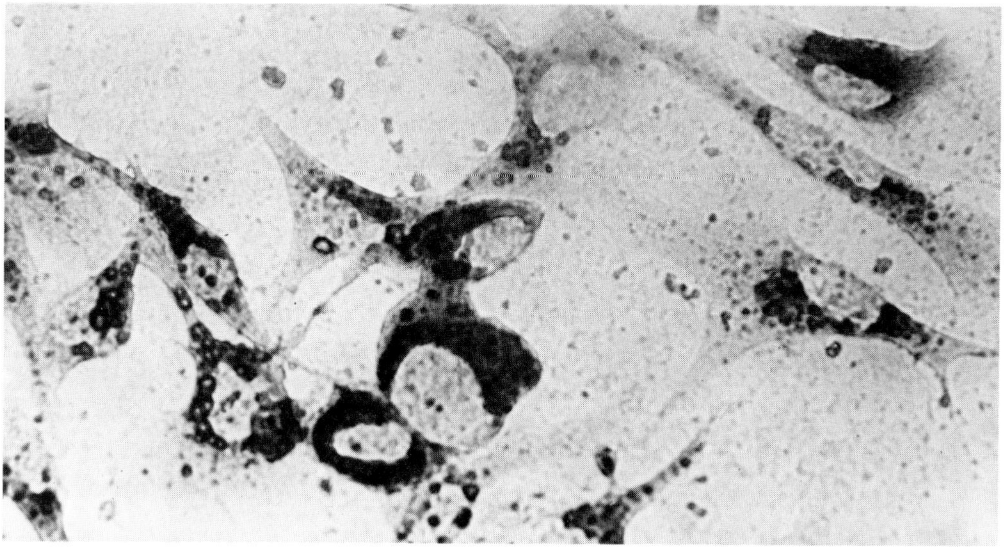

Figure 4. Pike gonad (PG) cells infected with SVCV for 24 hours and stained with an indirect immunoperoxidase technique. Dark-staining cytoplasmic antigen is localized near the cell nuclei. From Faisal and Ahne (1980). Reprinted with the permission of Springer-Verlag.

Dixon and Hill (1984) were unable to develop a satisfactory ELISA technique for SVCV and Egtved virus, although they were successful with IHNV and the eel rhabdoviruses eel virus American (EVA) and eel virus European unknown (EVEX). They attributed the difference to the fact that high-titered neutralizing serum could be produced against IHNV (1 : 50,000) and the eel viruses (1 : 200,000 to 1 : 300,000), whereas titers against SVCV and Egtved virus were very low (only 1 : 3500 and 1 : 2000, respectively). They postulated that a titer of at least 1 : 50,000 would be needed for a successful ELISA with SVCV. (The situation brings to mind the recent development of highly sensitive direct binding assay for IHNV by Hsu and Leong [1985]; the methodology can be applied to SVCV.)

In the initial report on the indirect immunoperoxidase technique for SVCV, Faisal and Ahne (1980) stated that in cell cultures held at 20°C positive results were first evident at 2 hours postinfection and markedly so at 24 hours (Fig. 4). When the indirect fluorescent antibody technique was used, reactivity was first evident at 6 hours postinfection. Frozen sections of spleens and kidneys also gave a positive reaction: the infectivity of those tissues was very high—10^8 $TCID_{50}/g$—whereas other tissues showed no demonstrable infectivity (Faisal and Ahne 1980).

In later work with infected cell cultures, also incubated at 20°C, Faisal and Ahne (1983) found first reactivity with indirect immunoperoxidase and indirect fluorescent antibody technique at about 12 hours postinoculation. For both techniques, sensitivity was highest at about 48 hours. In contrast, titratable infectivity was first found at 24 hours. When applied to frozen sections, the two serologic techniques were about equally sensitive. When Faisal and Ahne (1984) infected carp with waterborne virus, the first evidence of virus appeared 7 days after infection, and only in the kidneys; assays at 3.5-day intervals showed that tissue reactivity to both immunoperoxidase and fluorescent antibody was consistently greatest in the kidneys from days 7 to 14. Thereafter, reactivity was generally highest in the liver and spleen from days 14 to 24, whereas that of kidneys declined; during days 21 to 24 no evidence of reactivity was found in the kidneys, nor could virus be isolated. The conclusion to be drawn is that, lacking data on the age of natural infections, one should assay livers as well as kidneys for serologic evidence of SVCV.

Also, when SVC is suspected, concurrent attempts at isolation and serum neutralization tests may save time in reaching definitive identification.

Transmission and Incubation

Under natural conditions, most transmission is probably horizontal, although circumstantial evidence suggests that vertical transmission may also take place in some situations. An example of this possibility was revealed in the assay of ovarian and seminal fluids of 5- to 10-year-old asymptomatic carp by Békési and Csontos (1985); as noted earlier, they found virus in 3 of 491 ovarian fluids (but in none of 211 seminal fluids). Ahne (1979) was similarly unable to find virus in seminal fluids from experimentally infected carp. An assay of the eggs themselves, before and during incubation, has yet to be reported.

During outbreaks, and for as much as several months thereafter, virus is shed in fecal casts and possibly also in urine (Ahne 1979; Baudouy et al. 1980a, b, c; Pfeil-Putzien 1977). Quantification showed that the fecal casts contain as much as 10^3–10^4 PFU/g (Ahne 1979). In constrast to findings with ovarian fluids, it is not known if asymptomatic survivors shed virus in feces or urine for more than 3 months. One can speculate that peak virus shedding occurs seasonally at or near spawning time, when susceptible young will be abundant.

The virus is readily transmitted experimentally, but the demonstrably effective injection methods— intraperitoneal, intracerebral, and intrapneumatic (by way of the swim bladder)—differ drastically from natural methods. Varović and Fijan (1973) found that 75-g carp held at about 17°C were most susceptible to intracerebral inoculation: it yielded an LD_{50} of $10^{2.6}$ $TCID_{50}$/g of body weight, compared with $10^{3.7}$ $TCID_{50}$/g for intraperitoneal injection, and $10^{4.5}$ $TCID_{50}$/g for swim bladder inoculation; however, survival time was only about 9 days in carp infected by way of the swim bladder compared with about 13 days in fish infected by other routes. The authors suggested that transmission by intraperitoneal or intrapneumatic injection gave results typical of the disease, but that the intracranial route should be used if susceptibility was a critical factor.

Surprisingly, Varović and Fijan (1973) found that as much as 10^6 $TCID_{50}$/g of body weight given by mouth caused no signs of disease.

Certain parasites potentially play a role in natural transmission. Experimentally, the carp louse *Argulus foliaceus* has been found to ingest the virus while feeding on SVC-infected carp, and 6 to 7 days later transfer it as it feeds on a healthy fish (Ahne 1979, 1985a; Pfeil-Putzien 1977, 1978). The leech *Piscicola geometra* has been similarly implicated (Ahne 1979, 1985a). The ciliate *Ichthyophthirius multifiliis*, tested under similar conditions, did not transfer virus from infected to healthy carp. It is emphasized that both the carp louse and leech carried the virus passively, and that the multiplication that usually takes place in true vectors did not occur. In fact, the SVCV titers in the parasites declined and disappeared within a matter of 14 days (Ahne 1979).

Waterborne transmission has been effected experimentally by contact between healthy and infected carp, and by the immersion of susceptible fish in culture-grown virus. Although infectivity levels of waterborne SVCV have not been reported for natural outbreaks, immersion of carp in quantified virus effectively transmitted the infection in a way that approximates transmission under husbandry. Not surprisingly, the gills were the primary portal of entry for SVCV.

Ahne (1977, 1978) reported a 20% mortaltity in 25- to 30-g carp that were immersed in the virus at a concentration of 4×10^4 PFU/mL for 2 hours at 13°C. In contrast, carp that were injected intraperitoneally with only 10^3 PFU virus sustained 90% mortality. Ahne followed the virus dynamics for 1 month. After immersion, virus was found in gills at 2 hours; it was probably replicated there because an eclipse occurred before infectivity was next found at 4 days, in the liver, kidneys, spleen, and gut. Virus was found on day 5 in circulating blood, where it reached a titer of 5×10^7 PFU/mL on day 6. Virus appeared in fecal material on day 11. Mortality began on day 20 and continued for 10 days.

Baudouy et al. (1980a, b, c) used a scheme of transmission by immersion that was similar to that of Ahne and reported similar results. However, they stressed aspects of temperature as related to seasonal change and outbreaks of SVC. They immersed 50-g carp for 2 hours in 5×10^3 PFU at 11°C and then held the fish for 50 days, gradually warming the water to 16°C, thus simulating springtime among fish that had overwintered and faced infection under unfavorable temperature. Infectivity reached 100% by day 2, but was found only in the gills. Fish became viremic on day 3, when the first virus was shed. No fish survived the 50-day period of observation. The authors stated that the course of the disease—as estimated from virus dynamics—was three to four times faster than occurred during the simulation of autumnal temperatures (Baudouy et al. 1980a).

In a related study, Baudouy et al. (1980b) held infected fish at three temperatures. Fish held at 18°C or above were infected, but developed an immune response and none died. Some of the fish held at 11 to 18°C became diseased and died, but some developed an immune response and survived. Carp held at 10°C or lower did not mount an immune response, and all died. Since virus replication in vitro was similar at 6°C and 20°C, the authors concluded that the temperature effect was principally on the host—immunosuppression being inferred.

Fall-winter-spring water temperatures were simulated by lowering the temperature from 11°C to 5°C during a 2-month period and then elevating it to 20°C during the next 5 months (Baudouy et al. 1980c). Again, 100% infection was achieved with waterborne virus, as determined by gently stroking gill surfaces with the orfice of a 1-mL pipette and assaying infectivity of the material thus aspirated by capillarity. Virus was first detected on day 2 and all fish were infected by day 3. At least 80% of the fish showed infectivity until day 120; a month-long rapid increase in mortality then occurred and virus detection by gill assay declined to zero. The work showed that virus transmitted in fall could result in disease and low mortality during the time of declining temperature of winter. However, mortality became high as the

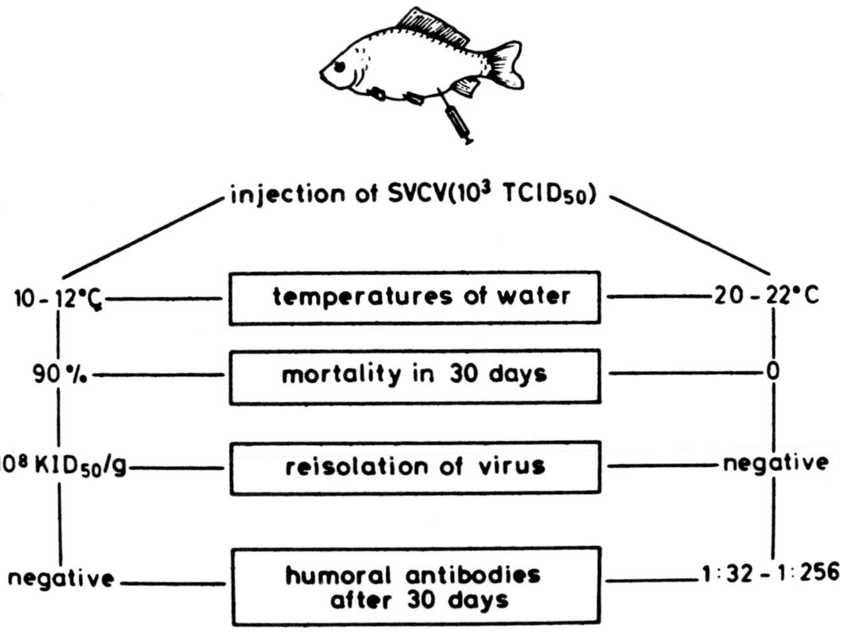

Figure 5. Water temperature determines whether injection of carp with SVCV results in disease and also mortality or survival and development of protective antibody. From Ahne (1984). Reprinted with the permission of Akadémiai Kaido, Budapest.

temperature rose above 7°C and killed most of the fish by about day 160. Interestingly, virus shedding peaked during days 60 to 120, which simulated yearly lows.

Overall, the various observations of temperature effects are in agreement. Fijan et al. (1971) noted that at 17 to 26°C, 3 of 10 carp injected with virus died and the average time to death was 13 days; at 16 to 17°C, mortality was 90% and the incubation time was only 10 days; at 11 to 15°C the percent mortality was the same but low-temperature retardation was evident in that incubation time to average death was 17 days. Ahne (1980) found that at 10 to 12°C infection dominated and mortality reached 90% within 30 days, whereas at 20°C the host immune response dominated—no mortality occurred and antibody was produced (Fig. 5).

Source

Survivors of SVC are the most important source of virus that initiates outbreaks. Once an epizootic is under way, virus is shed by victim fish and serves to sustain the disease in a population. By virtue of numbers propagated, the source of virus is the common carp, but other fish species—especially but not exclusively cyprinids—might also be sources. Ahne (1979) reported that none of several kinds of aquatic animals that he tested (snails, insects, isopods, and frogs) sustained long-term survival of the virus. It is not known whether SVCV retains viability during passage through fish-eating birds. Considering their mobility, such birds could also serve as transfer agents of the virus.

Host and Geographic Range

The common carp, in its various scaled and partly scaled forms, is the principal host of SVCV; the virus kills both the adults and young. Partly from their own investigations and partly from reports of others, Shchelkunov et al. (1984) concluded that the virus has also been isolated from other cyprinids: the silver carp, bighead carp (*Aristichthys nobilis*), and crucian carp (*Carassius carassius*). Incongruously, the closely related goldfish (*C. auratus*) was found to be refractory (Ahne 1973), whereas fry of the unrelated sheatfish (*Silurus glanis*) suffered a natural infection that resulted in a more than 90% loss (Fijan et al. 1984).

Experimentally, guppies (*Lebistes reticulata*) were susceptible to strain 10/3 virus and fry of northern pike (*Esox lucius*) were comparably vulnerable to the 5/30 reference strain SVCV (Ahne 1973, 1985b). In an interesting study wholly unrelated to fish health, Bussereau et al. (1975) found that SVCV, like PFR, multiplied in the fruit fly *Drosophila melanogaster* and there conferred sensitivity to carbon dioxide. This effect is identical to that of sigma virus, the rhabdovirus that occurs naturally in the fruit fly and is transmitted genetically in it.

SCVC occurs throughout Europe and the Soviet Union wherever carp are propagated. Bucke and Finlay (1979) found the virus in the United Kingdom, but it has not been reported from North America, despite repeated introductions of the common carp. However, the common carp is seldom propagated in North America; instead, it is typically a feral fish that is generally held in low esteem and seldom warrants or receives virologic examination.

The common carp has been under husbandry in China for several thousand years, but western knowledge of fish diseases in that country is meager, and SVCV has not yet been reported. Fish cell culture has been started only recently and communication has been poor. Now, however, some reports on fish health are being published in the English language. Accordingly, it seems inevitable that the literature on SVCV will eventually include reports of the virus in Chinese cultural facilities.

Immunity

Carp can mount an immune response to SVCV, but the response is influenced by temperature, route of immunization, quantity and strain of virus, and age and condition of the host. Factors in a protective response are only incompletely known, but probably include antibody, interferon, and cellular immunity. Because neutralizing antibody can be easily quantified, most investigators of SVCV immunity have worked with it; however, even though titers are generally low, overall immune responses have been protective.

Sulimanović (1973) published an early account of an immune response to SVCV. He omitted specific information on virus quantification and temperature; however, he inoculated live virus per os and induced protection against a later intraperitoneal challenge. He noted that the challenge resulted in a gain in indirect hemagglutination (HAI) titer.

Fijan and coworkers (Fijan, Petrinec, Stancl, Dorson, and Le Berre 1977; Fijan, Petrinec, Stancl, Kezić, and Teskeredzić 1977), also using live virus, determined the effects of temperature and routes of inoculation. Adult carp injected intraperitoneally produced neutralizing antibody slowly and unevenly when they were held at about 14°C, but produced more antibody more rapidly and more uniformly when they were held at 25°C. Booster injections increased neutralizing antibody only in fish that showed a low primary response. The neutralizing activity was predominantly located in the IgM fraction of serum, and clearly distinguished SVCV from PFR (Fijan, Petrinec, Stancl, Dorson, and Le Berre 1977). When virus was given to yearlings intraperitoneally or by mouth—presumably at 25°C—solid protection was provided to all for 9 months, but only the injected fish were protected at 11 months. Neutralizing antibody was present in injected fish for only 2 months, and remained in only 1 of 118 fish given virus orally. Understandably, the authors considered that neutralizing antibody was not a major factor in protection.

In addition to antibody, SVCV induces synthesis of interferon. Baudouy et al. (1977) immersed carp in 10^2 or 10^4 PFU/mL for 2, 7, or 24 hours. They reported, not surprisingly, that mortality was about 50% at 12°C, regardless of exposure time; among survivors, levels of antibody were highest at about 2 months. When Baudouy (1978) injected carp intraperitoneally with 10^5, 10^6, or 10^7 PFU virus but held them at the more favorable 20°C, interferon was evident at 24 hours—9 of 10 sera effected reduction of 80% or more of 100 PFU of IPNV—and the highest activity reflected the greater quantities of virus injected. At 72 hours and 96 hours, however, the levels of interferon had declined. Neutralizing antibody, present in most fish, reached a maximum at 2.5 months, and declined somewhat in 1:200 dilutions at 4 months. The work did not include tests for protection.

After carp were exposed by immersion in virus at 15°C, survivors had no measurable neutralizing antibody but nevertheless were protected against an intraperitoneal challenge (Hill 1977). Moreover, the fish that survived the intraperitoneal challenge developed significant levels of neutralizing antibody. Again, neutralizing antibody seemed not to be directly involved in protection.

Studies of immunoprophylaxis of SVC were begun in Czechoslovakia in 1976, where inactivated virus was administered parenterally in large-scale trials. Significant benefit was obtained as early as 7 days after vaccination, and protection persisted for as long as 11 months (Tesarčík and Macura 1981; Tesarčík et al. 1978). Macura et al. (1983), who summarized 5 years of research on etiology, pathology, and virus characteristics, included some details of the commercial SVC vaccine. For that product, virus is inactivated with 0.05% beta propiolactone—apparently at 4°C—after which the pH is adjusted to 7.2 and the preparation lyophilized. The authors clearly acknowledged the limitations of the product, which must be injected into individual fish.

Tesarčík et al. (1984) eventually reviewed the vaccination studies in Czechoslovakia in an English-language paper. There seemed to be little doubt that the vaccine was protective, but the results of field or pilot-scale testing were typically given in qualitative rather than quantitative terms. As an alternative to injecting thousands of individual fish, adult females were vaccinated; however, no colostrum-like maternal transfer of protection against SVC was observed.

In Yugoslavia, Fijan and Matašin (1980) showed that immersion of groups of 30 common carp fry for 1 hour in 1×10^5 $TCID_{50}$/mL provided protection after 4 weeks at 21°C. Challenge consisted of an intraperitoneal injection of 1×10^4 $TCID_{50}$, which killed 50% of the nonvaccinated controls. Hyperosmotic infiltration was neither beneficial nor detrimental. That methodology has generally been phased out of fish vaccination.

Fijan (1984) discussed prospects and constraints of vaccination of European pond fishes. Unfortunately, few methods of vaccine administration can be incorporated into husbandry technology when temperatures are favorable for a good immune response, or when the young have reached the age of immune competence. Oral and spray delivery methods merit rigorous pilot-scale testing.

The immune response of carp to SVCV has found application in epizootiologic studies. In Germany, Wizigmann et al. (1980, 1983), testing for neutralizing antibody, found that sera from healthy carp showed no neutralizing activity at a dilution of 1:2. Considering activity at a 1:6 or more dilution as presumptive evidence, they surveyed fish from 51 farms and found neutralizing antibody in 472 of 1963 serum samples. Titers were generally low; they were 1:6 in 400 sera, 1:7 to 1:23 in 60, and 1:24 to 1:114 in the remaining 12. SVCV was considered to be widespread in the Bavarian populations of carp.

Control

Several control measures are known for SVC, but each measure has certain features that limit its applications or effectiveness. Avoidance—the almost universal control measure—is limited in part by incomplete knowledge of sources of the virus, and by the requirements of large-scale husbandry. Virologic methods have not yet been developed to detect carriers of SVC. Although neutralizing antibody can be identified in sera from brood stock, its significance is not known. Does neutralizing antibody indicate freedom from the virus, or latent infection and a carrier state? The surface water from streams or rivers that is typically used in carp husbandry nearly always harbors wild fish that may carry and shed virus.

The inability to identify carriers of SVCV also limits the development of specific-pathogen–free brood stock. Health history is now the sole means of recognizing populations that are free of the virus.

A sound biological basis exists for using temperature manipulation as a control measure—by simply rearing the fish at about 20°C. Seasonal climatic changes and energy costs pose significant limitations to all but small-scale operations. Cooling waters or effluents from nuclear or fossil fuel generating plants offer cost advantages, providing that the warmed water does not foster other fish pathogens.

Genetic selection for resistance to SVC has decided application, but control is less than complete and the practice is not widely used. In the USSR a program of multifaceted selection has been in progress since 1963, and survival differences between resistant and susceptible stocks of about 30% were claimed (Kirpichnikov and Faktorovich 1972; Kirpichnikov et al. 1972). That differential was maintained into the 1980s, and on a 1982 visit to the USSR, I learned that the Krasnodor strain of carp was highly touted for its resistance to SVC. In Great Britain, Hill (1977) found that domestic (farmed) strains of carp were significantly more resistant than wild fish. Mortality was 60 to 90% in wild carp, whereas farmed fish sustained "very few mortalities."

As with realistic and cost-effective therapy or prophylaxis of any viral infection of fishes, prospects for the chemical control of SVC are dim to the point of near nonexistence. On the other hand, immunization undoubtedly has the greatest potential for cost-effective and realistic control of SVC. Effective immunoprophylaxis has been developed, but requirements for administration are severely limiting: the fish must have reached the age of immune competence, and temperature (about 20°C or higher) must be favorable for an immune response; and the vaccine must be economical—hence inactivated preparations are probably ruled out. Above all, the vehicle of administration must be compatible with husbandry practices—e.g., it must be a product that is effective when incorporated in feeds, or one that can be used as a spray when fry are normally handled for pond stocking.

References

Ahne, W. 1973. Zellkulturen aus verschiedenen Süsswasserteleosteergeweben und Untersuchung über die Ätiologie der Schwimmblasenentzündung der Karpfen. Ph.D. dissertation. Ludwig-Maximilians Univ., Munich.

Ahne, W. 1976. Untersuchungen über die Stabilität des karpfenpathogenen Virusstammes 10/3. Pages 121–127 in H. H. Reichenbache-Klinke, ed. Fisch und Umwelt, Vol. 2. Gustav Fischer Verlag, Stuttgart.

Ahne, W. 1977. Evidence for the systemic character of *Rhabdovirus carpio* infection. Bull. Off. Int. Epizoot. 87:435–436.

Ahne, W. 1978. Uptake and multiplication of spring viremia of carp virus in carp, *Cyprinus carpio* L. J. Fish Dis. 1:265–268.

Ahne, W. 1979. Untersuchungen über die acute Form der infektiösen Bauchwassersucht bei Cypriniden (*Cyprinus carpio, Ctenopharyngodon idella*) Habilitationsschrift zur Erlangung der Venia legendi. Ludwig-Maximilians Univ., Munich.

Ahne, W. 1980. *Rhabdovirus carpio*—Infektion beim Karpfen (*Cyprinus carpio*): Untersuchungen über Reaktionen des Wirtsorganismus. Fortschr. Veterinärmed. 30:180–183.

Ahne, W. 1982a. Vergleichende Untersuchungen über die Stabilität von vier fischpathogenen Viren (VHSV, PFR, SVCV, IPNV). Zentralbl. Veterinärmed. [B] 29:457–476.

Ahne, W. 1982b. Untersuchungen zur Tenazität der Fischviren. Fortschr. Veterinärmed. 35:305–309.

Ahne, W. 1984. Important viral diseases in European fish cultures. Symp. Biol. Hung. 23:3–15.

Ahne, W. 1985a. *Argulus foliaceus* L. and *Philometra geometra* L. as mechanical vectors of spring viremia of carp virus (SVCV). J. Fish Dis. 8:241–242.

Ahne, W. 1985b. Viral infection cycles in pike (*Esox lucius* L.). Z. Angew. Ichthyol. 1:90–95.

Ahne, W., and K. Wolf. 1977. Spring viremia of carp. U.S. Fish Wildl. Serv., Fish Dis. Leaflet 51.

Ahne, W., and K. Wolf. 1980. Viruserkrankungen der Fische. Pages 56–105 in H. H. Reichenbach-Klinke, ed. Krankheiten und Schädigungen der Fische. Gustav Fischer Verlag, Stuttgart.

Antychowicz, J. 1973. A summary account of swim-bladder inflammation of cyprinids (SBI). Pages 145–155 in W. A. Dill, ed. Symposium on the major communicable fish diseases in Europe and their control. EIFAC (Eur. Inland Fish. Advis. Comm.) Tech. Paper 17, Suppl. 2.

Arshaniza, N. M., and O. N. Bauer. 1973. The epizootiology, diagnosis and prophylaxis of swim-bladder inflammation of cyprinids (SBI). Pages 140–144 in W. A. Dill, ed. Symposium on the major communicable fish diseases in Europe and their control. EIFAC (Eur. Inland Fish. Advis. Comm.) Tech. Paper 17, Suppl. 2.

Arshaniza, N. M., O. N. Bauer, and V. L. Vladimirov. 1968. Air bladder disease of carps; its aetiology, epizootiology and control. Bull. Off. Int. Epizoot. 69:999–1006.

Bachmann, P. A., and W. Ahne. 1973. Isolation and characterization of agent causing swim bladder inflammation in carp. Nature 244:235–237.

Bachmann, P. A., and W. Ahne. 1974. Biological properties and identification of the agent causing swim bladder inflammation in carp. Arch. Gesamte Virusforsch. 44:261–269.

Baudouy, A. M. 1978. Relation hôte-virus au cours de la virémie printanière de la carpe. C. R. Acad. Sci. [D] (Paris) 286:1225–1228.

Baudouy, A. M., G. Merle, and M. Danton. 1977. Résultats concernant l'étude expérimentale de la relation hôte-virus: virémie printanière de la carpe. Bull. Off. Int. Epizoot. 87:437–438.

Baudouy, A. M., M. Danton, and G. Merle. 1980a. Virémie printanière de la carpe: résultats de contaminations expérimentales effectuées au printemps. Ann. Rech. Vet. 11:245–249.

Baudouy, A. M., M. Danton, and G. Merle. 1980b. Virémie printanière de la carpe: étude expérimentale de l'infection évoluant à différentes températures. Ann. Virol. (Paris) 131E:479–488.

Baudouy, A. M., M. Danton, and G. Merle. 1980c. Experimental infection of susceptible carp fingerlings with spring viremia of carp virus under wintering environmental conditions. Pages 23–27 in W. Ahne, ed. Fish diseases. Third COPRAQ Session. Springer-Verlag, Berlin.

Bauer, O. N., and K. A. Faktorovich. 1969. Research on infectious diseases of carp in the USSR. Bul. VÚR Vodňany 3:30–43.

Békési, L., and L. Csontos. 1985. Isolation of spring viraemia of carp virus from asymptomatic broodstock carp, *Cyprinus carpio* L. J. Fish Dis. 8:471–472.

Békési, L., and E. Szabó. 1977. *Rhabdovirus carpio* elsö izolálása hazai pontyállományokból. Magy. Állatorv. Lapja 32:369–371.

Békési, L., and E. Szabó. 1979. A rhabdovirus isolated from carps in Hungary. Experimental infection of carps and resistance of the virus. Acta Microbiol. Acad. Sci. Hung. 26:193–197.

Bootsma, R., and Ebregt. 1983. Spring viremia of carp. Pages 81–86 in D. P. Anderson, M. Dorson, and P. Dubourget, eds. Antigens of fish pathogens. Collection Fondation Marcel Merieux, Lyon, France.

Bootsma, R., N. Fijan, and J. Blommaert. 1977. Isolation and preliminary identification of the causative agent of carp erythrodermatitis. Vet. Arh. 47:291–302.

Bucke, D., and J. Finlay. 1979. Identification of spring viraemia in carp (*Cyprinus carpio* L.) in Great Britain. Vet. Rec. 104:69–71.

Bussereau, F., P. de Kinkelin, and M. Le Berre. 1975. Infectivity of fish rhabdoviruses for *Drosophila melanogaster*. Ann. Microbiol. (Paris) 126A:389–395.

Clark, H. F., and E. Z. Soriano. 1974. Fish rhabdovirus replication in non-piscine cell culture: new system for the study of rhabdovirus-cell interaction in which the virus and cell have different temperature optima. Infect. Immun. 10:180–188.

Clerx, J. P. M., and M. C. Horzinek. 1978. Comparative protein analysis of non-salmonid fish rhabdoviruses. J. Gen. Virol. 40:287–295.

Clerx, J. P. M., M. C. Horzinek, and A. D. M. E. Osterhaus. 1978. Neutralization and enhancement of infectivity of non-salmonid fish rhabdoviruses by rabbit and pike immune sera. J. Gen. Virol. 40:297–308.

Cohen, J., and G. Lenoir. 1974. Ultrastructure et morphologie de quatre rhabdovirus de poissons. Ann. Rech. Vét. 5:443–450.

Csaba, G., É. Kovács-Gayer, L. Békési, M. Bucsek, J. Szakolczai, and K. Molnár. 1984. Studies into the possible protozoan aetiology of swim bladder inflammation in carp fry. J. Fish Dis. 7:39–56.

de Kinkelin. P., and M. Le Berre. 1974. Rhabdovirus des poissons. II. Propriétés in vitro du virus de la virémie printanière de la carpe. Ann. Microbiol. (Paris) 125A:113–124.

de Kinkelin, P., M. Dorson, and A. M. Hattenberger-Baudouy. 1982. Interferon synthesis in trout and carp after viral infection. Dev. Comp. Immunol. Suppl. 2:167–174.

de Kinkelin, P., J. Bernard, and A. M. Hattenberger-Baudouy. 1984. Immunization against viral diseases occurring in cold water. Pages 167–197 in Symposium on fish vaccination, Paris, Office International Epizooties, February 20–22, 1984.

Deuter, A., P. J. Enzmann, and K. Bigott. 1982. Comparison of two fish pathogenic rhabdoviruses: L-(^{75}Se) seleno-methionine-labelled proteins. Zentralbl. Bakteriol. Mikrobiol. Hyg. [A] 253:10–11.

Dixon, P. F., and B. J. Hill. 1984. Rapid detection of fish rhabdoviruses by the enzyme-linked immunosorbent assay (ELISA). Aquaculture 42:1–12.

Enzmann, P. J., B. Maier, and K. Bigott. 1981. Biochemical data on the RNA of VHS-V and RVC indicating a possible serological relationship. Bull. Eur. Assoc. Fish Pathol. 1:37–39.

Faisal, M., and W. Ahne. 1980. Use of immunoperoxidase technique for detection of fish virus antigens. Pages 186–192 in W. Ahne, ed. Fish diseases. Third COPRAQ Session. Springer-Verlag, Berlin.

Faisal, M., and W. Ahne. 1983. Quantitative Nachweis des Spring Viremia of Carp Virus (SVCV)–Antigens mit Hilfe der Immunofluoreszenz– und Immunoperoxidaseplaquestechnik. Zentralbl. Veterinärmed. [B] 30:466–470.

Faisal, M., and W. Ahne. 1984. Spring viraemia of carp virus (SVCV): comparison of immunoperoxidase, fluorescent antibody and cell culture isolation techniques for detection of antigen. J. Fish Dis. 7:57–64.

Faktorovich, K. A. 1968. Histological methods of diagnosis of carp dropsy. Bull. Off. Int. Epizoot. 69:1481–1497.

Fijan, N. 1966. Experimental transmission of infectious dropsy. Bull. Off. Int. Epizoot. 65:731–738.

Fijan, N. 1972. Infectious dropsy in carp—a disease complex. Pages 39–51 in L. E. Mawdesley-Thomas, ed. Diseases of fish. Symposia of the Zoological Society of London. Academic Press, London.

Fijan, N. 1973. Spring viremia of carp (SVC)—a review. Pages 119–123 in W. A. Dill, ed. Symposium on the major communicable fish diseases in Europe and their control. EIFAC (Eur. Inland Fish. Advis. Comm.) Tech. Paper 17, Suppl. 2.

Fijan, N. 1975. Advances in knowledge of viral diseases of fish. Pages 127–133 in L. A. Page, ed. Wildlife diseases. Plenum Press, New York.

Fijan, N. 1984. Vaccination of fish in European pond culture: prospects and constraints. Symp. Biol. Hung. 23:233–241.

Fijan, N., and Ž. Matašin. 1980. Spring viraemia of carp: preliminary experiments on vaccination by exposure to virus in water. Vet. Arh. 50:215–220.

Fijan, N., Z. Petrinec, Đ. Sulimanović, and L. O. Zwillenberg. 1971. Isolation of the viral causative agent from the acute form of infectious dropsy of carp. Vet. Arh. 41:125–138.

Fijan, N., Z. Petrinec, Z. Stancl, M. Dorson, and M. Le Berre. 1977. Hyperimmunization of carp with *Rhabdovirus carpio*. Bull. Off. Int. Epizoot. 87:439–440.

Fijan, N., Z. Petrinec, Z. Stancl, N. Kezić, and E. Teskeredzić. 1977. Vaccination of carp against spring viraemia: comparison of intraperitoneal and peroral application of live virus to fish kept in ponds. Bull. Off. Int. Epizoot. 87:441–442.

Fijan, N., Ž. Matašin, Z. Jeney, J. Oláh., and L. O. Zwillenberg. 1984. Isolation of *Rhabdovirus carpio* from sheatfish (*Silurus glanis*) fry. Symp. Biol. Hung. 23:17–24.

Gupta, K. C., and P. Roy. 1980. Alternate capping mechanisms for transcription of spring viremia of carp virus: evidence for independent mRNA initiation. J. Virol. 33:292–303.

Gupta, K. C., and P. Roy. 1981. Synthesis of capped and uncapped methylated oligonucleotides by the virion transcriptase of spring viremia of carp virus, a rhabdovirus. Proc. Natl. Acad. Sci. USA 78:4758–4762.

Gupta, K. C., D. H. L. Bishop, and P. Roy. 1979. 5′-Terminal sequences of spring viremia of carp virus RNA synthesized in vitro. J. Virol. 30:735–745.

Havelka, J. 1973. Infectious dropsy of cyprinids (IDC) in Czechoslovakia. Pages 124–132 in W. A. Dill, ed. Symposium on the major communicable fish diseases in Europe and their control. EIFAC (Eur. Inland Fish. Advis. Comm.) Tech. Paper 17, Suppl. 2.

Hill, B. J. 1977. Studies on SVC virulence and immunization. Bull. Off. Int. Epizoot. 87:455–456.

Hill, B. J., B. O. Underwood, C. J. Smale, and F. Brown. 1975. Physico-chemical and serological characterization of five rhabdoviruses infecting fish. J. Gen. Virol. 27:369–378.

Hill, B. J., R. F. Williams, and J. Finlay. 1981. Preparation of antisera against fish virus disease agents. Dev. Biol. Stand. 49:209–218.

Hofer, B. 1904. Handbuch der Fischkrankheiten. Verlag der Allgemein Fischerei-Zeitung, B. Heller, Munich.

Hsu, Y. L., and J. C. Leong. 1985. A comparison of detection methods for infectious hematopoietic necrosis virus. J. Fish Dis. 8:1–12.

Kirpichnikov, V. S., and K. A. Faktorovich. 1972. Increase in the resistance of carp to dropsy by means of breeding. Communication II. The course of the selection and evaluation of the breeding groups selected. Sov. Genet. (English translation of Genetika) 8:592–600.

Kirpichnikov, V. S., K. A. Faktorovich, and V. S. Suleimanyan. 1972. Increasing the resistance of carp to dropsy by breeding. Communication I. Methods of breeding for resistance. Sov. Genet. (English translation of Genetika) 8:306–312.

Kiuchi, A., and P. Roy. 1984. Comparison of the primary sequence of spring viremia of carp virus M protein with that of vesicular stomatitis virus. Virology 134:238–243.

Körting, W. 1982. Protozoan parasites associated with swimbladder inflammation (SBI) in young carp. Bull. Eur. Assoc. Fish Pathol. 2:25–28.

Kovács-Gayer, É. 1983. Histopathological studies on protozoan swimbladder inflammation of common carp fry. Parasitol. Hung. 16:39–46.

Lenoir, G. 1973. Structural proteins of spring viremia virus of carp. Biochem. Biophys. Res. Commun. 51:895–898.

Lenoir, G., and P. de Kinkelin. 1975. Fish rhabdoviruses: comparative study of protein structure. J. Virol. 16:259–262.

Liebmann, H. 1973. Swim-bladder inflammation of cyprinids (SBI). Pages 134–136 in W. A. Dill, ed. Symposium on the major communicable fish diseases in Europe and their control. EIFAC (Eur. Inland Fish. Advis. Comm.) Tech. Paper 17, Suppl. 2.

McAllister, P. E. 1979. Fish viruses and viral infections. Pages 401–470 in H. Fraenkel-Conrat and R. R. Wagner, eds. Comprehensive virology, Vol. 14. Plenum Press, New York.

Macura, B., L. Dedek, and J. Tesařčik. 1973. The positive cytopathic effect from a case of spring viremia of carp. (Pozitivní cytopatický efekt z případu jarní viremie kaprů). Bul. VÚRH Vodňany 9:7–14. In Czechoslovakian, English abstract.

Macura, B., J. Tesarčík, and J. Řehulka. 1983. Survey of methods of specific immunoprophylaxis of carp spring viremia in Czechoslovakia. Pr. VÚRH Vodňany 12:50–56.

Negele, R. D. 1977. Histopathological changes in some organs of experimentally infected carp fingerlings with *Rhabdovirus carpio*. Bull. Off. Int. Epizoot. 87:449–450.

Osadčaja, E. F. 1964. Isolation of cytopathic agents from carp with acute form of rubella. (Vyldlenie tsitopatogennykh agentov ot karpov pri ostroi formi krasnukhi.) Veterinarija 41:29. Translated from the Russian by N. Fijan.

Otte, E. 1963. The contemporary views on the etiology of the infectious dropsy of carp. (Die heutigen Ansichten über die Ätiologie der infektiösen Bauchwassersucht der Karpfen.) Wien. Tierärztl. Monatsschr. 50:995–1005. Translated from the German by S. F. Snieszko.

Pfeil, C. 1978. Modell der Schnelldiagnose einer Infektion mit dem *Rhabdovirus carpio* durch die indirekte Immunofluoreszenz. Pages 18–19 in Fisch und Umwelt 5. Gustav Fischer Verlag, Stuttgart.

Pfeil-Putzien, C. 1977. New results in the diagnosis of spring viraemia of carp caused by experimental transmission of *Rhabdovirus carpio* with carp louse (*Argulus foliaceus*). Bull. Off. Int. Epizoot. 87:457. Abstract.

Pfeil-Putzien, C. 1978. Experimentelle Übertragung der Frühjahrsvirämie (spring viremia) der Karpfen durch Karpfenläuse (*Argulus foliaceus*). Zentralbl. Veterinärmed. [B] 25:319–328.

Pilcher, K. S., and J. L. Fryer. 1980. The viral diseases of fish: a review through 1978. Part 1: Diseases of proven viral etiology. CRC Crit. Rev. Microbiol. 7:287–363.

Roegner-Aust, S. 1953. Zur Frage einer Virus-Ätiologie bei verschiedenen Fischkrankheiten. Münch. Beitr. Abwasser-Fisch. Flussbiol. 1:120–145.

Roegner-Aust, S., G. Brunner, and R. Jaxtheimer. 1950. Elektronenmikroskopische Untersuchungen über den Erreger der infektiösen Bauchwassersucht der Karpfen. Bakterium?—Virus? Allg. Fisch-Ztg. 75:420–422.

Roegner-Aust, S., and F. Schleich. 1951. Zur Ätiologie einiger Fischkrankheiten. Z. Naturforsch. 6b:448–451.

Roy, P. 1976. The transcription process of the RNA of spring viremia of carp. Fed. Proc. 35:812. Abstract.

Roy, P. 1981. Phosphoproteins of spring viremia of carp virus. Virology 112:274–281.

Roy, P., and J. P. Clewley, 1978a. Phosphoproteins of spring viremia carp virus and other rhabdoviruses. Pages 116–125 in B. W. J. Mahy and R. D. Barry, eds. Negative strand viruses and host cells. Academic Press, New York.

Roy, P., and J. P. Clewley. 1978b. Spring viremia of carp virus RNA and virion-associated transcriptase activity. J. Virol. 25:912–916.

Roy, P., K. C. Gupta, and A. Kiuchi. 1984. Characterization of spring viremia of carp virus mRNA species and the 3′ sequence of the viral RNA. Virus Res. 1:189–202.

Schäperclaus, W., editor. 1979. Fischkrankheiten, 4th ed. Akademie-Verlag, Berlin, D.D.R.

Shchelkunov, I. S., L. N. Yukhimenko, T. I. Shchelkunova, I. D. Trombitskii, and A. P. Mantsu. 1984. O vedelenii virusa ot belogo tolstolobika c Sindromom krasnukhi. Rybnoe khozyistvo. Ehkspress-informatsia seriya: rybokhozyaistvennoyo ispolezovanie vnutrennkh vodoyomov. Moskva 4:3–7.

Sokol, F., and H. Koprowski. 1975. Structure-function relationships and mode of replication of animal rhabdoviruses. Proc. Natl. Acad. Sci. USA 72:933–936.

Sokol, F., H. F. Clark, T. J. Wiktor, M. L. McFalls, D. H. L. Bishop, and J. F. Obijeski. 1974. Structural phosphoproteins associated with ten rhabdoviruses. J. Gen. Virol. 24:433–445.

Sokol, F., K. B. Tan, M. L. McFalls, and P. Madore. 1974. Phosphate acceptor amino acid residues in structural proteins of rhabdoviruses. J. Virol. 14:145–151.

Stankiewicz, E. B. 1977. Neuropathological picture in fish affected by carp septicaemia (*Septicaemia haemorrhagica cyprinorum*). Acta Ichthyol. Piscatoria 7:35–45.

Sulimanović, Đ. 1973. Immunity of carp to *Rhabdovirus carpio* and determination of antibodies by indirect hemagglutination. Vet. Arh. 43:153–161.

Svobodová, Z., and J. Tesarčík. 1973. The red blood picture in erythrodermatitis, aerosystitis, and spring viremia of carp in the third vegetation period. Acta Vet. Brno 42:265–269.

Tesarčík, J. F. 1977. Studies in the relations between air bladder inflammation and infectious dropsy in carps. Pr. VÚRH Vodňany 10:75–90.

Tesarčík, J., and B. Macura. 1981. Field carp vaccination against spring viraemia on the fish farms of the state fisheries (I.). (Provozní vakcinace kaprů proti jarní virémii na závodech Státního rybářství [I.].) Bul. VÚRH Vodňany 17:3–11. In Czechoslovakian, English abstract.

Tesarčík, J., B. Macura, and L. Dedek. 1977. The isolation of a virus agent from a case of spring viraemia of carps in the Czech Socialist Republic. (Izolace virového agens z případu jarní virémie kaprů v ČSR.) Bul. VÚRH Vodňany 13:10–14. In Czechoslovakian, English abstract.

Tesarčík, J., B. Macura, and J. Rehulka. 1978. Evaluation of a broader clinical experiment with the application of a biopreparation for the control of the spring viraemia of carp in 1977. (Vyhodnocení širšího klinického pokusu s aplikací biopreparátu proti jarní virémii kaprů v roce 1977.) Bul. VÚRH Vodňany 14:3–6. In Czechoslovakian, English abstract.

Tesarčík, J., B. Macura, L. Dedek, L. Valíček, and B. Smíd. 1977. Isolation and electron microscopy of rhabdovirus from the acute form of infectious dropsy of carp (spring viremia of carp). Zentralbl. Veterinärmed. [B] 24:340–343.

Tesarčík, J., B. Macura, J. Řehulka, M. Hrdoňka, and V. Konášová. 1984. Summarized results of pilot vaccination of carp against spring viremia in the Czech Socialist Republic. Pr. VÚRH Vodňany 13:68–74.

Tets, V. I., and G. S. Yakovleva. 1962. Obtaining carp tissue culture and its application in studying the etiology of rubella of fish. (Poluchenie kulturi tkanei karpa i ee primenenie pri izuchenii etiologii krasnuchi ryb.) Nauchno-Tekh. Byull. GosNIORH 15:73–77. Translated from the Russian by N. N. Fijan.

Tomašec, I. 1973. Infectious dropsy of carp (IDC). Pages 107–111 in W. A. Dill, ed. Symposium on the major communicable fish diseases in Europe and their control. EIFAC (Eur. Inland Fish. Advis. Comm.) Tech. Paper 17, Suppl. 2.

Tomašec, I., Z. Brudnjak, N. Fijan, and L. Kunst. 1964. Weiterer Beitrag zur Aetiologie der infektiösen Bauchwassersucht des Karpfens. Bull. Int. Acad. Jugosl. Sci. 16:35–44.

Tomašec, I. I., and N. N. Fijan. 1965. The etiology of infectious dropsy of carp. Ann. N.Y. Acad. Sci. 126(Art. 1):606–615.

Varović, K., and N. Fijan. 1973. Osjetljivost šarana prema *Rhabdovirus carpio* pri raznim načinima inokulacije. Vet. Arh. 43:271–276.

Whipple, H. E., editor. 1965. Viral diseases of poikilothermic vertebrates. Ann. N.Y. Acad. Sci. 126(Art. 1):1–680.

Wizigmann, G., C. Pfeil-Putzien, C. Baath, H. Dangschat, and G. Koppel. 1980. Serologische Untersuchungen über das Vorkommen von Antikörpern gegenüber *Rhabdovirus carpio* bei Karpfen in bayerischen Teichwirtschaften. Fisch Umwelt 8:31–36.

Wizigmann, G., H. Dangschat, C. Baath, and C. Pfeil-Putzien. 1983. Untersuchungen über Virusinfektionen bei Süsswasserfischen in Bayern. Tierärztl. Umsch. 38:250–258.

18 Viral Hemorrhagic Septicemia

Synonyms: Historic (some names shown here in translation)—abdominal dropsy of trout, Egtved disease (named for the small village of Egtved, Denmark, where the disease was first recognized), entero-hepatic-renal syndrome, epizootic exophthalmia, hemorrhagic virus septicemia, infectious anemia, infectious kidney swelling and liver degeneration, infectious renal hypertrophy and liver degeneration, new disease of trout, pernicious anemia, popeye sickness, trout plague, and virus disease of rainbow trout.

The name viral hemorrhagic septicemia was adopted by international agreement (Altara 1963).

Definition

Viral hemorrhagic septicemia (VHS) is an acute to chronic viscerotropic disease of rhabdoviral etiology that occurs in Europe among certain fish in husbandry; it commonly causes significant mortality. Outbreaks typically involve fingerling to yearling rainbow trout and begin when water temperatures are 10°C or lower. Although not usual, the disease also afflicts adult rainbow trout and several other salmonid species, and has even caused mortality among certain nonsalmonid fishes under husbandry.

History

The early history of Egtved virus, the etiologic agent of VHS, is conjectural. The agent has probably existed in Europe for eons. One might suppose that it was maintained without undue consequences in a native salmonid such as the brown trout, Atlantic salmon, or Danube salmon (*Hucho hucho*). The response of the Danube salmon to the virus has yet to be determined.

The assumed pristine relationship of balance between virus and host could have been tilted in favor of the pathogen as salmonid husbandry developed and host population densities were greatly increased. However, a major change occurred in the biology of Egtved virus when the North American rainbow trout was introduced into Europe. According to MacCrimmon (1971), the rainbow trout was introduced into France in 1879 and into Germany in 1882. Thereafter, introductions into other countries were made from European as well as North American sources. In a classic example of a new host-pathogen relationship, the virus proved to be markedly virulent.

Schäperclaus (1938) described "kidney swelling" (Nierenschwellung) of rainbow trout as a serious new disease but one of undetermined etiology. His illustrations of several afflicted trout are suggestive of signs of VHS, and one caption indicates that the photograph was taken in 1931. In his book *Fischkrankheiten*, Schäperclaus (1954) presented details of the disease that he then termed "infectious kidney swelling and liver degeneration." Based on the absence of bacteria and parasites, and on the fact that he was able to transmit the disease with bacteria-free filtrates of infective organ homogenates, he considered a virus to be the cause.

As the 1950s closed, the disease was well recognized by fish health specialists as a presumed viral condition, but it was known by several names. Deufel (1959) used the term coined by Schäperclaus and reported confirmation of transmission by intraperitoneal injection of bacteria-free filtrates of infective organ homogenates. Tack (1959) also reported experimental transmission of "trout plague" or "Egtved disease" and noted the high susceptibility of the brook trout *Salvelinus fontinalis*. Reichenbach-Klinke (1959), in a review of the status of fish diseases in Bavaria, noted that the "new trout disease" also afflicted grayling and whitefish.

The early 1960s saw two reports on the histopathology of VHS: in Munich, Schmidt (1960) concentrated on the liver and compared normal tissue with that taken from clinically affected trout; and in Italy, Ghittino (1962) distinguished between the changes of infectious kidney swelling and liver degeneration, and those of lipoid liver degeneration (a then current nutritional problem among hatchery trout). Deufel (1960) elaborated on earlier work and by means of differential filtration determined the size of the presumed virus to be about 100 to 150 μm.

The most significant report of 1960–1964 was that of Jensen (1963) at Denmark's State Serum Laboratory. Using primary monolayer cultures of trout cells, he isolated a cytopathic agent from two outbreaks. The infectivity was filterable, serially transferable, and able to produce typical Egtved disease in experimentally inoculated rainbow trout. Moreover, the agent could be reisolated from test fish that died. The cause of VHS had been found. At the University of Bern, an attempt was made to visualize the virus, but the ensuing report of viruslike particles (Zwillenberg and Zwillenberg 1964) was at best inconclusive. Soon thereafter, electron microscopy clearly established Egtved virus to be a rhabdovirus, like vesicular stomatitis virus (Zwillenberg et al. 1965).

Also during the early 1960s, an international meeting of fish health specialistis was held in Torino, Italy, under the auspices of the Permanent Commission for the Study of Diseases of Fish of the Office of International Epizooties. A resolution adopted at the meeting called for the acceptance of one name, viral hemorrhagic septicemia, and the abandonment of the many other names of the disease (Altara 1963).

During a 1964 international conference on viral diseases of poikilothermic vertebrates, sponsored by the New York Academy of Sciences, six major or minor reports were presented that dealt with VHS. The collected papers were published as an edited annal (Whipple 1965).

At that conference, Jensen (1965) gave details of his isolation of Egtved virus—an isolant designated F1, the reference strain. Also, Rasmussen (1965) gave a comprehensive summary of his 10 years of research in Denmark, during which he routinely transmitted the disease to several species of salmonids and determined the role of temperature; he presented data that convincingly undermined the then widely held belief that the disease had a nutritional cause. Other papers reported VHS in various countries: France (Bellet 1965; Besse et al. 1965), Italy (Ghittino 1965), and Poland (Miaczynski 1965).

Research on VHS in Denmark increased with the efforts of Vestergard Jorgensen (1968), who reported on the preparation of rabbit antiserum and its use in neutralization tests and immunodiffusion reactions. As a result of Swiss-Danish collaboration, Zwillenberg et al. (1965) demonstrated the rhabdoviral form of Egtved virus, the first such fish pathogen. In another collaboration, Yasutake and Rasmussen (1968) described and illustrated histopathologic changes in fingerling trout that had been injected with culture-grown virus.

Other reports documented occurrences of the disease elsewhere. Brown trout with VHS were found in Italy (Ghittino 1968a); and the disease was reported by Tesarčík et al. (1968) in Czechoslovakia, by Zwillenberg et al. (1965) in Switzerland, and by Hastein et al. (1968) in Norway. Reviews of VHS appeared in English (Ghittino 1965), Italian (Ghittino 1967), and French (Ghittino 1968b). Brief reports by Pfitzner (1966) and Zwillenberg et al. (1968) dealt with the response of certain cyprinids to experimental infections.

During the early 1970s, the first reports appeared on the characterization or biophysical and biological properties of the virus. Notable information came in companion reports from France's Laboratory of Ichthyopathology (de Kinkelin 1972; de Kinkelin and Scherrer 1970). Inactivation of the virus by common chemical or physical factors was described by Frost and Wellhausen (1974) and by Vestergard Jorgensen (1973b). Clark and Soriano (1974) showed that temperature played a major role in the replication of Egtved virus by mammalian and reptilian cells.

In Denmark, research on immunology and serology of VHS continued. Neutralizing antibody was found in infected trout (Vestergard Jorgensen 1971) and serologic comparisons of 76 isolants from four European countries revealed a second serotype that was designated 61 by Vestergard Jorgensen (1972) but that later became strain II. Fluorescent antibody was used to demonstrate Egtved virus in infected trout tissues (Vestergard Jorgensen and Meyling 1972), and then to demonstrate specific antibody, as well as the virus (Vestergard Jorgensen 1974b). It was shown by de Kinkelin and Dorson (1973) that the virus stimulated production of interferon in trout and that the interferon exerted a protective effect before neutralizing antibody could be produced. The possible relatedness of Egtved and infectious hematopoietic necrosis virus (IHNV) was examined by McAllister et al. (1974), who found the two agents to be serologically distinct.

Yasutake (1970), in continuing histopathologic investigations, found distinctive differences among the three salmonid viruses then recognized: IHNV, infectious pancreatic necrosis virus (IPNV), and Egtved virus (VHSV).

Examination of different aspects of transmission led to the finding that the virus was shed by infected fish and that the disease could be spread by waterborne contact (Vestergard Jorgensen 1973a). All indications were that the virus was eluted from eggs during incubation and that vertical transmission did not occur (Vestergard Jorgensen 1970). In contrast to IPNV, Egtved virus lost its infectivity during passage through sea gulls (Eskildsen and Vestergard Jorgensen 1973).

Minor reports dealt with dual infections of cell cultures with Egtved virus and IPNV (Wolf and Vestergard Jorgensen 1970), and with the finding that VHS did not produce inclusion bodies distinctive enough for reliable diagnostics (Hofmann 1971). The coho salmon was found to be refractory to VHS (de Kinkelin et al. 1974)—a discovery that would later prove to be of value in hybridizing rainbow trout.

Continuing his run of reviews, Ghittino (1973) presented a panel review on VHS during a 1972 international symposium on the major communicable fish diseases in Europe, held in Amsterdam under the auspices of the European Inland Fisheries Advisory Commission of the Food and Agriculture Organization of the United Nations.

Compared with the early 1970s, publications on VHS increased only modestly during 1975–1979. Among the reports were the first on the molecular biology of the virus. Five structural proteins were identified (Hill et al. 1975; Lenoir and de Kinkelin 1975; McAllister and Wagner 1975). Two years later, the characteristics of the virus RNA were described (McAllister and Wagner 1977; Robin and Rodrigue 1977).

Matters dealing with serologic methods received attention: indirect fluorescent antibody techniques were touted as providing identification in only 2 days, versus the 7 or more days generally required for cultures to show CPE (Meier and Vestergard Jorgensen 1975; Pfeil and Wiedemann 1977); and the unusually effective role that complement played in antibody reactions of Egtved virus was described (Dorson and Torchy 1979; Dorson et al. 1979).

Following Vestergard Jorgensen's earlier discovery of a second strain of Egtved virus, de Kinkelin and Le Berre (1977) identified isolant 23/75 as distinctive; it later became recognized as strain III.

The virus was found in new locations, and the responses of several species of fish were tested. Pastoret et al. (1976) made two isolations in Belgium, and Meier and Vestergard Jorgensen (1980) described an acute outbreak of VHS among northern pike fry in a Swiss rearing facility. Golden trout (*Salmo aguabonita*) proved to be fully susceptible (Ahne et al. 1976), whereas hybrids of coho salmon and rainbow trout were only moderately affected (Ord et al. 1976). Chinook salmon fingerlings proved to be refractory even to virus given by injection (Ord 1975).

Vestergard Jorgensen (1976) described his efforts to develop a vaccine. Even after being passed some 90 times in cell culture, the virus retained a low level of residual pathogenicity, however, it provided partial protection in field testing. Other aspects of protection were reported by Chilmonczyk (1978), who found that chemically inactivated virus induced lymphostimulation in trout that survived VHS.

Dorson et al. (1975) induced interferon in vivo and demonstrated several of its physicochemical properties, and de Kinkelin et al. (1977) again reported that interferon was produced long before neutralizing antibody was demonstrable.

There were single reports on miscellaneous topics: the pathogenesis of experimental infections was described (Pfeil-Putzien and Hoffmann 1979); the length of time that virus could be recovered from refrigerated fish or from those that had been frozen was determined (Wiedemann 1977); histopathologic changes and hematology in experimentally infected trout were described (Hoffmann et al. 1979); the morphologic variability of the virus and the kinds of fragments produced during disintegration were illustrated (Olberding and Frost 1975); and major aspects of rhabdoviral infections of salmonids were discussed and standardized methods for effecting experimental infections were offered (de Kinkelin et al. 1979).

The first case of fish with VHS and concurrent IPN was reported by Schlotfeldt and Frost (1975).

The number of publications on VHS during the early 1980s was nearly double that in the late 1970s. However, not all reports of the 1980s conveyed new information; some were simply variations of presentations that had been published earlier.

Nearly one-fifth of the 1980–1984 publications on VHS were case-type reports of new or unusual outbreaks or of experimental infections in new or atypical hosts: the virus was isolated from trout in the Soviet Union, presumably in the Ukraine (Osadchaya et al. 1981); several natural or experimental infections in pike fry were described (Ahne 1980a; Meier 1981; Meier and Vestergard Jorgensen 1980; Wizigmann et al. 1980); a natural outbreak in grayling was reported by Wizigmann et al. (1980); an outbreak of high mortality in brown trout was reported by Bovo et al. (1982); the response of brown trout fry to three strains of the virus was reported (Vestergard Jorgensen 1980); and de Kinkelin and Castric (1982) found Atlantic salmon fry to be susceptible to the virus administered by injection but not by bath. Perhaps the most unusual newly identified hosts were sea bass (*Dicentrarchus labrax*) and turbot (*Scophthalmus maximus*), the young of which were susceptible to waterborne as well as to injected virus (Castric and de Kinkelin 1984).

Topics relating to immunization and protection accounted for some papers of the period. A general overview of theoretical and practical aspects of fish immunization by de Kinkelin et al. (1984) placed special emphasis on VHS. Two live virus preparations were tested: in Denmark, attenuated virus designated Reva showed promise in field trials (Vestergard Jorgensen 1981, 1982a); and in France, a comparable low-virulence preparation designated F25 (for its upper 25°C temperature tolerance) provided partial protection (Bernard et al. 1983; de Kinkelin and Bearzotti 1981). These Danish and French preparations both had low virulence. Philosophical discussions of VHS vaccination were published by Enzmann (1983) and Enzmann and Konrad (1984). Continuing the work with interferon, de Kinkelin et al. (1982) reported the passive transfer of protection with interferon that had been produced in vivo.

Under the broad heading of molecular biology, Bernard et al. (1984) identified membrane lipids of Egtved virus. In a comparison of the polypeptide synthesis of wild Egtved virus and the more resistant F25 strain, de Kinkelin et al. (1980) found that, among other differences, the F25 strain lacked the L and G proteins. Evidence for possible relatedness of an mRNA of spring viremia of carp virus (SVCV) and Egtved virus was presented by Enzmann et al. (1981) and again by Enzmann and Maier (1983).

Egtved virus was included in a comparison of four fish viruses for their stability when exposed to a variety of physical and chemical agents (Ahne 1982a, b; Ahne and Held 1980). Maisse et al. (1980), who compared the response of Egtved virus and IPNV to ultraviolet irradiation, found the rhabdovirus to be markedly the more sensitive to inactivation.

As expected, interest in serologic methods of detection and identification continued, and in 1981 Ahne reviewed techniques that were in current use in fish virology. Enzmann (1981), using the indirect fluorescent antibody technique, identified the virus in ascitic fluid and tissue smears from acute cases in only 2 to 3 hours—the proviso being that adequate levels of virus be present. Faisal and Ahne (1980), using indirect fluorescent antibody and immunoperoxidase techniques on inoculated cell cultures, obtained results within 8 hours postinoculation. The newest approach was made by Dixon and Hill (1984), who were unsuccessful in attempting to derive answers with ELISA. They attributed the failure to their inability to obtain antisera with high neutralizing activity.

Several papers appeared in which electron microscopy was used: the morphogenesis and ultrastructure of the virus in cell culture was described by Baroni et al. (1982), and the role of gill pillar cells in virus entry and replication was described and illustrated by Chilmonczyk (1980) and again by Chilmonczyk and Monge (1980).

During the period 1980–1984, outbreaks of VHS were documented among trout in brackish water and in seawater. Bovo and Giorgetti (1983) found the virus in an epizootic that occurred 5 months after trout were transferred from freshwater to brackish water. Horlyck et al. (1984) reported an outbreak in brackish water that took place some 18 months after trout were transferred from a farm registered free of VHS. The first case of VHS in full-strength seawater resulted in 80% mortality and began just 1 month after the fish were transferred from freshwater (Castric and de Kinkelin 1980).

Several publications of the period reported on some of the less frequently studied clinical features of VHS. Hoffmann (1980) quantified the sequence of hematologic changes seen in experimentally infected rainbow trout. Histologic findings and hematologic values of naturally infected trout were described and illustrated, and Egtved virus was visualized in the anterior kidney by Amlacher et al. (1980). Two years later, in a similar report, hematologic and histologic features were revealed by electron microscopy (Amlacher et al. 1982). Wizigmann and Hoffmann (1982) followed the sequence of histologic changes in experimentally infected trout and related the changes to the appearance of Egtved virus antigen, as determined by fluorescent antibody. Serum enzymes and isoenzymes of trout with VHS were found to differ in certain specific activities (Georgiev and Kamenov 1980).

Indications of the prevalence of VHS were evident in two reports: Wizigmann et al. (1983) summarized 5 years of diagnostic work in Bavaria, in which VHS was found in about 16% of 1413 cases and neutralizing antibody was present in the fish at 60% of the farms examined. In Switzerland, Meier (1984) reported an overall 1.5% incidence of VHS in 1100 lots of imported trout examined during a similar 5-year span.

Other miscellaneous topics on VHS were covered: Ahne (1980b) was able to superinfect CHSE-214 cells that were persistently infected with IPNV, but the yield of Egtved virus was only one-tenth that of normal cells. Vestergard Jorgensen (1982b) isolated Egtved virus from free-living rainbow trout—probably hatchery escapees—but only during the yearly low temperatures. Neukirch (1984) investigated the entry, multiplication, and spread of infection, but was able to demonstrate virus only in the gills immediately after waterborne exposure; his findings added to the considerable body of evidence that gills are a prime portal of entry. In an elegant presentation, Elger and Hentschel (1983) described and illustrated their study of a glomerulonephritis in fish with presumptively identified chronic VHS; however, attempts to isolate the virus were not successful.

A well-referenced review of VHS was given by de Kinkelin (1983) at an international symposium on antigens of fish pathogens at Talloires, France, in May 1982.

In the most recent period of record here, 1985, most reports addressed the susceptibility of various hosts to VHS. Meier (1985) demonstrated an age-related development of resistance in northern pike: fry were highly susceptible but 2-year-olds were refractory. Ahne (1985) infected northern pike fry and showed that, when these fry were fed to yearlings, the virus retained its infectivity and was capable of producing clinical signs of VHS. Ahne and Thomsen (1985) reported a natural outbreak of VHS among whitefish, and the successful induction of disease and mortality in whitefish fry. Work with hybrids of coho salmon and rainbow trout was reported by Dorson and Chevassus (1985), who found that triploid rainbow trout were susceptible to Egtved virus but that triploid hybrids were wholly refractory.

Neukirch (1985) republished much of the material on uptake and multiplication of VHS in trout that had been published earlier (Neukirch 1984).

Olesen and Vestergard Jorgensen (1986) developed a microculture plaque reduction assay in which complement was used. The system measured trout antibody titers that were as high as 1 : 10,240 within 6 weeks postinfection and remained at detectable levels as long as 1 year. The method was more sensitive than conventional neutralization tests.

Bernard et al. (1985) followed virus multiplication and production of interferon and neutralizing antibody in trout infected partly with virulent virus and partly with strains attenuated in cell culture passage. Virulent virus produced the higher yield of progeny virus, interferon, and neutralizing antibody, and hence provided the greater protection. The G protein induced neutralizing antibody, but differences in G protein between strains could not be correlated with virulence.

Signs and Pathologic Changes

Viral hemorrhagic septicemia commonly occurs as an uncomplicated health problem, but some outbreaks are compounded by bacterial infection. Moreover, a second viral infection, notably IPN, may also be present. Accordingly, the emphasis here is on confirmed natural and experimental VHS in typical hosts—fingerling to yearling rainbow trout—in which Egtved virus is the only pathogen known to be present.

Early descriptions of pathologic findings date back 30 years or more but, retrospectively, it is recognized that the viral cause of pre-1963 cases is subject to uncertainty. Accordingly, and with exceptions to be noted, the following account is based on personal observations and on reports such as those of Amlacher et al. (1980, 1982), Ghittino (1973), Hoffmann (1980), Hoffmann et al. (1979), Rasmussen (1965), Yasutake (1970, 1975), and Yasutake and Rasmussen (1968).

Behavior

About 30 years ago, the disease now known as VHS was described as having three forms: chronic-subacute, acute-hyperactive, and latent. In time, that designation gave way to stages of infection: acute, chronic, and nervous—terminology that held sway for traditional rather than virologic reasons. The value of clinical differentiation of the infection is not clear.

In typical outbreaks, afflicted individuals do not feed, and their behavior ranges from lethargic to hyperactive—within even a few minutes. Some victims that seem unable or unwilling to exert themselves move to pond or raceway edges, where currents are presumably weaker. Even along the shallower edges, the weaker fish slowly drift downstream, but feebly maintain normal upstream orientation and balanced posture until they lodge against the shore or outlet grid. However, obviously weak fish sometimes respond to stimulation and swim away. In earthen ponds, victims seek refuge in edge vegetation, but may exhibit near-agonal tail-chasing activity, followed by rapid respiration and obvious prostration.

External Signs

Victims are noticeably darker than normal, and may have exophthalmia in one or both eyes. Gills are pale, but may be lightly flecked with petechiae. Hemorrhage may occur in the orbits and fin bases.

Internal Signs

The possible range of internal signs is great. In worst-case victims, the internal organs and body cavity may show profuse bleeding, and changes in the liver and kidneys are obvious. In contrast, the more typical victims show less dramatic pathology. The gastrointestinal tract is devoid of food. A few petechiae may be scattered in the swim bladder and peritoneum, and occasionally in the visceral adipose tissue and musculature of the abdominal wall. The liver is usually pale and off-color or mottled with hyperemic areas. Kidneys are more intensely red than normal (noticeable swelling is the exception) and the spleen is usually enlarged and reddish.

Histopathologic Findings

As the name indicates, VHS is termed a hemorrhagic disease. However, actual hemorrhaging is usually limited; degenerative changes and necrosis are the more common signs, present in many tissues.

Kidneys are the principal target, and damage to hematopoietic tissue is more severe than that to excretory tissue (Fig. 1). Foci of macrophages present are attributable to hyperplasia. In more advanced disease, macromelanophore centers are destroyed and the pigments disseminated. Leukopenia is evident, as is vascular and sinusoidal congestion.

Focal necrosis and degeneration occur in the liver, another prime target (Fig. 2). Hepatic cells are vacuolated, pyknotic, and karyolytic, but distinctive or diagnostic inclusions are lacking. Hofmann (1971) stressed that occasional inclusions are not

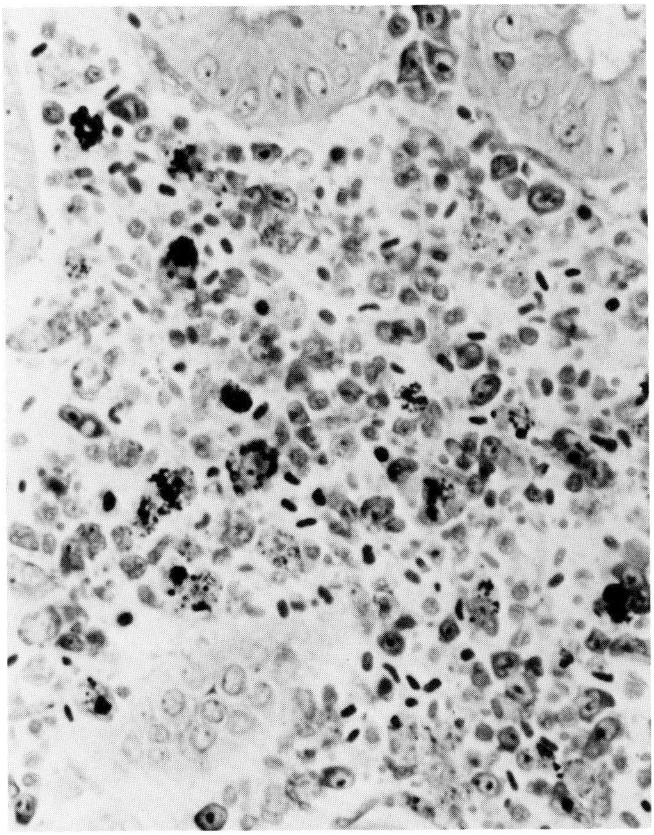

Figure 1. Histologic section through the kidney of a rainbow trout with experimental VHS. Hematopoietic tissue shows degeneration and necrosis. May-Grünwald-Giemsa. From Yasutake (1975). Reprinted with the permission of the University of Wisconsin Press.

readily distinguishable from normal nucleoli. Sinusoids are dilated and congested with blood cells.

Extravasation of blood can be found in the skeletal muscle of some victims, but muscle fibers and bundles are not damaged (Fig. 3). Changes in the hematopoietic tissue of the spleen are similar to those in the kidneys. Pancreatic tissues—in contrast to typical findings in IHN and IPN—show fewer and less destructive changes, and the intestine and gills appear normal.

Essentially the same signs and pathologic changes as those in rainbow trout were reported in five species: Atlantic salmon (de Kinkelin and Castric 1982), brown trout (Ghittino 1968b), golden trout (Ahne et al. 1976), and sea bass and turbot (Castric and de Kinkelin 1984).

Changes were more severe in young northern pike than in rainbow trout. Ahne (1980a) reported increased muscle and skin hemorrhages, and Meier and Vestergard Jorgensen (1980) described the hemorrhages as more "violent." Wizigmann et al. (1980) found damaged pancreatic islet tissue in pike. Kidney tubules showed evidence of degeneration, and their lumens contained hyaline amorphous masses. Similar renal changes and, additionally, vacuolated gut epithelium were found in whitefish.

Hematologic changes associated with VHS have been reported by Amlacher et al.

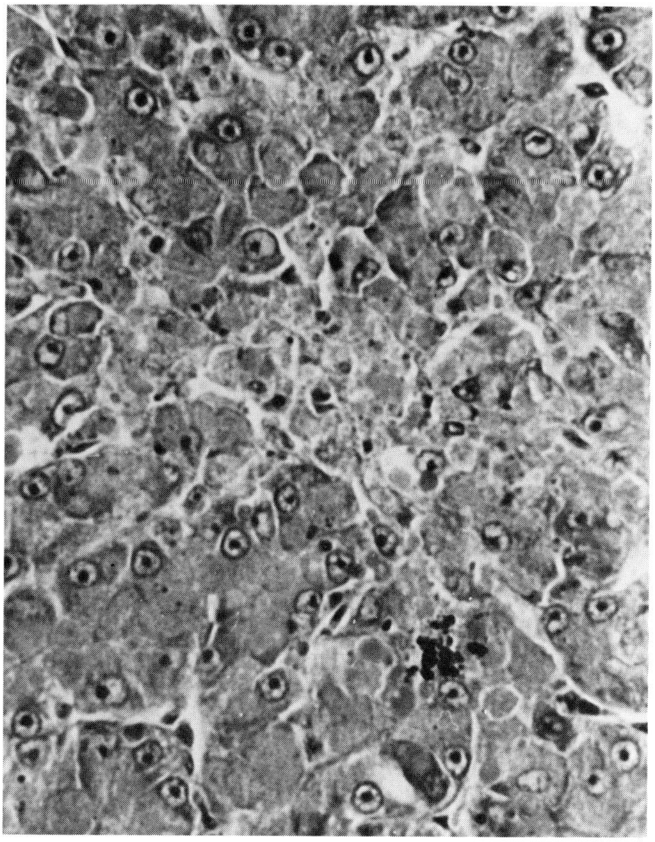

Figure 2. Histologic section through the liver of a rainbow trout with experimentally induced VHS. Focal degeneration and necrosis are evident. H & E. From Yasutake (1975). Reprinted with the permission of the University of Wisconsin Press.

(1980, 1982), Deufel and von Pollnitz (1977), Hoffman (1980), and Hoffman et al. (1979). Collectively, the findings reflect damage to hematopoietic tissue and attendant anemia, leukopenia, and thrombocytopenia; thus they are wholly in agreement with the histopathologic changes. Values for thrombocytes, red blood cells, hematocrits, and hemoglobin are depressed. There is an increase in damaged erythrocytes and granulocytes and a marked increase in immature erythrocytes, particularly late in the infection. Necrobiotic bodies—products of degenerating or necrotic cells—have been found in blood smears of VHS victims, as wall as in those of IHN victims (W. T. Yasutake, personal communication).

In studies on the hematology of rainbow trout with VHS, Hoffmann et al. (1979) transmitted the disease with strain F1 Egtved virus grown in culture and recorded observations on 16 occasions during the 3 months that followed. The disease resulted in a marked reduction of thrombocytes and erythrocytes, and consequently in a depression of hematocrit values. Leukocyte number showed a slight peak during the first 2 weeks, but then declined. Numbers of immature erythrocytes increased during the second month, presumably reflecting a compensatory response of the hematopoietic tissue.

Hoffmann (1980) described hematologic changes that complemented those re-

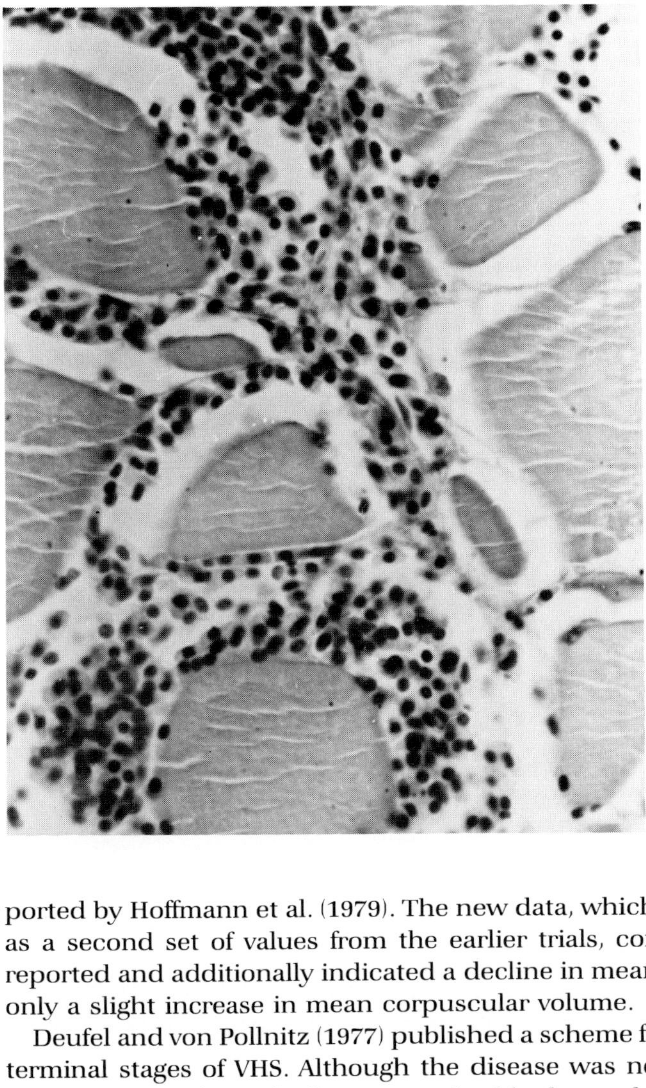

Figure 3. Histologic section of rainbow trout skeletal muscle showing extravasation of blood accompanying VHS. H & E. From Yasutake (1975). Reprinted with the permission of the University of Wisconsin Press.

ported by Hoffmann et al. (1979). The new data, which appeared to have been taken as a second set of values from the earlier trials, combined the values previously reported and additionally indicated a decline in mean corpuscular hemoglobin but only a slight increase in mean corpuscular volume.

Deufel and von Pollnitz (1977) published a scheme for diagnosing rainbow trout in terminal stages of VHS. Although the disease was not confirmed virologically, the general hematologic findings agreed with those obtained by workers who used experimental infections with VHS strain F1 virus. Deufel and von Pollnitz found a reduction in the numbers of erythrocytes, leukocytes, and thrombocytes and an increase in the numbers of erythroblasts and degenerating erythrocytes. Comparable findings were reported by Flondro et al. (1984), who examined two lots of rainbow trout that had clinical signs of VHS but in which the presence of the virus was not confirmed. In diseased fish, mean hematocrit values dropped from a normal value of about 43% to 12%, hemoglobin from about 10 g/dL to only 4 g/dL, and erythrocyte numbers from about 940,000/μL to about 250,000/μL.

Georgiev and Kamenov (1980) found that serum enzyme activities of yearling and 3-year-old rainbow trout said to have VHS differed markedly from those of healthy fish. Concentration of glutamic-oxaloacetic transaminase, glutamic-pyruvic transaminase, and lactic acid dehydrogenase were markedly higher in the diseased fish, particularly in the 3-year-olds. In contrast, the alkaline phosphatase levels decreased 42% in the diseased yearlings and 86% in the 3-year-olds.

Etiology

Although VHS had long been considered a viral disease, Jensen (1963) was the first to demonstrate its etiology. He isolated the agent on primary cultures of immature ovaries of rainbow trout, passaged it six times in vitro, filtered the material through a Seitz asbestos pad, inoculated rainbow trout, and produced mortality and signs of VHS. The apparent causal agent was reisolated from several of the dead fish. Rivers' postulates were thus fulfilled, and so acknowledged by all later investigators. He later provided more details of his studies (Jensen 1965). He used primary cultures and the RTG-2 cell line and showed that Egtved virus was sensitive to ether and to glycerol, and it was labile at room temperature.

Size and Shape

Egtved virus is a regular rhabdovirus (Fig. 4) whose diameter was given by Zwillenberg et al. (1965) as 60 nm in thin section and 70 nm in negatively stained preparations; in either kind of preparation, the overall length of the particles was 180 nm. Slightly larger values for particles in thin sections—75 × 240 nm—were given by de Kinkelin and Scherrer (1970). Cohen and Lenoir (1974), who compared four rhabdoviruses of fishes, gave the mean dimensions of Egtved virus as 60 × 177 nm. In addition, they reported that the particles were extremely fragile—a feature also noted by later investigators (Baroni et al. 1982; Hill et al. 1975; Olberding and Frost 1975)—and had only ill-defined surface spikes. Morphogenesis and ultrastructure of the virus in vitro were described and illustrated by Baroni et al. (1982).

Biophysical Properties

Major contributions to characterization of Egtved virus were made by Ahne (1982a, b) and de Kinkelin and Scherrer (1970).

The virus is ether, heat, and acid labile (at pH 3). It is stable at pH 5 to 10, and also stable through several freeze-thaw cycles. It is replicated by six commonly used fish cell lines—BF-2, CHSE-214, EPC, FHM, PG, and RTG-2—in which maximal titers of somewhat more than 10^7 PFU/mL are produced. Like IHNV and SVCV, Egtved virus has been grown in the mammalian cell lines BHK-21 and WI-38 and the reptilian cell line GL-1; however, titers were consistently lower than in fish cells (Clark and Soriano 1974).

In fish cell lines, the range of replication extends from 4°C to about 20°C, but yields are highest at 10 to 15°C (Fig. 5). The virus loses infectivity at temperatures above 20°C, but a thermoresistant vaccine strain that has been developed grows at 25°C.

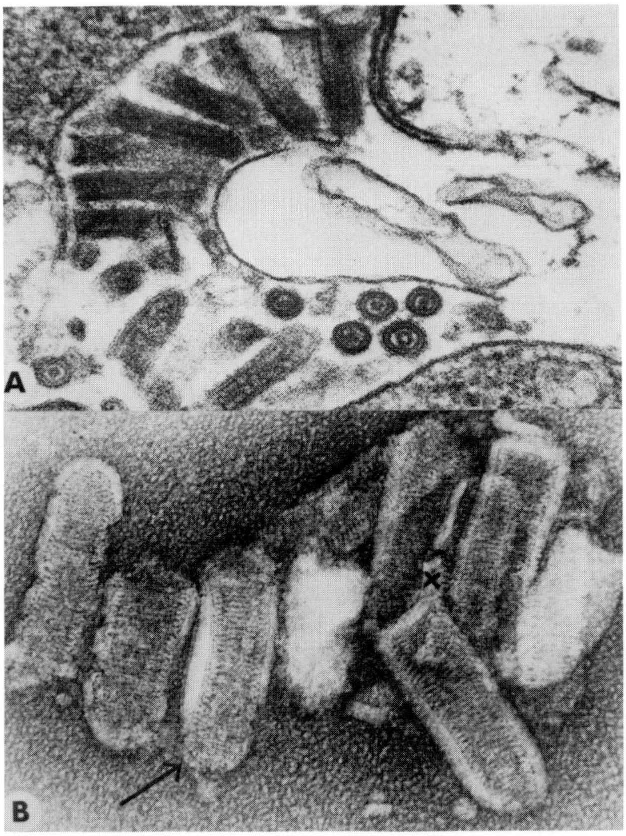

Figure 4. Egtved virus in RTG-2 cells. *A.* Extracellular virions in longitudinal section and cross section, in a thin section of an infected culture. *B.* Negatively stained virions showing typical rhabdovirus morphology and evidence of fragility (arrow). From Wolf (1966). Reprinted with the permission of Academic Press, Inc.

Comparative one-step growth curves in FHM and RTG-2 cells have been carried out at 14°C (Fig. 6). In the FHM cells, new virus appears at about 7 hours; growth is exponential for 5 to 6 hours and reaches a plateau at about 24 hours. Growth in RTG-2 cells is slower: new virus appears at 10 hours, develops exponentially during the next 10 to 12 hours, and reaches a plateau at about 48 hours. It is important for in vitro replication of Egtved virus that the pH of cultures be 7.6 to 7.8.

Like SVCV and PFR, Egtved virus is sensitive to deleterious chemicals and environmental and physical factors (Table 1). Because it is a pathogen of cold-water fishes, the virus is somewhat more sensitive than SVCV or PFR to high temperatures. Low temperatures thus favor the retention of infectivity, and the presence of protein (such as 2% or more of serum) is additionally protective. In cell culture medium containing serum, infectivity persists for many months at 4°C or lower. Freezing at −20°C or lower maintains infectivity for several years, and lyophilization can be expected to maintain the virus even longer.

No doubt because VHS results in great economic losses to commercial trout culturists in much of Europe, research on this disease has been directed largely toward practical biological aspects of the organism, rather than toward molecular biology. Consequently, less is known about the biophyscial properties of Egtved virus than about those of IHNV and SVCV.

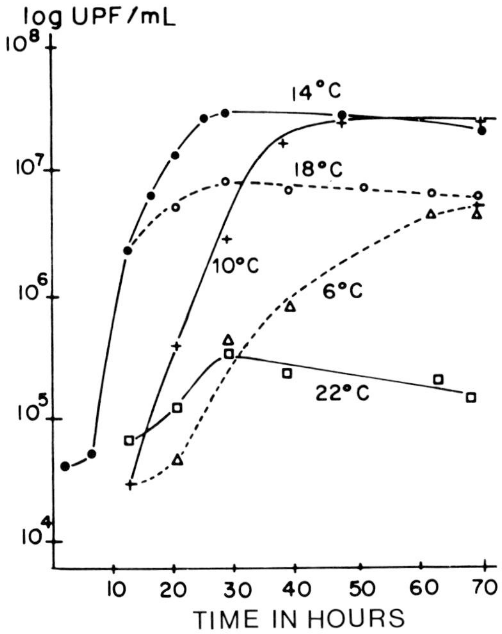

Figure 5. Growth rates of Egtved virus in FHM cells at five temperatures. The ultimate yields at 10°C and 14°C were the same, but peak yield was reached earlier at 14°C. From de Kineklin and Scherrer (1970). Reprinted with the permission of the Institute National de la Recherche Agronomique, Paris.

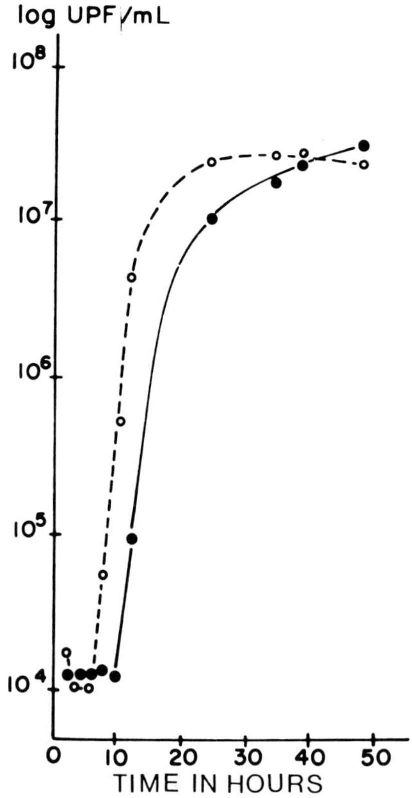

Figure 6. Comparative one-step growth curves of Egtved virus at 14°C in FHM cells (broken line) and RTG-2 cells (solid line). From de Kinkelin and Scherrer (1970). Reprinted with the permission of the Institute National de la Recherche Agronomique, Paris.

Table 1. Effects of selected chemicals and environmental and physical factors on infectivity of Egtved virus

Agent or factor	Effect	
	Inactivation (%)	Period
Formalin, 3%	100	5 min
NaOH, 2%	100	5 min
Chlorine, 500 ppm	99	<2 min
Actomar (I_2), 0.01%	100	5 min
Tap water at 10°C	90	14 days
Stream water at 10°C	90	14 days
Mud at 4°C, pH 7.4	>99	10 days
Drying at 15°C	>99	14 days
Gamma irradiation 1.7×10^5 rads[a]	100	—
UV irradiation at 254 nm at 5 cm	100	10 min

Source: Adapted from Ahne (1982a, b); Ahne and Held (1980).
[a]In tap water.

The RNA has a buoyant density in cesium sulphate of 1.69 g/mL, and the base composition is 29.3% cytosine, 23.6% adenine, 14.5% uridine, and 32.6% guanine—values expected for single-stranded RNA (Robin and Rodrigue 1977). In a sucrose gradient of 5 to 25% the RNA sediments at 38 to 40S—as would be expected for a single-stranded molecule (Hill et al. 1975).

The structural proteins have been resolved and identified, and molecular weights determined (Table 2). The consensus of investigators is that Egtved virus is similar to rabies virus (de Kinkelin et al. 1980; Hill et al. 1975; Lenoir and de Kinkelin 1975; McAllister and Wagner 1975). Egtved virus lacks a nonstructural (NS) protein, but McAllister and Wagner (1977) found that it contains an RNA-dependent RNA polymerase and that this enzyme has an optimal temperature for in vitro transcription of about 15°C—close to the optimum for in vitro replication of the virus.

Diagnosis

A clinical, albeit presumptive, diagnosis of VHS is indicated if mortality breaks out among rainbow trout—particularly fingerlings or yearlings—or other young sus-

Table 2. Molecular weights ($\times 10^3$) of structural proteins of Egtved virus

Polymerase L	Glycoprotein G	Nucleocapsid N	Matrix		Reference
			M_1	M_2	
190	80	38	25	19	Lenoir and de Kinkelin (1975)
>150	90–86	62	38	33	Hill et al. (1975)
157	74	41	22	19	McAllister and Wagner (1975)
190	80	38	25	19	de Kinkelin et al. (1980)

ceptible species during or after seasonal low temperatures. Affected fish should show most of the major signs and behavior changes described.

Epizootiologic features also should be consistent: e.g., VHS is now known only in Europe and in facilities that have a history of the disease or are exposed to potential or known sources of the virus.

Diagnosis of VHS is reinforced by finding the histopathologic changes previously described. The possibility of IHN is readily excluded, because histologic changes in the stratum compactum do not occur in VHS, and because IHN has not yet been reported from Europe.

Confirmation of a diagnosis of VHS requires that the virus be identified serologically, following isolation or demonstration of specific antigen in sections or smears from target tissues. Diagnostic material from suspected cases in non-European locations should be promptly referred to a qualified reference laboratory.

Isolation

Egtved virus can be readily isolated during epizootics and for a short time afterward; thereafter, isolation from survivors is problematic. Unlike IPNV, Egtved virus is not shed in feces; accordingly, the use of feces (a favored nondestructive approach) is fruitless. Whereas IPNV is shed from carriers throughout the year, Egtved virus (like IHNV) is silent or latent much of the year. Both of these salmonid rhabdoviruses are shed with eggs. Vestergard Jorgensen (1982b) isolated Egtved virus from organ homogenates of feral rainbow trout, but only during winter when water temperature was low; attempts at isolation during spring and autumn were unproductive. Low temperature is apparently required for the virus to break latency. Although the virus has been isolated from urine, isolation from blood or peritoneal washings has not been attempted.

Kidneys and spleen are the organs recommended for isolating Egtved virus from epizootic specimens. Brain tissue is suggested if survivors are to be assayed. These recommendations are based on the findings of early investigators of VHS, on histopathologic evidence, and on quantification of virus in tissues of victim fish (Table 3).

Table 3. Infectivity of different organs of rainbow trout that died during an outbreak of viral hemorrhagic septicemia

Organ	Range of titers[a]
Kidneys	5.5–6.5
Spleen	4.5–6.8
Stomach	3.2–4.2
Muscle	2.5–4.5
Liver	2.2–3.2
Hindgut	1.8–4.2
Foregut	–2.2

Note: Determinations made after 8 months storage at −20°C; specimens provided by C. J. Rasmussen.

[a]Infectivity expressed as exponent of $\log_{10}$ $TCID_{50}$/g.

Suspect organs are pooled, homogenized, diluted 1:10 or more with physiologic saline or balanced salt solution, and decontaminated by membrane filtration or with combined centrifugation and treatment with antibiotics—preferably bacteriocidal. Inocula can be added to cultures in medium, but sensitivity is greater if the inocula are adsorbed onto drained cell sheets. Adsorption should be at 12 to 15°C and for 30 to 60 minutes. The presence of 50 μg/mL DEAE dextran during adsorption increases sensitivity two- to threefold (Campbell and Wolf 1969).

The origins of cells that are susceptible to Egtved virus in vitro include fish, reptiles, and mammals, but fish cells are preferred because of their tolerance of low temperatures and consistently higher virus yield. Primary cultures of cyprinid or salmonid origin have been used, but most workers prefer established cell lines: BF-2, CHSE-214, EPC, FHM, PG, RTG-2, or STE-137. The RTG-2 line offers the advantage of yielding plaques of four salmonid viruses that are morphologically distinctive (Wolf and Quimby 1973). If this line is used for isolation, plaque morphology may reveal the presence of a second virus (Fig. 7). At the time of inoculation, cultures should be actively dividing, about 80 to 90% confluent. Moreover, the pH should be 7.6 to 7.8 and the temperature of incubation 15 ± 3°C (Jensen 1965; Vestergard Jorgensen 1974a). Incubation should be continued until CPE appears or, if it does not, for about 12 days (Vestergard Jorgensen 1968). Blind passage follows.

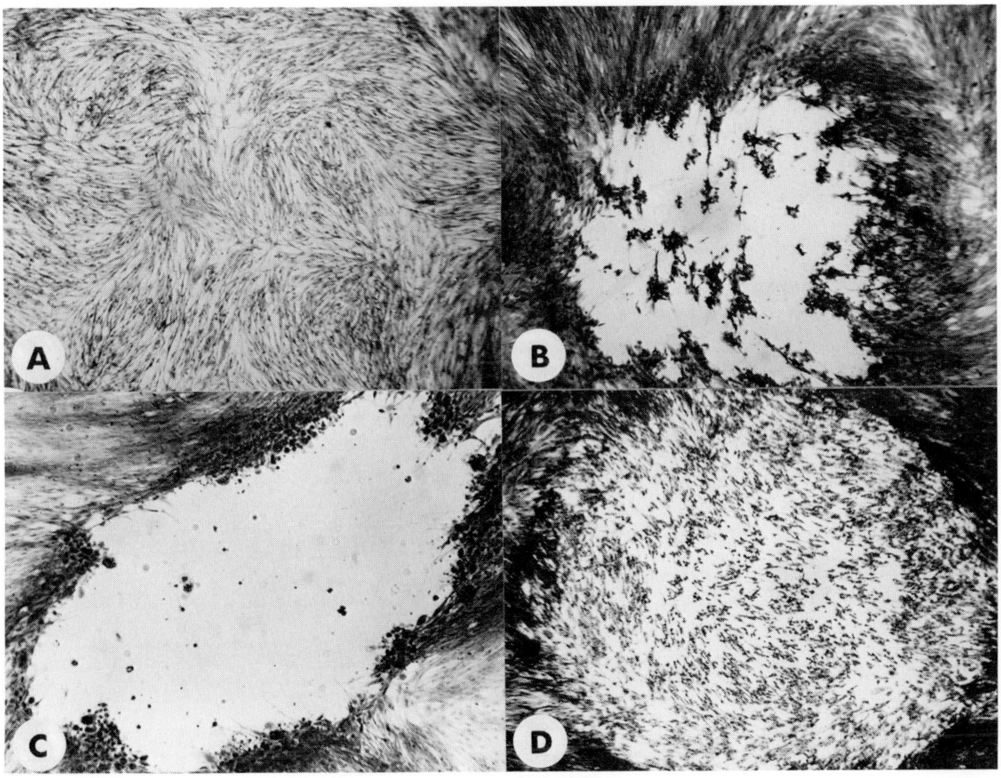

Figure 7. Plaque characteristics of four viruses of salmonids growing on RTG-2 cells. *A.* Infectious pancreatic necrosis virus. *B.* Infectious hematopoietic necrosis virus. *C. Herpesvirus salmonis. D.* Egtved virus.

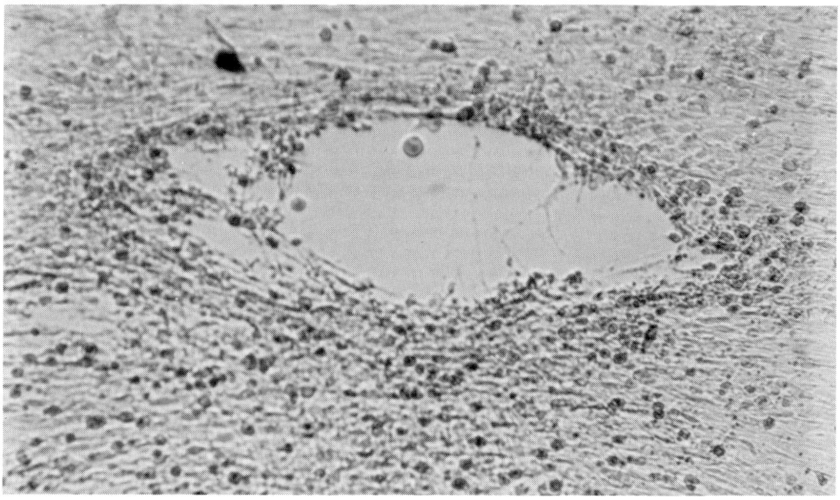

Figure 8. Focus of Egtved virus caused CPE in living, unstained RTG-2 cells. Rounding of infected cells is typical.

The CPE of Egtved virus first appears in the form of focal areas of rounded cells (Fig. 8). The foci then enlarge and lysis ensues, but affected cells do not show the margination of chromatin typically shown by IHNV.

Identification

Egtved virus is a serologically distinct pathogen; it is not related to the other prominent rhabdoviruses of fishes—IHNV, SVCV, PFR, and those found in eels. However, Egtved virus has yet to be compared with the rhabdoviruses from halibut and the Eurasian perch (*Perca fluviatilis*). On the other hand, the rhabdovirus from cod with the ulcus syndrome has recently been serologically identified as an isolant of strain F1 Egtved virus (P. E. Vestergard Jorgensen, personal communication, 1986).

The three strains or subtypes of the virus that are recognized react strongly with homologous antiserum, but cross-react only weakly in neutralization assays with antisera against heterologous strains. Strain I is the original or F1 isolant of Jensen (1963). Strain II is represented by isolant 61 and three others reported by Vestergard Jorgensen (1972). Strain III is an isolant from brown trout reported by de Kinkelin and Le Berre (1977) and originally designated 23/75. Inasmuch as several strains of the virus are known, a polyvalent antiserum encompassing all three strains should be used in serologic identification and detection procedures.

The most commonly used procedures for identifying Egtved virus are neutralization tests, fluorescent antibody techniques, immunoperoxidase staining, and plaque reduction. Dorson et al. (1979) designed a complement fixation test in which virus is used for quantification of specific anti–Egtved virus antibody. The test has not become popular, but the discovery that normal trout serum can be used as a source of complement has proven advantageous. The use of complement has been found to effect a 10-fold increase in sensitivity of Egtved virus neutralization by trout antiserum (Dorson and Torchy 1979).

A satisfactory ELISA has yet to be developed for Egtved virus. Dixon and Hill (1984) were successful with eel virus American, eel virus European unknown, and IHNV, but not with Egtved virus and SVCV. They attributed the problem to inherently low-titered antiserum—1 : 2000 for Egtved virus—and postulated that an Egtved virus antiserum would have to have a 50% plaque neutralization titer of 1 : 50,000 or greater to be suitable for ELISA.

Serologic methods used in examining the virology of eight pathogens of fish in husbandry were reviewed by Ahne (1981).

Vestergard Jorgensen (1968) was the first to undertake preparation of Egtved virus antiserum and to use it in neutralization and immunodiffusion tests. Both tests provided identification, but probably because gel diffusion lacks sensitivity, no other uses of the immunodiffusion test have been reported. On the other hand, although neutralization is considered to be the most sensitive and most specific test, it is also the slowest, sometimes requiring 2 or more days to produce results. Olesen and Vestergard Jorgensen (1986), who described their use of complement in a 50% plaque neutralization test in microculture plates, quantified record fish serum titers as high as 1 : 10,240; however, 5 days of incubation time were required.

Fluorescent antibody techniques have been widely used for identification and detection of Egtved virus. In the first of such applications, Vestergard Jorgensen and Meyling (1972) used the direct method to demonstrate the virus in frozen tissue sections and in infected cell cultures. Cell cultures showed specific fluorescence 8 hours postinfection, whereas CPE first became evident after 16 to 24 hours. The time advantage was significant. In addition, it was discovered that Egtved strains I and II were both reactive in direct tests with antiserum prepared against strain I, the F1 isolant.

Vestergard Jorgensen (1974b) used an indirect fluorescent antibody technique to demonstrate Egtved virus and its corresponding antibody. The advantage of having one conjugate, an anti–rabbit IgG, function in several pathogen-antibody reactions was evident, but elapsed time for development of fluorescence was not mentioned. Meier and Vestergard Jorgensen (1975) also used an indirect technique on infected cell cultures. First results appeared at 24 hours and fluorescence peaked at 48 hours; results were thus obtained sooner than by plaque assay but sensitivity was slightly reduced.

Pfeil and Wiedemann (1977), who used experimentally infected trout in a comparison of indirect fluorescent antibody technique with virus isolation, found that the rates were 85% for virus isolation and 62% with the fluorescence technique; however, elapsed time was 7 days for isolation and only 2 days for fluorescent antibody.

Fluorescence techniques continue in use. For example, in a 5-year survey of several fish viruses, Wizigmann et al. (1983) used an indirect technique on cell cultures and to demonstrate specific Egtved virus antibody, and the direct method on frozen tissue sections.

The most rapid fluorescent antibody results were reported by Enzmann (1981), who used the indirect procedure on smears of infected tissues, on isolated white blood cells, and on ascitic fluid. He claimed results in only 2 to 3 hours—providing that the solid-phase material had "about 10^5–10^6 units measured by infectivity." The approach certainly seems appropriate for epizootic material.

The immunoperoxidase method of demonstrating viral antigen is underexploited; the sole report of application with Egtved virus was that by Faisal and Ahne (1980), who found that results from this method and the fluorescent antibody technique were both obtained within about 8 hours—much sooner than CPE, which required about 24 hours. A decided advantage of immunoperoxidase was that results could be read with conventional light microscopy. Although untried as yet, the logical application of immunoperoxidase is in the demonstration of Egtved virus in imprints or smears of infected kidneys and spleens, in the buffy coat of suspect specimen fish, or in ascitic fluids.

Ahne (1981) tabulated data pertaining to several methods of preparing sera immune to Egtved and other fish viruses. Based on maximum 50% plaque reduction titer (1 : 32,000), the most potent antiserum was produced in rabbits by de Kinkelin. In a personal communication to Ahne, he reported using three intramuscular and intravenous inoculations of pelleted virus during a 1-month period of immunization. No details are given about virus titer or the use of adjuvant.

Transmission and Incubation

Multiple factors are involved in the transmission of the virus and the production of disease, but temperature is of paramount importance. Transmission readily occurs in the range of 1°C to about 12°C, but temperatures much above 15°C are nonpermissive. Age is another factor in rainbow trout and other highly susceptible species: afflictions are more severe and mortalities higher in the young than in older fish. Furthermore, strains or isolants of wild virus differ in virulence, and the route of infection also has an effect. As an example, the oral route has not been infective for salmonids—probably because of the acid lability of the virus (de Kinkelin et al. 1979). Nevertheless, Ahne (1980a) successfully infected young northern pike by feeding them infected trout. It should be noted, however, that the oral route does not exclude the considerable possibility of virus transmission by way of the exposed gills.

Natural infections occur by horizontal transmission of waterborne virus, and it is generally conceded that vertical transmission is extremely rare or does not occur. In any event, vertical transmission has yet to be documented. Vestergard Jorgensen (1970) reported isolating virus from freshly taken trout eggs; however, after the eggs had been incubated in flowing water for several hours, virus could no longer be demonstrated. Culture-grown virus added to eggs could no longer be demonstrated after 10 days of incubation. Under husbandry, virus may be shed with eggs, but it is soon eluted by water flow during incubation. By all accounts, egg-borne infectivity has never been implicated in outbreaks of VHS.

Vectors of VHS are neither known nor even suspect. It is widely recognized that gulls and other fish-eating birds can carry VHS victims from one facility to another. Unlike IPNV, however, Egtved virus is inactivated during passage through the gut of a gull as a result of being confronted with the high acidity of the anterior digestive tract and the 40°C avian body temperature.

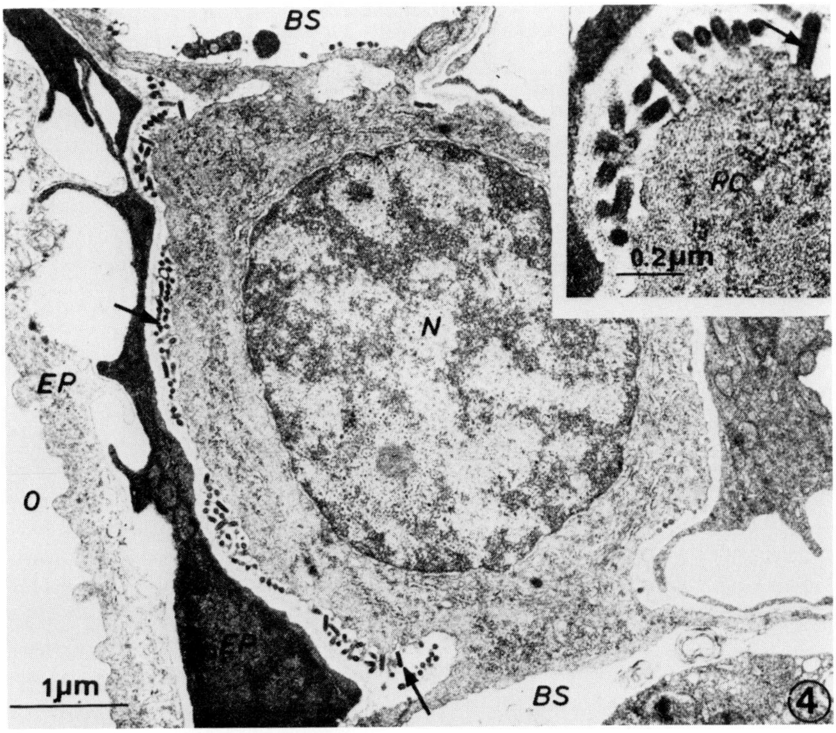

Figure 9. Gill pillar cell infected with Egtved virus. The insert shows higher magnification of virus budding. BS, blood space; EP, epithelial cell; N, nucleus; O, outside of fish. From Chilmonczyk and Monge (1980). Reprinted with the permission of Alan R. Liss Inc., New York.

It has been amply demonstrated that waterborne virus can be carried downstream to infect susceptibles; however, the identity of the organ or tissue source (except for urine) is not clear. Virus has long been seen in kidney tubules and assumed to be shed in urine, but Neukirch (1985) and Neukirch and Glass (1984) provided the first experimental proof. They found virus in catheterized urine specimens from experimentally infected trout, beginning at 3 days postinfection and continuing for as long as 30 days.

Inasmuch as partial cannibalization of dead or dying fish—enucleation of eyes, breaching of the abdominal wall, or the consumption of soft tissues—is common, it could conceivably be a source of virus. Virus has also been visualized in pillar cells of the gills (Fig. 9) and might be released during the typical rapid postmortem breakdown and decomposition of gills. Surprisingly, Egtved virus has not been reported in the seminal fluid or feces. It has been sought but not found in the feces of trout with VHS (Neukirch 1984, 1985; Neukirch and Glass 1984). In general, information on the subject of Egtved virus shedding is scarce.

Incubation time depends on temperature, dose, and route. Early workers transmitted the disease by injecting extracts of infected tissues (Deufel 1960; Schäperclaus 1954), but it remained for Rasmussen (1965) to show a relation between incubation time and route of delivery. He found that the incubation time was about 2 weeks after intraperitoneal injection (temperature at 2 to 11°C), but 1 to 2 weeks longer when contact transmission (the least rigorous method of infection) was used. At the low temperature of 1°C, Yasutake and Rasmussen (1968) reported an incubation time of 3 to 4 weeks following intraperitoneal injection of culture-grown virus.

Vestergard Jorgensen (1973a), who held fish at 9°C, noted that incubation time was only 5 days after virus was injected, and 8 to 11 days when fish were immersed in virus. Ahne et al. (1976) reported a slightly longer incubation time of 1 to 2 weeks at 8 to 10°C after virus injection.

Standardized methods of infecting young trout with Egtved virus that were advocated by de Kinkelin et al. (1979) provide predictable incubation times and results. Virus-free trout should weigh 0.4 to 5.0 g. In numbers of 50 to 100, depending on weight, they are to be held in 6-L containers supplied with three changes per hour of aerated water (9 to 12°C, pH 7.2, total hardness about 250 ppm calcium carbonate). Infection by immersion requires that the fish be held in 5×10^4 PFU/mL virus in 3 L of water for 3 hours; the flow of water is then to be restored. Alternatively, comparable results are obtained by an intramuscular injection of 5 to 10 PFU per fish. With both procedures mortality begins on day 5. The same general approach is used in a method that most closely matches natural infection, but infection is achieved by holding one VHS carrier with every 10 healthy trout. This procedure gives a 7-day incubation time to first mortality.

Ghittino (1965) was able to transmit Egtved virus simply by brushing the gills with a homogenate of infected tissues. Later work showed that virus could be recovered from gill tissues immediately after exposure of the fish to waterborne infectivity (Neukirch 1984, 1985). Accordingly, gills are considered to be a prime portal of entry—but not necessarily the only one—through which Egtved virus is transmitted.

Source

During an outbreak, victim fish shed virus and serve as sources of infection that help sustain an epizootic. Thereafter, some survivors become carriers and shed virus under conditions that are not yet completely known, but which almost certainly include low temperature. It is a reasonable supposition that low temperature is required to ensure continuity of the virus from one generation of fish to another.

Historically, Egtved virus was probably first found in one or more of Europe's salmonids—Atlantic salmon, Danube salmon, or brown trout. Among these, the brown trout is usually considered to be the prime or initial source species, because of its widespread original habitat, abundance, and comparatively high resistance to VHS. Although less common, the Atlantic salmon, Danube salmon, graylings, and whitefishes must remain suspect as the possible original source or sources.

Host and Geographic Range

Viral hemorrhagic septicemia is a problem disease and cause of mortality among certain species of fish in husbandry—notably the rainbow trout cultured in fresh-

Table 4. Response of fish species tested for susceptibility to Egtved virus

Response and species	Reference
Susceptible	
Atlantic salmon (*Salmo salar*)	Rasmussen (1965)
Brook trout (*Salvelinus fontinalis*)	Rasmussen (1965)
Brown trout (Salmo trutta)	Rasmussen (1965)
Golden trout (*Salmo aguabonita*)	Ahne et al. (1976)
Grayling (*Thymallus thymallus*)	Wizigmann et al. (1980)
Lake trout (*Salvelinus namaycush*)	Ghittino (1973)
Pike (*Esox lucius*)	Meier and Vestergard Jorgensen (1980)
Rainbow trout (*Salmo gairdneri*)	Jensen (1963)
Sea bass (*Dicentrarchus labrax*)	Castric and de Kinkelin (1984)
Turbot (*Scophthalmus maximus*)	Castric and de Kinkelin (1984)
Whitefish (*Coregonus* sp.)	Ahne and Thomsen (1985)
Hybrid (rainbow trout × coho salmon) (*Salmo gairdneri* × *Oncorhynchus kisutch*)	Ord et al. (1976)
Refractory	
Common carp (*Cyprinus carpio*)	Zwillenberg et al. (1968)
Chinook salmon (*Oncorhynchus tshawytscha*)	Ord et al. (1976)
Chub (*Leuciscus* [*Squalius*] *cephalus*)	Tack (1959)
Coho salmon (*Oncorhynchus kisutch*)	de Kinkelin et al. (1974)
Goldfish (*Carassius auratus*)	Pfitzner (1966)
Eurasian perch (*Perca fluviatilis*)	Zwillinberg et al. (1968)
Roach (redeye minnow) (*Leuciscus* [*Rutilus*] *rutilus*)	Tack (1959)
Tench (*Tinca vulgaris*)	Zwillenberg et al. (1968)
Hybrid (triploid rainbow trout × coho salmon) (*S. gairdneri* × *Oncorhynchus kisutch*)	Dorson and Chevassus (1985)

water at low temperature. Egtved virus can also result in mortality, of at least the young, in Atlantic salmon and several species of trout (Salmonidae) and the closely related graylings (Thymallidae) and whitefishes (Coregonidae). The virus has also been found to cause mortality in pike (Esocidae); furthermore, under experimental seawater conditions, the fry of turbot (Bothidae) and sea bass (Percichthyidae) proved to be susceptible, both to injection and to waterborne infectivity (Table 4).

Species reported to be refractory include chinook salmon and coho salmon, both of which were challenged by intraperitoneal injection of virus. Triploid hybrids of coho salmon × rainbow trout were also refractory, but they were challenged by the less severe waterborne method (Table 4).

In somewhat older literature, the Eurasian perch and several species of cyprinids were reported to be refractory (Table 4), but the response of those species merits more critical evaluation.

Standard methods of challenge should include representatives of all three known strains of Egtved virus, because they are known to differ in virulence. Several age groups of each fish species should also be tested, but especially fry and fingerlings because the young are the more vulnerable among susceptible species. Virus should be administered by intraperitoneal injection, as well as by the waterborne or immersion method. Levels of virus should be comparable to those that produce mortality in rainbow trout fingerlings, and fish should be held at a temperature that favors virulence of the agent (8 ± 3°C).

Viral hemorrhagic septicemia is only known to occur in Europe. The disease has

Table 5. Countries in which viral hemorrhagic septicemia has been reported

Country	Reference
Belgium	Pastoret et al. (1976)
Bulgaria	Georgiev and Kamenov (1980)
Czechoslovakia	Tesarčík et al. (1968)
Denmark	Jensen (1963)
Federal Republic of Germany	Schlotfeldt and Frost (1975)
France	Besse et al. (1965)
German Democratic Republic	Amlacher et al. (1982)
Italy	Ghittino (1965)
Norway	Hastein et al. (1968)
Poland	Miaczynski (1965)
Sweden	Vestergard Jorgensen (1972)
Switzerland	Zwillenberg et al. (1965)
USSR	Osadchaya et al. (1981)

not been found in other parts of the world where significant husbandry of trout is practiced—Australia, Japan, New Zealand, North America, or South America.

Within Europe, VHS has been reported from some countries in the west and central areas, and has been identified in the western portion of the Soviet Union (Table 5). One cannot assume, however, that VHS does not occur in countries where it has not yet been reported. Although Vestergard Jorgensen (1972) reported that fish with VHS had been found in Sweden, de Kinkelin (1983) noted that VHS had not been found in either Sweden or Finland. Insofar as can be determined from the literature, the British Isles have thus far managed to avoid the problem.

Immunity

As awareness of VHS increased, commercial trout producers in Europe recognized that survivors of the disease were resistant to reinfection. The logical conclusion was that factors of the immune response were involved; in time, research was conducted and confirming evidence was soon produced.

Work on the immune response to VHS began in Denmark; surprisingly, however, survivors seemingly lacked neutralizing antibody. Moreover, even after trout received monthly injections of attenuated virus for a year, titers of neutralizing antibody were low (Vestergard Jorgensen 1971). Work in France showed that trout produced interferon in response to Egtved virus and that production peaked at 3 days postinfection, presumably at the time it could be present to function in resistance (de Kinkelin and Dorson 1973). Danish work showed still another factor: in trout, antibody reactive in the fluorescent antibody technique was produced 5 months earlier than neutralizing antibody (Vestergard Jorgensen 1974b). Laboratory trials later showed that immersion of trout in attenuated virus resulted in an appreciable level of protection against waterborne virulent challenge (Vestergard Jorgensen 1976). Prospects for a practical vaccine seemed realistic.

As a test for the immune state, Chilmonczyk (1978) showed that inactivated Egtved virus strain I caused lymphostimulation in survivors of VHS outbreaks and in about half of the trout vaccinated by immersion. It could not be evoked by strain III virus, however, nor was it present in controls. Recognizing a need for standardized VHS infections, de Kinkelin et al. (1979) developed three procedures for inducing the disease experimentally with predictable results (see preceding section on transmission and incubation). More important, their method of ensuring contact between healthy trout and a standard percentage of carriers (i.e., confining 1 carrier trout with 10 healthy trout) most closely matches natural transmission. The experimental procedures were helpful in later studies of vaccine development.

Strain I virus in high passage obtained from Denmark was adapted in stepwise passage in EPC cells to grow at 25°C (de Kinkelin et al. 1980). Immunization trials with the temperature-tolerant F25 virus were carried out at 10°C. Trout fry were immersed in 5×10^5 PFU/mL virus for 30 minutes, then challenged at 20 days. The vaccine strain was not completely avirulent; mortality during immunization ranged from 2 to 15%, but virulence did not increase during in vivo passage. The vaccine strain provided measurable protection: survival was at least 30% greater in immunized trout than in nonimmunized controls. In addition, protection was provided against all three strains of Egtved virus and it lasted more than 100 days (de Kinkelin and Bearzotti 1981).

Tests with high passages of the Danish vaccine (designated Reva) showed that it retained a low level of virulence, but nevertheless still provided protection under laboratory conditions and, to some extent, under field conditions (Vestergard Jorgensen 1981, 1982a). Virulence of the high-passage Reva strain was stable: when tested after 20 passages in trout fry, mortality for 3-g trout was still about 16% (the pre–in vivo passage level) and only 5% for 100-g trout.

For immunization with the Reva vaccine, fish were immersed for 60 minutes in static oxygenated water held at 10°C and containing about 2×10^4 PFU/mL of the virus. Portions were held at various temperatures and challenged with virulent virus 9 weeks later. Vaccine virus persisted in fish for 98 days at 5°C, but for only 8 days at 15°C. Challenge by immersion in 5×10^4 PFU/mL of virulent virus killed 100% of the nonimmunized control fish, 80% of the immunized fish kept at 20°C, 60% of those held at 15°C and at 10°C, but only 25% of those held at 5°C. Field results were reported as being similar to the laboratory results in fish held at 5°C (Vestergard Jorgensen 1982a).

Continuation of the vaccine work in France showed that the temperature-tolerant F25 virus was, in comparison with wild virus, deficient in glycosylated polypeptide but that its sensitivity to host defense was increased. Immunization with F25 showed protection as early as 24 hours postinoculation, and that effect was postulated to be the result either of interferon or of saturation of primary target cells that effectively prevented superinfection. The F25 vaccine stimulated synthesis of neutralizing antibody that eventually provided late protection (Bernard et al. 1983).

Various aspects of immunity were included by de Kinkelin (1983) in his outline-type review of VHS, but coverage was more detailed in a review of immunization against IHN, IPN, and VHS presented at a Paris Symposium on Fish Vaccination (de Kinkelin et al. 1984). Overall, prospects for a practical vaccine are promising, although an opposing view has been expressed by Enzmann (1983) and Enzmann and Konrad (1984).

In principle, the vaccine most likely to be cost effective—an aspect that has yet to be addressed—would be an attenuated live virus product intended for immersion or spray delivery to fry or small fingerlings. Virulence of existing products should be further reduced and ways found to increase protection at temperatures more favorable for trout growth. Existing products seem not inclined to revert to greater virulence but, ideally, vaccine strains ought to be readily identifiable as such.

The ethics of application require that no live virus vaccines be applied in watersheds where wild-type virus is not known to be present.

Control

Control of VHS is best effected by avoidance. Ideally, that requires a water supply that is free of the virus and can be kept free of it. Secondarily, the stock to be propagated—eggs or young fish—should come from a certified virus-free source. Since the disease is not transmitted vertically, eggs can be taken with disciplined hygienic precautions from carriers and carefully disinfected in 100 ppm iodine at about pH 6.5 for 10 minutes, with reasonable confidence that hatching can occur and the young grow without developing VHS.

General considerations for control of VHS and two other fish virus diseases of problem proportions in Europe have been presented through the Office of International Epizooties of the Permanent Commission for the Study of Fish Diseases. The several measures include diagnostic methods, control methods, certification procedures, sanitation, and a view to the future that anticipates an important role for vaccination in controlling VHS (Ghittino et al. 1980).

Providing that the water supply is free of Egtved virus, facilities with enzootic VHS can be reclaimed for virus-free production. An eradication program in progress in Denmark since 1965 has been highly effective and singularly cost effective. The best account of the program was given by Vestergard Jorgensen (1974a), who implemented it. Steps in the program included removing infected stock, emptying and disinfecting holding facilities, repopulating with virus-free stock, registering VHS-free producers, and conducting clinical inspections at least once a year (Vestergard Jorgensen 1974a, 1980).

Where VHS cannot be avoided, measures can be applied to reduce losses. If marketing and stock availability permit, refractory salmonids—chinook or coho salmon, or triploid hybrids of coho salmon and rainbow trout—can be reared. Alternatively, the stock at risk can be vaccinated, as previously described here.

References

Ahne, W. 1980a. Experimentelle Egtved virus Infektion beim Hecht (*Esox lucius* L.). Tierärztl. Umsch. 35:225–229.

Ahne, W. 1980b. IPN-virus permanent infizierte Fischzellen: Untersuchungen über Zellvermehrung und Egtvedvirusreplikation. Zentralbl. Bakteriol. Mikrobiol. Hyg. [A] 246:304–307.

Ahne, W. 1981. Serological techniques currently used in fish virology. Dev. Biol. Stand. 49:3–27.

Ahne, W. 1982a. Vergleichende Untersuchungen über die Stabilität von vier fischpathogenen Viren (VHSV, PFR, SVCV, IPNV). Zentralbl. Veterinärmed. [B] 29:457–476.

Ahne, W. 1982b. Untersuchungen zur Tenazität der Fischviren. Fortschr. Veterinärmed. 35:305–309.

Ahne, W. 1985. Viral infection cycles in pike (*Esox lucius* L.). J. Appl. Ichthyol. 1:90–91.

Ahne, W., and C. Held. 1980. Untersuchungen über die viruzide Wirkung von Actomar K_{30} auf fischpathogene Viren. Tierärztl. Umsch. 35: 308–318.

Ahne, W., and I. Thomsen. 1985. Occurrence of VHS in virus in wild white fish (*Coregonus* sp.). Zentralbl. Veterinärmed. [B] 32:73–75.

Ahne, W., R. D. Negele, and B. Ollenschlager. 1976. Vergleichende Infektionsversuche mit Egtved-Viren (Stamm F1) bei Regenbogenforellen (*Salmo gairdneri*) und Goldenforellen (*Salmo aguabonita*). Berl. Münch. Tierärztl. Wochenschr. 89:161–164.

Altara, J., chairman. 1963. Resolutions on Item A, viral hemorrhagic septicemia. Permanent Commission for the Study of Diseases of Fish, Office International des Epizooties. Bull. Off. Int. Epizoot. 59:298–299.

Amlacher, E., J. Ude, C. Rudolph, and G. Ernst. 1980. Direct electron microscopical visualization of the presumptive virus of viral hemorrhagic septicaemia (VHS) in the rainbow trout *Salmo gairdneri* Richardson and additional histopathological and haematological observations. J. Fish Dis. 3:55–62.

Amlacher, E., C. Rudolph, J. Ude, and H. J. Eisengarten. 1982. Virale haemorrhagische Septikämie (VHS) der Regenbogenforelle (*Salmo gairdneri* Richardson)—Pathologie, elektronenmikroskopischer Virusnachweis und peripheres Blutbild. Arch. Exp. Veterinärmed. 36:193–201.

Baroni, A., R. Galesso, and G. Bovo. 1982. Ultrastructural observations on the virus of viral haemorrhagic septicaemia in culture. J. Fish Dis. 5:439–444.

Bellet, R. 1965. Viral hemorrhagic septicemia (VHS) of the rainbow trout bred in France. Ann. N.Y. Acad. Sci. 126:461–467.

Bernard, J., P. de Kinkelin, and M. Bearzotti-Le Berre. 1983. Viral hemorrhagic septicemia of rainbow trout: relation between the G polypeptide and antibody production in protection of the fish after infection with the F25 attenuated variant. Infect. Immun. 39:7–14.

Bernard, J., B. Kerouault, and P. de Kinkelin. 1984. High-performance thin-layer chromatography of lipids from different fish cell lines and from viral hemorrhagic septicaemia virus. Ann. Virol. (Paris) 135E:177–182.

Bernard, J., M. Bearzotti-Le Berre, and P. de Kinkelin. 1985. Viral haemorrhagic septicaemia in rainbow trout: attempt to relate interferon production, antibody synthesis and structure of the virus with the mechanism of virulence. Ann. Virol. (Paris) 136E:13–26.

Besse, P., J. C. Levaditi, J. C. Guillon, and P. de Kinkelin. 1965. Occurrence of viral diseases in the rainbow trout hatcheries in France. First histopathological results. Ann. N.Y. Acad. Sci. 126:543–546.

Bovo, G., and G. Giorgetti. 1983. Episodio di setticemia emorragica virale in trote iridea (*Salmo gairdneri*) allevate in acqua salmastra. Atti Soc. Ital. Sci. Vet. 37:686–689.

Bovo, G., E. Zanin, and G. Giorgetti. 1982. Grave episodio di setticemia emorragica virale (SEV) in avannotti di trota fario (*Salmo trutta*) d'allevamento. Atti Soc. Ital. Sci. Vet. 36:631–634.

Campbell J. B., and K. Wolf. 1969. Plaque assay and some characteristics of Egtved virus (virus of viral hemorrhagic septicemia of rainbow trout). Can. J. Microbiol. 15:635–637.

Castric, J., and P. de Kinkelin. 1980. Occurrence of viral hemorrhagic septicaemia in rainbow trout *Salmo gairdneri* Richardson reared in sea water. J. Fish Dis. 3:21–27.

Castric, J., and P. de Kinkelin. 1984. Experimental study of the susceptibility of two marine fish species, sea bass (*Dicentrarchus labrax*) and turbot (*Scophthalmus maximus*), to viral hemorrhagic septicaemia. Aquaculture 41:203–212.

Chilmonczyk, S. 1978. Stimulation spécifique des lymphocytes de truites arc-en-ciel (*Salmo gairdneri*) résistantes à la septicème virale. C. R. Acad. Sci. [D] (Paris) 287:387–389.

Chilmonczyk, S. 1980. Some aspects of trout gill structure in relation to Egtved virus infection and defence mechanism. Pages 18–22 in W. Ahne, ed. Fish diseases. Springer-Verlag, Berlin.

Chilmonczyk, S., and D. Monge. 1980. Rainbow trout gill pillar cells: demonstration of inert particle phagocytosis and involvement in viral infection. J. Reticuloendothel. Soc. 28:327–332.

Clark, H. F., and E. Z. Soriano. 1974. Fish rhabodvirus replication in non-piscine cell culture: new system for the study of rhabdovirus-cell interaction in which the virus and cell have different temperature optima. Infect. Immun. 10:180–188.

Cohen, J., and G. Lenoir. 1974. Ultrastructure et morphologie de quatre rhabdovirus de poissons. Ann. Rech. Vet. 5:443–450.

de Kinkelin, P. 1972. Le virus d'Egtved II. Purification. Ann. Rech. Vet. 3:199–208.

de Kinkelin, P. 1983. Viral haemorrhagic septicaemia. Pages 51–62 in D. P. Anderson, M. Dorson, and P. Dubourget, eds. Antigens of fish pathogens: development and production for vaccines and serodiagnostics. Collection Fondation Marcel Merieux, Lyon, France.

de Kinkelin, P., and M. Bearzotti. 1981. Immunization of rainbow trout against viral hemorrhagic septicemia (VHS) with a thermoresistant variant of the virus. Dev. Biol. Stand. 49:431–439.

de Kinkelin, P., and J. Castric. 1982. An experimental study of the susceptibility of Atlantic salmon fry, *Salmo salar* L., to viral haemorrhagic septicaemia. J. Fish Dis. 5:57–65.

de Kinkelin, P., and M. Dorson. 1973. Interferon production in rainbow trout (*Salmo gairdneri*) experimentally infected with Egtved virus. J. Gen. Virol. 19:125–127.

de Kinkelin, P., and M. Le Berre. 1977. Isolement d'un rhabdovirus pathogène de la truite fario (*Salmo trutta*, L. 1766). C. R. Acad. Sci. [D] (Paris) 284:101–104.

de Kinkelin, P., and R. Scherrer. 1970. Le virus d'Egtved I. Stabilité, développement et structure du virus de la souche danoise F_1. Ann. Rech. Vét. 1:17–30.

de Kinkelin, P., M. Le Berre, A. Meurillon, and M. Calmels. 1974. Septicémie hémorragique virale: démonstration de l'état réfractaire du saumon coho (*Oncorhynchus kisutch*) et de la truite fario (*Salmo trutta*). Bull. Fr. Piscic. 253:166–176.

de Kinkelin, P., A. M. Baudouy, and M. Le Berre. 1977. Réaction de la truite fario (*Salmo trutta* L. 1766) et arc-en-ciel (*Salmo gairdneri* Richardson, 1836) à l'infection par un nouveau rhabdovirus. C. R. Acad. Sci. [D] (Paris) 284:401–404.

de Kinkelin, P., S. Chilmonczyk, M. Dorson, M. Le Berre, and A. M. Baudouy. 1979. Some pathogenic facets of rhabdoviral infection of salmonid fish. Pages 357–375 in P. Bachmann, ed. Mechanisms of viral pathogenesis and virulence. Proceedings of the 4th Munich symposium on microbiology, Munich, June 6–7, 1979.

de Kinkelin, P., M. Bearzotti-Le Berre, and J. Bernard. 1980. Viral hemorrhagic septicemia of rainbow trout: selection of thermoresistant virus variant and comparison of polypeptide synthesis with the wild-type virus strain. J. Virol. 36:652–658.

de Kinkelin, P., M. Dorson, and A. M. Hattenberger-Baudouy. 1982. Interferon synthesis in trout and carp after viral infection. Dev. Comp. Immunol. Suppl. 2:167–174.

de Kinkelin, P., J. Bernard, and A. M. Hattenberger-Baudouy. 1984. Immunization against viral diseases occurring in cold water. Pages 167–197 in P. de Kinkelin and C. Michel, eds. Symposium on fish vaccination, Paris, France, February 20–22, 1984.

Deufel, J. 1959. Über das Virus der "infektiösen Nierenschwellung und Leberdegeneration" und einige Bekämpfungsversuche. Allg. Fisch-Ztg. 84:269.

Deufel, J. 1960. Grössenbestimmung des Virus der infektiösen Nierenschwellung und Leberdegeneration der Forellen mittels Ultrafiltration. Arch. Fischereiwiss. 10:169–232.

Deufel, J., and C. von Pollnitz. 1977. Zytodiagnostik von Krankheiten bei Regenbogenforellen (*Salmo gairdneri* Richardson). Fisch Umwelt 4:15–24.

Dixon, P. F., and B. J. Hill. 1984. Rapid detection of fish rhabdoviruses by the enzyme-linked immunosorbent assay (ELISA). Aquaculture 42:1–12.

Dorson, M., and B. Chevassus. 1985. Etude de la réceptivité d'hybrides triploïdes truite arc-en-ciel × saumon coho à la nécrose pancréatique infectieuse et à la septicémie hémorragique virale. Bull. Fr. Peche Piscic. 296:29–34.

Dorson, M., and C. Torchy. 1979. Complement dependent neutralization of Egtved virus by trout antibodies. J. Fish Dis. 2:345–347.

Dorson, M., A. Barde, and P. de Kinkelin. 1975. Egtved virus induced rainbow trout serum interferon: some physicochemical properties. Ann. Microbiol. (Paris) 126B:485–489.

Dorson, M., C. Torchy, and C. Michel. 1979. Rainbow trout complement fixation used for titration of antibodies against several pathogens. Ann. Rech. Vét. 10:529–534.

Elger, M., and H. Hentschel. 1983. Glomerular disease in cultured rainbow trout, *Salmo gairdneri* Richardson, suffering from presumptive chronic viral hemorrhagic septicaemia. J. Fish Dis. 6:211–229.

Enzmann, P. J. 1981. Rapid identification of VHS-virus from trout by immunofluorescence. Dev. Biol. Stand. 49:57–62.

Enzmann, P. J. 1983. Considerations on the effectiveness of VHS-vaccination. Bull. Eur. Assoc. Fish Pathol. 3:54–55.

Enzmann, P. J., and M. Konrad. 1984. Die Virale hämorrhagische Septikämie der Regenbogenforelle (VHS) und ihre Bekämpfung in epidemiologischer Sicht. Tierärztl. Umsch. 39:886–892.

Enzmann, P. J., and B. Maier. 1983. Serologische Verwandtschaft zwischen zwei fischpathogenen Rhabdoviren. Fisch Umwelt 12:13–21.

Enzmann, P. J., B. Maier, and K. Bigott. 1981. Biochemical data on the RNA of VHS-V and RVC indicating a possible serological relationship. Bull. Eur. Assoc. Fish Pathol. 1:37–39.

Eskildsen, U. K., and P. E. Vestergard Jorgensen. 1973. On the possible transfer of trout pathogenic viruses by gulls. Riv. Ital. Piscicol. Ittiopatol. 8:104–105.

Faisal, M., and W. Ahne. 1980. Use of immunoperoxidase technique for detection of fish virus antigens. Pages 186–192 in W. Ahne, ed. Fish diseases. Springer-Verlag, Berlin.

Flondro, F., E. Ropel, and Z. Żdan. 1984. Zastosowanie badania hematologicznego w diagnostyce wirusowej posocznicy lososiowatych-VHS. Med. Weter. 40:50–54.

Frost, J. W., and S. Wellhausen. 1974. Studies on the effect of temperature and glycerol on the inactivation of the virus of haemorrhagic septicaemia (VHS) of rainbow trout by use of the microtiter system. Zentralbl. Veterinärmed. [B] 21:625–631.

Georgiev, G. S., and Y. Kamenov. 1980. Studying the activity of some serum enzymes and of their isoenzymes in rainbow trout during viral hemorrhagic septicemia. Med. Vet. 33:143–146.

Ghittino, P. 1962. Differential-Diagnose der bei Regenbogenforellen vorkommenden "lipoiden Leberdegeneration" und "infektiösen Nierenschwellung und Leberdegeneration" auf Grund histopathologisch-anatomischer Untersuchungen. Z. Fisch. 11:549–555.

Ghittino, P. 1965. Viral hemorrhagic septicemia (VHS) in rainbow trout in Italy. Ann. N.Y. Acad. Sci. 126:468–478.

Ghittino, P. 1967. La setticemia emorragica virale (SEV) della trota iridea di allevamento. Riv. Ital. Piscicol. Ittiopatol. 2:90–102.

Ghittino, P. 1968a. Grave enzoozia di Setticemia Emorragica Virale in trote fario di allevamento (Salmo trutta). Riv. Ital. Piscicol. Ittiopatol. 3:17–19.

Ghittino, P. 1968b. Septicémie hémorragique virale de la truite arc-en-ciel. Pages 19–37 in Les maladies contagieuses des poissons incluses dans le Code Zoosanitaire International de l'O.I.E. Office International des Epizooties, Paris.

Ghittino, P. 1973. Viral hemorrhagica septicemia (VHS). Pages 4–41 in W. A. Dill, ed. Symposium on the major communicable fish diseases in Europe and their control. EIFAC (Eur. Inland Fish. Advis. Comm.) Tech. Paper 17, Suppl. 2.

Ghittino, P., N. Fijan, and P. de Kinkelin. 1980. Control methods for major viral diseases of fish in Europe. Bull. Off. Int. Epizoot. 92:967–978.

Hastein, T., G. Holt, and J. Krogsrud. 1968. Hemorrhagisk virusseptikemi (Egtvedsyke) hos regnbueorret i Norge. Nord. Vet. Med. 20:708–711.

Hill, B. J., B. O. Underwood, C. J. Smale, and F. Brown. 1975. Physico-chemical and serological characterization of five rhabdoviruses infecting fish. J. Gen. Virol. 27:369–378.

Hoffmann, R. 1980. Untersuchungen des roter Blutbildes bei viraler hämorrhagischer Septikämie (VHS) der Regenbogenforellen (*Salmo gairdneri*). Fisch Umwelt 8:37–48.

Hoffmann, R., C. Pfeil-Putzien, H. Dangschat, and M. Vogt. 1979. Untersuchungen zur Pathogenese der viralen hämorrhagischen Septikämie (VHS) bei Regenbogenforellen (*Salmo gairdneri*). II. Hämatologische und histologische Untersuchungen. Berl. Münch. Tierärztl. Wochenschr. 92:180–185.

Hofmann, W. 1971. On the diagnostic importance of inclusion bodies in viral hemorrhagic septicemia (VHS) of trout. (Zur diagnostischen Bedeutung der Einschlusskörperchen bei der hämorrhagischen Virusseptikämie (HVS) der Forellen. D.T.W. 78:14–17. Translated from the German by R. Howland.

Horlyck, V., S. Mellergard, I. Dalsgaard, and P. E. Vestergard Jorgensen. 1984. Occurrence of VHS in Danish maricultured rainbow trout. Bull. Eur. Assoc. Fish Pathol. 4:11–13.

Jensen, M. H. 1963. Preparation of fish tissue cultures for virus research. Bull. Off. Int. Epizoot. 59:131–134.

Jensen, M. H. 1965. Research on the virus of Egtved disease. Ann. N.Y. Acad. Sci. 126:422–426.

Lenoir, G., and P. de Kinkelin. 1975. Fish rhabdoviruses: comparative study of protein structure. J. Virol. 16:259–262.

McAllister, P. E., and R. R. Wagner. 1975. Structural proteins of two salmonid rhabdoviruses. J. Virol. 15:733–738.

McAllister, P. E., and R. R. Wagner. 1977. Virion RNA polymerases of two salmonid rhabdoviruses. J. Virol. 22:839–843.

McAllister, P. E., J. L. Fryer, and K. S. Pilcher. 1974. An antigenic comparison between infectious hematopoietic necrosis virus (OSV strain) and the virus of haemorrhagic septicaemia of rainbow trout (*Salmo gairdneri*) (Denmark strain) by cross-neutralization. J. Wildl. Dis. 10:101–103.

MacCrimmon, H. R. 1971. World distribution of rainbow trout (*Salmo gairdneri*). J. Fish. Res. Board Can. 28:663–704.

Maisse, G., M. Dorson, and C. Torchy. 1980. Inactivation de deux virus pathogènes pour les salmonidés (virus de la nécrose pancréatique infectieuse et de la septicémie hémorragique virale) par les rayons ultraviolets. Bull. Fr. Piscic. 278:34–40.

Meier, W. 1981. Viral hemorrhagic septicemia in non-salmonid fishes. Bull. Eur. Assoc. Fish Pathol. 1:15–17.

Meier, W. 1984. Some aspects of the Swiss import examination in a 5-year period (1978–1982). Bull. Eur. Assoc. Fish Pathol. 4:24–25.

Meier, W. 1985. Virale hämorrhagische Septikämie: Empfänglichkeit und epizootiologische Rolle des Hechts (*Esox lucius* L.). J. Appl. Ichthyol. 4:171–177.

Meier, W., and P. E. Vestergard Jorgensen. 1975. A rapid and specific method for the diagnosis of viral haemorrhagic septicaemia (VHS) of rainbow trout. Riv. Ital. Piscicol. Ittiopatol. 10:11–16.

Meier, W., and P. E. Vestergard Jorgensen. 1980. Isolation of VHS virus from pike fry (*Esox lucius*) with hemorrhagic symptoms. Pages 8–17 in W. Ahne, ed. Fish diseases. Springer-Verlag, Berlin.

Miaczynski, B. 1965. Viral diseases and diseases of uncertain etiology in fish in Poland. Ann. N.Y. Acad. Sci. 126:620–628.

Neukirch, M. 1984. An experimental study of the entry and multiplication of viral haemorrhagic septicaemia virus in rainbow trout, *Salmo gairdneri* Richardson, after water-borne infection. J. Fish Dis. 7:231–234.

Neukirch, M. 1985. Uptake, multiplication and excretion of viral haemorrhagic septicaemia virus in rainbow trout (*Salmo gairdneri*). Pages 295–300 in A. E. Ellis, ed. Fish and shellfish pathology. Academic Press, London.

Neukirch, M., and B. Glass. 1984. Some aspects of virus shedding by rainbow trout (Salmo gairdneri Rich.) after waterborne infection with viral haemorrhagic septicaemia (VHS) virus. Zentralbl. Bakteriol. Mikrobiol. Hyg. [A] 257:433–438.

Olberding, K. P., and J. W. Frost. 1975. Electron microscopical observations of the structure of the virus of viral haemorrhagic septicaemia (VHS) of rainbow trout (*Salmo gairdneri*). J. Gen. Virol. 27:305–312.

Olesen, N. J., and P. E. Vestergard Jorgensen. 1986. Detection of neutralizing antibody to Egtved virus in rainbow trout (*Salmo gairdneri*) by plaque neutralization test with complement addition. J. Appl. Ichthyol. 2:33–41.

Ord, W. 1975. Resistance of chinook salmon (*Oncorhynchus tschawytscha*) fingerlings experimentally infected with viral hemorrhagic septicemia virus. Bull. Fr. Piscic. 257:149–152.

Ord, W. M., M. Le Berre, and P. de Kinkelin. 1976. Viral hemorrhagic septicemia: comparative susceptibility of rainbow trout (*Salmo gairdneri*) and hybrids (*S. gairdneri* × *Oncorhynchus kisutch*) to experimental infection. J. Fish Res. Board Can. 33:1205–1208.

Osadchaya, E. F., M. G. Nakonenchnaya, and V. L. Rogozhkina. 1981. Vydelenie ot bolnykh forelei vozbuditelya virusnoi gemorragicheskoi septitsemii. (Isolation of the agent of hemorrhagic septicemia from diseased trout.) Pages 3–14 in V. A. Musselius and I. S. Shestrin, eds. Proceedings of scientific reports on fish diseases and water toxicology, Issue 32. Ministry of Fish Husbandry of the USSR, All Union Research Fisheries Production Society, All Union Research Institute of Fish Husbandry, Moscow, USSR. In Russian.

Pastoret, P. P., A. Chauveheid, G. Burtonboy, and F. Schoenaers. 1976. Isolement du virus de la septicémie hémorragique virale de la truite (SHV) en Belgique. Ann. Méd. Vét. 120:411–418.

Pfeil, C., and H. Wiedemann. 1977. Virusnachweis durch indirekte Immunofluoreszenz und Virusanzüchtung bei experimentell erzeugter viraler hämorrhagischer Septikämie (VHS) der Regenbogenforelle (*Salmo gairdneri*). D.T.W. 84:125–164.

Pfeil-Putzien, C., and R. Hoffmann. 1979. Untersuchungen zur Pathogenese der viralen hämorrhagischen Septikämie (VHS) bei Regenbogenforellen (*Salmo gairdneri*). I. Virologische Untersuchungen. Berl. Münch. Tierärztl. Wochenschr. 92:162–164.

Pfitzner, I. 1966. Beitrag zur Ätiologie der "hämorrhagischen Virusseptikämie der Regenbogenforellen." Zentralbl. Bakteriol. [Orig. A] 201:306–320.

Rasmussen, C. J. 1965. A biological study of the Egtved disease (INUL). Ann. N.Y. Acad. Sci. 126:427–460.

Reichenbach-Klinke, H. 1959. Fischkrankheiten in Bayern in den Jahren 1957 und 1958. Allg. Fischztg. 84:226–228.

Robin, J., and A. Rodrigue. 1977. Purification and biochemical properties of Egtved viral RNA. Can. J. Microbiol. 23:1489–1491.

Schäperclaus, W. 1938. Die Schädigungen der deutschen Fischerei durch Fischparasiten und Fischkrankheiten. Allg. Fischztg. 41:256–259, 267–270.

Schäperclaus, W. 1954. Fischkrankheiten, 3rd ed. Akademie-Verlag, Berlin, D.D.R.

Schlotfeldt, H. J., and J. W. Frost. 1975. Doppelinfektion von Regenbogenforellen mit dem Virus der viralen hämorrhagischen Septikämie (VHS) und dem Virus der infektiösen Pankreasnekrose (IPN). Berl. Münch. Tierärztl. Wochenschr. 88:455.

Schmidt, B. 1960. Histologische Untersuchungen an der Leber von gesunden und an "infektiöser Nierenschwellung und Leberdegeneration" erkrankten Forellen. Inaugural Dissertation. Ludwig-Maximillians-Universität, Munich.

Tack, E. 1959. Beiträge zur Erforschung der Forellensuche. Arch. Fischereiwiss. 10:20–30.

Tesarčík, J., J. Cibulka, and J. Kalivoda. 1968. Výskyt virové hemorrhagické septikémie (anaemia infectiosa salmonum) u pstruhů amerických duhových (Parasalmo gardnerii Richardson, 1836). Bul. VÚR Vodňany 3:1–8.

Vestergard Jorgensen, P. E. 1968. Serological identification of Egtved virus (virus of viral haemorrhagic septicaemia of rainbow trout). A preliminary report. Bull. Off. Int. Epizoot. 69:985–989.

Vestergard Jorgensen, P. E. 1970. The survival of viral hemorrhagic septicemia (VHS) virus associated with trout eggs. Riv. Ital. Piscicol. Ittiopatol. 5:13–14.

Vestergard Jorgensen, P. E. 1971. Egtved virus: demonstration of neutralizing antibodies in serum from artificially infected rainbow trout (*Salmo gairdneri*). J. Fish. Res. Board Can. 28:875–877.

Vestergard Jorgensen, P. E. 1972. Egtved virus: antigenic variation in 76 virus isolates examined in neutralization tests and by means of the fluorescent antibody technique. Symp. Zool. Soc. Lond. 30:333–340.

Vestergard Jorgensen, P. E. 1973a. Artificial transmission of viral haemorrhagic septicaemia (VHS) of rainbow trout. Riv. Ital. Piscicol. Ittiopatol. 8:101–102.

Vestergard Jorgensen, P. E. 1973b. Inactivation of IPN and Egtved virus. Riv. Ital. Piscicol. Ittiopatol. 8:107–108.

Vestergard Jorgensen, P. E. 1974a. A study of viral diseases in Danish rainbow trout, their diagnosis and control. Ph.D. dissertation. Royal Danish Veterinary and Agricultural University, Copenhagen.

Vestergard Jorgensen, P. E. 1974b. Indirect fluorescent antibody techniques for demonstration of trout viruses and corresponding antibody. Acta Vet. Scand. 15:198–205.

Vestergard Jorgensen, P. E. 1976. Partial resistance of rainbow trout (*Salmo gairdneri*) to viral haemorrhagic septicaemia (VHS) following exposure to non-virulent Egtved virus. Nord. Vet. Med. 28:570–571.

Vestergard Jorgensen, P. E. 1980. Egtved virus: the susceptibility of brown trout and rainbow trout to eight virus isolates and the significance of the findings for the VHS control. Pages 3–7 in W. Ahne, ed. Fish diseases. Springer-Verlag, Berlin.

Vestergard Jorgensen, P. E. 1981. Experimental VHS-vaccine in field trials. Bull. Eur. Assoc. Fish Pathol. 1:13–14.

Vestergard Jorgensen, P. E. 1982a. Egtved virus: temperature-dependent immune response of trout to infection with low-virulence virus. J. Fish Dis. 5:47–55.

Vestergard Jorgensen, P. E. 1982b. Egtved virus: occurrence of inapparent infections with virulent virus in free-living rainbow trout, *Salmo gairdneri* Richardson, at low temperature. J. Fish Dis. 5:251–255.

Vestergard Jorgensen, P. E., and A. Meyling. 1972. Egtved virus: demonstration of virus antigen by the fluorescent antibody technique in tissues of rainbow trout affected by viral haemorrhagic septicaemia and in cell cultures infected with Egtved virus. Arch. Virusforsch. 36:115–122.

Whipple, H. E., Editor. 1965. Viral diseases of poikilothermic vertebrates. Ann. N.Y. Acad. Sci. 126:1–680.

Wiedemann, H. 1977. Methodische Untersuchungen zur Virusisolierung bei experimenteller viraler hämorrhagischer Septikämie (VHS) der Forellen. Berl. Münch. Tierärztl. Wochenschr. 90:340–343.

Wizigmann, G., and R. Hoffmann. 1982. Vergleichende lichtmikroskopische und immunhistologische Untersuchungen bei der viralen hämorrhagischen Septikämie (VHS) der Regenbogenforelle. Zentralbl. Veterinärmed. [B] 29:782–788.

Wizigmann, G., C. Baath, and R. Hoffmann. 1980. Isolierung des Virus der viralen hämorrhagischen Septikämie (VHS) aus Regenbogenforellen-, Hecht- und Äschenbrut. Zentralbl. Veterinärmed. [B] 27:79–81.

Wizigmann, G., H. Dangschat, C. Baath, and C. Pfeil-Putzien. 1983. Untersuchungen über Virusinfektionen bei Süsswasserfischen in Bayern. Tierärztl. Umsch. 38:250–258.

Wolf, K. 1966. The fish viruses. Adv. Virus Res. 12:35–101.

Wolf, K., and M. C. Quimby. 1973. Fish viruses: buffers and methods for plaquing eight agents under normal atmosphere. Appl. Microbiol. 25:659–664.

Wolf, K., and P. E. Vestergard Jorgensen. 1970. Salmonid viruses: double infection of RTG-2 cells with Egtved and infectious pancreatic necrosis viruses. Arch. Virusforsch. 29:337–342.

Yasutake, W. T. 1970. Comparative histopathology of epizootic salmonid virus diseases. Pages 341–350 in S. F. Snieszko, ed. A symposium on diseases of fishes and shellfishes. Am. Fish. Soc., Spec. Publ. 5.

Yasutake, W. T. 1975. Fish viral diseases: clinical, histopathological, and comparative aspects. Pages 247–271 in W. E. Ribelin and G. Migaki, eds. The pathology of fishes. Univ. Wisconsin Press, Madison.

Yasutake, W. T., and C. J. Rasmussen. 1968. Histopathogenesis of experimentally induced viral hemmorrhagic septicemia in fingerling rainbow trout (*Salmo gairdneri*). Bull. Off. Int. Epizoot. 69:977–984.

Zwillenberg, L. O., and H. H. L. Zwillenberg. 1964. Elektronenmikroskopische Untersuchungen und Regenbogenforellen mit infektiöser Nierenschwellung und Leberdegeneration. Arch. Virusforsch. 14:319–331.

Zwillenberg, L. O., M. H. Jensen, and H. H. L. Zwillenberg. 1965. Electron microscopy of the virus of viral haemorrhagic septicaemia of rainbow trout (Egtved virus). Arch. Virusforsch. 17:1–19.

Zwillenberg, L. O., I. Pfitzner, and H. H. L. Zwillenberg. 1968. Infektionsversuche mit Egtved-Virus an Zellkulturen und Individuen der Schleie (*Tinca vulgaris* Cuv.) sowie an anderen Fischarten. Zentralbl. Bakteriol. Hyg. I Abt. Orig. 208:218–226.

SECTION II DISEASES AND AGENTS OF LOW VIRULENCE

19 Bluegill Hepatic Necrosis Reovirus

Synonym: oyster reovirus 13_{p2}

Definition

Bluegill hepatic necrosis is a subacute disease condition with attendant mortality that can be experimentally induced in fry of *Lepomis macrochirus* by intraperitoneal injection of a reovirus or exposure to waterborne infectivity. The virus was originally isolated from pooled normal-appearing juvenile American oysters (*Crassostrea virginica*). The mollusk, however, was refractory to experimental infection.

History

During 1976–1977 disease surveys were conducted in mollusk hatcheries on Long Island Sound, New York, and virus was isolated twice—in March 1977 at one hatchery and in August 1977 at another. On both dates, the source mollusks were healthy-appearing American oysters, and virus was detected in only one of several pooled samples of whole-oyster tissues. Decontaminated extracts of oyster tissues were inoculated on drained cultures of the BF-2 cell line of bluegill origin and incubated at 15°C for 7 days. Evidence of virus appeared when BF-2 cultures developed syncytia after being inoculated with first-passage material. Tests showed that the virus was replicated by several fish cell lines but not by avian or mammalian cells, and that the agent was neither infectious pancreatic necrosis virus nor the virus from the mollusk *Tellina*, but instead a new reovirus-like agent. For reasons that were not explained, the virus was designated 13_{p2} (Meyers 1979). In the following year, Meyers and Hirai (1980) described and illustrated the results of rotational enhancement of 13_{p2} images and concluded that the virus was an icosahedron with $T = 3$ symmetry, and therefore conformed morphologically to reovirus.

Next, 13_{p2} was tested for pathogenicity to juvenile oysters and to bluegills (because the virus had been replicated best by cultures of bluegill cells). Results of that work showed that oysters were resistant but that bluegills developed focal necrotic hepatitis, which sometimes resulted in death (Meyers 1980).

Meyers (1983) reported that the virus produced a nonvirulent infection in fingerling and adult rainbow trout. Virus titer increased in the fish and histopathologic examination showed multifocal hepatitis and occasional pancreatitis. Moreover, infections were accompanied by the production of antivirus precipitating and neutralizing antibodies. Confirmation of the reovirus nature of 13_{p2} was reported by Winton et al. (1987), who found that it was similar to, but distinct from, the reoviruses of golden shiner, chum salmon, and channel catfish—all isolants that were taken from fish.

Signs and Pathologic Changes

Of 16 bluegill fingerlings inoculated intraperitoneally with 2×10^4 $TCID_{50}$ virus, 7 died within 2 to 7 weeks postinoculation. Death occurred 3 to 7 days after the onset of signs that included anorexia, reclusion, listlessness, loss of equilibrium, and terminal prostration (Meyers 1980).

External lesions did not develop among bluegills exposed to waterborne infectivity, nor among those that had been injected with virus. Neither did waterborne virus induce internal signs. In contrast, the livers of fish that died after injection of virus were white instead of the normal tan.

Histologic examination of 23 fish that had been injected with virus yielded a 94% incidence of hepatic lesions. Moreover, the liver was the only organ showing change. Two of four fish exposed by immersion showed multifocal necrotic hepatitis—a condition not found in three controls that were examined.

Features of the hepatic lesions included large syncytia and coagulation necrosis of hepatocytes but little or no infiltration of inflammatory cells. Normal cord pattern of the livers was disorganized to the point of being almost unrecognizable. Nevertheless, hepatocyte regeneration and diffuse infiltration by phagocytes was also evident. Hepatotropism of the virus was also evident in rainbow trout that had been inoculated intraperitoneally. Although no clinical signs of disease were evident, adult and juvenile specimens showed about a 60% incidence of multifocal granulomatous hepatitis (Meyers 1983).

Etiology

The role of the virus in nature is wholly unknown.

Size and Shape

Particles are icosahedral, doubly encapsidated, and measure about 75 nm.

Biophysical Properties

The virus is stable in chloroform, acid (pH 3), and alkali (pH 9), but labile at pH 2. Treatment at pH 3 results in a 1000-fold increase in titer. Replication occurs at temperatures of 15 to 23°C, but not at 30°C. Highest titers (~ 10^7 $TCID_{50}$/mL) are replicated in the BF-2 and BB cell lines, but other lines of fish cells also support growth: AS, GE-4 (guppy embryo), WF-2 (walleye fry), and ASH (Atlantic salmon heart). The FHM, RTG-2, and RTS (rainbow trout spleen) cell lines were refractory, as were primary chick kidney cultures and six cell lines of mammalian origin.

The virus does not hemagglutinate human O+ erythrocytes and is not neutralized by avian reovirus antiserum or antiserum against reovirus types 1 and 3. Also, it does not produce CPE in avian or mammalian kidney cells that replicate other animal reoviruses (Meyers 1979).

Winton et al. (1987) showed that, like the three reoviruses of fish origin, 13_{p2} had an RNA genome composed of 11 segments of double-stranded RNA and five structural proteins, which, although similar, were distinctive for each virus.

Source

The normal or natural host of 13_{p2} is not known. It could be either a vertebrate or an invertebrate, probably living in an aquatic environment. That environment could be freshwater, brackish, estuarine, or strictly marine. Considering the temperature of replication and the resistance of homeotherm cells, the sole restriction appears to be that the virus originates in a poikilotherm animal.

References

Meyers, T. R. 1979. A reo-like virus isolated from juvenile American oysters (*Crassostrea virginica*). J. Gen. Virol. 43:203–212.

Meyers, T. R. 1980. Experimental pathogenicity of reovirus 13_{p2} for juvenile American oysters *Crassostrea virginica* (Gmelin) and bluegill fingerlings *Lepomis macrochirus* (Rafinesque). J. Fish Dis. 3:187–201.

Meyers, T. R. 1983. Serological and histopathological responses of rainbow trout, *Salmo gairdneri* Richardson, to experimental infection with the 13_{p2} reovirus. J. Fish Dis. 6:277–292.

Meyers, T. R., and K. Hirai. 1980. Morphology of a reo-like virus isolated from juvenile American oysters (*Crassostrea virginica*). J. Gen. Virol. 46:249–253.

Winton, J. R., C. N. Lannan, J. L. Fryer, R. P. Hedrick, T. R. Meyers, J. A. Plumb, and T. Yamamoto. 1987. Morphological and biochemical properties of four members of a novel group of reoviruses isolated from aquatic animals. J. Gen. Virol. 68:353–364.

20 Fish Pox

Synonyms: carp pox, cutaneous warts, *Epithelioma papulosum*, hyperplastic epidermal disease, *Papillosum cyprini*, plaque warty hyperplasia, variola

Definition

Fish pox is a chronic skin disease that occurs among several species of propagated cyprinids and typically appears as transient focal lesions of benign, nonnecrotizing epidermal hyperplasia. The disease has an associated herpesvirus that is probably causal.

Although the disease described here involves a herpesvirus and not a poxvirus, the terms fish pox and carp pox are firmly entrenched in the literature. The common carp (*Cyprinus carpio*) is probably the most widely propagated species of fish and consequently the most frequently reported victim of the viral disease. However, it is now evident that cyprinids other than carp are susceptible to the herpesvirus, and hence the broadening of the name to fish pox.

History

B. Hofer, the dean of European fish pathologists, included "pox disease of carp," *Epithelioma papulosum*, in his 1904 textbook of fish diseases. He noted that the disease was recognized in the Middle Ages and mentioned in Swiss naturalist K. Gessner's 1563 book of fishes. Inasmuch as fish viruses were unknown, it is understandable that Hofer attributed the skin lesions to secondary effects of myxosporidan parasitism in the kidneys.

The possibility of a viral etiology was broached by Lowenthal (1907), who described histologic features of the lesions and, more specifically, the presence of intranuclear inclusions. Convinced that he was dealing with an infectious disease, he related the inclusions to those of rabies and several pox diseases then recognized.

Although the evidence could not be considered conclusive as judged by today's standards, much stronger indications of virus involvement were reported during the early 1950s, when Roegner-Aust and her associates at the Bavarian Biological Research Institute examined tissue smears and palladium-shadowed lesion extracts with electron microscopy (Roegner-Aust 1953; Roegner-Aust and Deubner 1952; Roegner-Aust and Schleich 1951). Illustrations in those reports showed abundant round particles in the size range of 70 to 220 nm, occurring both in aggregates and singly. The investigators stated that such particles occurred rarely or not at all in preparations from normal skin.

Roegner-Aust (1953) reviewed the relevant literature on the condition and pointed out that previous works proposed avitaminosis, calcium deficiency, and environmental factors as being involved. Also considered were hereditary factors or genetic predisposition—possibilities supported by the work of Probst (1951), also at the Bavarian Institute.

Meanwhile, at the New York Aquarium, Nigrelli (1948) described a neoplastic condition in the bluegill (*Lepomis macrochirus*) that was "similar to carp pox." His 1952 review of virus and tumors discussed the two conditions as "hyperplastic epidermal disease or fish pox."

At the Robert Koch Institute in Berlin, Grützner (1956) made the first application of tissue culture to fish virology. She cultured explants of embryonic tissue from the guppy (*Lebistes reticulatus* [*Poecilia reticulata*]) and inoculated them with extracts of carp pox lesions. After several passages, some of the cultures showed giant cells containing three to eight nuclei and a higher incidence of nuclear polymorphism than was present in control cultures. The effects were attributed to virus.

Four years later, work at the Zoological Institute at Stuttgart-Hohenheim eventually led to the unequivocal demonstration of intracellular herpesvirus in the lesions. In the first of several reports on thin-section electron microscopy, Schubert and Meyer (1960) described and illustrated electron-dense intracellular forms that measured 300 to 500 nm and were believed to be organisms now termed chlamydia. Later work showed the forms to be membrane-bound inclusions, and the earlier interpretation was recanted when unequivocal herpesvirus was demonstrated (Schubert 1964, 1966). Later reviews uniformly accepted the herpesvirus interpretation (Ghittino et al. 1968; Riedmüller [nee Roegner-Aust] 1965; Schäperclaus 1969).

Poxlike lesions of a superficial and transient character were reported by Tesarčík and Mares (1967) from northern pike (*Esox lucius*) in Czechoslovakia, and Mawdesley-Thomas and Bucke (1967) illustrated gross lesions and a histologic section of hyperplastic epithelium from the roach (*Rutilus rutilus*) in England. Neither report documented the presence of virus. In Italy, Ghittino et al. (1968) briefly reviewed the disease and proposed that the lesions be designated as "plaque warty-hyperplasia of the epidermis."

From Czechoslovakia, Lucký (1970) reported what he considered to be "pox" in the sheatfish or European catfish (*Silurus glanis*). He noted that among reports of various European authors at least 11 species of fish other than carp had been diagnosed as having the disease. Lucký's illustrations of hyperplastic skin of the sheatfish generally conformed to growths typical of those occurring on the common carp; however, the most critical evidence—demonstration of herpesvirus particles—was lacking.

In 1974, Hines et al. in Israel reported the differential susceptibility of a strain of carp that had been developed in Yugoslavia. The authors were apparently not aware of the earlier work by Probst (1951) showing a genetic implication (they did not cite it).

A curious feature of the apparent occurrence of the disease was noted by Wolf (1973). Although common carp were introduced to North America about 100 years ago and were widely established in natural waters, none had been found with pox lesions. That situation probably reflects the fact that carp are regarded in North America as a coarse fish, unwanted in many waters, and did not warrant the attention of fish health personnel.

This situation changed in the late 1970s. Continuing their interest in fish diseases—more specifically in tumors or neoplasms—Sonstegard and Sonstegard (1978) found the disease in carp from several of the Great Lakes. They were able to transmit the tumor experimentally and showed that herpesvirus was present in thin sections of the lesions. They were unable to isolate the virus in any of the commonly used fish cell lines.

Hungarian investigators who carried out light and electron microscopic studies of the disease and also tried to isolate the virus in cell culture were also unsuccessful (Racz et al. 1980).

As reported by Fijan et al. (1983), studies were begun in Yugoslavia in 1969 on the etiology of pox disease in carp. Isolation of the herpesvirus from hyperplastic tissue was attempted but not achieved; however, the effort resulted in the development and establishment of the cell line named *Epithelioma papulosum cyprini* (EPC).

Over the years, the EPC cell line has found widespread use in fish virology. Presumably, the tissue of origin was infected by herpesvirus, the genome of which could be integrated in EPC cell nuclei. Such integration has not been shown, but cultures from sources in the United States and Europe carry a coronavirus that induces syncytia at temperatures of 30°C or more (Hetrick and Lidgerding 1982).

In 1984, at an international seminar on fish pathology held in Tokyo, Sano and colleagues at Tokyo University of Fisheries reported repeated isolation of herpesvirus from tumor tissues of Asagi carp, a koi or colored strain of *Cyprinus carpio*. Oncogenicity of the virus in experimentally infected carp was claimed, but that finding was clouded by tumor development in the control. A brief illustrated version of the work appeared later (Sano, Fukuda, Furukawa, Hosoya, and Moriya 1985).

Sano, Fukuda, and Furukawa (1985) provided greater details of the herpesvirus isolation, some characteristics of the virus, and its oncogenicity after long incubation. Fulfillment of Rivers' postulates was not claimed, but photographic documentation left no doubt that a herpesvirus was isolated. The virus is the probable cause of the neoplasia. The name suggested by the authors was *Herpesvirus cyprini* or cyprinid herpesvirus 1 (CHV).

Signs and Pathologic Changes

Most accounts indicated that the usual infection, being superficial, had little effect on the well-being of the host, but conflicting opinions also existed. Heavier infections clearly affected the health of the host: Bauer (1957) noted that fish with many lesions showed growth retardation; Schäperclaus (1969) described two outbreaks in which heavily infected fish grew poorly, and cited similar observations by earlier investigators; and Sano, Fukuda, and Furukawa (1985) reported that some mortality occurred among Asagi carp that were experimentally inoculated with virus.

External Signs

The lesions are superficial, and usually smooth and firm. The largest measure several centimeters in greatest dimension and several millimeters in thickness. The lesions can be somewhat glassy or waxy, but are sometimes opaque. The color can be slightly milky, creamy to brown (depending on pigmentation of the locus), or tinged with pink from vascularity (Fig. 1).

The lesions are transient and may persist for months before they slough. The affected areas sometimes develop dark pigmentation that is esthetically undesirable in either ornamental or table fish.

Internal Signs

The neoplasms are not invasive, nor do they metastasize, but fish culturists and investigators have over the years noted that carp with neoplasms often had spinal deformities, osteomalacia (Knochenweiche), abnormal limberness, emaciation, or growth retardation (Hines et al. 1974; Mann 1951; Schäperclaus 1969). Mann (1951), who analyzed the soft vertebrae, found abnormally low levels of ash, calcium, and phosphorus. However, a positive correlation between the presence of the neoplasm or virus and abnormal metabolism—particularly in bone—has not been established.

Histopathologic Findings

The growths consist of epidermal cells with well-formed intercellular connections, but the usual stratified layers are disrupted. In places, the cells are elongated and radiate outward to form papillae involving highly proliferated epidermal cells. Mitoses are common, especially in the outer layers. Normal-looking mucous cells, normally a surface feature, may occur at depth between papillae. Some of the epithelial cells show Cowdry type A intranuclear inclusions, and chromatin is marginated. The growths do not penetrate the basal lamina. They are vascularized and usually show no evidence of necrosis.

Virus occurs in cells that are in an advanced state of degeneration. The endoplasmic reticulum is enlarged and coarse, the mitochondria are swollen, and cristae are indistinct. Ribosomes are abundant. The nucleus shows only a small amount of peripheral chromatin.

Figure 1. Cyprinid species showing lesions of fish pox. *Top.* Common carp (*Cyprinus carpio*) with glistening translucent patches. From Wolf (1973). Reprinted with the permission of Academic Press, Inc. *Bottom.* Ide (*Leuciscus* [*Idus*] *idus*) with opaque discrete and fused nodules. From McAllister et al. (1985). Reprinted with the permission of the Wildlife Disease Association.

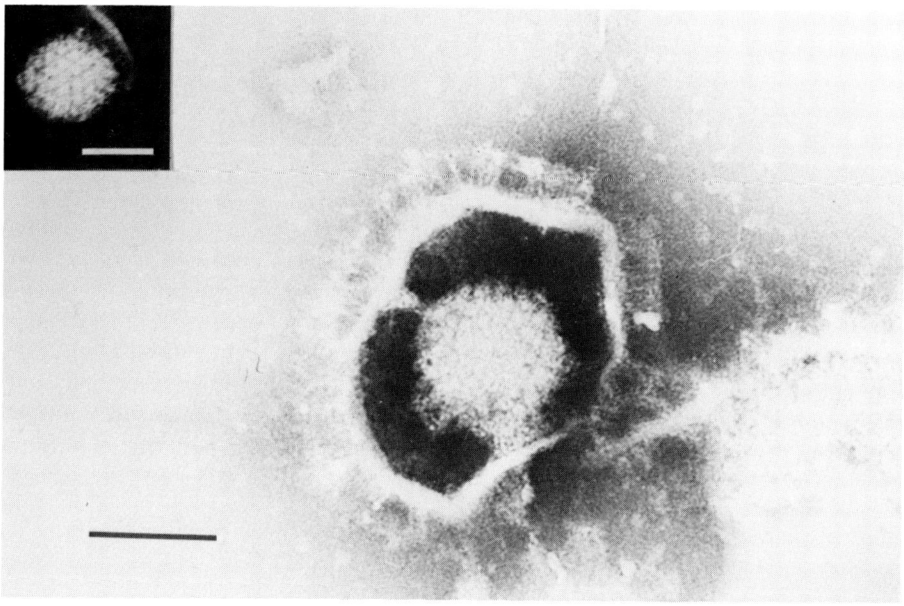

Figure 2. Negatively stained enveloped herpesvirus particle of a fish pox isolant. Bar = 100 nm. From Sano, Fukuda, Furukawa, Hosoya, and Moriya (1985). Reprinted with the permission of the Japanese Society of Fish Pathology.

Etiology

Although critical proof—fulfillment of Rivers' postulates by reisolation of virus from experimentally induced tumors—is lacking, the finding of herpesvirus in hyperplastic cells (Racz et al. 1980; Schubert 1964; Sonstegard and Sonstegard 1978), and especially the work of Sano and colleagues (Sano et al. 1984; Sano, Fukuda, and Furukawa 1985; Sano, Fukuda, Furukawa, Hosoya, and Moriya 1985), support the reasonable assumption of viral etiology.

Size and Shape

Sano, Fukuda,and Furukawa (1985) give the size of enveloped virions in FHM cells as 190 ± 27 nm and that of the nucleoid as 113 ± 9 nm (Fig. 2). In naturally infected cells, Schubert (1964) reported values of 140 and 110 nm, respectively, and Racz et al. (1980) a size of about 150 nm for the virions. Morphologically the virus conforms to herpesvirus.

Biophysical Properties

The virus is ether, pH, and heat labile and its replication is inhibited by iododeoxyuridine. It is grown in FHM, MCT, and EPC cell lines (in order of preference) at

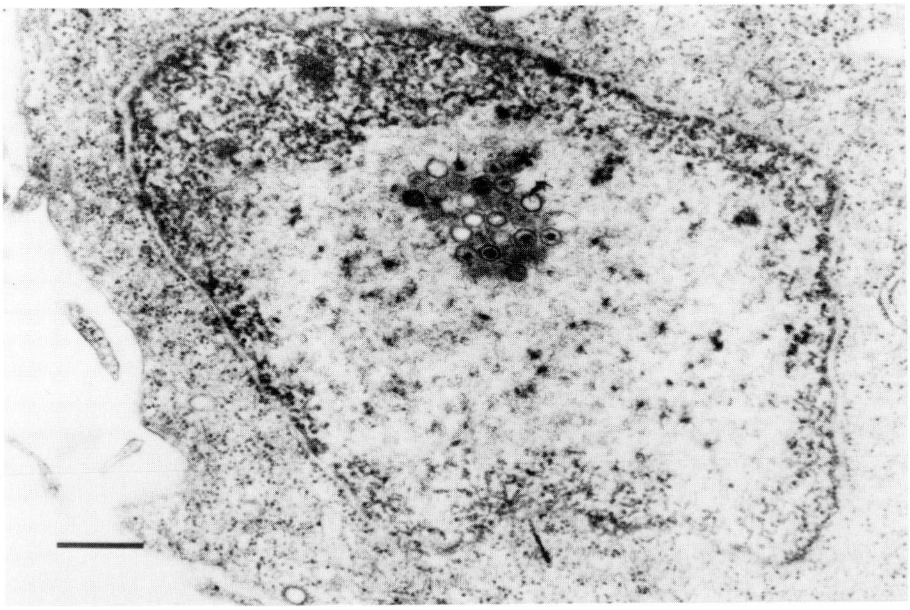

Figure 3. Thin-section electron micrograph of an infected FHM cell showing margination of the chromatin, disruption of the nuclear membrane, and empty capsids within the intranuclear inclusion. Bar = 500 nm. From Sano, Fukuda, Furukawa, Hosoya, and Moriya (1985). Reprinted with the permission of the Japanese Society of Fish Pathology.

temperatures of 15 to 20°C. The maximal titer in FHM and MCT cells at 20°C is nearly 10^6 $TCID_{50}$/mL. Syncytia are not produced, but infected cells show margination of chromatin and some develop Cowdry type A intranuclear inclusions (Fig. 3). In vitro CPE begins at about 5 days and consists of focal vacuolation, pyknosis, and slow sloughing. Overgrowth of areas of CPE is occasionally observed.

Diagnosis

Diagnosis may be made on the basis of finding the distinctive epidermal lesions in carp and other cyprinids. (Because of the general host specificity of herpesvirus, noncyprinid species are currently excluded.) The reliability of the diagnoses may be improved by histologic examination that establishes epithelial hyperplasia, and especially intranuclear inclusions. A definitive diagnosis can be made by electron microscopy and the demonstration of herpesvirus.

Isolation

Sano and coworkers (Sano, Fukuda, Furukawa, Hosoya, and Moriya 1985) reported making isolations from papilloma tissues of eight Asagi and one common carp in 1981–1983; however, few details were provided. Inocula were prepared "employing antibiotics and filtration methods," and CPE occurred in 2 to 3 weeks in EPC and FHM cells incubated at 20°C.

Sano, Fukuda, and Furukawa (1985) used FHM cells for isolation and, because this cell line gave a high titer, they also used it for characterization work. Other workers who attempted isolation of the virus in FHM cells were not successful (McAllister et al. 1985; Racz et al. 1980; Sonstegard and Sonstegard 1978).

Isolation of the virus may be facilitated if, in addition to cell lines, primary cultures of normal skin epithelium of the same cyprinid host species are used.

Identification

Serologic methods of identification have not been reported. A presumptive identification is possible by the isolation or electron microscopic demonstration of herpesvirus in the subject hyperplastic tissues.

Transmission and Incubation

Fijan (personal communication, 1972) achieved transmission routinely by abrading the skin of healthy carp and rubbing it with lesions of carp pox (e.g., Fig. 1). Sonstegard and Sonstegard (1978) reported two methods of transmitting the disease: first, by simple cohabitation and, second, by abrading epithelium of normal carp and rubbing the traumatized area with tumor tissue. At 10°C, the tumors develop in about 2 months.

In early attempts at transmission, Sano and coworkers (Sano, Fukuda, Furukawa, Hosoya, and Moriya 1985) inoculated 40 Asagi and 40 common carp by intraperitoneal injection or by scarification with culture-grown virus. Although tumors developed in 96 to 100% of the Asagi and 18% of the common carp, essentially the same percentages of tumors developed in the controls.

In later work (Sano, Fukuda, and Furukawa 1985), 3 of 10 common carp given an intraperitoneal injection of virus and held at 15°C developed tumors within 6 months, whereas 10 control fish injected with Hanks' balanced salt solution remained free of the lesions. Five months elapsed before the first tumors became visible.

According to Hines et al. (1974) and McAllister et al. (1985), outbreaks of fish pox have occurred during the low temperatures of winter or early spring; the lesions sloughed or disappeared as the water warmed in the summer. Temperature appears to be a permissive factor in the long incubation and months-long persistence of the infection.

Source

Considering the typical host specificity of herpesviruses, the source of the fish pox agent is without doubt a cyprinid host. Monogenetic trematodes, leeches, copepods, and myxosporidans have each been proposed as possible or probable vectors, but proof is lacking. Hines et al. (1974) found more infected carp in ponds stocked at high density than in ponds stocked at low density.

Host and Geographic Range

Histologically similar lesions of epidermal hyperplasia have been reported from other species of fish, but herpesvirus particles have been found only in carp and several other cyprinids. Sonstegard and Sonstegard (1978) reported the disease in the wild and in laboratory infections of hybrids of carp and goldfish (*Carassius auratus*), but they were unable to transmit the disease to goldfish. That apparent degree of host specificity is surprisingly high, even for a herpesvirus.

Reports of poxlike epidermal hyperplasias of esocid, percid, osmerid, and centrarchid species notwithstanding, actual involvement of herpesvirus must await virologic or electron microscopic documentation. As a case in point, electron microscopy of hyperplastic epidermal disease of the bluegill has failed to show virus.

The uncertainty of herpesvirus involvement in species other than the common carp is compounded by the fact that cells of some hosts can integrate a herpesvirus genome into their own. The Lucké adenocarcinoma of frogs is an example. The Lucké tumor is present throughout the year, but virus can be visualized only after frogs are held at the low temperature of hibernation. That feature seemingly is tied to the biology of the host and the virus, for it ensures the synthesis and release of infectious virus and transmission to the next generation of frogs at the time of the early spring spawning. Thereafter, in warm weather, the virus cannot be visualized.

Immunity

The lesions are infiltered with leukocytes, suggesting an inflammatory response; other aspects of either cellular or humoral immunity are unknown.

Control

Because obviously infected fish are unsightly, the disease is a problem where carp are produced for table use, or exotic cyprinids for ornamental use. Sloughing of lesions does occur spontaneously and might be accelerated by exposing affected specimens to elevated temperatures.

In Israel the practice of crossbreeding instead of inbreeding carp strains is credited with bringing about a sharp decline in the prevalence of the disease. Also, among susceptible strains, low-density stocking of ponds is accompanied by a reduced prevalence (Hines et al. 1974).

References

Bauer, O. N. 1957. Diseases of carp in fish ponds in Leningrad, Velikie Luki, and Novgorod Oblast's. Pages 68–88 in G. K. Petrushevskii, ed. Parasites and diseases of fishes. Bulletin of the All-Union Scientific Research Institute Fresh-water Fisheries. Translated from the Russian by The Israel Program for Scientific Fisheries for The National Science Foundation and the U.S. Department of the Interior.

Fijan, N., D. Sulimanović, M. Berzotti, D. Muzinić, L. O. Zwillenberg, S. Chilmonczyk, J. F. Vautherot, and P. de Kinkelin. 1983. Some properties of the *Epithelioma papulosum cyprini* (EPC) cell line from carp *Cyprinus carpio*. Ann. Virol. (Paris) 134E:207–220.

Ghittino, P., S. Dalforno, A. Provana, and M. L. Codegone. 1968. Iperplasia verrucosa a placca dell'epidermide in ciprinidi (cosi detto "valiolo dei pesci"). Riv. Ital. Piscicol. Ittiopatol. 3:53–54.

Grützner, L. 1956. Überprüfung einiger Anwendungsmöglichkeiten der Gewebekultur von *Lebistes reticulatus* (Peters) und *Macropodus opercularis* (Linne) in der Virusforschung. Zentralbl. Bakteriol. Parasitenkd. Infektionskr. Hyg. Abt. I Orig. 165:81–96.

Hetrick, F., and B. Lidgerding. 1982. Persistent infection of a fish cell line with a coronavirus. Page 137 in E. Kurstak, chairman. Fourth international conference on comparative virology, Banff Springs Hotel, Banff, Alberta, Canada, October 17–22, 1982. Abstract W4–6.

Hines, R. S., G. W. Wohlfarth, R. Moav, and G. Hulata. 1974. Genetic differences in susceptibility to two diseases among strains of the common carp. Aquaculture 3:187–197.

Hofer, B. 1904. Handbuch der Fischkrankheiten. Verlag Allg. Fischerei-Zeitung, B. Heller, Munich.

Lowenthal, W. 1907. Einschlussartige Zell- und Kernveränderungen in der Karpfenpocke. Z. Krebsforsch. 5:197–204.

Lucký, Z. 1970. Pathological changes with pox (*Epithelioma papulosum*) in the sheat fish (*Silurus glanis*). Acta Vet. Brno, Suppl. 1 1970:81–86.

McAllister, P. E., B. C. Lidgerding, R. L. Herman, L. C. Hoyer, and J. Hankins. 1985. Viral diseases of fish: first report of carp pox in golden ide (*Leuciscus idus*) in North America. J. Wildl. Dis. 21:199–204.

Mann, H. 1951. Chemische Untersuchungen über Knochenweiche bei Karpfen. Arch. Fischereiwiss. 3:105–113.

Mawdesley-Thomas, L. E., and D. Bucke. 1967. Fish pox in the roach (*Rutilus rutilus* L.). Vet. Rec. 81:56.

Nigrelli, R. F. 1948. Hyperplastic epidermal disease in the bluegill sunfish, *Lepomis macrochirus* Rafinesque. Zoologica 33:133–137.

Nigrelli, R. F. 1952. Virus and tumors in fishes. Ann. N.Y. Acad. Sci. 54:1076–1092.

Probst, E. 1951. Ein Beitrag zur Pockenkrankung des Karpfens. Allg. Fisch-Ztg. 76:479–480.

Racz, F., L. Békéski, and E. Szabo. 1980. Neuere Beiträge zur Ätiologie der Karpfenpocken. Fisch Umwelt 8:49–58.

Riedmüller, S. 1965. Elektronenmikroskopische Untersuchungen zur Pockenkrankheit und infektiösen Bauchwassersucht des Karpfens. Allg. Fisch-Ztg. 18:39–41.

Roegner-Aust, S. 1953. Zur Frage einer Virusätiologie bei verschiedenen Fischkrankheiten. Münch. Beitr. Abwasser-Fisch-Flussbiol. 1:120–145.

Roegner-Aust, S., and B. Deubner. 1952. Weitere Untersuchungen über die Pockenkrankheit des Karpfens. Z. Naturforsch. 7:572–573.

Roegner-Aust, S., and F. Schleich. 1951. Zur Ätiologie einiger Fischkrankheiten. Z. Naturforsch. 6:448–451.

Sano, T., H. Fukuda, and M. Furukawa. 1984. *Herpesvirus cyprini*: biological and oncogenic properties. Pages 55–56 in International seminar on fish pathology for the 20th anniversary of the Japanese Society of Fish Pathology, Tokyo, Japan, September 8–10, 1984. Abstract.

Sano, T., H. Fukuda, and M. Furukawa. 1985. *Herpesvirus cyprini*: biological and oncogenic properties. Fish Pathol. 20:381–388.

Sano, T., H. Fukuda, M. Furukawa, H. Hosoya, and Y. Moriya. 1985. A herpesvirus isolated from carp papilloma in Japan. Pages 307–311 in A. E. Ellis, ed. Fish and shellfish pathology. Academic Press, London.

Schäperclaus, W. 1969. Virusinfektionen bei Fischen. Pages 1067–1141 in H. Rohrer, ed. Handbuch der Virusinfektionen bei Tieren, Vol. 2. Gustav Fischer Verlag, Jena.

Schubert, G. 1964. Elektronenmikroskopische Untersuchungen zur Pockenkrankheit des Karpfens. Z. Naturforsch. 19:675–682.

Schubert, G. H. 1966. The infective agent in carp pox. Bull. Off. Int. Epizoot. 65:1011–1022.

Schubert, G., and G. Meyer. 1960. Licht- und elektronenmikroskopische Untersuchungen zur Pockenkrankheit des Karpfens. Pages 472–477 in Verh. Dtsch. Zool. Ges., Münster/Westf., May 18–25, 1959.

Sonstegard, R. A., and K. S. Sonstegard. 1978. Herpesvirus-associated epidermal hyperplasia in fish (carp). Pages 863–868 in G. de-The, W. Henle, and F. Rapp, eds. Proceedings of the international symposium on oncogenesis and herpesviruses III. Int. Agency Res. Cancer, Sci. Publ. 24, Lyon, France.

Tesarčík, J., and J. Mares. 1967. Případ puchýřnatosti (epithelioma papulosum) u generačnií štiky (*Esox lucius* L.) Bul. VÚR Vodňany 1:1–5.

Wolf, K. 1973. Herpesviruses of lower vertebrates. Pages 495–520 in A. S. Kaplan, ed. The herpesviruses. Academic Press, New York.

21 Golden Shiner Virus Disease

Definition

Golden shiner virus disease is a reovirus infection, generally of low virulence, that occurs during high temperature (late summer and fall) in the commercially propagated bait minnow *Notemigonus crysoleucas*.

History

Golden shiner virus (GSV) was first encountered in the southern United States in 1977 during virologic examinations of host populations in which low levels of mortality were evident. The virus showed several properties similar to those of reovirus, and others similar to those of the virus of infectious pancreatic necrosis (IPNV). However, it was not neutralized by IPNV antiserum (Plumb et al. 1979). In confirmation of that early finding, Schwedler and Plumb (1980) prepared antiserum to GSV and showed that GSV and IPNV were unrelated serologically.

Field observations suggested that GSV was inherently an agent of low virulence, but Schwedler and Plumb (1982a) showed that it could result in high mortality if test fish were stocked at high density. Those authors also determined temperatures at which replication occurred in vitro and storage was favorable (Schwedler and Plumb 1982b).

Uncertainty about the virus grouping of GSV disappeared when Winton et al. (1987) compared biophysical properties of the RNAs and structural proteins of three viruses from three species of fish and found that they were each distinctive but typical members of the Reoviridae.

Signs and Pathologic Changes

Affected fish are listless and swim near the surface; they dive when disturbed but soon return to the top. Dying fish sometimes swim in a spiral. When viewed from above, many victims show a marked hemorrhagic dorsal surface and petechial hemorrhages on ventral surfaces, the cornea, visceral fat, and intestinal mucosa (Plumb et al. 1979).

Histopathologic changes have not been reported.

Etiology

Circumstantially, GSV has been implicated in episodes of low mortality among golden shiners, but only during seasonally high water temperatures. Limited mortality has been produced experimentally by intraperitoneal injection of the virus but clinical signs of disease did not appear.

Size and Shape

The GSV is a doubly encapsidated, nonenveloped particle having a mean diameter of about 75 nm.

Biophysical Properties

As determined by polyacrylamide gel electrophoresis, the genome of double-stranded RNA consists of 11 segments that occur in three size classes. Three large segments have molecular weights of $2.6–2.1 \times 10^6$, three are in the range of $1.9–1.5 \times 10^6$, and five are $1.1–0.3 \times 10^6$. Five structural proteins were separated: the two largest had molecular weights of 137,000 to 125,000, one was in the range of 75,000 to 70,000, a fourth was 50,000 to 40,000, and the fifth was 34,000 to 30,000 (Winton et al. 1987).

The virus is ether, pH, and heat stable; the preferred storage temperature is −70°C. Infectivity persists for at least 6 months at 4°C, but is lost within a week at 30°C and at −15°C (Plumb et al. 1979; Schwedler and Plumb 1982b). Replication occurs in FHM cells to a maximum titer of about 10^7 $TCID_{50}$/mL at 20 to 30°C, but not at 15 or 35°C. The BB, CCO, and RTG-2 cell lines are refractory. In FHM cells the virus causes focal syncytia, vacuolation followed by rounding, and local lysis. Focal areas of affected cells are sometimes overgrown by normal-looking cells. In a one-step growth curve at 30°C, a maximum level of cell-associated virus occurs at about 8 hours postinfection and CPE appears 2 hours later. Released virus peaks at about 16 hours.

Superior results and fewer problems with overgrowth are claimed for EPC cells (R. P. Hedrick, personal communication, 1987).

Diagnosis

The disease can be considered if, during hot weather, large golden shiners show clinical signs and mortality, and if other pathogens can be excluded as causal agents. Virus isolation and identification are suggested.

Isolation

Until specific infective tissues are identified, homogenates should be prepared from pooled visceral organs, diluted with a physiologic saline solution, decontaminated with antibiotics or by membrane filtration, and used as inocula. Incubation should be at 25 to 30°C, and cultures should be examined at least daily to avoid overlooking transient areas of CPE that become overgrown.

Identification

Virus is identified serologically by serum neutralization test with specific anti-GSV serum. A low level of cross-reactivity occurs with the reoviruses from the channel catfish and chum salmon; however, the viruses are considered to be serologically distinct (Amend et al. 1984; Hedrick et al. 1984). In addition, the three viruses differ in electrophoretic mobilities of structural proteins and RNA segments and in the range of cell lines that support growth (Hedrick et al. 1984; Winton et al. 1987).

Transmission and Incubation

Although vertical transmission is not excluded, fish-to-fish transmission best accounts for natural infections. Incubation time is probably about 1 to 2 weeks.

Source

Until shown to be otherwise, golden shiners are assumed to be the source of GSV.

Host and Geographic Range

The disease was first found among golden shiners in several southern states, where they are raised and shipped to other areas of the United States. Later, the virus was found in California in golden shiners and in normal-appearing grass carp (R. P. Hedrick, personal communication, 1987).

Control

Practical control measures are not known. Avoidance is the only realistic approach.

References

Amend, D. F., T. McDowell, and R. P. Hedrick. 1984. Characteristics of a previously unidentified virus from channel catfish (*Ictalurus punctatus*). Can. J. Fish. Aquat. Sci. 41:807–811.

Hedrick, R. P., R. Rosemark, D. Aronstein, J. R. Winton, T. McDowell, and D. F. Amend. 1984. Characteristics of a new reovirus from channel catfish (*Ictalurus punctatus*). J. Gen. Virol. 65:1527–1534.

Plumb, J. A., P. R. Bowser, J. M. Grizzle, and A. J. Mitchell. 1979. Fish viruses: a new double-stranded RNA icosahedral virus from a North American cyprinid. J. Fish. Res. Board Can. 36:1390–1394.

Schwedler, T. E., and J. A. Plumb. 1980. Fish viruses: serologic comparison of the golden shiner and infectious pancreatic necrosis viruses. J. Wildl. Dis. 16:597–599.

Schwedler, T. E., and J. A. Plumb. 1982a. Golden shiner virus: effects of stocking density on incidence of viral infection. Prog. Fish-Cult. 44:151–152.

Schwedler, T. E., and J. A. Plumb. 1982b. In vitro growth kinetics and thermostability of the golden shiner virus. J. Wildl. Dis. 18:441–446.

Winton, J. R., C. N. Lannan, J. L. Fryer, R. P. Hedrick, T. R. Meyers, J. A. Plumb, and T. Yamamoto. 1987. Morphological and biochemical properties of four members of a novel group of reoviruses isolated from aquatic animals. J. Gen. Virol. 68:353–364.

22 Lymphocystis Disease

Synonym: lymphocystis

Definition

Lymphocystis is a common, chronic, and benign infection caused by an iridovirus that results in uniquely hypertrophied cells, typically in the skin and fins of only the more advanced orders of fishes.

History

Lymphocystis literature spans more than 100 years and includes unusual events not found in the history of other fish diseases. Earlier writers reviewed details of the discovery of lymphocystis and the first decades of its history (Nigrelli and Ruggieri 1965; Nigrelli and Smith 1939; Rašín 1927; Weissenberg 1914). Accordingly, only selected highlights are included here.

As cited in Rašín (1927), Weissenberg (1914), and Woodcock (1904), the earliest reports were made during the latter part of the 19th century: Lowe, McIntosh, and Sandeman independently encountered peculiar skin tumors in pleuronectid fishes and, like Woodcock, arrived at erroneous identifications. For example, Sandeman mistook lymphocystis cells for parasite eggs, and Woodcock concluded that the cells were a previously unknown sporozoan parasite.

Woodcock and the marine zoologist Johnstone were interested in microsporidan infections and corresponded with each other. Johnstone sent a specimen flounder and a drawing of the head showing cysts that were thought to be another "myxosporidian infection." Woodcock sectioned the cysts and interpreted them to be "enormously hypertrophied cells." The sections showed no evidence of bacteria, and since knowledge of animal virology was then primitive, Woodcock's interpretation was understandable. The cells seemed to be located in lymph spaces and to be surrounded by lymphocytes; accordingly, the presumed sporozoan cause was named *Lymphocystis johnstonei*.

One year later, Woodcock's interpretation was supported by Johnstone himself. Other reports favoring a parasite orientation followed.

The year 1914 was especially significant because Weissenberg investigated the condition and became the first to conclude that lymphocystis was a viral infection. He used experimental methods and laid the groundwork for investigators who followed. Reasoning from his findings and from the comparatively scarce facts existing about virology, he stated that the giant cells were not protozoans but infected connective tissue cells. He equated the cytoplasmic inclusions with those of smallpox, rabies, and insect polyhedrosis viruses, and additionally proposed that the virus size might be determined by filtration. He also found that the disease could be transmitted by cohabitation. According to Rašín (1927), Weissenberg's interpretation of lymphocystis was independently supported by Claussen and by Joseph, who used the term "ultravisible virus."

Rašín undertook studies of the pathogenesis of lymphocystis in 1924 and 1925. He transmitted the disease by implantation of infective tissue, by injection of tissue homogenate, and by application of lesion material to scarified skin (Rašín 1927, 1928). His work was soon confirmed by Babiak (1927) and Schmidt (1929). Although Rašín made a highly significant contribution, it went largely unnoticed for years. Almost 25 years later, Weissenberg saw Rašín's publications and found support for the postulated viral etiology. He was impressed with Rašín's findings and made specific reference to his "brilliant experimental work" (Weissenberg 1951b).

The early literature on lymphocystis came from Europe, but as noted in the annotated bibliography of Nigrelli and Ruggieri (1965), Mavor and Feinberg found the disease in North America in 1918 and considered the cells to be a protozoan, which they named *Lymphocystis vitrel*. That bibliography also indicated that in 1926 Johnstone abandoned his belief in the parasite etiology and adopted Weissenberg's interpretation.

From the 1930s through the 1950s, much of the lymphocystis literature dealt with the occurrence of the disease in additional hosts and locations, but several details of Weissenberg's personal life during that time merit retelling. Because he was a Jew in Germany, his career was disrupted. Although he had served as a German military physician for 4 years during World War I, the Nazi Ministry withdrew his academic position in 1936. He left Germany, literally a pauper, in 1937. Coming to the United States, he resumed his investigations and as R. Weissenberg, he published 16 reports during the next 30 years (Weissenberg 1938, 1939a, b, 1945, 1949, 1951a, b, c, d, 1952, 1955, 1956, 1960, 1965a, 1965b; R. Weissenberg et al. 1937); in addition he published four reports as J. R. Weissenberg (Weissenberg 1951, 1960a, b, c). Clearly, he was the leading authority on the subject.

At the New York Aquarium, Nigrelli became interested in lymphocystis and published a histologic study with Smith in 1939. He added other case history reports (Nigrelli 1940, 1943, 1950), and later briefly wrote of being unable to transmit the disease either by cohabitation or by feeding, but that the method of implantation was successful (Nigrelli 1952). In Germany, Grützner (1956) similarly reported experimental transmission. Although her in vitro work did not result in the production of lymphocystis cells, she was the first to apply tissue culture to a fish virus. She also examined imprints of lesions by electron microscopy. Her micrographs showed round particles that were 270 to 460 nm in diameter, some of which were most certainly virus.

The 1960s produced new information on lymphocystis. Weissenberg continued his efforts in microscopy with emphasis on the osmiophilic granules of the inclusion substance (J. R. Weissenberg 1960a, b, c; R. Weissenberg 1960). As a measure of the high esteem in which he was held, he was guest of honor at the 1964 "Conference on Viral Diseases of Poikilothermic Vertebrates," sponsored by the New York Academy of Sciences and the U.S. Fish and Wildlife Service. He presented two papers of his own and coauthored a third (Walker and Weissenberg 1965; R. Weissenberg, 1965a, b). The most memorable presentation of the entire conference was his review of his 50 years of research on lymphocystis (Weissenberg 1965a), at the conclusion of which he received a standing ovation. Walker and Weissenberg (1965) correlated osmiophilic granules with virus particles, but this paper also served to mark the limitations of light microscopy in virology and the end of an era.

Walker's interests in lymphocystis dated back to 1947, and he later published brief descriptions of the disease in walleyes (*Stizostedion vitreum vitreum*), (Walker 1957, 1958). A preview of things to come was given as an oblique comment in an abstract (Walker 1961), where he noted having seen lymphocystis virus in electron micrographs. The telling confirmation of Weissenberg's thesis was published the following year: virus was abundantly present in lymphocystis cells (Gelber 1963; Walker 1962). Reinforcing data soon followed (Walker 1963; Walker and Wolf 1962). In a different vein, Walker (1965) reported on acridine orange staining that confirmed earlier Feulgen staining.

In West Virginia, Wolf chanced to collect an adult largemouth bass (*Micropterus salmoides*) with prominent lymphocystis lesions. Rašín's implantation transmission procedure was again effective. Weissenberg's filtration results were confirmed and extended, Rašín's desiccation of tissue without loss of infectivity was confirmed, and the virus was found to be glycerol and ether labile (Wolf 1962). The work led to a visit by Weissenberg, then in his early 80s but still surprisingly agile and seemingly tireless. He provided deeply appreciated interest, counsel, and encouragement. At the 1964 Conference on Viral Diseases, Wolf and Carlson (1965) published a one-step in vivo growth curve; and Dunbar and Wolf (1966) later described and illustrated the detailed sequence of changes in the infected cells. Almost anticlimatically, lymphocystis virus was isolated and Rivers' postulates were fulfilled by Wolf et al. (1966).

Infected cell cultures were also used for an ultrastructural study of the virus (Zwillinberg and Wolf 1968). The first marine fish cell culture system for lymphocystis was reported by Sigel et al. (1966) at the University of Miami. Lopez et al. (1969) published research on nucleic acid and protein synthesis in lymphocystis cells.

At Lehigh University, Midlige and Malsberger (1968) described in vitro multiplication and morphology of the virus, and Pritchard and Malsberger (1968) reported a cytochemical study of lymphocystis cells in vivo that paid particular attention to the hyaline capsule. Other case and host record reports appeared during the 1960s (McCosker 1969; Nordenberg 1962, 1966; Ryder 1961; Templeman 1965).

During the 1970s, most of the literature on lymphocystis consisted of reports that can be categorized as new host records, surveys, reports of occurrence, aspects of pathology, electron microscopy, or miscellaneous minor topics.

The following investigators reported new hosts: Aneer and Ljungberg (1976), Howse (1972), Huizinga (1977), Lawler et al. (1974, 1975, 1977), McCosker and Nigrelli (1971), McCosker et al. (1976), Miyazaki and Egusa (1972), Murchelano and Bridges (1976), and Paperna (1973).

Reports of occurrence and results of surveys extended the known range to much of the world: Australia (Ashburner 1975); Baja California and Panama (McCosker and Nigrelli 1971); Baltic Sea (Aneer and Ljungberg 1976; Vitnish and Baranova 1976); Bering Sea (Alpers et al. 1977; McCain et al. 1978); California coast (Stephens et al. 1970); Chesapeake Bay (Krantz 1970); Georgia coast (Smith 1970, 1972; Stickney and White 1974); Gulf of Mexico (Christmas and Howse 1970; Howse 1973; Overstreet and Howse 1977); Irish Sea (Perkins et al. 1972; Shelton and Wilson 1973); Japan (Miyazaki and Egusa 1972); North Sea (Mann 1970); Scotland (Roberts 1970); Tanzania (Paperna 1973); and Wisconsin (Amin 1979).

Papers with a theme of pathology ranged from pathogenesis (Roberts 1975) to infection of organs such as the eye (Dukes and Lawler 1975; Lawler and Dukes 1973); and internal tissues (Howse et al. 1977; Huizinga and Cosgrove 1973); effects of tagging (Clifford and Applegate 1970), and histochemistry (Howse and Christmas 1970).

Electron microscopy was the principal topic in the reports of Dolowy et al. (1976), Howse and Christmas (1971), Koch et al. (1976), and Madeley et al. (1978).

Judging from the literature that appeared, the early 1980s saw a decided shift in interest. Case reports of lymphocystis disease continued to appear, but the emphasis of research was on biophysical properties of the virus. Walker and Hill (1980) confirmed earlier studies of in vitro replication, and Robin and Berthiaume (1981) showed that small incomplete particles were present with larger and infectious virions. Other workers reported analyses of virus proteins (Flugel et al. 1982) or details of isotope uptake by infected host cells (Spitzer et al. 1982). The genome itself came under study by Darai et al. (1983) and proximate analysis of the composition of the virion was reported by Robin et al. (1983). Then Robin et al. (1984) gave the results of their analyses of virus proteins, and Berthiaume et al. (1984) published a definitive description of structural features of the virus.

Signs and Pathologic Changes

Lymphocystis is a terminal infection of individual cells, and the effect on the host is generally slight. The virus is typically dermatotropic and superficial; the condition is much like that of warts, in that the lesions are macroscopic and occur mostly at the periphery of the vascular system (Fig. 1).

Behavior

Infected fish behave normally but extensive growths can slightly affect swimming. Among fish in confinement, and conceivably in the wild, the protruding clusters of cells lead to aggression and cannibalization of the lesions. Vaughan (1979) reported, however, that susceptibility to predation was no greater in bluegills with experimen-

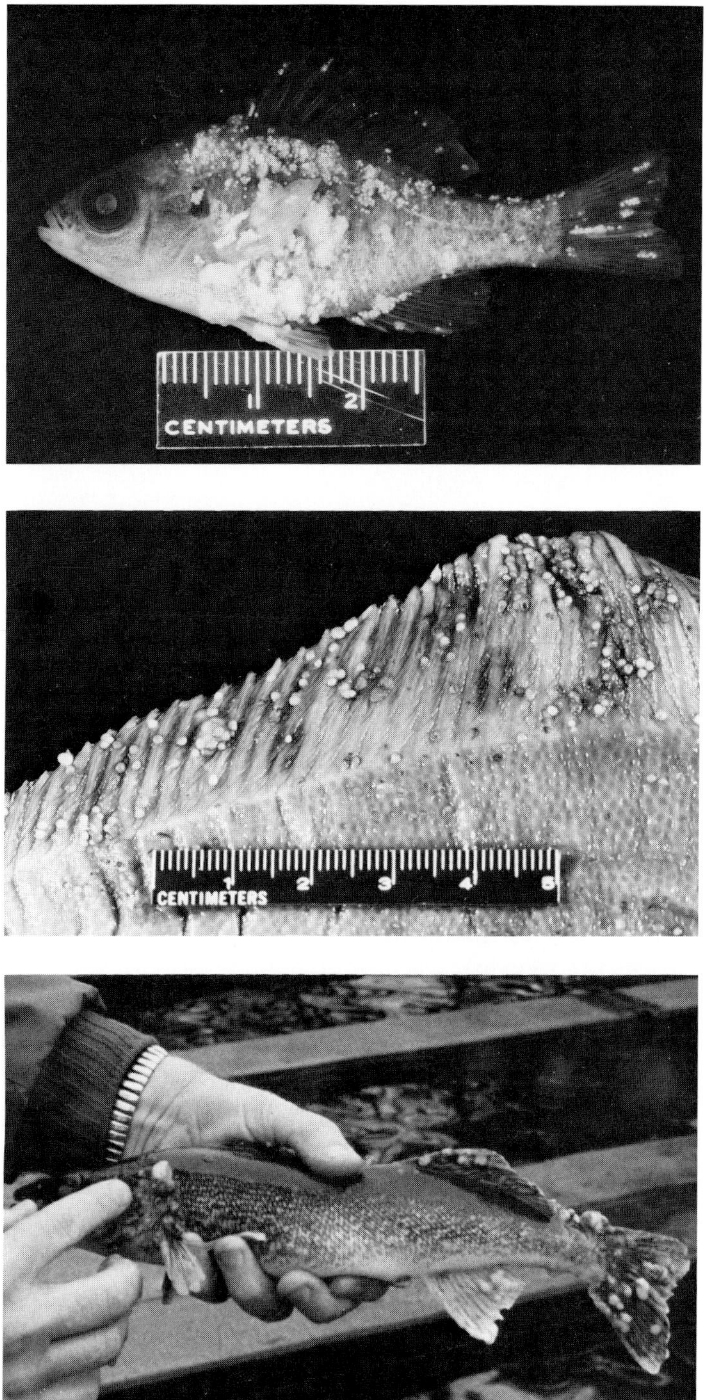

Figure 1. Lymphocystis disease in a centrarchid (top), pleuronectid (middle), and percid (bottom).

tal infections than in noninfected fish. Understandably, the size, location, and texture of lymphocystis lesions increase vulnerability to gill netting.

External Signs

At maturity the lesions are irregularly elevated masses of pebbled texture. The color is light cream to grayish, but covering epithelial tissue may be normally pigmented. Vascularity sometimes gives large clusters of cells a reddish hue. Considerable variation occurs in size, location, and distribution of the masses; infected cells also occur singly, but that is the exception. The tendency of lymphocystis cells to occur in clusters is puzzling and has not been explained.

Internal Signs

Lymphocystis cells are sometimes in muscle, peritoneum, and membranes covering internal organs, particularly among marine fish. Considering prediliction of the virus for connective tissue cells, the internal occurrence is not surprising. If more attention were paid to the viscera, the incidence of internal lymphocystis would undoubtedly be greater than is now known.

Histopathologic Findings

Lymphocystis cells are uniquely and enormously hypertrophied fish cells that at maturity are commonly 100 μm or more in greatest dimension and sometimes exceed 1 mm. The hypertrophy commonly results in a 50,000- to 100,000-fold increase in cell volume.

Major features of the cell are its thick hyaline capsule, a central enlarged nucleus, and prominent basophilic DNA cytoplasmic inclusions (Fig. 2). Isolated cells are spherical to oval and without processes or pseudopodia. In clusters, compression due to enlargement results in more angular forms.

The capsule is commonly 5 to 10 μm thick and laminar; its substance has been found to consist of sulfated acid mucopolysaccharide that stains positively with periodic acid–Schiff, (PAS), Bismark brown, and Hale's colloidal iron. The capsule is digestible with pronase and gives a positive methylene blue extinction test. Howse and Christmas (1970) showed that capsules stain dark purple with aldehyde fuchsin and that they are strongly alcianophilic at acid pH, thereby indicating functional sulfate groups. Metachromasia with toluidine blue is also indicative of sulfate groups. Where the capsules of adjacent lymphocystis cells touch, they sometimes give the appearance of being fused, and in old lesions the cells are embedded in a hyaline matrix.

Lymphocystis cells produced in vitro appear much like those that are the result of natural infection: they are greatly enlarged and have a hyaline capsule and an enlarged nucleus (Fig. 3).

The nucleus and the nucleolus of infected cells are enlarged more or less propor-

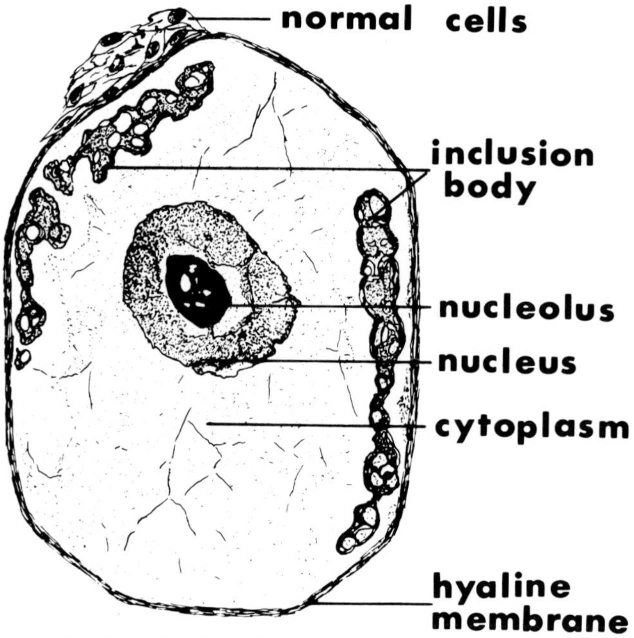

Figure 2. Sketch of a mature centrarchid-type lymphocystis cell showing its major features and (for size comparison) normal fibroblastic cells. Length of the infected cell is 140 μm.

tionally. The nuclear membrane is folded, invaginated, and otherwise grossly irregular. The in vitro cell nucleus differs in that its margin is smooth and regular. In natural lymphocystis the nucleus is typically centrally located, a feature that distinguishes lymphocystis from epitheliocystis (in which the nucleus is typically peripheral).

The cytoplasmic inclusion bodies of mature lymphocystis are basophilic and Feulgen positive, and are the sites of virus synthesis and maturation. The inclusions are of two morphologic types, differing according to host species. As typified by those in centrarchid cells, the inclusions may be massive, cordlike, vesicular to fenestrated, and located between the nucleus and capsule in discontinuous basketlike array (Fig. 4). The percid-type inclusion is netlike and distributed throughout the cytoplasm in dendritic fashion.

Developmental stages of infected cells have been described and illustrated by several investigators (Alexandrowicz 1951; Nigrelli and Smith 1939; Weissenberg 1914, 1920, 1939a, b, 1945, 1965b). A month-long time course at 25°C, showing 11 developmental stages (Fig. 5) was described by Dunbar and Wolf (1966).

Marine virus and cells have also been developed in vitro by Lopez et al. (1969). However, an associated dose-response phenomenon was evident. When a low multiplicity of infection (MOI) was used, the affected cells underwent hypertrophy and accumulated large amounts of protein, RNA, and DNA. On the other hand, when a much higher MOI was used, the cells showed increased rates of macromolecular synthesis, but became pyknotic and did not accumulate the products. Walker and Hill (1980) confirmed these findings and noted a progressive decline in infectivity with in vitro transfer, until on the fourth passage CPE failed to occur. The extinction

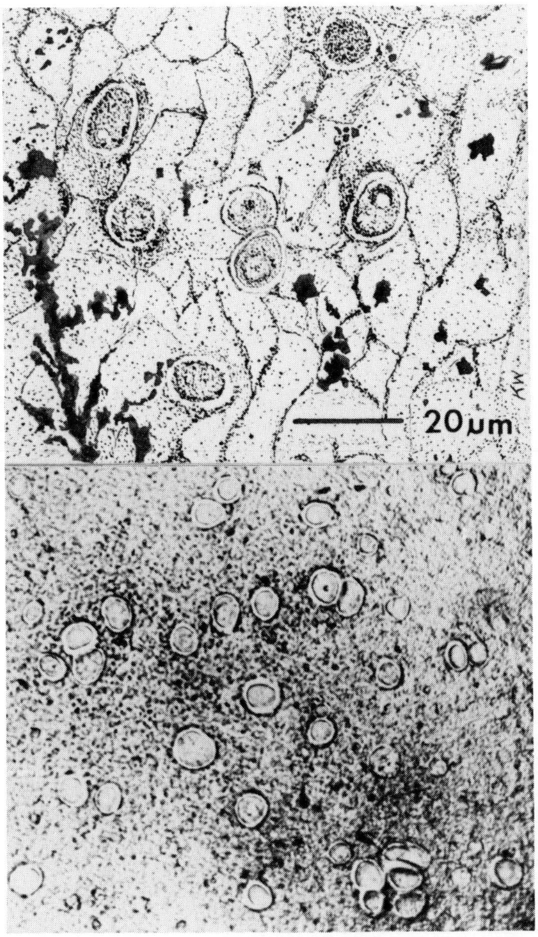

Figure 3. Living lymphocystis cells sketched by R. Weissenberg (1920) in a percid fin (top) are similar to centrarchid cells in vitro (bottom). From K. Wolf, M. Gravell, and R. G. Malsberger, "Lymphocystis virus: isolation and propagation in centrarchid fish cell lines," *Science* 151 (1966): 1004–1005, © AAAS.

of infectivity could have been a result of cropping cells before the growth cycle was complete.

Whereas light microscopy has revealed much about the infected cell and the sequence of changes, electron microscopy has mostly been used to examine the virus, and relatively few details are known about the cells' fine structure. Supporting the original findings of Walker (1962), agreement is general that virus is replicated at surfaces of the inclusion bodies and that the inclusions themselves are not membrane bound. The enlarged nucleus has a folded membrane and shows evidence of being polyploid. The inner surface of the capsule has microvilli. Midlige and Malsberger (1968) examined BF-2 cells at intervals after infection: cells were unchanged, except that cellular processes were retracted at 4 days; inclusions were present on day 8; virus was first found on day 15; and at 28 days virus was abundant and estimated to number 100,000/cell. Howse et al. (1977) described some fine-structure changes of lymphocystis cells in the epicardium of a marine sciaenid fish.

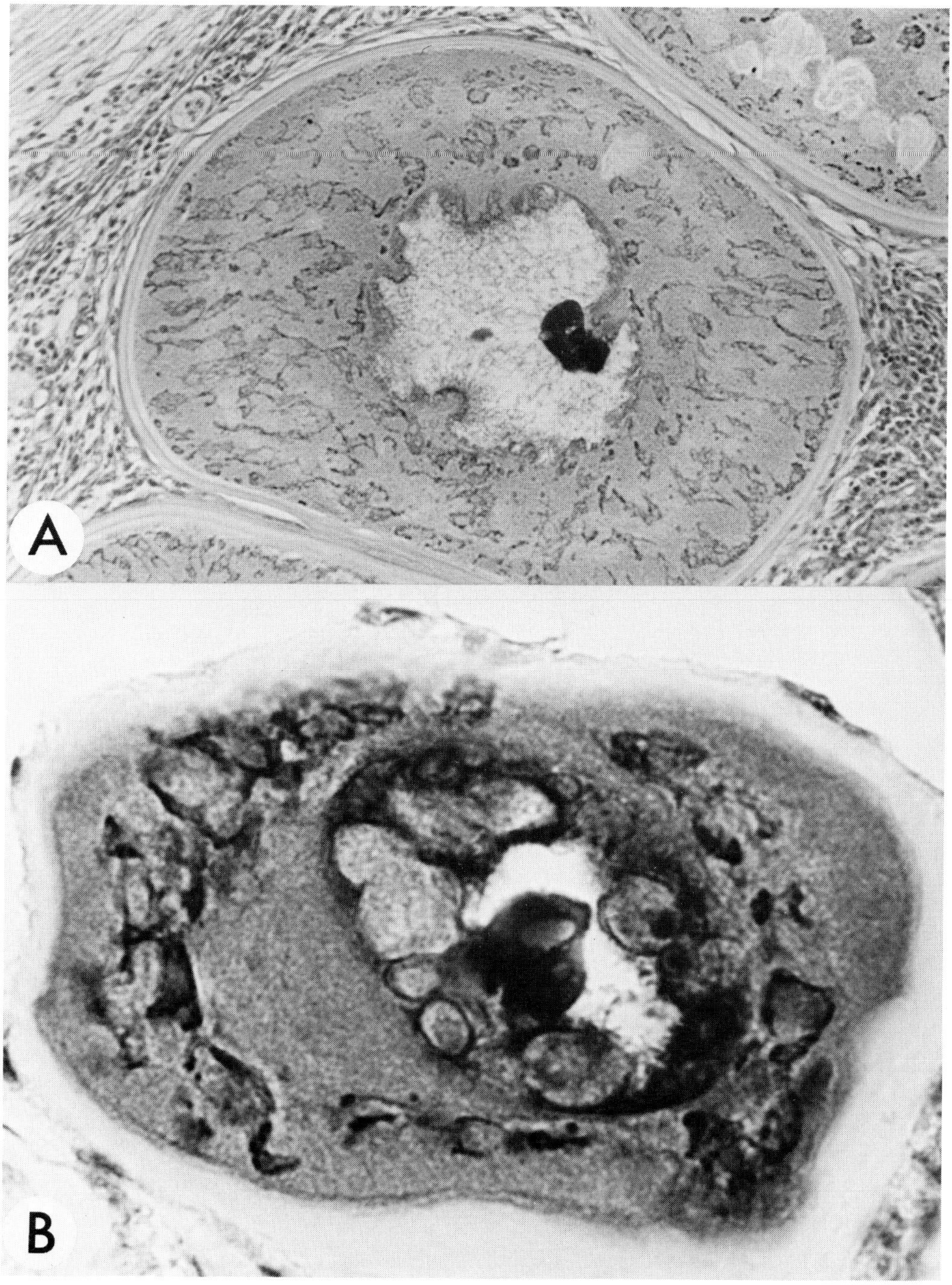

Figure 4. Representative lymphocystis cells show differences principally in the form and distribution of cytoplasmic inclusion substance. *A.* Percid type. *B.* Centrarchid type.

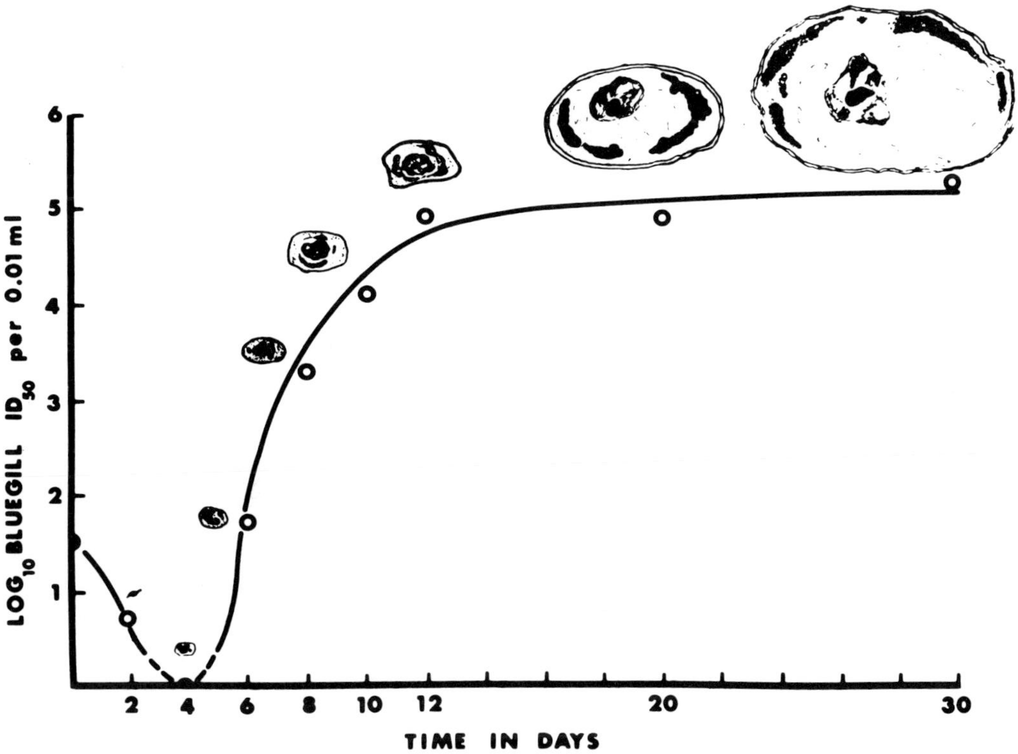

Figure 5. A comparison of lymphocystis cell development (after Dunbar and Wolf 1966) and replication of infective virus in a one-step growth cycle in bluegills at 25°C. Sketches of cells are proportional to mean cell size. From Wolf and Carlson (1965). Reprinted with the permission of the New York Academy of Sciences.

Clinical Findings

The impact of lymphocystis on its host is slight. For walleyes in Saginaw Bay, Lake Michigan, Hile (1954) reported that infected fish weighed about 6% less than noninfected fish of the same length. Among white crappies (*Pomoxis annularis*), infected fish weighed 3 to 5% less than noninfected fish of the same length in one population (Witt 1957) and 3.9% less in another (Greene and Murphy 1974). On the other hand, Petty and Magnuson (1974), who removed lesions from infected bluegills, found that the victims still weighed slightly more than normal fish of the same length. The differences were only marginally significant.

Etiology

Lymphocystis disease is caused by an iridovirus; virus of centrarchid origin has been deposited with the American Type Culture Collection as VR-342.

Size and Shape

Lymphocystis is the largest of the iridoviruses (Kelly and Robertson 1973), and although most determinations of its size are in the range of 200 ± 50 nm, the extremes are as small as 130 nm and as large as 330 nm. In part, size differences could reflect host origin or the presence of incomplete particles of 100 to 150 nm as reported by Robin and Berthiaume (1981). Other differences are undoubtedly due to the kind of preparation used—whether thin section or negatively stained.

Major features of the virus consist of a dense core within two unit membranes of the capsid, surface subunits, and an outermost fringe of fuzzy material (Figs. 6 and 7). Walker (1962) described the nucleoid as a ball of osmiophilic thread and noted that adjacent virions were separated from each other due to the presence of an outer fibrous material. Those findings were confirmed by later workers (Berthiaume et al. 1984; Howse and Christmas 1971). Madeley et al. (1978) noted that the nucleoid of degenerating particles showed tubular material that was suggestive of being coiled, and that the surface of the capsid had knoblike subunits. Berthiaume et al. (1984) described the nucleoid as a tangled ball of osmiophilic threads or granules; they confirmed the presence of globular surface subunits as well as the outer fringe of filaments (Fig. 7).

Biophysical Properties

The virus is ether, glycerol, and heat labile, and does not hemagglutinate. Its infectivity persists through three cycles of freezing and thawing, and is readily preserved by freezing. Lyophilized and simply desiccated preparations have retained infectivity for more than 10 years (Wolf et al. 1979). The virus has specific tropism for fibroblastic cells and is best replicated by cultures derived from the host or a closely related species.

The virus is composed of 42% protein, 17% lipid, and 1.6% DNA; the remainder consists of unidentified materials that could be sugars (Robin et al. 1983). When virus that had been isolated from fish was separated on polyacrylamide gel, as many as 33 polypeptides were identified; their molecular weights ranged from 14,000 to 220,000 (Flugel et al. 1982). Culture-grown virus was similarly separated into 22 polypeptides whose molecular weights ranged from 30,700 to a comparable high of 210,000 (Robin et al. 1984).

Contour length of DNA from pleuronectid fishes averaged 49 ± 23 μm, a value that corresponded to a molecular weight of 93,000 ± 44,000. However, restriction enzyme analysis and denaturation, followed by annealing, gave a value of only about 65,000. The annealing process yielded a contour length of 34.25 μm, suggesting terminal redundancy (Darai et al. 1983).

Isotope labeling has shown that lymphocystis virus stimulates increased uptake of DNA precursor but that overall, normal DNA synthesis is switched off in favor of virus replication (Sigel et al. 1971).

A one-step growth curve in bluegills at 25°C includes an eclipse at 3 to 5 days postinfection. Thereafter, growth is exponential until about day 12, when a maximal

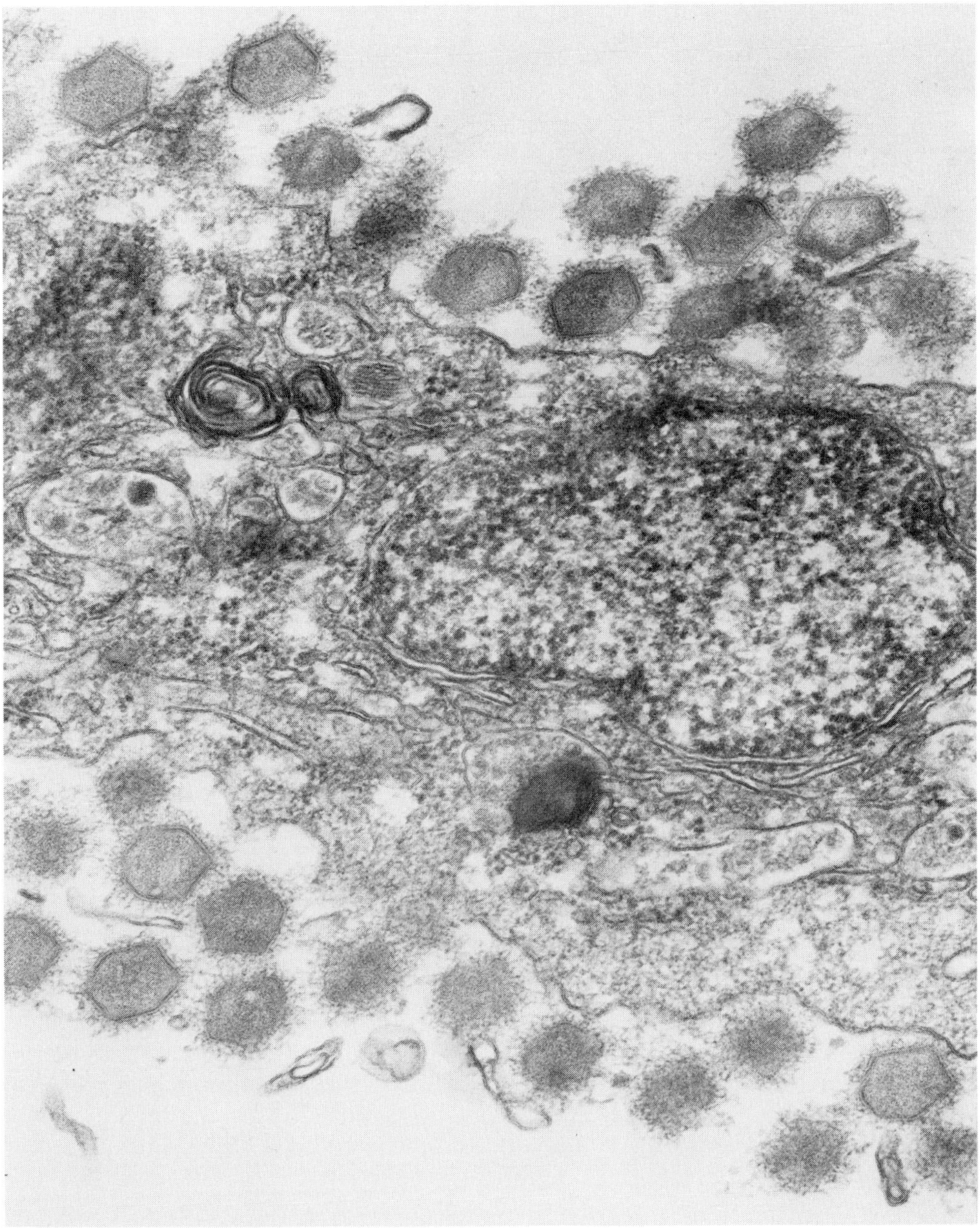

Figure 6. Lymphocystis virus in a section of infected centrarchid tissue. The angularity of the particles is evident, as is the outer fringe of the fuzzy filaments. ×57,000. Courtesy of P. D. Lunger.

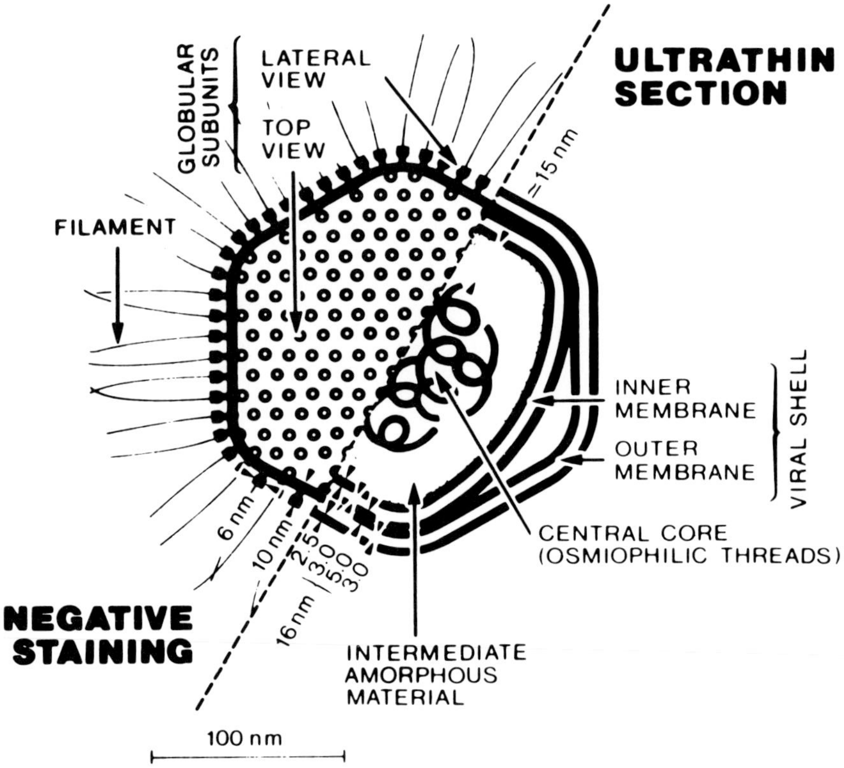

Figure 7. Proposed ultrastructure of negatively stained lymphocystis virus. From Berthiaume et al. (1984). Reprinted with the permission of Academic Press, Inc.

titer of about 10^7 ID_{50}/g is reached. Cell size continues to increase for 2 more weeks, but infectivity does not (Fig. 5).

Diagnosis

The usual cases of lymphocystis disease can be diagnosed by visual examination and the finding of individual or clustered hypertrophied cells on the skin or fins of fishes known to be susceptible. They might be confused with superficial cysts of *Ichthyophthirius multifiliis* or of trematodes, but those cysts show internal movement. Diagnosis can be confirmed by histologic section.

In some cases of lymphocystis in walleyes, lesions coexist with and resemble dermal sarcomas. Lymphocystis cells have in fact been found adjoining and even embedded in sarcomatous tissue. Careful examination distinguishes between the two growths, and histologic sections clearly reveal the giant lymphocystis cells. Electron microscopy shows large icosahedral virions in the hypertrophied cells and C-type particles in cells of the neoplasm, but the two viruses have never been found in the same cell (Walker 1957, 1958, 1969; Yamamoto et al. 1976).

Isolation

If lesions are mature, virus is abundant and can be readily isolated if the cell culture system is susceptible. Virus is readily released during mechanical homogenization. Sonication can be used, but cautiously, because excessive power input can inactivate the virus. Decontamination can be effected with antibiotic treatment or by filtration, but porosity should be no smaller than 0.45 μm. Incubation should be in a temperature range that is favored by the host and extended for at least several weeks. Maximal titers can be expected to reach 10^6–10^7 $TCID_{50}$/mL after in vitro cells reach maturity. Wolf and Quimby (1973) have described plaquing procedures for quantification of lymphocystis virus.

Identification

Histologic sections provide a rare instance where a virologic identification can be made confidently without isolation and serologic testing. The large cells, with their distinctive capsule, enlarged and centrally located nucleus and nucleolus, and cytoplasmic inclusions, are unique. Histologic sections of senescent lymphocystis cells may lack a nucleus and inclusion substance, but the mass of eosinophilic cytoplasm is homogeneous. In contrast, epitheliocystis cells have an eccentric nucleus and a massive granular basophilic inclusion body.

Serologic identification of lymphocystis virus has been made by using an indirect fluorescent antibody test. When used against homologous virus, rabbit antiserum shows reactivity with inclusion bodies in infected cell cultures (Walker and Hill 1980). The response of heterologus isolants of lymphocystis virus has not been reported. Antigenic and serologic comparisons of virus from an array of unrelated hosts will prove to be interesting, as will still other comparisons with the large icosahedral fish viruses—the agent of viral erythrocytic necrosis and the iridovirus of goldfish.

Transmission and Incubation

The occurrence of lymphocystis in nature, in husbandry, among exhibit fish, and as a result of experimentation indicates that contact transmission is the principal means by which virus is spread. External surfaces, including the gills, are the chief portal of entry; the oral route seems not to be involved, and no evidence exists for vertical transmission.

Factors such as high population density and external trauma enhance transmission. Incidences of virtually 100% have occurred among centrarchids and percids in intensive freshwater pond husbandry and, according to Paperna et al. (1982), among sparids at high density in seawater holding tanks. External trauma has been implicated in transmission during spawning (Ryder 1961), from combat behavior (McCosker 1969), and from such human activities as netting, marking, or tagging (Clifford and Applegate 1970; Olson 1958).

Circumstantially, transmission has seemingly been enhanced by several external parasites: copepods (Nigrelli 1950), leeches (Petty and Magnuson 1974), and perhaps isopods (Lawler et al. 1974). However, Lawler et al. were unable to demonstrate experimental transmission with the isopods. Virus transmission also failed when a ciliate, *Ichthyophthirius multifiliis*, and monogenetic trematodes moved from infected to healthy bluegills (K. Wolf, unpublished data).

Experimental transmission has been achieved by contact, by implantation, by application of infectious material to gills or scarified or traumatized skin, or internally by injection.

Incubation time and lesion duration depend on host species, virus, and temperature. In cool-water hosts, incubation and duration are lengthy (visible lesions may persist for a year). In warm-water species, the cycle is often completed in several weeks. Based on development of visual lesions, lymphocystis infection at the cellular level is a relatively slow process: 9 to 14 days in percids in aquaria (Weissenberg 1914), 13 to 15 days in anabantids (Rašín 1927, 1928), 6 weeks in centrarchids (Weissenberg 1945), 6 weeks at 12°C and 10 days at 25°C in centrarchids (Wolf 1962), a month in pleuronectids at 10°C (Roberts 1975), 5 to 7 days in sciaenids at 25°C, 9 to 11 days at 20°C, and 21 days at 15°C (Cook 1972).

R. Walker (personal communication), who compared lymphocystis development in a percid and in a centrarchid, reported that homologous virus in the percid required 40 days to produce a cell size that occurred in the centrarchid in only 10 days.

Source

The source of virus is infected fish whose mature lesions have been breached, abraded, sloughed, or otherwise damaged.

Host and Geographic Range

Lymphocystis occurs in freshwater and marine species but only among the more highly evolved fishes. Infection is common among some perciform, pleuronectiform, and tetraodontiform fishes—three of the highest orders—but also occurs in at least one representative of five lower orders (Table 1).

Nigrelli and Ruggieri (1965) compiled a list of 49 susceptible species representing 20 families and 5 orders. Lawler et al. (1977) added 47 species from 12 families and 1 order. Still other susceptibles have been reported and more will be found.

There are puzzling features about lymphocystis. In mixed natural populations it appears to be selective for one or two species and yet, under cohabitation, it is able to infect several species of a family. That ability is particularly marked within the Centrarchidae, Chaetodontidae, Cichlidae, Pleuronectidae, and Sciaenidae. Experimentally, the host range can be extended even further if virus is injected. One can effect interfamilial infections, and one case exists of an experimental interordinal infection (Weissenberg 1951d). However, even within a particular family some genera have proved to be susceptible and others refractory.

Factors in natural occurrence undoubtedly involve the particular habits and eco-

Table 1. A list of orders and families of fishes with one or more species susceptible to lymphocystis virus

Order number[a]	Order and family[a]	Common name	Reference[b]
20	Clupeiformes		
	Clupeidae	Herrings	2
26	Salmoniformes		
	Osmeridae	Smelts	1
32	Ophidiiformes		
	Ophidiidae	Cusk-eels	2
34	Lophiiformes		
	Ogcocephalidae	Batfishes	2
36	Cyprinodontiformes		
	Cyprinodontidae	Killifishes	1
47	Scorpaeniformes		
	Scorpaenidae	Scorpion fishes	2
	Hexagrammidae	Greenlings	1
48	Perciformes		
	Centropomidae	Snooks	2
	Percicthyidae	Temperate basses	2
	Serranidae	Sea basses	1, 2
	Pseudochromidae	Dottybacks	2
	Centrarchidae	Sunfishes	1, 2
	Percidae	Perches	1, 2
	Lutjanidae	Snappers	2
	Sparidae	Porgies	1, 2
	Sciaenidae	Drums	2
	Mullidae	Goatfishes	1
	Scatophagidae	Scats	1
	Chaetodontidae	Butterfly fishes	1, 2
	Cichlidae	Cichlids	1, 2
	Pomacentridae	Damselfishes	1, 2
	Labridae	Wrasses	1, 2
	Clinidae	Clinids	2
	Blenniidae	Combtooth blennies	1, 2
	Eliotrididae	Sleepers	1
	Gobiidae	Gobies	2
	Siganidae	Rabbitfishes	2
	Anabantidae	Labyrinth fishes	1
49	Pleuronectiformes		
	Bothidae	Lefteye flounders	2
	Pleuronectidae	Righteye flounders	1, 2
	Soleidae	Soles	1
50	Tetraodontiformes		
	Balistidae	Leatherjackets	
	Ostraciidae	Boxfishes	1
	Diodontidae	Porcupine fishes	2

[a]Phylogenetic sequence of orders and families according to Nelson (1984).
[b]1, Nigrelli and Ruggieri (1965); 2, Lawler et al. (1977).

logical niche of the host. Other factors might also play a role, but it seems likely that the virus itself has properties that suit it to some hosts and not to others. The difficulty in trying to better understand such relationships is that the available information is a mix of field observations and experimental results with fishes whose health history is unknown.

In a study of seasonal distribution of the disease in a lake, Witt (1957) found that lymphocystis occurred only in one centrarchid species, even though other members of the family were abundant. On the other hand, experimental infections among species and genera of centrarchids (*Micropterus* to *Lepomis*) were readily accomplished (Wolf 1962). Smith (1970) found that two genera of sciaenids (*Stellifer* and *Cynoscion*) were infected in a Georgia river, and Christmas and Howse (1970) reported the same occurrence in *Micropogon* and *Cynoscion* in estuarine waters of the Gulf of Mexico. Cook (1972) transmitted virus from one sciaenid genus to two others (from *Micropogon* to *Cynoscion* and *Pogonias*), but was unable to infect a member of the sciaenid genus *Leiostomus*, or even one other member of the genus *Cynoscion*. The virus also failed to infect a freshwater centrarchid. Lawler et al. (1974) reported one successful intergeneric infection of sciaenids (*Bairdiella* to *Larimus*), and one that was reciprocally refractory (*Bairdiella-Micropogon*). Among cichlids, Nigrelli and Ruggieri (1965) transmitted virus from *Aequidens* to *Hemichromis* and to three species of *Tilapia*.

R. Weissenberg (1951d) was partly successful with interordinal infections—Salmoniformes to Cyprinodontiformes. In two of three trials, he infected cyprinodonts with percid material (*Stizostedion* to *Fundulus*); however, early cell degeneration occurred and he concluded that conditions for the *Stizostedion* virus were unfavorable in the distantly related host.

In laboratory experiments, with susceptible hosts, freshwater centrarchids and cichlids and marine sciaenids lend themselves well to being held at room temperature.

The geographic range of lymphocystis is probably global. Most accounts of the disease are from Europe, the British Isles, and North America. It was found by Ashburner (1975) in Australia and by Paperna (1973) in Africa. Alpers et al. (1977) and McCain et al. (1978) reported the infection in the Bering Sea, and Miyazaki and Egusa (1972) documented it in Japan. Weissenberg (1965b) and McCosker and Nigrelli (1971) found the disease in fishes of Central America. According to T. Sano (personal communication), lymphocystis is also known from Hawaii, the South Pacific, and Indochina. Thakur and Nasar (1977) reported it in India. There is no reason to doubt that it will eventually be found in Indonesia, Asia, and elsewhere.

Immunity

Comparatively little information is available on the immune response to lymphocystis virus, but Roberts (1975) and Russell (1974) have shown evidence for the existence of both a cellular and a humoral response. Moreover, their works help explain some of the apparent inconsistencies concerning the virus and the disease.

The very location of the virus in the host works against its being effectively recognized as an antigen, and yet the usual occurrence of the disease is among the young, and large-scale recurrence is not known; accordingly, one could argue for an age-dependent resistance. Resistance probably develops with age, but it differs markedly from that in viruses that can infect but cause no disease in adults, e.g., channel catfish virus and IPNV. Weissenberg (1945), for example, was able to reinfect some specimens as many as three times during a 4-month period. Experimentally at least, older fish can usually be infected. Wolf (1962) noted that a second infection could be initiated in bluegills that bore lymphocystis lesions. However, lesions from the second inoculum were always smaller than those from the first—evidence for possible immune factors.

The usual location of lymphocystis virus—superficial focal lesions at the periphery of the vascular system and behind the heavy hyaline capsule—inhibits antigen recognition by the host. That recognition occurs when antigen is released.

At Stirling University in Scotland, Russell (1974), who worked with naturally infected pleuronectids, called specific attention to features of the cellular response and demonstrated a humoral response. He described and illustrated macrophages, plasma cells, or plasma cell precursors (immune cells) around lymphocystis cells. The immune cells were most common around the contents of infected cells that had burst—an event that occurs late in the duration of the lesion. He also showed that serum from infected fish reacted reciprocally in a precipitin test with homologous and heterologous lesion homogenates. Normal skin and fin homogenates were unreactive, as were sera from noninfected specimens.

At the same laboratory at Stirling University, Roberts (1975) carried out similar studies and confirmed Russell's findings. He found no immune response until late in the disease cycle, when mature infected cells burst or were breached and masses of antigen were released. Cell-mediated and humoral responses occurred. Regression of the lesions began when precipitin reactions were first noted.

Lymphocystis seems to persist longest in cool-water and cold-water fishes. The duration of infection may be due to the direct effect of temperature on virus replication, but the known reduction of efficiency of the immune response at low temperature might also be a factor.

Control

To the host, lymphocystis is seldom, if ever, life-threatening; but it can affect human economic interests. Lesions can be unsightly and the marketability of obviously infected fish is depressed or lost, as sometimes occurs among food and aquarium fishes.

Avoidance is a universal control measure, but its applicability is limited to propagated species intended for market. The literature contains suggestions that food fish with lymphocystis that are captured should be removed from the population; however, that measure probably has greater psychological benefit to the fisherman than to the lowering of disease prevalence in wild stocks.

Theoretically at least, captive propagated brood stock with external lesions could be culled. The success of such a measure would hinge on the absence of internal and submacroscopic lesions, and on the absence of virus in the facility and its water supply. Like surgical removal or cautery of visible lesions, culling is more applicable to the hobbyist or the aquarist than to the aquaculturist.

In the United States, some commercial producers of yellow perch (*Perca flavescens*) and largemouth bass have had populations of fingerlings with infections that deterred prospective buyers. Suggested control measures included holding the fish until lesions were shed, marketing the apparently clean specimens, and disinfecting the ponds and resuming operations with clean stock.

At present, there is no known method of therapy or of immunization, and the possible existence of resistant fish strains is wholly unexplored.

References

Alexandrowicz, J. S. 1951. Lymphocystis tumours in the red mullet (*Mullus surmuletus* L.). J. Mar. Biol. Assoc. U.K. 30:315–332.

Alpers, C. E., B. B. McCain, M. S. Myers, and S. R. Wellings. 1977. Lymphocystis disease in yellowfin sole (*Limanda aspera*) in the Bering Sea. J. Fish. Res. Board Can. 34:611–616.

Amin, O. M. 1979. Lymphocystis disease in Wisconsin fishes. J. Fish Dis. 2:207–217.

Aneer, G., and O. Ljungberg. 1976. Lymphocystis disease in Baltic herring (*Clupea harengus* var. *membras* L). J. Fish Biol. 8:345–350.

Ashburner, L. D. 1975. Lymphocystis in paradise fish (*Macropodus opercularis*) in Australia. Aust. Vet. J. 51:448–449.

Babiak, P. 1927. Příspěvek k vývoji buněk lymphocystis. Dissertation. Institute for General Zoology and Parasitology of the College of Veterinary Medicine, Brno, Czechoslovakia.

Berthiaume, L., R. Alain, and J. Robin. 1984. Morphology and ultrastructure of lymphocystis disease virus, a fish iridovirus, grown in tissue culture. Virology 135:10–19.

Christmas, J. Y., and H. D. Howse. 1970. The occurrence of lymphocystis in *Micropogon undulatus* and *Cynoscion arenarius* from Mississippi estuaries. Gulf Res. Rep. 3:131–154.

Clifford, T. J., and R. L. Applegate. 1970. Lymphocystis disease in tagged and untagged walleyes in a South Dakota lake. Prog. Fish-Cult. 32:177.

Cook, D. W. 1972. Experimental infection studies with lymphocystis virus from Atlantic croaker. Pages 329–335 in J. W. Avault, E. Boudreaux, and E. Jaspers, eds. Proceedings of the third annual workshop, World Mariculture Society, St. Petersburg, Florida. January 26–28, 1972.

Darai, G., K. Anders, H.-G. Koch, H. Delius, H. Gelderblom, C. Samalecos, and R. M. Flugel. 1983. Analysis of the genome of fish lymphocystis disease virus isolated directly from epidermal tumors of Pleuronectes. Virology 126:466–479.

Dolowy, W. C., E. A. Koch, R. H. Spitzer, S. Greenberg, and E. R. Brown. 1976. Subviral particles within and associated with the lymphocystis virus of fish. Proc. Am. Assoc. Cancer Res. 17:3. Abstract.

Dukes, T. W., and A. R. Lawler. 1975. The ocular lesions of naturally occurring lymphocystis in fish. Can. J. Comp. Med. 39:406–410.

Dunbar, C. E., and K. Wolf. 1966. The cytological course of experimental lymphocystis in the bluegill. J. Infect. Dis. 116:466–472.

Flugel, R. M., G. Darai, and H. Elderblom. 1982. Viral proteins and adenosine triphosphate phosphorhydrolase activity of fish lymphocystis disease virus. Virology 122:48–55.

Gelber, E. 1963. Lymphocystis disease. Rensselaer Eng. 16:8–10.

Greene, D. S., and C. E. Murphy. 1974. Food and feeding habits of the white crappie (*Poxomis annularis* Rafinesque) in Benbrook Lake, Tarrant County, Texas. Tex. J. Sci. 25:35–51.

Grützner, L. 1956. Überprüfung einiger Anwendungsmöglichkeiten der Gewebekultur von *Lebistes reticulatus* (Peters) und *Macropodus opercularis* (Linne') in der Virusforschung. Zentralbl. Bakteriol. Parasitenkd. Infektionskr. Hyg. Abt. I Orig. 165:81–96.

Hile, R. 1954. Fluctuations in growth and year-class strength of the walleye in Saginaw Bay. U.S. Fish Wildl. Serv. Fish. Bull. 56:7–59.

Howse, H. D. 1972. Snook (*Centropomus*: Centropomidae): new host for lymphocystis, including observations on the ultrastructure of the virus. Am. Midl. Nat. 88:476–479.

Howse, H. D. 1973. Wart-like growths are unsightly but not dangerous to man or fish. Mar. Briefs 2:3.

Howse, H. D., and J. Y. Christmas. 1970. Lymphocystis tumors: histochemical identification of hyaline substances. Trans. Am. Microsc. Soc. 89:276–282.

Howse, H. D., and J. Y. Christmas. 1971. Observations on the ultrastructure of lymphocystis virus in the Atlantic croaker *Micropogon undulatus* (Linnaeus). Virology 44:211-214.

Howse, H. D., A. R. Lawler, W. E. Hawkins, and C. A. Foster. 1977. Ultrastructure of lymphocystis in the heart of the silver perch, *Bairdiella chrysura* (Lacepede), including observations on normal heart structure. Gulf Res. Rep. 6:39–57.

Huizinga, H. W. 1977. Lymphocystis disease in the green terror. Trop. Fish Hobbyist 25:47–60.

Huizinga, H. W., and G. E. Cosgrove. 1973. Surface and visceral lymphocystis disease in a white crappie, *Pomoxis annularis*. J. Wildl. Dis. 9:349–351.

Kelly, D. C., and J. S. Robertson. 1973. Icosahedral cytoplasmic deoxyriboviruses. J. Gen. Virol. 20(Suppl.):17–41.

Koch, E. A., W. C. Dolowy, R. H. Spitzer, S. Greenberg, and E. R. Brown. 1976. Postulated developmental forms in the life cycle of the lymphocystis virus. Cancer Biochem. Biophys. 1:163–166.

Krantz, G. E. 1970. Lymphocystis in striped bass, *Roccus saxatilis*, in Chesapeake Bay. Chesapeake Sci. 11:137–139.

Lawler, A. R., and T. W. Dukes. 1973. Lymphocystis in the eye. Vet. Rec. 93:297.

Lawler, A. R., H. D. Howse, and D. W. Cook. 1974. Silver perch, *Bairdiella chrysura*: new host for lymphocystis. Copeia 1974:266–269.

Lawler, A. R., H. D. Howse, and D. W. Cook. 1975. Lymphocystis infections in the silver perch, *Bairdiella chrysura*. J. Miss. Acad. Sci. 19:183.

Lawler, A. R., J. T. Ogle, and C. Donnes. 1977. *Dascyllus* spp: new hosts for lymphocystis, and a list of recent hosts. J. Wildl. Dis. 13:307–312.

Lopez, D. M., A. R. Beasley, L. S. Dietrich, and M. M. Sigel. 1968. Biochemical and morphological aspects of the early phase of lymphocystis infection in tissue culture. Bacteriol. Proc. 1968:176. Abstract.

Lopez, D. M., M. M. Sigel, A. R. Beasley, and L. S. Dieterich. 1969. Biochemical and morphologic studies of lymphocystis disease. Natl. Cancer Inst. Monogr. 31:223–236.

McCain, B. B., M. S. Myers, W. D. Gronlund, S. R. Wellings, and C. E. Alpers. 1978. The frequency, distribution, and pathology of three diseases of demersal fishes in the Bering Sea. J. Fish Biol. 12:267–276.

McCosker, J. E. 1969. A behavioral correlate for the passage of lymphocystis disease in three blennioid fishes. Copeia 1969:636–637.

McCosker, J. E., and R. S. Nigrelli. 1971. New records of lymphocystis disease in four eastern Pacific fish species. J. Fish. Res. Board Can. 28:1809– 1810.

McCosker, J. E., M. D. Lagios, and T. Tucker. 1976. Ultrastructure of lymphocystis virus in the quillback rockfish, *Sebastes maliger*, with records of infection in other aquarium-held fishes. Trans. Am. Fish. Soc. 105:333–337.

Madeley, C. R., D. A. Smail, and S. I. Egglestone. 1978. Observations on the fine structure of lymphocystis virus from European flounders and plaice. J. Gen. Virol. 40:421–431.

Mann, H. 1970. Über den Befall der Plattfische der Nordsee mit Lymphocystis. Ber. Dtsch. Wiss. Komm. Meeresforsch. 21:219–223.

Midlige, F. H., Jr., and R. G. Malsberger. 1968. In vitro morphology and maturation of lymphocystis virus. J. Virol. 2:830–835.

Miyazaki, T., and S. Egusa. 1972. A histopathological observation on lymphocystis disease in the sea bass *Lateolabrax japonicus* (Cuvier and Valenciennes). Fish Pathol. 6:83–89. In Japanese.

Murchelano, R. A., and D. W. Bridges. 1976. Lymphocystis disease in the winter flounder, *Pseudopleuronectes americanus*. J. Wildl. Dis. 12:101–103.

Nelson, J. S. 1984. Fishes of the world, 2nd ed. Wiley, New York.

Nigrelli, R. F. 1940. Mortality statistics for specimens in the New York Aquarium, 1939. Zoologica 25:525–552.

Nigrelli, R. F. 1943. Causes of diseases and death of fishes in captivity. Zoologica 28:203–216.

Nigrelli, R. F. 1950. Lymphocystis disease and ergasilid parasites in fishes. J. Parasitol. 36:36. Abstract.

Nigrelli, R. F. 1952. Virus and tumors in fish. Ann. N.Y. Acad. Sci. 54:1076–1092.

Nigrelli, R. F., and G. D. Ruggieri. 1965. Studies on virus diseases of fishes. Spontaneous and experimentally induced cellular hypertrophy (lymphocystis disease) in fishes of the New York Aquarium, with a report of new cases and an annotated bibliography (1874–1965). Zoologica 50:83–96.

Nigrelli, R. F., and G. M. Smith. 1939. Studies on lymphocystis disease in the orange filefish, *Ceratacanthus schoepfii* (Walbaum), from Sandy Hook Bay, N.J. Zoologica 24:255–264.

Nordenberg, C.-B. 1962. Das Vorkommen der Lymphocystiskrankheit bei Scholle und Flunder im Öresund. K. Fysiogr. Sallsk. Lund Forh. 32:17–26.

Nordenberg, C.-B. 1966. Lymphocystis-sjuka hos flatfisk i Öresund 1962–1965. Sartryck Sven. Fiskeri Tidskr. 1966:40–41.

Olson, D. E. 1958. Statistics of a walleye sport fishery in a Minnesota lake. Trans. Am. Fish. Soc. 87:52–72.

Overstreet, R. M., and H. D. Howse. 1977. Some parasites and diseases of estuarine fishes in polluted habitats of Mississippi. Ann. N.Y. Acad. Sci. 298:427–462.

Paperna, I. 1973. Lymphocystis in fish from east African lakes. J. Wildl. Dis. 9:331–335.

Paperna, I., I. Sabnai, and A. Colorni. 1982. An outbreak of lymphocystis in *Sparus aurata* L. in the Gulf of Aqaba, Red Sea. J. Fish Dis. 5:433–437.

Perkins, E. J., J. R. S. Gilchrist, and O. J. Abbott. 1972. Incidence of epidermal lesions in fish of the north-east Irish Sea area, 1971. Nature 238:101–103.

Petty, L. L., and J. J. Magnuson. 1974. Lymphocystis in age 0 bluegills (*Lepomis macrochirus*) relative to heated effluent in Lake Monona, Wisconsin. J. Fish. Res. Board Can. 31:189–l193.

Pritchard, H. N., and R. G. Malsberger. 1968. A cytochemical study of lymphocystis tumor cells in vivo. J. Exp. Zool. 169:371–380.

Rašín, K. 1927. Příspěvek k pathogenesi lymphocystis johnstonei Woodcock I. Biol. Spisy Vysoké Školy Zvěrolékařské Brno, ČSR 6:1–28.

Rašín, K. 1928. Příspěvek k pathogenesi lymphocystis johnstonei Woodcock II. Biol. Spisy Vysoké Školy Zvěrolékařské Brno, ČSR 7:1–14.

Roberts, R. J. 1970. Lymphocystis in young plaice, and some other aspects of disease in plaice husbandry. J. Fish Biol. 2:383. Abstract.

Roberts, R. J. 1975. Experimental pathogenesis of lymphocystis in the plaice (*Pleuronectes platessa*). Pages 431–441 in L. A. Page, ed. Wildlife diseases. Plenum Press, New York.

Robin, J., and L. Berthiaume. 1981. Purification of lymphocystis disease virus (LDV) grown in tissue culture. Evidences for the presence of two types of viral particles. Rev. Can. Biol. 40:323–329.

Robin, J., C. Larivière-Durand, and J. Bernard. 1983. The chemical composition of lymphocystis disease virus of fish. Rev. Can. Biol. Exp. 42:173–176.

Robin, J., L. Berthiaume, and A. Laperriere. 1984. Study of the structural proteins of lymphocystis disease virus (LDV), strain Leetown NFH, grown in tissue culture. (Etude des polypeptides de structure du virus de la maladie lymphokystique des poissons [lymphocystis disease virus] multiplié en culture cellulaire.) Ann. Virol. (Paris) 135E:67–80.

Russell, P. H. 1974. Lymphocystis in wild plaice, *Pleuronectes platessa* (L.), and flounder, *Platichthys flesus* (L.) in British coastal waters: a histopathological and serological study. J. Fish Biol. 6:771–778.

Ryder, R. A. 1961. Lymphocystis as a mortality factor in a walleye population. Prog. Fish-Cult. 23:183–186.

Schmidt, A. 1929. Vliv teploty na vývoj buněk lymphocystis. Dissertation. Institute for General Zoology and Parasitology, College of Veterinary Medicine, Brno, Czechoslovakia.

Shelton, R. G. J., and K. W. Wilson. 1973. On the occurrence of lymphocystis, with notes on other pathological conditions, in the flatfish stocks of the north-east Irish Sea. Aquaculture 2:395–410.

Sigel, M. M., A. R. Beasley, and M. C. Launer. 1966. Cytological and cytochemical (AO) studies on lymphocystis in tissue culture. In Vitro 2:153–154. Abstract.

Sigel, M. M., D. M. Lopez, A. R. Beasley, and L. S. Dieterich. 1971. Virus-cell interaction in lymphocystis disease of fish. Pages 124–135 in M. Sanders, ed. Viruses affecting man and animals. W. H. Green, St. Louis.

Smith, F. G. 1970. A preliminary report on the incidence of lymphocystis disease in the fish of the Sapelo Island, Georgia, area. J. Wildl. Dis. 6:469–471.

Smith, F. G. 1972. Lymphocystis disease in *Cynoscion nothus*, *Cynoscion regalis* and *Stellifer lanceolatus* from Georgia estuaries. Proc. Annu. Conf. Southeast. Assoc. Game Fish Comm. 26:598–608.

Spitzer, R. H., E. A. Koch, R. B. Reid, and S. W. Downing. 1982. Metabolic-morphologic characteristics of the integument of teleost fish with mature lymphocystis nodules. Cell Tissue Res. 222:339–357.

Stephens, J. S., Jr., R. K. Johnson, G. S. Key, and J. E. McCosker. 1970. The comparative ecology of three sympatric species of California blennies of the genus *Hypsoblennius* gill (Teleostomi, Blenniidae). Ecol. Monogr. 40:227–233.

Stickney, R. R., and D. B. White. 1974. Lymphocystis in tank-cultured flounder. Aquaculture 4:307–308.

Templeman, W. 1965. Lymphocystis disease in American plaice of the eastern Grand Bank. J. Fish. Res. Board Can. 22:1345–1356.

Thakur, N. K., and S. A. K. Nasar. 1977. The occurrence of lymphocystis in *Anabas testudineus* (Bloch). Curr. Sci. 46:150–151.

Vaughn, G. E. 1979. Comparative vulnerability of bluegills with and without lymphocystis disease to predation by largemouth bass. Prog. Fish-Cult. 41:163–164.

Vitnish, M., and T. Baranova. 1976. Lymphocystis disease in Baltic flounder (*Platichthys flesus* L.). Int. Counc. Explor. Sea ICES-CM-1976-P6:1–12.

Walker, D. P., and B. J. Hill. 1980. Studies on the culture, assay of infectivity and some *in vitro* properties of lymphocystis virus. J. Gen. Virol. 51:385–395.

Walker, R. 1947. Lymphocystis disease and neoplasia in fish. Anat. Rec. 99:559–560.

Walker, R. 1957. Warty walleyes. N.Y. State Conserv. 12:28–29.

Walker, R. 1958. Lymphocystis warts and skin tumors of walleyed pike. Rensselaer Rev. Grad. Stud. 14:1–5.

Walker, R. 1961. Fine structure of a virus tumor of fish. Am. Zool. 1:395–396.

Walker, R. 1962. Fine structure of lymphocystis virus of fish. Virology 18:503–505.

Walker, R. 1963. The capsule of virus induced lymphocystis cells of fish. Am. Zool. 3:490. Abstract.

Walker, R. 1965. Viral DNA and cytoplasmic RNA in lymphocystis cells of fish. Ann. N.Y. Acad. Sci. 126:386–395.

Walker, R. 1969. Virus associated with epidermal hyperplasia in fish. Natl. Cancer Inst. Monogr. 31:195–207.

Walker, R., and R. Weissenberg. 1965. Conformity of light and electron microscopic studies on virus particle distribution in lymphocystis tumor cells of fish. Ann. N.Y. Acad. Sci. 126:375–385.

Walker, R., and K. Wolf. 1962. Virus array in lymphocystis cells of sunfish. Am. Zool. 2:566. Abstract.

Weissenberg, J. R. 1951. Positive result of a filtration experiment supporting the view that the agent of the lymphocystis disease of fish is a true virus. Anat. Rec. 111:582. Abstract.

Weissenberg, J. R. 1960a. Further studies on the lymphocystis disease of fish. Anat. Rec. 137:400. Abstract.

Weissenberg, J. R. 1960b. Further studies on the intracellular microorganism observed in the lymphocystis disease of fish. Anat. Rec. 138:387. Abstract.

Weissenberg, J. R. 1960c. Paired structures of the inclusion bodies in the lymphocystis virus disease of perches. Bacteriol. Proc. 1960. Abstract M180.

Weissenberg, R. 1914. Über infektiöse Zellhypertrophie bei Fischen (Lymphocystiserkrankung). Sitzber. Kgl. Preuss. Acad. Wiss. Physik.-Math. 30:792–804.

Weissenberg, R. 1920. Lymphocystisstudien (infektiöse Hypertrophie von Stutzgewebzellen bei Fischen) I. Die riefen Geschwülste bei Kaulbarsch und Flunder. Lymphocystisgenese beim Kaulbarsch. Arch. Mikrobiol. Anat. 94:55–134.

Weissenberg, R. 1938. Studies on virus diseases of fish. I. Lymphocystis disease of the orange filefish (*Aleutera schoepfii*). Am. J. Hyg. 28:455–462.

Weissenberg, R. 1939a. Studies on virus diseases of fish. II. Lymphocystis disease of *Fundulus heteroclitus*. Biol. Bull. 76:251–255.

Weissenberg, R. 1939b. Studies on virus diseases of fish. III. Morphological and experimental observations on the lymphocystis disease of the pike perch *Stizostedion vitreum*. Zoologica 24:245–254.

Weissenberg, R. 1945. Studies on virus diseases of fish. IV. Lymphocystis disease in Centrarchidae. Zoologica 30:169–184.

Weissenberg, R. 1949. Studies on lymphocystis tumor cells of fish. I. The osmiophilic granules of the cytoplasmic inclusions and their interpretation as elementary bodies of the lymphocystis virus. Cancer Res. 9:537–542.

Weissenberg, R. 1951a. Studies on lymphocystis tumor cells of fish. II. Granular structures of the inclusion substance as stages of the developmental cycle of the lymphocystis virus. Cancer Res. 11:608–613.

Weissenberg, R. 1951b. Some results of morphological studies on the developmental cycle of the lymphocystis virus of fish with reference to the experimental work of K. Rašín. Anat. Rec. 111:530–531. Abstract.

Weissenberg, R. 1951c. Four additions to the list of host fishes in which lymphocystis tumors have been observed as the result of spontaneous viral infection. Anat. Rec. 111:581. Abstract.

Weissenberg, R. 1951d. Experimental lymphocystis infection of the killifish *Fundulus heteroclitus* with emulsion of lymphocystis tumors of the perch *Stizostedion vitreum*. Anat. Rec. 111:582. Abstract.

Weissenberg, R. 1952. Parallel features in the parasitism and life cycle of fish microsporidia and fish viruses. Proc. Soc. Protozool. 3. Abstract 5. P. 3.

Weissenberg, R. 1955. The third spontaneous case of lymphocystis virus disease of fish from the Pacific Coast of North America. Anat. Rec. 122:434–435. Abstract.

Weissenberg, R. 1956. Granular components of the basophilic lattice in the lymphocystis virus inclusion bodies of *Pleuronectes*. Arch. Gesamte Virusforsch. 7:1–17.

Weissenberg, R. 1960. Some remarkable osmiophilic structures of the inclusion bodies in the lymphocystis virus disease of the European flounder. Arch. Gesamte Virusforsch. 10:253–266.

Weissenberg, R. 1965a. Fifty years of research on the lymphocystis virus disease of fishes (1914–1964). Ann. N.Y. Acad. Sci. 126:362–374.

Weissenberg, R. 1965b. Morphological studies on lymphocystis tumor cells of a cichlid from Guatemala *Cichlasoma synspilum* Hubbs. Ann. N.Y. Acad. Sci. 126:396–413.

Weissenberg, R., R. F. Nigrelli, and G. M. Smith. 1937. Lymphocystis in the hogfish, *Lachnolaimus maximus*. Zoologica 22:303–305.

Witt, A., Jr. 1957. Seasonal variation in the incidence of lymphocystis in the white crappie from the Niangua arm of the Lake of the Ozarks, Missouri. Trans. Am. Fish. Soc. 85:271–279.

Wolf, K. 1962. Experimental propagation of lymphocystis disease of fishes. Virology 18:249–256.

Wolf, K., and C. P. Carlson. 1965. Multiplication of lymphocystis virus in the bluegill (*Lepomis macrochirus*). Ann. N.Y. Acad. Sci. 126:414–419.

Wolf, K., and M. C. Quimby. 1973. Fish viruses: buffers and methods for plaquing eight agents under normal atmosphere. Appl. Microbiol. 25:659–664.

Wolf, K., M. Gravell, and R. G. Malsberger. 1966. Lymphocystis virus: isolation and propagation in centrarchid fish cell lines. Science 151:1004–1005.

Wolf, K. E., M. C. Quimby, C. P. Carlson, and W. J. Owens. 1979. Lymphocystis virus: infectivity of lesion preparations stored after lyophilization or simple desiccation. J. Fish Dis. 2:259–260.

Woodcock, H. M. 1904. Note on a remarkable parasite of plaice and flounders. Trans. Liverpool Biol. Soc. 18:143–152.

Yamamoto, T., R. D. Macdonald, D. C. Gillespie, and R. K. Kelly. 1976. Viruses associated with lymphocystis disease and dermal sarcoma of walleye (*Stizostedion vitreum vitreum*). J. Fish. Res. Board Can. 33:2408–2419.

Zwillenberg, L. O., and K. Wolf. 1968. Ultrastructure of lymphocystis virus. J. Virol. 2:393–399.

23 Walleye Herpesvirus

Synonyms: *Herpesvirus vitreum*, percid herpesvirus 1

Definition

The walleye herpesvirus, *Herpesvirus vitreum*, is an agent found in and presumably the cause of transient skin lesions of epidermal hyperplasia that occur in spawning walleyes, *Stizostedion vitreum vitreum*, in certain Canadian watersheds.

History

The walleye herpresvirus was first reported by Kelly, Nielsen, and Yamamoto (1980) of the Freshwater Institute of Winnipeg, Canada. The brief characterization described the agent as a "herpes-like virus"; the same designation was used in a review of skin lesions of the walleye, which also gave epizootiologic data on the associated epidermal hyperplasia (Kelly, Nielsen, Campbell, and Yamamoto 1980). The original isolant proved to be a bona fide herpesvirus, and its more detailed characterization—including electron micrographs and a one-step growth curve—was published by Kelly et al. (1983). Nothing further appeared until Yamamoto et al. (1985), at the University of Alberta, reviewed the virus-associated skin neoplasms of walleyes in certain parts of Canada. They noted that the epidermal hyperplasia harbored herpesvirus and that the agent was routinely isolated. The walleye herpesvirus is therefore represented by multiple isolants and in time might also be found elsewhere and in other percid fishes.

Signs and Pathologic Changes

Fish with epidermal hyperplasia are found during their spawning run, but population incidence has not been reported. Presumably, the fish appear to be otherwise normal. The lesions are transient and are not to be found in late spring or summer. Moreover, the hyperplasia is seemingly benign, at least in adults, because some affected specimens that were tagged have been recaptured a year later.

Histopathologic examination has not been reported.

Etiology

Rivers' postulates have not been fulfilled, but it seems reasonable to assume that the walleye virus causes the hyperplasia.

Size and Shape

Negatively stained capsids measure 100 nm in diameter, and enveloped particles have a size range of 190 to 230 nm. The capsomere number is 162. Well-defined peplomers are present on the surface of the envelope.

Biophysical Properties

The agent is ether labile, and its replication is inhibited by 10^{-3} M phosphonacetate, by 10^{-5} M bromodeoxyuridine, and by 10^{-5} M 9-(2-hydroxyethoxymethyl) guanine (Acyclovir). However, Acyclovir inhibits replication only in the WO cell line of walleye ovary origin, and not in the walleye fibroblast line WC-1 described by Kelly, Miller, et al. (1980).

Replication occurs at 4 to 15°C but not at 20°C. A one-step growth curve in WO cells at 13°C shows some cell clumping after 24 hours. The appearance of prominent syncytia by 36 hours is followed by disintegration of nuclear membranes and lysis. Exponential replication occurs from 12 to about 24 hours, and peak infectivity is reached about 12 hours later. Maximal yield is 8×10^5 PFU/mL. Titers are slightly higher and plaques smaller in WC-1 cells than in WO cells.

As would be expected in a herpesvirus, the walleye agent is not replicated by nonpercid fish lines.

Electron microscopy shows naked capsids—some with and some without nucleoids—in nuclei of epithelial cells in lesion tissue. Envelopment occurs at the nuclear membrane, and complete virions are abundant in intercellular spaces.

By conventional determination, infectivity is retained by membranes of 100 nm mean porosity. A 90-minute adsorption time is recommended for plaquing, presumably to obtain maximal titers.

Virus was frozen at −70°C by Kelly et al. (1983)—presumably a temperature favorable for retaining infectivity. Virus survival or lability at temperatures other than −70°C has not been reported.

Herpesvirus vitreum is distinctly different from the C-type particle or presumed retrovirus that Walker (1969) reported in hyperplastic epidermal cells of walleyes from New York.

Host and Geographic Range

The virus is thus far known only from Bad Carrot River, a tributary of Hanson Lake, Saskatchewan, Canada. Its effect on young walleyes is completely unknown. The species is relatively difficult to obtain and maintain under experimental conditions. It would be interesting to determine the effect of *H. vitreum* on young and adult yellow perch (*Perca flavescens*) and other common percid species.

References

Kelly, R. K., H. R. Miller, O Nielsen, and J. W. Clayton. 1980. Fish cell culture: characteristics of a continuous fibroblastic cell line from walleye (*Stizostedion vitreum vitreum*). Can. J. Fish. Aquat. Sci. 37:1070–1075.

Kelly, R. K., O. Nielsen, J. S. Campbell, and T. Yamamoto. 1980. Current status of lymphocystis, dermal sarcoma/fibroma, and related diseases in walleye, *Stizostedion vitreum vitreum*, in western Canada. In Proceedings of the joint 4th biennial fish health section and fish disease workshops, Seattle, Washington, June 24–26, 1980.

Kelly, R. K., O. Nielsen, and T. Yamamoto. 1980. A new herpes-like virus (HLV) of fish (*Stizostedion vitreum vitreum*). In Vitro 16:255. Abstract.

Kelly, R. K., O. Nielsen, S. C. Mitchell, and T. Yamamoto. 1983. Characterization of *Herpesvirus vitreum* isolated from hyperplastic epidermal tissue of walleye, *Stizostedion vitreum vitreum* (Mitchill). J. Fish Dis. 6:249–260.

Walker, R. 1969. Virus associated with epidermal hyperplasia in fish. Natl. Cancer Inst. Monogr. 31:195–207.

Yamamoto, T., R. K. Kelly, and O. Nielsen. 1985. Morphological differentiation of virus associated skin tumors of walleye (*Stizostedion vitreum vitreum*). Fish Pathol. 20:361–372.

PART II

VIRAL INFECTIONS OF INDETERMINATE PATHOGENICITY

24 Atlantic Cod Ulcus Syndrome

Synonym: ulcer disease of Atlantic cod

Definition

The ulcus syndrome is a subacute to chronic progressive ulcerating skin condition of Atlantic cod, 1.5 to 2.5 years old, taken from certain organically polluted coastal waters of Denmark. The condition involves virus, and attendant high mortality is indicated. The ulcus syndrome shows a seasonal aspect in that prevalence is highest during autumn.

The disease described here, or one that causes lesions having a similar appearance (Fig. 1), has been recognized as a pathologic skin condition of Atlantic cod (*Gadus morhua*) in western European waters for more than 50 years. During much of that time, bacteria—particularly vibrios—were thought to be responsible, but recent work has shown evidence of two viruses, and indications are that one could be the etiologic agent.

History

For more than 50 years, Atlantic cod from certain Baltic areas have been known to have a variety of skin lesions, but only within the past decade or so has a specific and progressive pathologic condition—the ulcus syndrome—been described. Jensen and Larsen (1976) presented a sequence of five progressive stages of the disease and noted a relation between occurrence of the disease and the presence of *Vibrio anguillarum* and carbohydrate pollution. Later, a bacteriologic study showed that, in addition to *V. anguillarum*, an aeromonad was frequently isolated; however, about 20% of the ulcers were free of cultivable bacteria (Larsen and Jensen 1977). The ulcus syndrome persisted in polluted water, but Larsen et al. (1978) were emphatic in stating that no simple relation could be found between *V. anguillarum* and prevalence of the disease.

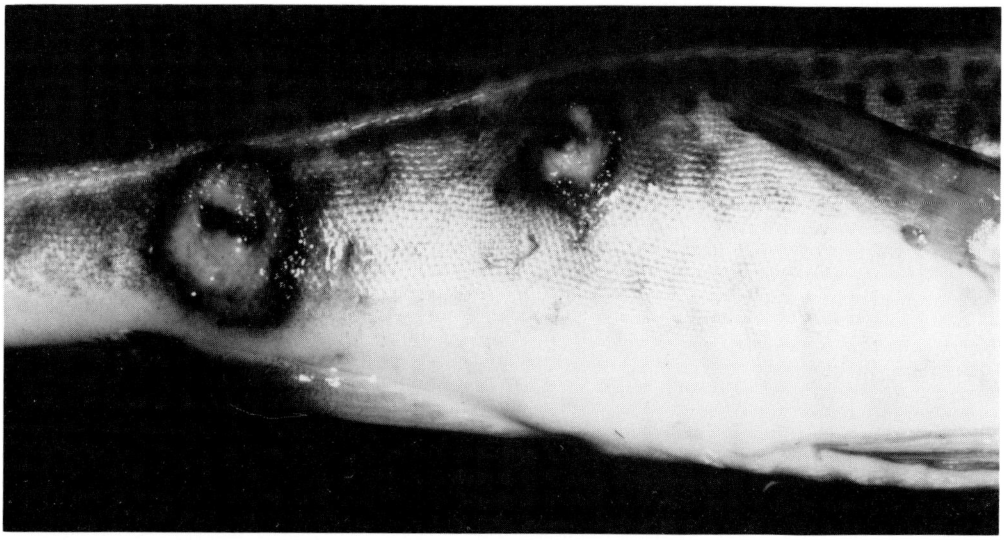

Figure 1. Late ulcerative lesions (stage 4) of cod ulcus syndrome. From Jensen and Larsen (1979). Reprinted with the permission of Danske Dyrlaegforening Alhambravej, Copenhagen.

During the period 1979 to 1982, a series of five related papers appeared: Jensen and Larsen (1979) described and illustrated gross and histopathologic features; Larsen and Jensen (1979) reviewed bacteriologic findings and concluded that *V. anguillarum* strains were common opportunists and not the primary pathogen; Jensen et al. (1979) reported the isolation of two viruses—a rhabdovirus and an iridovirus—from a group of cods with ulcus syndrome; Jensen and Larsen (1982) showed laboratory transmission and implicated the iridovirus as the more suspect cause; and Larsen and Jensen (1982) summarized results of the 1976 to 1979 studies on prevalence and pollution.

In two companion papers, Jensen (1983) reviewed the ulcus syndrome, and Larsen (1983) summarized the bacteriologic aspects of the disease. In addition, Stork (1983) reported a stongly positive correlation between polychlorinated biphenyl (PCB) residue concentrations in fish tissues and the presence of the ulcus syndrome.

Considerable recycling or redundancy is evident in the literature on Atlantic cod ulcus syndrome. New data are needed, particularly on the role of virus.

Signs and Pathologic Changes

Ulcus syndrome is a specific, progressive, chronic condition that is capable of causing mortality. Jensen and Larsen (1976, 1979) partitioned the progress of the disease into five stages: (1) papulovesicular, (2) erosive, (3) early ulcerative, (4) late ulcerative, and (5) healing.

Behavior

Victim fish were taken by pound net and seine; behavior was not mentioned.

External Signs

Distinctive signs accompany each of the five stages recognized by Jensen and Larsen (1976, 1979).

Stage 1. In the papulovesicular stage, multiple dermal papules and occasional vesicles are distributed laterally on the body, but never on fins or head. The papules are 2 to 8 mm in diameter and 1 to 3 mm high, and some have a slightly hemorrhagic apex. Papules progress to slightly larger hemispherical vesicles containing serous fluid.

Stage 2. In the erosive stage, small craterlike holes with slightly elevated margins appear in the skin. The central epithelium and connective tissue have been lost. Bottoms of lesions are gray to yellow with a pinkish hue.

Stage 3. The early ulcerative stage is marked by the appearance of small red concave ulcers with a margin of hyperplastic tissue and sometimes with centers showing a pale gray necrosis.

Stage 4. The late ulcerative stage is typified by the presence of one or more ulcers 2 to 8 cm in diameter and of varied depth and severity. Some ulcers penetrate the abdominal wall. Ulcers may develop from papules or result from papular fusion (Fig. 1).

Stage 5. The healing stage, marked by small white patches or scars devoid of pigment cells and scales, is observed in autumn at temperatures of 7°C or lower. Healing is most common after stages 1 and 2.

Internal Signs

Inasmuch as internal features of diseased fish have not been reported, dramatic changes presumably were not obvious.

Histopathologic Findings

Stage 1. The epithelium is rough but intact and appears to be normal at low magnification. At higher magnification and with electron microscopy, edema is evident in the stratum spongiosum. Scale pockets show fluid accumulation and infiltration, predominantly by granulocytes. Hyperemia and hemorrhaging are also present. Bacteria are found in the loose connective tissue. Alterations do not occur in tissue below the stratum compactum.

Stage 2. Edema and cell infiltration extend downward and into intramuscular and intermuscular septa. The epidermis is breached and absent from the centers of open papules. Edges of the epithelium are thickened. Loose connective tissue is strongly infiltrated.

Stage 3. The stratum compactum is markedly edematous and infiltrated but usually intact. Capillaries are distended and granulation tissue has begun to form in the stratum compactum and its overlying loose connective tissue. In some cases, the stratum compactum has been breached and the underlying musculature exposed. Well-developed granulation tissue originating from the subcutis is richly vascularized and infiltrated.

Stage 4. The later ulcerative stage is much like stage 3, except that the granulation tissue is thicker and the stratum compactum is always breached. Also, the underlying musculature is edematous, necrotic, and infiltrated.

Stage 5. In the healing process, infiltrating cells are replaced by fibroblasts and the lesion is covered by epithelium.

Etiology

The cause of the ulcus syndrome has not been conclusively demonstrated, but there is adequate reason to reject a bacterial etiology. In a sample of about 100 affected cod, Larsen and Jensen (1977) found bacteria in 83% of the ulcers but *V. anguillarum* in only about half of the fish. In a later examination of 350 fish, bacteria (predominantly *V. anguillarum*) were found in only about half of the ulcers; the remaining fish were described as having "sterile skin lesions" (Larsen and Jensen 1979). Equally telling, young cod failed to develop the ulcus syndrome when they were exposed to *V. anguillarum* by bath, by swabbing on scarified skin, or by intraperitoneal injection (Jensen and Larsen 1982).

The possibility of a viral etiology was first indicated when an iridovirus and a rhabdovirus were isolated from pooled papules collected from seven cod. Examination of an additional 46 affected individuals showed iridovirus in only one. Twenty apparently healthy cod showed no evidence of either virus (Jensen et al. 1979). Considered alone, the data on virus isolations only weakly support the hypothesis of viral etiology. Rivers' postulates have not been fulfilled; however, results of experimental transmission tend to support the concept of a viral etiology. A 100% transmission was achieved by cohabitation of 6 healthy fish with 4 that bore stage 1 lesions. Also, the injection of culture-grown iridovirus apparently sometimes transmitted the disease (Jensen and Larsen 1982). It seems reasonable to reject the rhabdovirus as a cause of the ulcus syndrome; only one isolation was made, and the culture grown rhabdovirus did not induce disease.

Size and Shape

The cytoplasmic iridovirus has a mean diameter of 145 to 150 nm and an internal nucleoid of about 100 nm (Fig. 2). The rhabdovirus has the typical size and shape of viruses in this group (Fig. 2). Mean particle dimensions are 55 × 176 nm.

Biophysical Properties

Neither virus produced CPE in the original cultures; visible effects became evident only on second or third passage and then only in certain cell lines (see below). Neither virus produced CPE during four passages in BF-2, FHM, or RTG-2 cells.

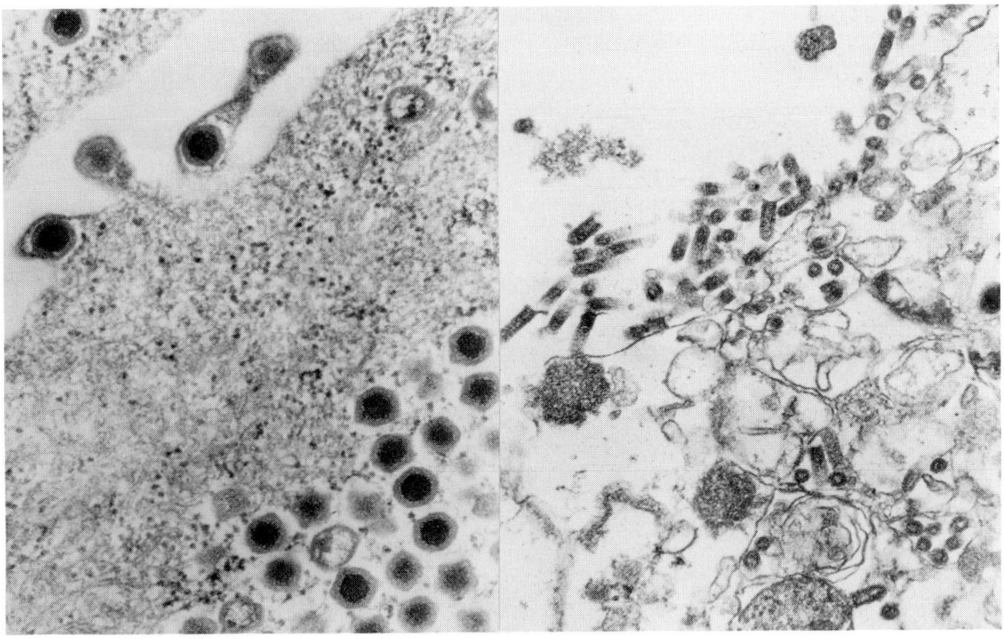

Figure 2. Viruses isolated from an Atlantic cod with ulcus syndrome. *Left.* Iridovirus in an EPC cell. ×40,250. *Right.* Rhabdovirus in a PS cell. ×33,850. From Jensen et al. 1979. Reprinted with the permission of Danske Dyrlaegforening Alhambravej, Copenhagen.

Iridovirus.

The EPC cell line gave good replication, and maximal titers were about 10^5 $TCID_{50}$/mL. At 15°C, CPE became evident as early as 20 hours postinoculation, and lysis ensued. Infected cells became rounded and developed a cytoplasmic inclusion that stained green with acridine orange and thus gave presumptive evidence of double-stranded nucleic acid. Maximal infectivity in the PS cell line from pike sarcoma was only 10^2 $TCID_{50}$/mL.

Rhabdovirus.

Ljungberg's PS line was preferred and maximal titers were about 10^5 $TCID_{50}$/mL. Cytopathic effects, consisting of swelling and sloughing of infected cells, became evident at 15°C in 3 to 4 days. In the EPC cell line, CPE was weak even after 7 days and the maximal titer was only 10 $TCID_{50}$/mL.

Diagnosis

On the basis of information now available, a pathologic diagnosis of the ulcus syndrome is the most that can be accomplished. Such a diagnosis would have to exclude traumatic lesions that might occur on the head, and rely heavily on the presence of papules, vesicles, and ulcers on the body. Failure to find *V. anguillarum* in some of the lesions would support the diagnosis, as would isolation of virus. At present, the ulcus syndrome could be considered as more of a research entity than as an established clinical disease. Accordingly, experimental transmission by cohabitation or by intracardial injection of crude homogenates of pooled papules and vesicles might provide welcome confirmation and additional data.

Isolation

The susceptible cell lines EPC and especially PS are suggested, and they should be incubated at 15°C. Serial 10-fold dilutions of pooled papular-vesicular homogenates should be tested and several blind transfers made before considering results negative. Time permitting, homogenates of internal organs ought also be examined, but thus far limited tests of such material have failed to show virus (N. J. Jensen, personal communication).

Identification

Identification of the iridovirus is at best presumptive and is based on morphology, host species, and response of the EPC and PS cell lines. The rhabdovirus has since been identified serologically as the F1 or strain I of Egtved virus (P. E. Vestergard Jorgensen, 1986 personal communication). Accordingly, an unanswered question arises: why did Jensen et al. (1979) not find CPE when FHM and RTG-2 cells were inoculated with the rhabdovirus?

Transmission and Incubation

Only limited success has been achieved in experimental transmission. Jensen and Larsen (1982) confined 6 healthy cod 25 to 35 cm long with 4 that bore stage 1 lesions. Within 9 to 11 days at 12 to 14°C, all 6 showed initial papulovesicular lesions (but no mention was made of controls). The same authors found a 36% transmission when 11 fish were first scarified and then rubbed with crude homogenates of stage 1 lesions. Incubation time was 9 to 12 days, and 11 controls remained normal. Intraperitoneal injection of a 0.45-μm membrane filtrate of lesion homogenate showed only 13% transmission among 31 recipients, whereas 21 controls remained normal. Again, the incubation period was about the same (8 to 12 days).

Jensen and Larsen (1982) conducted seven trials with cultures of viruses. In two trials, rhabdovirus was injected intraperitoneally or intracardially; 54 fish were used but none developed lesions, and 53 controls remained healthy. In a third trial, in which 20 fish were injected intraperitoneally with a mixture of rhabdoviruses and iridoviruses, 2 developed the syndrome and 20 controls remained normal. In a fourth trial, the icosahedral virus was injected intraperitoneally, but all of 20 inoculated fish and 26 controls remained healthy. The best results were achieved in the fifth to seventh trials, in which icosahedral virus was injected intracardially: 21 of 113 test fish (19%) developed stage 1 lesions; however, 1 of 96 control fish also showed signs.

Limiting factors undoubtedly play a role in preventing consistent results from experimental transmission. Foremost is the fact that the test fish might have had prior exposure to the causal agent. Environmental factors seem to be involved. For one, the peaking of the ulcus syndrome in autumn could be interpreted as either following peak summer temperatures or occurring at a time of declining temperature. Pollution too is a factor: Larsen and Jensen (1982) showed 4 years of data in which prevalence peaked in areas heavily polluted with organic effluents from sugar and cellulose processing plants. In contrast, fewer than 1% affected fish were taken from the nonpolluted control area.

Source

The source is unknown but assumed to be the cod or a closely related species.

Host and Geographic Range

Thus far, male and female Atlantic cod 1.5 to 2.5 years old, taken from certain of Denmark's coastal waters, are involved. Perhaps cod and related gadids elsewhere will eventually be shown to have the condition.

References

Jensen, N. J. 1983. The ulcus syndrome in cod (*Gadus morhua*): a review. Rapp. P.-V. Reun. Cons. Int. Explor. Mer 182:58–64.

Jensen, N. J., and J. L. Larsen. 1976. Research project concerning the possible relationship between pollution factors, especially carbohydrate pollution and marine vibrios. (Forskningsprojekt vedrorende marine vibrio populationers forhold til forureningskomponenter, isaer kulhydratforurening i marine biotoper.) Dan. Vetinaertidsskr. 59:521–524.

Jensen, N. J., and J. L. Larsen. 1979. The ulcus-syndrome in cod (*Gadus morhua*). I. A pathological and histopathological study. Nord. Vet. Med. 31:221–228.

Jensen, N. J., and J. L. Larsen. 1982. The ulcus-syndrome in cod (*Gadus morhua*). IV. Transmission experiments with two viruses isolated from cod and *Vibrio anguillarum*. Nord. Vet. Med. 34:136–142.

Jensen, N. J., B. Bloch, and J. L. Larsen. 1979. The ulcus-syndrome in cod (*Gadus morhua*). III. A preliminary virological report. Nord. Vet. Med. 31:436–442.

Larsen, J. L. 1983. Bacteriological aspects of the ulcus syndrome in cod. Rapp. P.-V. Reun. Cons. Int. Explor. Mer 182:65–71.

Larsen, J. L., and N. J. Jensen. 1977. An Aeromonas species implicated in ulcer-disease of the cod (*Gadus morhua*). Nord. Vet. Med. 29:199–211.

Larsen, J. L., and N. J. Jensen. 1979. The ulcus-syndrome in cod (*Gadus morhua*). II. A bacteriological investigation. Nord. Vet. Med. 31:289–296.

Larsen, J. L., and N. J. Jensen. 1982. The ulcus-syndrome in cod (*Gadus morhua*). V. Prevalence in selected Danish marine recipients and a control site in the period 1976–1979. Nord. Vet. Med. 34:303–312.

Larsen, J. L., N. J. Jensen, and N. O. Christensen. 1978. Water pollution and the ulcer syndrome in the cod (*Gadus morhua*). Vet. Sci. Commun. 2:207–216.

Stork, T. 1983. PCB levels correlated with the ulcus-syndromes in cod (*Gadus morhual*) in the North Sea. International Council for Exploration of the Sea, council memorandum 1983/ E:50.

25 Atlantic Menhaden Spinning Disease

Definition

Atlantic menhaden spinning disease is a name used to describe die-offs of epizootic proportion that occur almost annually among some populations of young *Brevoortia tyrannus* along northeastern coastal waters of the United States during spring or summer. Infectious pancreatic necrosis virus (IPNV) has been isolated from specimens of some recent epizootics, but quantification of virus—a critical factor in assessing causality—has not been reported. In addition, IPNV—typically a readily isolated virus—has been sought but not found in victims of other die-offs.

Limited laboratory trials with young menhaden that were injected intraperitoneally with IPNV or immersed in it showed that the fish underwent an acute IPN-like disease and died within about a week after the onset of signs. The evidence therefore suggests that IPNV could be an associated presence or even a prime pathogen in some die-offs, but that it is not the universal cause of spinning disease in Atlantic menhaden.

History

According to Newman (1980), large die-offs of menhaden have been known for decades and were recognized in the Chesapeake Bay as early as the 18th century. Historically, environmental factors—bacteria and parasites—have been investigated, but no single entity has been convincingly assigned a causal role. Sindermann (1970) noted that virus was suspected as an etiologic agent, and a virus was successfully isolated by Stephens et al. (1980). The isolant had the principal biophysical characteristics of IPNV, was neutralized by polyvalent anti-IPNV serum, and produced an IPN-like disease when young menhaden were injected with or immersed in the virus. The signs that developed were similar to those of spinning disease.

Other outbreaks of spinning disease were examined virologically. In two cases, IPNV was isolated directly on CHSE-214 cells, but specimens from two other outbreaks failed to yield virus (F. M. Hetrick, personal communication).

No additional reports on the possible virology of spinning disease have appeared.

Signs and Pathologic Changes

Newman (1980), who described his findings in naturally afflicted specimens, noted that gross signs include abnormal swimming in a corkscrew pattern. Some fish have hyperemic fin bases and hemorrhage in the eye and about the vent. Internally, the spleen is consistently congested; congestion of the brain, sometimes accompanied by unilateral hemorrhage of the subarachnoid, is a frequent finding; and blood occurs in the lumen of the digestive tract in many specimens.

Histologically, meningitis of varying severity is common, and focal necrosis of pancreatic acinar cells is an occasional finding. At the ultrastructural level, brain tissues show demyelinating lesions accompanied by slight inflammation.

In contrast to histologic findings of IPNV in salmonids, visceral lesions of spinning disease in clupeids are less severe and more focal. The gut usually contains food, and hemorrhage is seldom seen in the anterior visceral mass.

Etiology

In critical evaluation, the role of virus in menhaden spinning disease is uncertain. It has not been found in all or even in a majority of the fish kills. Trials by two different investigators resulted in disease and 100% mortality. Virus was reisolated from pooled internal organ homogenates of all victims 5 to 7 and 10 to 13 cm long examined by Stephens et al. (1980), and from blood, pancreas, and some internal organ tissues assayed individually from larger fish. Virus quantification—an index of the role of an IPNV—was not reported.

Host and Geographic Range

Young-of-the-year Atlantic menhaden are the usual victims of naturally occurring outbreaks, but older specimens have sometimes been involved. Experimentally, another clupeid—the American shad (*Alosa sapidissima*)—has proved to be susceptible to immersion in virus of menhaden origin (Newman 1980).

In the Chesapeake Bay, a focus of investigations, most mortalities have occurred along about 40 km of shoreline between Annapolis and Baltimore, Maryland. Die-offs have also been reported farther north and from as far south as North Carolina (F. M. Hetrick, personal communication).

References

Newman, M. W. 1980. IPN virus disease of clupeid fishes. International Council for Exploration of the Sea, Special Meeting on Diseases of Commercially Important Marine Fish and Shellfish. No. 24.

Sindermann, C. J. 1970. Principal diseases of marine fish and shellfish. Academic Press, New York.

Stephens, E. B., M. W. Newman, A. L. Zachary, and F. M. Hetrick. 1980. A viral aetiology for the annual spring epizootics of Atlantic menhaden *Brevoortia tyrannus* (Latrobe) in Chesapeake Bay. J. Fish Dis. 3:387–398.

26 Bluegill Virus

Definition

The bluegill virus (BGV) is a small enveloped ribovirus having provisional grouping with picornaviruses and togaviruses but of unknown pathogenicity for the source species *Lepomis macrochirus*. Although known since 1965, BGV is still considered to be an orphan virus in that its role in a fish is undetermined.

History

The BGV was first isolated by Wolf and Quimby during team efforts to culture the agent of epitheliocystis (Hoffman et al. 1969). The source specimen was an adult with a light case of epitheliocystis. Scrapings were made from its caudal fin in an attempt to harvest the superficial epitheliocystis cells. The scraped material was homogenized, decontaminated through a 0.45-μm membrane filter, inoculated into BF-2 cultures, and incubated at 23°C. A few apparently encapsulated giant cells that appeared after 2 days were suspected of being epitheliocystis cells. Although the transfer of the material was unavoidably delayed, the culture remained infectious: when it was transferred to fresh BF-2 cells, CPE developed 2 to 6 days later. Affected cells underwent exceptional elongation and became pyknotic, thus exhibiting the unique CPE of BGV. Since that effect was serially transferable, infected cell sheets were fixed and sent to L. O. Zwillenberg, Bern, Switzerland, for examination by electron microscopy.

Zwillenberg found abundant virus particles maturing at cell surfaces, but never intracellularly. In contrast, examination of natural epitheliocystis cells revealed chlamydia but no virus. As judged by its morphology and apparent method of maturation, the in vitro virus was considered to be similar to a myxovirus.

Young-of-the-year bluegills injected with the newly isolated virus showed no evidence of disease, and none died. The virus was put in frozen storage, and a stock was deposited with the American Type Culture Collection as VR-604.

During 1972, under the tutelage of Malsberger at Leigh University, Beckwith began a characterization of BGV, which was later published (Beckwith and Malsberger 1979).

In West Virginia, several perfunctory attempts over the years to find BGV in other bluegills failed, but during 1979 a virus was isolated from adult bluegills and presumptively identified as BGV (J. Camper, personal communication). The bluegills were normal-appearing brood stock at the Frankfort (Kentucky) National Fish Hatchery, and isolation was made during a routine health inspection that included virologic examination. In contrast to the single and serendipitous original finding of BGV, the new isolations were made from two of five pools of homogenized bluegill kidneys and spleens, each pool representing five donor fish. Virus was successfully transferred to fresh cultures several times and each time the transfer resulted in CPE. Infective cultures were sent to the National Fish Health Research Laboratory; there its viral identity was presumptively confirmed, but infectivity was soon lost. Virologic examination of additional brood stock fish failed to reveal more BGV. The role of BGV in the bluegills from Kentucky was not determined.

Etiology

Attempts to demonstrate pathogenicity of BGV for bluegills have thus far failed. Two or more factors might have been involved. First, the test fish were young-of-the-year and, of necessity, obtained from ponds that harbored feral fish. In essence, the health history of the stock could not be determined. Second, prior exposure to the virus could have occurred and thus stimulated an immune response. Additionally, fish of the size tested could have developed age or size resistance, the kind of pattern well known for most of the fish-pathogenic viruses.

Size and Shape

Virions are spherical to ovoid, 90 to 120 nm in diameter (Fig. 1). Particles that are electron dense are generally smaller than those that are electron lucent, but both kinds show a doubly contoured envelope (Beckwith and Malsberger 1979; Berthiaume et al. 1982). Subtle periodic surface projections occur on some of the virions. Infectivity passes membranes of 100 nm, but not 50 nm, porosity.

Biophysical Properties

Although BGV resembles several other RNA viruses in size and shape, it is recognized as being a representative of a novel virus family. Beckwith and Malsberger (1979) showed that BGV lacks essential attributes of a myxovirus, and the work of investigators at the University of Sherbrooke, Quebec, established that it could not be a bunyavirus or a coronavirus. Robin and Dery (1982) determined that the RNA is single stranded, that it sediments in sucrose as a single molecular species, and that its molecular weight is 2.17×10^6. In cesium sulfate, the RNA bands at 1.69 g/mL. Berthiaume et al. (1982) confirmed earlier electron microscopy and additionally

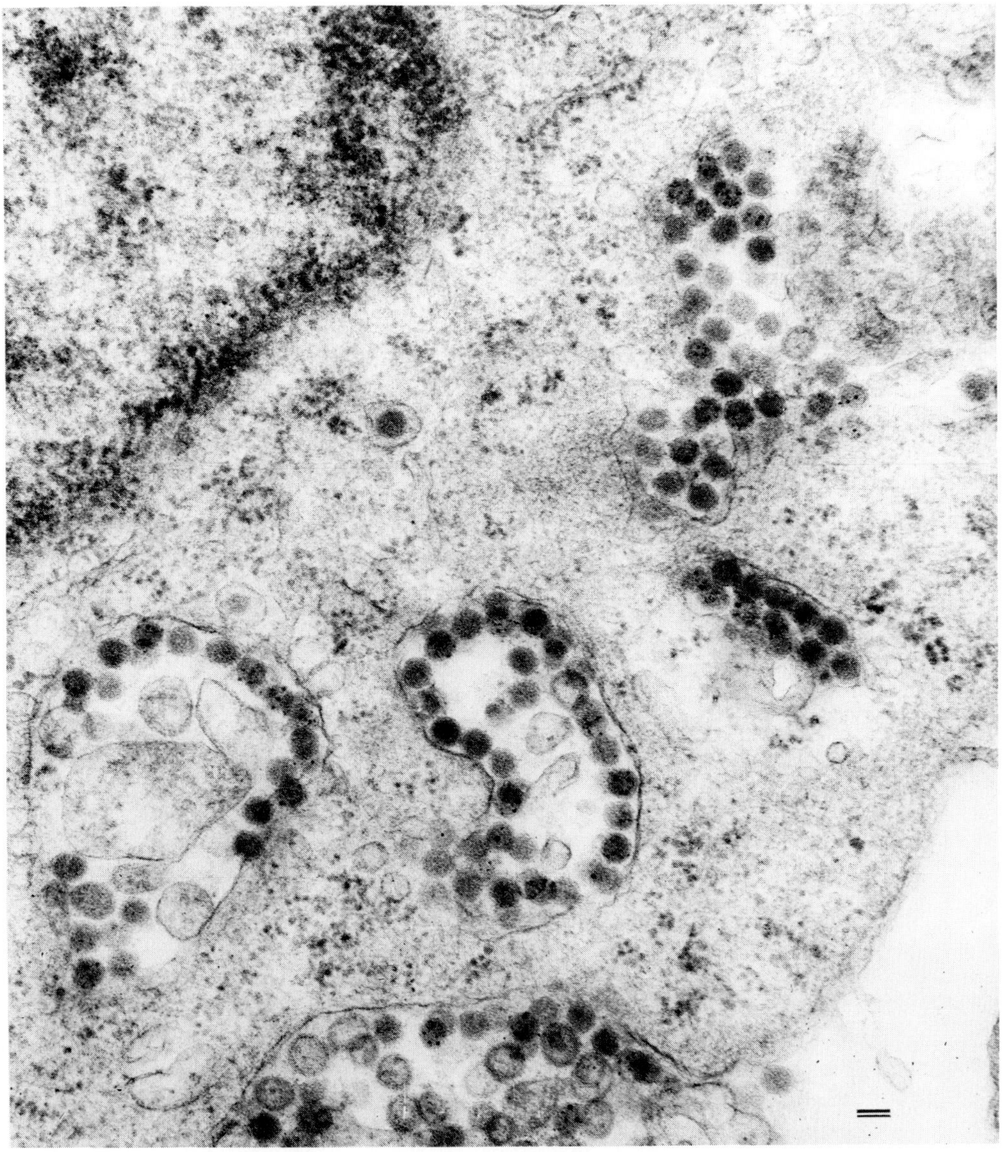

Figure 1. A BF-2 cell infected with bluegill virus. Note that the electron-dense virions are smaller than those that are electron lucent. Bar = 100 nm. Courtesy of R. W. Darlington.

showed that the genome is linear. Robin and Larivière-Durand (1983) reported that the RNA has class 1 positive-strand genome polarity and, moreover, that the nucleic acid is infectious.

Intact virus is ether sensitive, and acid and heat labile. Its infectivity is best maintained in medium containing 10% serum at −80°C; although infectivity declines, it persists for several weeks at 4°C. The effect of lyophilization is not known.

Beckwith and Malsberger (1979) tested cells of mammalian, reptilian, amphibian,

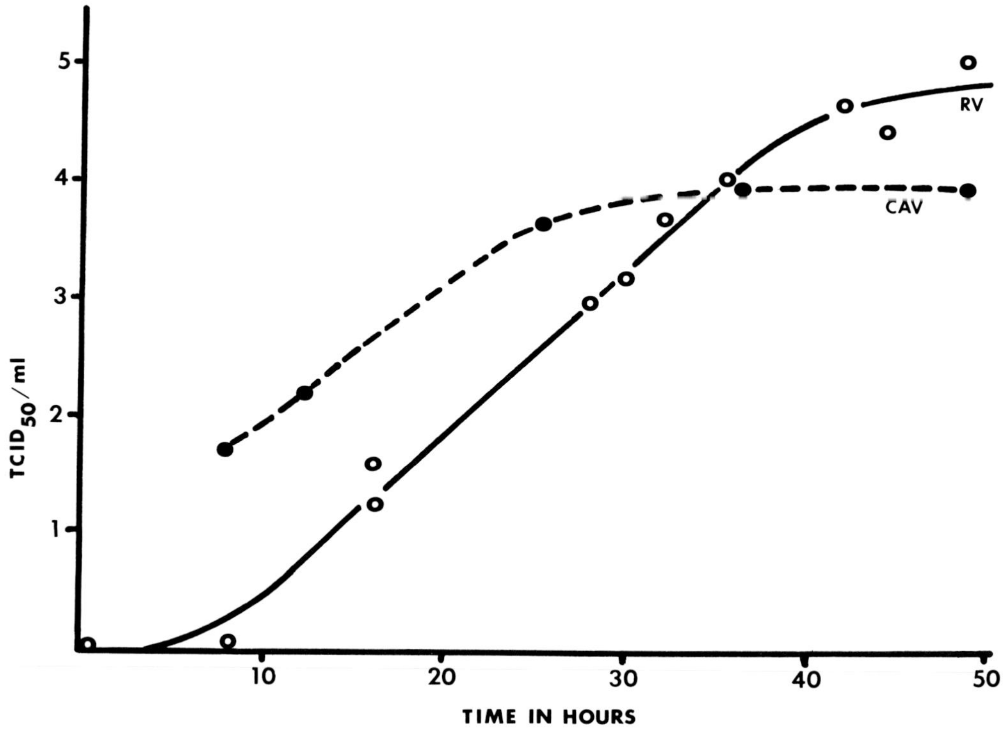

Figure 2. One-step growth curve of bluegill virus in BF-2 cells at 20°C. The multiplicity of infection is about 10. RV, released virus; CAV, cell-associated virus. After Beckwith and Malsberger (1979).

and teleostean origin, but found that BGV was replicated only in fish cells; moreover, the best yield—about 10^6 PFU/mL—was in centrarchid fish cells at 20°C. Wolf and Quimby (1973) plaqued BGV in 3 days under a two-phase agarose gel in normal atmosphere at 25°C and Robin et al. (1982) plaqued it in 7 days under methylcellulose in 5% carbon dioxide at 21°C.

The virus has been grown at 20 to 25°C, but would probably also replicate at temperatures as low as 15°C and as high as 30°C. At 20°C, a one-step growth curve requires about 48 hours (Fig. 2). New virus appears at 8 to 10 hours and becomes exponential until about 24 to 36 hours. Released virus exceeds cell-associated virus by a factor of 10 (Beckwith and Malsberger 1979).

Like lymphocystis virus, BGV appears to have an absolute requirement for cells of fibroblastic morphology in order to be synthesized. Cell lines of centrarchid origin, notably the BF-2 line, are generally susceptible. As had happened at the National Fish Health Research Laboratory, Beckwith and Malsberger (1979) found that at high passage level the lineage of BF-2 cells was refractory. The solution to that problem is to go back to lower passage cells or to clone the line for susceptible cells.

The distinctive feature of the CPE produced by BGV is its exceptionally elongated pyknosis, involving both nucleus and cytoplasmic processes. That effect is usually apparent on the second day of incubation, when foci of infection show a "jackstraw" pattern (Fig. 3). Pyknosis is inevitably followed by lysis.

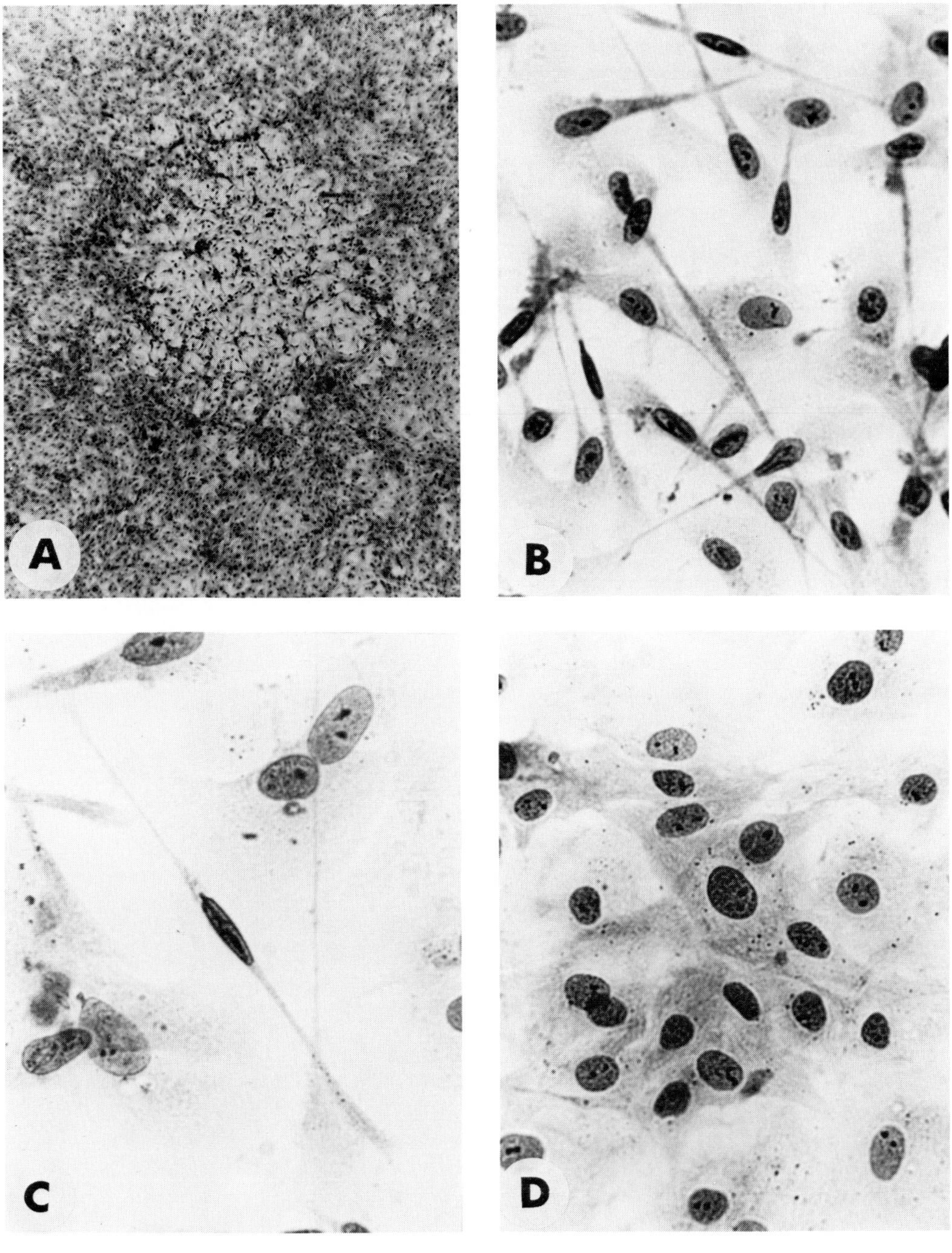

Figure 3. Response of BF-2 cells to the bluegill virus. *A.* Focal infection in the center of an otherwise normal cell sheet. *B.* At higher magnification, elongated pyknosis of infected cells is evident. *C.* Compared with noninfected cells, cells infected with bluegill virus become extremely elongated. The chromatin is marginated. *D.* Noninfected BF-2 cells.

Identification

Beckwith and Malsberger (1979), who prepared anti-BGV serum in rabbits, reported that at a dilution of 1 : 16 it neutralized 100 $TCID_{50}$ homologous virus.

Presumptive identification of BGV is made by finding a virus of similar morphology in a centrarchid fish and showing that it evokes the unique elongated CPE in fibroblastic cells of centrarchid origin.

Host and Geographic Range

Two isolations were made, both from the centrarchid bluegill. Since natural hybridization occurs between centrarchid sunfishes that inhabit the same environments, BGV could easily occur in related species.

The virus has been found only in the southeastern United States, but the host species is widely distributed and may harbor the virus in other geographic regions.

References

Beckwith, D. G., and R. G. Malsberger. 1979. Characterization and morphology of the bluegill virus. J. Gen. Virol. 43:489–501.

Berthiaume, L., J. Robin, and R. Alain. 1982. Electron microscopic study of the bluegill virus. Can. J. Microbiol. 28:398–402.

Hoffman, G. L., C. E. Dunbar, K. Wolf, and L. O. Zwillenberg. 1969. Epitheliocystis, a new infectious disease of the bluegill (*Lepomis macrochirus*). Antonie Van Leeuwenhoek. J. Microbiol. Serol. 35:146–158.

Robin, J., and C. Dery. 1982. The genome of bluegill virus. Can. J. Microbiol. 28:58–64.

Robin, J., and C. Larivière-Durand. 1983. Bluegill virus is a ribovirus of positive strand polarity. Arch. Virol. 77:119–125.

Robin, J., C. Larivière-Durand, and L. Berthiaume. 1982. Plaque assay of bluegill virus using a methylcellulose overlay. J. Virol. Methods 5:351–354.

Wolf, K., and M. C. Quimby. 1973. Fish viruses: buffers and methods for plaquing eight agents under normal atmosphere. Appl. Microbiol. 25:659–664.

27 Carp Iridovirus

Definition

The carp iridovirus is represented by two isolants that were found in homogenates of gill and kidney tissues during an investigation of the possible viral etiology of the disease of the common carp (*Cyprinus carpio*) known as gill necrosis (Popkova and Shchelkunov 1978; Shchelkunov and Shchelkunova 1981, 1984). The role of the iridovirus in gill necrosis has yet to be determined.

History

Carp gill necrosis, a disease of economic significance in husbandry of the common carp, has been known for about 30 years. The history and postulated causes—among them a viral etiology—were reviewed in a report by Schreckenbach et al. (1975). They concluded from their own work that the disease was noninfectious and instead arose as a result of ammonia toxicity. Miaczynski (1965), who reviewed the viral diseases and others of uncertain etiology that were known in Poland, included gill necrosis of carp and, although he offered no evidence, was the first to suggest a viral etiology.

The idea of virus involvement was investigated later by others. Popkova and Shchelkunov (1978) and Schreckenbach et al. (1975) noted that reports of virus isolation and experimental induction of gill necrosis had been published in the Soviet Union. However, the fate or disposition of the isolants themselves is uncertain. On the other hand, at least two isolations, which later received more detailed study, were made by Popkova and Shchelkunov (1978). The isolations seem to be of a single virus entity, the characterization of which was reported by Shchelkunov and Shchelkunova in Russian in 1981 and in an English translation in 1984.

Etiology

The role of the virus is not known, but most evidence suggests that it is not the cause of the gill necrosis disease. Not only has the virus not been isolated from all or even most clinical cases—indeed, Shchelkunov and Shchelkunova (1984) found virus in only 2 of 61 cases—but experimental transmission (by these invesitgators) was unsuccessful in 7 of 8 trials with 2-year-old common carp. They concluded that the virus was a pathogen of low virulence. (As with the virulent viruses of fishes, effects would have been more likely to be significant if specific-pathogen–free fry or fingerlings had been used.)

The carp iridovirus measures about 200 nm, it is ether stable, but labile in chloroform and at pH 3. Replication in FHM or EPC cells is accompanied by lytic CPE and a peak titer of about 10^5 $TCID_{50}$/mL.

Isolation

Virus was isolated from gill and kidney homogenates that were decontaminated and then adsorbed onto susceptible FHM monolayer cultures and incubated at 22 to 28°C. Blind passage should be made if CPE does not appear within 10 days.

Identification

Serologic identification has yet to be reported. The virus can be presumptively identified by host, source, morphology, and size.

Host and Geographic Range

Virus has been isolated only from common carp and only in the Soviet Union. Considering how widely this species is propagated, the virus will undoubtedly be found elsewhere. Also, the relationship to the goldfish iridovirus needs to be determined. The hosts are closely related and capable of hybridizing. Although they are now described as two agents, they might in reality be but one.

References

Miaczynski, T. B. 1965. Viral diseases and diseases of uncertain etiology in fish in Poland. Ann. N.Y. Acad. Sci. 126:620–628.

Popkova, T. I., and I. S. Shchelkunov. 1978. Isolation of virus from carp afflicted with gill necrosis. (Vedelenie virusa ot karpov, bol'nykh zhabernym nekrozom.) VNIIPRKh. Rybn. Khoz. 4:34–38.

Schreckenbach, K., R. Spangenberg, and S. Krug. 1975. Die Ursache der Kiemennekrose. Z. Binnenfisch. DDR 22:257–288.

Shchelkunov, I. S., and T. I. Shchelkunova. 1981. Resultati virusologichskich issoledovani bolnych nekrozom zhabr karpov. Pages 466–482 in J. Olah, K. Molnar, and Z. Jeney, eds.

Proceedings of an international seminar on fish, pathogens and environment in European polyculture, Fisheries Research Institute, Szarvas, Hungary, June 23–27, 1981.
Shchelkunov, I. S., and T. I. Shchelkunova. 1984. Results of virological studies on gill necrosis. Pages 31–43 in J. Olah, ed. Symposia biologica Hungarica, Vol. 23. Akademiai Kiado, Budapest.

28 Channel Catfish Reovirus

Definition

The channel catfish reovirus (CRV) is an agent of undetermined significance that has been isolated from *Ictalurus punctatus* during a program of surveillance of catfish farms for the herpesvirus of channel catfish (CCV).

History

The first isolation was made in 1982 when a virologic examination was carried out on juvenile fish that were experiencing chronic low mortality at a channel catfish farm in southern California. The virus was later isolated at a second California farm. The source of the virus was not determined; however, both farms used water from the same open canal (Amend et al. 1982, 1984).

Signs and Pathologic Changes

Mortality was not severe at either of the two sources of the virus. Of the specimens examined, the sole pathologic change was hyperplasia of gill lamellae with multifocal fusion, which Amend et al. (1984) regarded as minor and nonspecific. In fingerling catfish, neither immersion nor infection by intraperitoneal injection resulted in mortality (Amend et al. 1982). Instead, the virus was replicated without evidence of virulence at 26°C and for as long as a month. (Other temperatures were not used.)

Etiology

The role of the reovirus in causing disease and its association with the chronic low mortality that originally prompted virologic examination is wholly unknown.

Hedrick et al. (1984), who characterized CRV, showed it to be doubly encapsidated;

the inner form measures 55 nm and the outer one 75 nm. Intact and infective virions bands at 1.36 g/mL in cesium chloride, and RNA that has been labeled with tritiated uridine banded at 1.56 g/mL in cesium sulfate. When the RNA was electrophoresed in 9% polyacrylamide gel, it proved to be composed of 11 segments. Seven structural polypeptides were identified, ranging in molecular weight from 32,000 to 132,000.

Replication of CRV occurs at temperatures of 20 to 30°C, and the optimum is about 25°C. Maximal infectivity produced is about $10^{8.1}$ $TCID_{50}$/mL by BB cells and $10^{7.3}$ $TCID_{50}$/mL by CCO cells. The yield of virus by FHM cells is only two to three times greater than the input. The CHSE-214 cell line is susceptible to the extent that it shows focal CPE, but the change does not progress. Although one-step growth curves were not carried out, the data presented by Amend et al. (1984) suggested that at 25°C the replicative cycle was completed well within 48 hours.

Much like the herpesvirus of channel catfish disease, CRV strain F82-34 produces syncytia in CCO and BB cells; unlike the herpesvirus, however, the reovirus also produces CPE in CHSE-214 cells. Development of syncytia is followed by a rounding of cells and detachment from the substrate.

The virus is stable in storage at −20 and −80°C for 5 months. Reasonable stability is observed during 4 weeks at 4°C, but loss accelerates at higher temperatures. Surprisingly, some infectivity persists for several days at 45°C.

Identification

Serologic identification of CRV is made by neutralization test or fluorescent antibody technique with antiserum prepared in rabbits (Amend et al. 1984). Anti-CRV serum shows no cross-reactivity with channel catfish virus. However, CRV has been shown by neutralization indices to be related to, but distinguishable from, the other two reoviruses from fishes—golden shiner virus and chum salmon virus (Hedrick et al. 1984). Notably, neither of these viruses is replicated by the ictalurid cell lines CCO and BB that are preferred for culture of CRV.

Host and Geographic Range

The CRV is known only from channel catfish in southern California.

References

Amend, D. F., T. McDowell, and R. P. Hedrick. 1982. Isolation of a reovirus from channel catfish (*Ictalurus punctatus*). Fish Health News 11(4):iii–v.

Amend, D. F., T. McDowell, and R. P. Hedrick. 1984. Characteristics of a previously unidentified virus from channel catfish (*Ictalurus punctatus*). Can. J. Fish. Aquat. Sci. 41:807–811.

Hedrick, R. P., R. Rosemark, D. Aronstein, J. R. Winton, T. McDowell, and D. F. Amend. 1984. Characteristics of a new reovirus from channel catfish (*Ictalurus punctatus*). J. Gen. Virol. 65:1527–1534.

29 Chinook Salmon Paramyxovirus

Definition

The chinook salmon paramyxovirus is an enveloped helical RNA agent that has been isolated on two occasions from normal-looking *Oncorhynchus tshawytscha* during health monitoring examinations (Winton et al. 1985). Its role in disease has yet to be determined.

History

The first isolation was made in 1982 and the second in 1983, but CPE occurred only after a blind passage had been made.

Signs and Pathologic Changes

None are evident, externally or internally.

Etiology

The role of the virus is wholly unknown, but Winton et al. (1985) reported that infectivity trials have been initiated with four species of salmonids.

Size and Shape

In stained thin sections of infected CHH-1 cells, the virus is enveloped and ranges in size from 125 to 250 nm (Fig. 1).

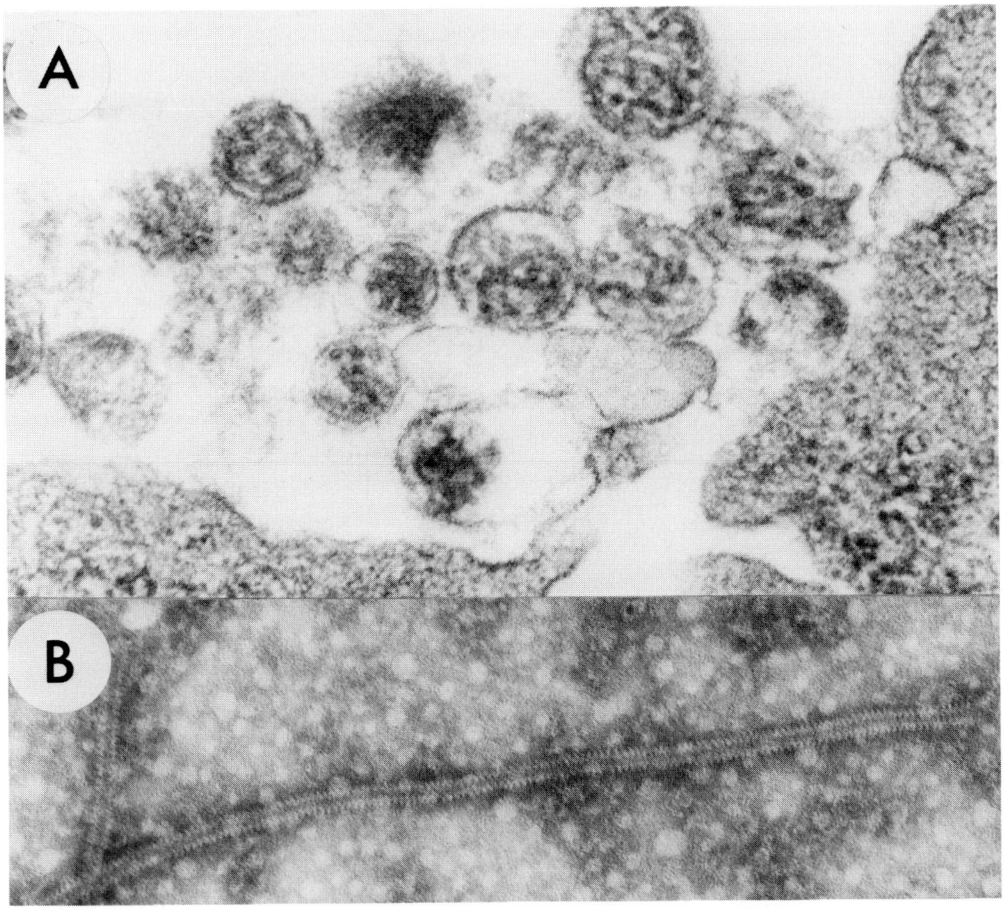

Figure 1. A. Electron micrograph of thin section of cell infected by chinook salmon paramyxovirus, showing enveloped virions containing a coiled genome. *B.* Negatively stained fragments of neucleocapsid. Courtesy of J. R. Winton.

Biophysical Properties

The virus is labile to chloroform, and is also heat labile, since it is completely inactivated within 6 hours at 56°C or in 24 hours at 37°C. Infectivity is stable for at least 7 days at 18°C and is not depressed by a single freeze-thaw cycle; it is stable within the range of pH 3 to pH 11 but is completely lost after a 30-minute exposure to pH 2. Although infectivity does not form a visible band, it is found in cesium chloride gradient fractions that have a density of 1.20 g/mL. The salmonid cell lines CHH-1, CHSE-214, KO-6, and CSE-119 replicate the virus in the temperature range of 6 to 21°C, but CPE is evident only at 15 to 21°C. The CPE begins as small foci of rounded and refractile cells, but eventually involves the entire cell sheet and the result is detachment (Fig. 2). The maximal titer ranges from 3×10^5 to 3×10^6 $TCID_{50}$/mL.

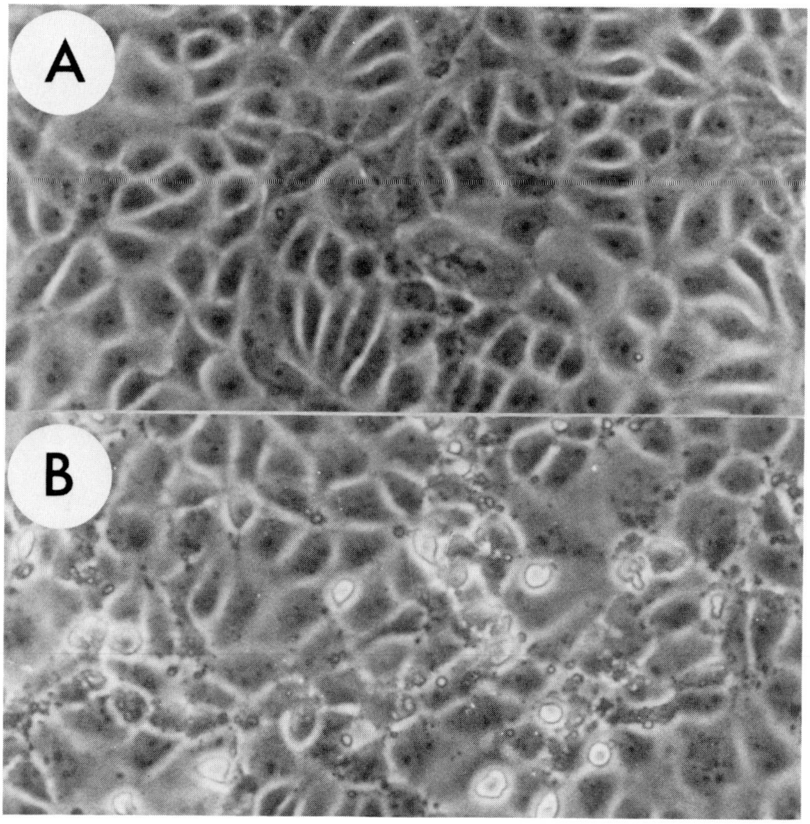

Figure 2. Response of CHH-1 cell line to the chinook salmon paramyxovirus. *A.* Noninfected control culture. *B.* Infected culture, showing refractile and rounded cells. Courtesy of J. R. Winton.

Replication is unaffected by iododeoxyuridine, and further evidence of an RNA genome is evident in the helical and coiled configuration of the nucleocapsid (Fig. 1).

Hemagglutination by this agent, the first fish virus to show that ability, is consistent with its placement in the paramyxovirus group. The virus rapidly agglutinates cells of rainbow trout and chinook, chum, and coho salmon. Turkey and chicken erythrocytes are also agglutinated, as are cells of human, rabbit, horse, guinea pig, and swine origin. Red blood cells from sheep, goats, and bovines are not agglutinated.

Isolation

Isolation was made in CHSE-214 cells that were inoculated with a homogenate of kidneys and spleens from five-fish pools of adult salmon. Incubation was at 18°C, but no evidence of virus was seen for 2 weeks. The material was blind passaged and an additional 2 weeks incubation elapsed before CPE was evident. Ovarian fluids from the same fish showed no virus (no other tissues were assayed).

Host and Geographic Range

The virus has been found only in returning adult chinook salmon in the Yaquina and Trask Rivers of coastal Oregon. Virus was sought but not found in progeny of the Yaquina River stock.

Reference

Winton, J. R., C. N. Lannan, D. P. Ransom, and J. L. Fryer. 1985. Isolation of a new virus from chinook salmon (*Oncorhynchus tshawytscha*) in Oregon, U.S.A. Fish Pathol. 20:373–380.

30 Chum Salmon Reovirus

Definition

The chum salmon reovirus is a member of the family Reoviridae that was first encountered in a pool of kidney and spleen homogenate from normal-appearing *Oncorhynchus keta* adults returning to Hokkaido Island, Japan. Originally considered to be of low virulence to several tested salmonids, the chum salmon virus has since been implicated in an epizootic among *O. masou*, also in Hokkaido. (T. Kimura, personal communication, 1987).

History

During 1978 a virologic examination was carried out on chum salmon returning to Tokushibetsu Hatchery in Hokkaido to meet American requirements for legal importation of salmonid eggs. Samples were sent to Oregon State University, where the virus was isolated.

Winton et al. (1980, 1981), who made the isolation, reported the details of initial characterization. They reported the size and shape of the organism, set forth the requirements for in vitro growth, and showed that the virus was not pathogenic for fry of three species of Pacific salmon. Further characterization—analyses of virus RNA and structural proteins—was reported by Winton et al. (1983). The biochemical components of the chum salmon isolant were compared with those of three other reoviruses from aquatic animals by Winton et al. (1987); they showed that the four viruses shared basic characteristics but were individually distinct and should be considered as members of a new genus.

Signs and Pathologic Changes

Winton et al. (1980, 1981) did not mention behavior aberrations or abnormal health conditions among the source salmon.

Fry (1 to 2 g) of three species of Pacific salmon— chum, chinook, and kokanee— proved to be infectible when the virus was injected intraperitoneally; however, neither signs of disease nor mortality occurred.

Fry were injected with 10^4 $TCID_{50}$ virus and held for 42 days at 12°C; during this time, specimen fry were periodically quantified for virus, and other specimens were fixed and stained for histopathologic examination.

Virus replication in fry of chinook salmon and kokanee was about 10 times greater than input during most of the period. At day 21 the amount of virus in chum salmon fry was about 100 times greater than input, but a slow decline then began, and by day 42 virus levels in all three species were comparable.

Histopathologic changes were found only in chum and chinook salmon fry and only in the liver. Areas of focal necrosis became evident in chum salmon fry on day 8 and were considered acute and numerous on day 21. Thereafter, however, healing took place and normal liver architecture returned. Lesser changes occurred in the liver of chinook salmon fry. Virus could not be demonstrated in electron micrographs of liver sections.

Etiology

Size and Shape

The chum salmon virus is a doubly encapsidated icosahedron about 75 nm in diameter.

Biophysical Properties

The genome is a double-stranded RNA with a buoyant density in cesium chloride of about 1.36 g/mL. Treatment with alpha chymotrypsin strips the outer capsid and results in a smaller particle size—50 to 55 nm—with an attendant 10–fold increase in infectivity (Winton et al. 1980, 1981).

The RNA consists of 11 segments: the 3 largest have molecular weights of 2.4 to 2.3 $\times 10^6$, 3 medium-size segments weigh 1.8 to 1.5 $\times 10^6$, and 5 small segments weigh 0.88 to 0.31 $\times 10^6$. Among five major and four minor proteins separated, the major ones had weights of about 137, 126, 72, 44, and 34 $\times 10^6$ (Winton et al. 1983, 1987).

Virus is replicated through the range of 10 to 20°C, but optimally at 15°C. Among 11 fish cell lines that support growth, yields are best—about 10^6 $TCID_{50}$/mL—in the CHSE-214 and CHH-1 line of chum salmon origin and lowest in the BB, FHM, and RTG-2 lines (the only lines tested that show no CPE).

The CPE in CHSE-214 consists of cell fusion that results in syncytia, and typical plaques show a band of nuclei concentrated near the inner margin of the plaque (Fig. 1).

Like other Reoviridae, the chum salmon agent is stable to chloroform and not inactivated at pH 3.0. This virus best conforms to the characteristics of the Orthoreoviridae; however, it differs in the number of RNA segments.

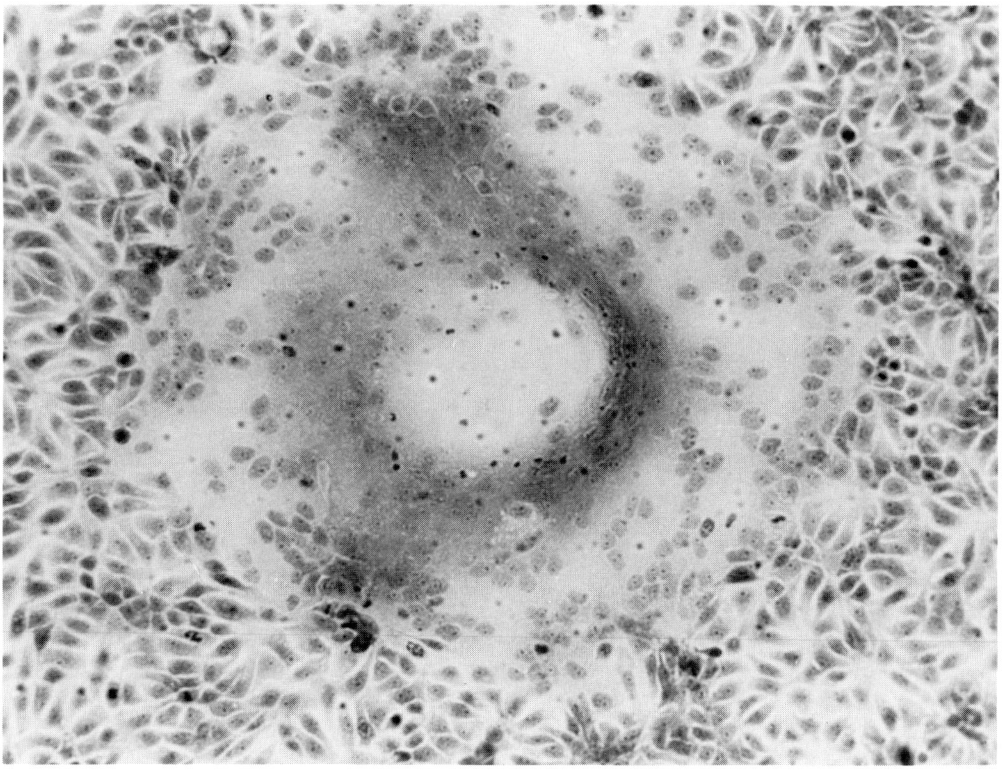

Figure 1. The CHSE-214 cells infected with the chum salmon reovirus develop plaquelike lesions in which marked cellular fusion and sheet contraction occur along the margins of focal CPE. From Winton et al. (1981). Reprinted with the permission of the Japanese Society of Fish Pathology, Tokyo.

Isolation

On original isolation, incubation was at 18°C, and 6 days elapsed before CPE became evident. Because CHSE-214 or CHH-1 cells yield the most virus and show CPE, they are recommended for isolation.

Identification

Chum salmon reovirus is readily identified serologically. Cross-neutralization tests with specific antisera against the reoviruses from the channel catfish and the golden shiner show that the three viruses are related but serologically distinct (Amend et al. 1984; Hedrick et al. 1984). The three fish reoviruses also differ in their growth characteristics and in electrophoretic mobilities of RNA segments and structural proteins (Hedrick et al. 1984; Winton et al. 1987).

References

Amend, D. F., T. McDowell, and R. P. Hedrick. 1984. Characteristics of a previously unidentified virus from channel catfish (*Ictalurus punctatus*). Can. J. Fish. Aquat. Sci. 41:807–811.

Hedrick, R. P., R. Rosemark, D. Aronstein, J. R. Winton, T. McDowell, and D. F. Amend. 1984. Characteristics of a new reovirus from channel catfish (*Ictalurus punctatus*). J. Gen. Virol. 65:1527–1534.

Winton, J. R., C. N. Lannan, J. L. Fryer, and T. Kimura. 1980. Isolation and characterization of a new reovirus from chum salmon. Pages 359–367 in Proceedings of the North Pacific aquaculture symposium, Anchorage, Alasaka, August 18–21, 1980; Newport, Oregon, August 25–27, 1980.

Winton, J. R., C. N. Lannan, J. L. Fryer, and T. Kimura. 1981. Isolation of a new reovirus from chum salmon in Japan. Fish Pathol. 15:155–162.

Winton, J. R., C. N. Lannan, and J. L. Fryer. 1983. Further characterization of a new reovirus of poikilothermic vertebrates. Pages 231–236 in R. C. Compans and D. H. L. Bishop, eds. Double-stranded RNA viruses. Elsevier, New York.

Winton, J. R., C. N. Lannan, J. L. Fryer, R. P. Hedrick, T. R. Meyers, J. A. Plumb, and T. Yamamoto. 1987. Morphological and biochemical properties of four members of a novel group of reoviruses isolated from aquatic animals. J. Gen. Virol. 68:353–364.

31 Eel Orthomyxovirus-like Agent

Synonyms: eel syncytial virus, EV-1, EV-2

Definition

The eel orthomyxovirus-like agent is a syncytium-inducing virus that has been isolated by various investigators during virologic examinations of subadult European eels (*Anguilla anguilla*) with stomatopapilloma— or as it was termed earlier, "cauliflower disease." The relationship of the virus to the neoplasm has not been determined.

History

Eels with distinctive neoplasms on and about the mouth have been recognized in Baltic and North Sea waters since the early 1900s. In the absence of visible pathogens, virus was considered as possibly causing the neoplasia, and in Germany the isolation of a small cytoplasmic icosahedron was reported by Pfitzner and Schubert (1969). Soon afterward in the United States, a virus that induced syncytia in FHM cells was isolated from internal organs and tumor homogenates of juvenile eels obtained from the same German area (Wolf and Quimby 1970). An American isolant was orally described to me as a small cytoplasmic icosahedron by the late R. W. Darlington. Designated EV-1, the virus plaque characteristics were described by Wolf and Quimby (1973).

McAllister et al. (1977) reviewed the eel virus literature and included early results obtained by Nagabayashi, who was then a visiting scientist in the National Fish Health Research Laboratory and was characterizing the virus. His micrographs of plaque purified virus (designated EV-2) showed no icosahedra, but instead pleomorphic virions with orthomyxovirus-like morphology. The characterization of EV-2 was published by Nagabayashi and Wolf (1979).

After several years during which no new information appeared, an isolant from an eel with stomatopapillomatosis, considered to be identical with EV-2, was described in Germany by Neukirch (1985). At about the same time, Ahne and Thomsen (1985) reported isolation of three different viruses from a single European eel with stomatopapilloma: an IPNV, a rhabdovirus, and a syncytium-inducing virus. Although the shape of the last of these was not then determined, the source fish and syncytium induction were reminiscent of EV-2. However, Ahne and Thomsen stated that their isolant was not neutralized by EV-2 antiserum.

Signs and Pathologic Changes

Although the cause is unknown, stomatopapilloma is of more than passing interest. The growths are exclusively external, rugose to verrucose, and pink to gray. The tissue is highly vascularized, soft, and easily damaged by handling. Growths consist of undifferentiated epithelial cells on a stroma of connective tissue. Schmid (1969) described the histologic findings and Schubert (1969) the ultrastructural details.

Etiology

The relationship of the virus to the neoplasm remains to be determined. All of the several isolations of a syncytium-inducing virus were made from specimen eels that had stomatopapillomatosis. However, Wolf and Quimby (1970) isolated virus from only 1 of 12 specimens that were assayed.

Size and Shape

Only the EV-2 isolant has been visualized. It is a pleomorphic particle that is 80 to 140 nm in diameter and has an outer surface studded with rod-shaped projections about 10 nm long (Nagabayashi and Wolf 1979).

Biophysical Properties

The EV-2 agent and the A_1B agent isolated by Neukirch (1985) have been characterized; both are ether and acid labile, have a buoyant density of 1.19 g/mL in sucrose, and hemagglutinate chicken erythrocytes but not those of human, rabbit, or rainbow trout. To demonstrate hemagglutination, one must first concentrate the virus (59,000 × g, 5 hours, 4°C) and use it at dilutions no greater than 1:64. Syncytia are produced by the EV-2 agent in FHM cells, optimally at 10 to 15°C (Fig. 1). Incubation at higher temperatures results only in pyknosis and lysis. The EPC, CHSE-214, and FHM lines form syncytia in the presence of A_1B. The BB, BF-2, and RTG-2 lines do not replicate EV-2.

Both isolants are inhibited by actinomycin D. The presence of 5 μg/mL greatly inhibits replication of EV-2 before, during, or after adsorption, but the effects are most pronounced when cells are treated 2 to 3 hours after infection.

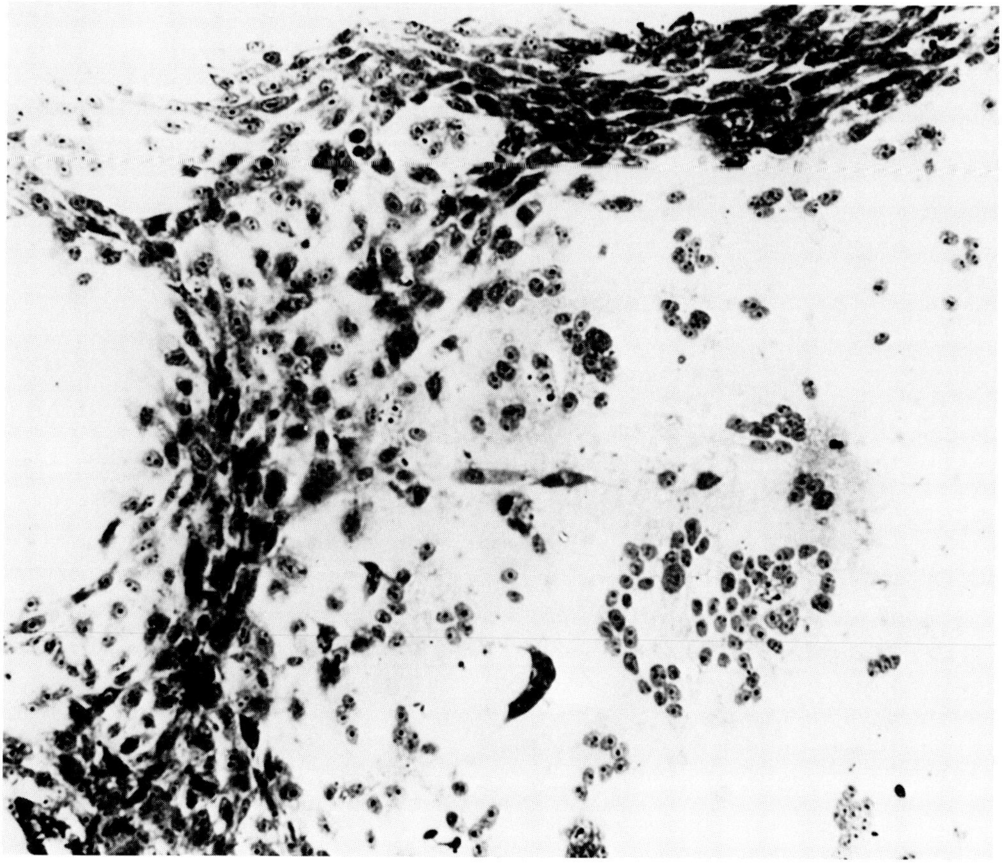

Figure 1. The EV-2 orthomyxovirus-like isolant at 10 to 15°C incubation in FHM cells induces the development of syncytia.

A one-step growth curve of EV-2 in FHM cells at 15°C showed that new infectivity appears at about 10 hours, peaks 10 hours later at about 10^5 PFU/mL, and then declines. Values for released and cell-associated virus are similar. The maximal titer for A_1B in FHM cells at 15°C is about 10^6 $TCID_{50}$/mL.

Heat lability of EV-2 and A_1B is marked: at 4°C the loss of infectivity is 99.9% for EV-2 by day 15 and even greater for A_1B by day 12. At 15°C, both isolants lose more than 99.9% of their infectivity within 4 days. Inactivation of EV-2 is complete at 50°C in 10 minutes, and it is not prevented by 1 M magnesium chloride.

Significant infectivity is also lost when EV-2 is lyophilized.

Isolation

In order of preference, tumor explants, tumor homogenates, or filtrates thereof should be assayed at 10 to 15°C on the FHM, EPC, and CHSE-214 cell lines. Plaque assay could show characteristic syncytia in the presence of mixed viral infection.

However, the use of polyvalent anti-IPNV and polyvalent anti–eel rhabdovirus sera would ensure selectivity for syncytium-forming virus.

Identification

Rabbit antiserum against EV-2 was used in an indirect fluorescent antibody technique that showed fluorescence in the cytoplasm of infected FHM cells but not in control cells. The antiserum was also used in direct immune electron microscopy that produced aggregates of virus particles.

Transmission and Incubation

Experimental transmission with EV-2 was attempted with elvers, but the results were inconclusive. During a 3-month period, half of the injected elvers died, but the virus was recovered from only 25% of the dead eels.

Source

The source of the virus is assumed to be eels.

Host and Geographic Range

In each of the three isolations, the eels had stomatopapillomatosis and came from waters near Hamburg, Germany. It is not known whether the syncytium-inducing virus is present in European eels that lack the neoplasm. The tumor has not been described from eels in North America nor in the extensive eel husbandry of Japan. However, about 20% of a sample of young eels obtained from Delaware's coastal waters yielded an agent with properties similar to those of EV-2, except that they were not neutralized by anti–EV-2 serum.

Immunity

Neukirch (1985) found that serum from about 20% of an unspecified number of eels contained antibodies that neutralized A_1B at from 1:4 to 1:24.

References

Ahne, W., and I. Thomsen. 1985. The existence of three different viral agents in a tumor bearing European eel (*Anguilla anguilla*). Zentralbl. Veterinärmed. [B] 32:228–235.

McAllister, P. E., T. Nagabayashi, and K. Wolf. 1977. Viruses of eels with and without stomatopapillomas. Ann. N.Y. Acad. Sci. 298:233–244.

Nagabayashi, T., and K. Wolf. 1979. Characterization of EV-2, a virus isolated from European eels (*Anguilla anguilla*) with stomatopapilloma. J. Virol. 30:358–364.

Neukirch, M. 1985. Isolation of an orthomyxovirus like agent from European eel (*Anguilla anguilla*). Bull. Eur. Assoc. Fish Pathol. 5:12–13.

Pfitzner, I., and G. Schubert. 1969. Ein Virus aus dem Blut mit Blumenkohlkrankheit behafteter Aale. Z. Naturforsch. 24b:790.

Schmid, O. J. 1969. Beitrag zur Histologie und Ätiologie der Blumenkohlkrankheit der Aale. Arch. Fischereiwiss. 20(Suppl. 1):16–23.

Schubert, G. 1969. Elektronenmikroskopische Untersuchungen an der Haut mit Blumenkohlkrankheit behafteter Aale. Arch. Fischereiwiss. 20(Suppl. 1):36–49.

Wolf, K., and M. C. Quimby. 1970. Virology of eel stomatopapilloma. Pages 94–95 in Progress in sport fishery research 1970. U.S. Fish Wildl. Serv. Resource Publ. 106.

Wolf, K., and M. C. Quimby. 1973. Fish viruses: buffers and methods for plaquing eight agents under normal atmosphere. Appl. Microbiol. 25:659–664.

32 Eel Rhabdoviruses

Isolants: EVA, EVEX, EVX, B_{12}, B_{44}, C_{26}, C_{30}, D_{13}

Definition

The rhabdoviruses of eels are a mixed collection of isolants that have originated mostly but not entirely among normal-looking eels (*Anguilla rostrata* or *A. anguilla*) of North American or European origin. Several of the isolants belong to the vesiculovirus subgroup of rhabdoviruses; the others have characteristics of the Lyssavirus subgroup.

The role of rhabdovirus in eels remains to be determined; most source animals are healthy, and virulence has yet to be shown. In contrast, eel isolants from the vesiculovirus subgroup have been found to be markedly pathogenic for rainbow trout fry. The pathologic changes in such trout fry were much like those affected by viral hemorrhagic septicemia, but the eel rhabdoviruses are serologically and biochemically different from the rhabdoviruses of salmonids and other fishes.

History

In Japan's Shizuoka Prefecture on Honshu Island, nearly 700 farms are devoted to intensive pond culture of eels. Because available stocks of the native eel, *Anguilla japonica*, have not been sufficient to meet the market demand, elvers of *A. anguilla* or *A. rostrata* have been imported from Europe or North America.

In Japan in 1974, an epizootic of hemorrhagic disease occurred in elvers 6 to 7 cm long in a pond that had been stocked 1 month earlier with fish from Cuba. Although mortality was nearly 60%, routine examinations revealed no pathogens; however, a rhabdovirus isolated in RTG-2 cells was not neutralizable by infectious hematopoietic necrosis virus (IHNV) or viral hemorrhagic septicemia virus (VHSV) antisera. Because the elvers came from an American eel, the virus was designated EVA by Sano (1976), the investigator from Tokyo University of Fisheries. Next, in efforts to confirm a foreign source of virus, a shipment of elvers imported from France was intercepted at Tokyo International Airport and sample specimens were taken for virologic examination. Again, a rhabdovirus was isolated in RTG-2 cells. Because of its European origin and unknown identity, it was named EVEX (Sano et al. 1976). A brief characterization of the isolant from the French elvers was published (Sano et al. 1977), but no comparisons were made with EVA. Summary information on both viruses was included in a review by McAllister et al. (1977).

The two viruses were sent to the Fish Diseases Laboratory at Weymouth, England, where comparisons of their characteristics and serology, and polypeptide analysis, showed that EVA and EVEX were related. Although details were not provided, the eel viruses were shown to be virulent for rainbow trout fry by Hill et al. (1980) and Sano et al. (1976). Methods for preparing antisera against the eel isolants and other fish rhabdoviruses were later given by Hill et al. (1981).

In the meantime, investigators at France's National Laboratory of Aquatic Animal Pathology found three viruses among apparently healthy elvers collected from French coastal waters (Castric and Chastel 1980). One isolant proved to be IPNV, but the other two were rhabdoviruses, designated B_{12} and C_{30}. The C_{30} isolant was serologically related to EVA and EVEX, but B_{12} differed not only serologically but also in its greater fragility, in its narrow low-temperature growth range, and in its total resistance to replication by RTG-2 cells.

Concerned about the possible impact of the eel viruses on other fishes of husbandry, Nishimura et al. (1981) tested fry of ayu, common carp, and rainbow trout, and elvers of Japanese eels. The work showed that only rainbow trout fry sustained mortality and that the pathologic changes evoked by EVA and EVEX were similar to those of IHNV. Only cell lines of heterologous fish origin had previously been used for the eel rhabdoviruses, but Sorimachi (1982) found that the EK-1 and EO-2 lines of eel origin gave the highest titers of EVEX and four other eel isolants. Dixon and Hill (1984) then developed an ELISA for rapid detection and identification of EVA and EVEX.

Castric et al. (1984), who used polyacrylamide gel electrophoroesis to determine polypeptide profiles of the eel rhabdoviruses, showed that five isolants—EVA, EVEX, C_{30}, B_{44}, and D_{13}—had polypeptides like those of vesicular stomatitis; and could therefore be considered as members of the vesiculovirus subgroup. Isolants B_{12} and C_{26} differed, and were considered to be in the Lyssavirus subgroup.

Ahne and Thomsen (1985) reported isolation of EVEX from an eel that also yielded two other viruses.

Signs and Pathologic Changes

Sano (1976) described the pathology of the elvers from which he had isolated EVA, but it is unclear whether the changes described were caused by the virus, for virulence of any rhabdovirus for eels has yet to be demonstrated experimentally.

In general, fish rhabdoviruses are recognized as being virulent for the young, but truly young eels have not been tested. The migration of leptocephali and elvers of *A. anguilla*, which spawns in the western Atlantic, to estuaries and streams of Europe takes as long as 2 years. Accordingly, the role of rhabdoviruses in eels might be determined when pelagic leptocephalus stages are challenged.

Elvers naturally infected with EVA have a hemorrhagic disease and show vascular congestion of abdominal surfaces and pectoral and anal fins (Sano 1976). The head is flexed downward, but neither behavioral changes nor internal changes in appearance have been reported. Histologic examination reveals intense hemorrhaging and degeneration of skeletal muscle. Gills are hemorrhagic, and blood or protein is present in Bowman's space and kidney tubules.

Isolants EVA and EVEX have been found to be nonvirulent for elvers (Castric and Chastel 1980; Nishimura et al. 1981). However, when those two isolants were used to challenge fry of ayu, common carp, and rainbow trout, the rainbow trout fry proved to be highly susceptible, especially to EVEX. At 20°C, EVA caused 25% mortality and EVEX caused 100%. The disease produced in trout fry was similar to IHN, and included hemorrhages in skeletal muscle, marked necrosis and hemorrhages in renal hematopoietic tissue, and focal necrosis of liver, spleen, and pancreas (Nishimura et al. 1981).

Etiology

The role of rhabdoviruses in eels could be that of pathogens in pelagic leptocephali and elvers, but the viruses might also be pathogens of other species and simply avirulent opportunists that are transmitted when eels feed.

The two subgroups of fish rhabdoviruses are classified on the basis of differences in polypeptide profiles (Table 1); in addition, they differ in serologic and biological properties (Table 2). Accordingly, if the rhabdoviruses are obligate pathogens of eels, they could evoke more than one kind of disease. As an example, two rhabdoviruses that occur in salmonids differ in polypeptide profile, serology, and biological properties. The diseases they produce are IHN and VHS—infections that have several features in common but other features that are peculiar to the virus.

The CPE of EVEX in RTG-2 cells consists of pyknosis, granulation of cytoplasm,

Table 1. Comparative molecular weights ($\times 10^3$) of major structural polypeptides of eel rhabdovirus subgroups

Protein	Virus subgroups: Vesiculovirus (C_{30})	Lyssavirus (B_{12})
L	165	165
G	65	75
N	50	45
NS	45	—
M	28	—
M_1	—	29
M_2	—	25

Source: Adapted from Castric et al. (1984).

Table 2. Comparison of some biological properties of eel rhabdovirus subgroups

Property	Subgroups	
	Vesiculovirus (C_{30})	Lyssavirus (B_{12})
Size	<185 nm	>200 nm
Growth in RTG-2 cells	Yes	No
Titer in EPC cells	10^7 PFU/mL	10^8 PFU/mL
Growth at 20°C	Yes	No
Neutralization by anti-EVEX serum	Yes	No

Source: Adapted from Castric et al. (1984).

and eventual lysis, but margination of nuclear chromatin does not occur (Sano et al. 1977).

Diagnosis

Notwithstanding the fact that most of the isolations have been made from seemingly healthy fish, eel rhabdovirus involvement can be considered if hemorrhagic disease and mortality occur among young eels at 20°C or higher, and other pathogens are not present.

Isolation

Virus can be present in eels of any age; internal organs, gills, and brains should therefore be pooled, homogenized, and inoculated as a decontaminated 1:50 to 1:100 dilution. Because both the vesiculovirus and Lyssavirus subgroups replicate in the EPC cell line, this line should be used for isolation. The vesiculovirus subgroup is also replicated in BF-2, EK-1, EO-2, FHM, and RTG-2 cells. An incubation temperature of 15°C is suggested, but additional cultures can be incubated at 20°C to hasten the response of vesiculoviruses. Cultures should be held for 5 to 7 days and, in the absence of CPE, blind passage should be made.

The presence of IPNV could obscure rhabdoviral CPE, but that effect can be prevented by using polyvalent anti-IPNV serum with the original inocula.

Identification

Reports on serologic identification of eel rhabdoviruses have concerned the preparation and use of antisera against vesiculovirus isolants EVA and EVEX (Castric and Chastel 1980; Dixon and Hill 1984; Hill et al. 1980, 1981). The results indicate that EVA and EVEX are closely similar, if not identical. Neutralization tests and ELISA show that these two isolants are wholly unrelated to the eel Lyssavirus isolants B_{12} and

C_{26}. Also, the eel rhabdoviruses were shown by Castric et al. (1984) and Dixon and Hill (1984) to be serologically different from the rhabdoviruses of carp (SVCV), perch, northern pike (PFRV), and salmonids (IHNV, VHSV). Serologic relationships to rhabdoviruses from Rio Grande cichlid, sheatfish, or halibut have yet to be assessed.

Transmission and Incubation

The general characteristics of rhabdoviruses suggest that horizontal transmission is most likely; however, no data on transmission or incubation have yet been developed.

Source

The sources of virus are not known. The possibilities include adult eels and fish of wholly unrelated families.

Host and Geographic Range

Thus far, the eel rhabdoviruses have been found only in elvers of stocks of the fish variously named *A. rostrata* or *A. anguilla*—generally considered to be North American and European races of one species. It remains to be determined whether the EVA-type virus has become established in *A. japonica* in Japan.

Control

It has not been determined whether a disease problem exists in eels; accordingly, a discussion of control measures is not appropriate.

References

Ahne, W., and I. Thomsen. 1985. The existence of three different viral agents in a tumour bearing European eel (*Anguilla anguilla*). Zentralbl. Veterinärmed. [B] 32:228–235.

Castric, J., and C. Chastel. 1980. Isolation and characterization attempts of three viruses from European eel, *Anguilla anguilla*: preliminary results. Ann. Virol. (Paris) 13E:435–448.

Castric, J., D. Rasschaert, and J. Bernard. 1984. Evidence of Lyssaviruses among rhabdovirus isolates from the European eel *Anguilla anguilla*. Ann. Virol. (Paris) 135E:35–55.

Dixon, P. F., and B. J. Hill. 1984. Rapid detection of fish rhabdoviruses by the enzyme-linked immunosorbent assay (ELISA). Aquaculture 42:1–12.

Hill, B. J., R. F. Williams, C. J. Smale, B. O. Underwood, and F. Brown. 1980. Physico chemical and serological characterization of two rhabdoviruses isolated from eels. Intervirology 14:208–212.

Hill, B. J., R. F. Williams, and J. Finlay. 1981. Preparation of antisera against fish virus disease agents. Dev. Biol. Stand. 49:209–218.

McAllister, P. E., T. Nagabayashi, and K. Wolf. 1977. Viruses of eels with and without stomatopapilloma. Ann. N.Y. Acad. Sci. 298:233–234.

Nishimura, T., M. Toba, F. Ban, N. Okamoto, and T. Sano. 1981. Eel rhabdovirus, EVA, EVEX and their infectivity to fishes. Fish Pathol. 15:173–184.

Sano, T., 1976. Viral diseases of cultured fishes in Japan. Fish Pathol. 10:221–226.

Sano, T., T. Nishimura, N. Okamoto, and H. Fukuda. 1976. Isolation of rhabdovirus from European eels (*Anguilla anguilla*) at Japanese port of entry. Fish Health News 5:5–6.

Sano, T., T. Nishimura, N. Okamoto, and H. Fukuda. 1977. Studies on viral diseases of Japanese fishes. VII. A rhabdovirus isolated from European eel, *Anguilla anguilla*. Bull. Jpn. Soc. Sci. Fish. 43:491–495.

Sorimachi, M. 1982. Susceptibility of fish cell lines to eel viruses. Fish Pathol. 17:119–123. In Japanese with English abstract.

33 Goldfish Iridovirus

Synonyms: GFV-1, GFV-2

Definition

The goldfish iridovirus (isolants GFV-1 and GFV-2) is an icosahedral cytoplasmic DNA virus that has been isolated from seemingly healthy *Carassius auratus* on two occasions. The role of the virus in the goldfish has not been determined, but experimental inoculation of *Fundulus heteroclitus* resulted in mortality (Shea 1983b).

History

The GFV isolants were found in 1978 at Northeastern University, Boston, Massachusetts, during the course of a search for fish viruses. In that search, primary cultures were initiated by trypsinization and explantation procedures with swim bladder and other tissues from several species of freshwater fish. Cultures of goldfish swim bladder explants were set up and incubated at 25°C. After 2 weeks, several cultures showed CPE that progressed during the next 2 weeks. Virus was suspected; accordingly, culture media were harvested, pooled, and stored at −70°C. Later testing showed that the suspect medium was infectious for the CAR cell line of goldfish origin. Further investigation showed the presence of icosahedral virions whose replication was inhibited by iododeoxyuridine. Initially termed the MA virus (initials of the person who made the isolation), the designation was later changed to goldfish virus, or GFV (J. Gabliks, personal communication).

News of the finding was not immediately published. In 1981, however, a manuscript describing early characterization of the virus was sent to me by J. Gabliks as first author. Information presented in the manuscript merited publication but a formal report did not soon appear.

In 1983, two reports on toxicologic studies appeared in which GFV-2 was used. Characterization of the virus was referred to by Shea (1983a, b) as being included in a paper in press by E. S. Berry and T. B. Shea. That report was then published under authorship that included J. Gabliks (Berry et al. 1983).

Signs and Pathologic Changes

The only evidence for pathogenicity was obtained with a group of 15 goldfish (6 to 8 cm long) injected intraperitoneally with 250 $TCID_{50}$ of third-passage virus. Eight fish that died 7 to 20 days later showed small hemorrhagic lesions in the peritoneal cavity. No controls died, and all attempts to isolate virus from the swim bladders of survivors failed. Two later trials were carried out with fifth- and seventh-passage virus, but no fish died during the ensuing 50 days and necropsy showed no signs of disease (J. Gabliks, personal communication). Shea (1983a) reported that GFV-2 was lethal to *Fundulus heteroclitus*, but gave no details.

Etiology

The two isolants of GFV are heat, pH, and chloroform labile. Three freeze-thaw cycles do not degrade infectivity. Enveloped virions are 180 nm in diameter and have an icosahedral capsid of 120 nm. Replication is cytoplasmic (Fig. 1). At 25°C in the CAR cell line, new infectivity first appears at 24 hours and peaks about 72 hours later. The maximal titer approaches 10^6 $TCID_{50}$/mL and most infectivity is cell associated (Berry et al. 1983).

Primary cultures of goldfish skin or swim bladder are susceptible but cultures of swim bladder from largemouth bass (*Micropterus salmoides*) and yellow perch (*Perca flavescens*) are not. The virus is also replicated by primary cultures of chick embryos.

Diagnosis

Diagnosis is not relevant—the virus was isolated from normal-looking fish.

Isolation

Initial isolations were made from primary cultures of goldfish swim bladder tissues that were incubated at 25°C. Homogenates of internal organs assayed on CAR might also show evidence of the virus if it is present. The FHM and EPC lines should also be tried.

Identification

Serologic identification has yet to be reported. A possible relationship to other fish iridoviruses—the agents isolated from carp gill necrosis and the cod ulcus syndrome—should be examined.

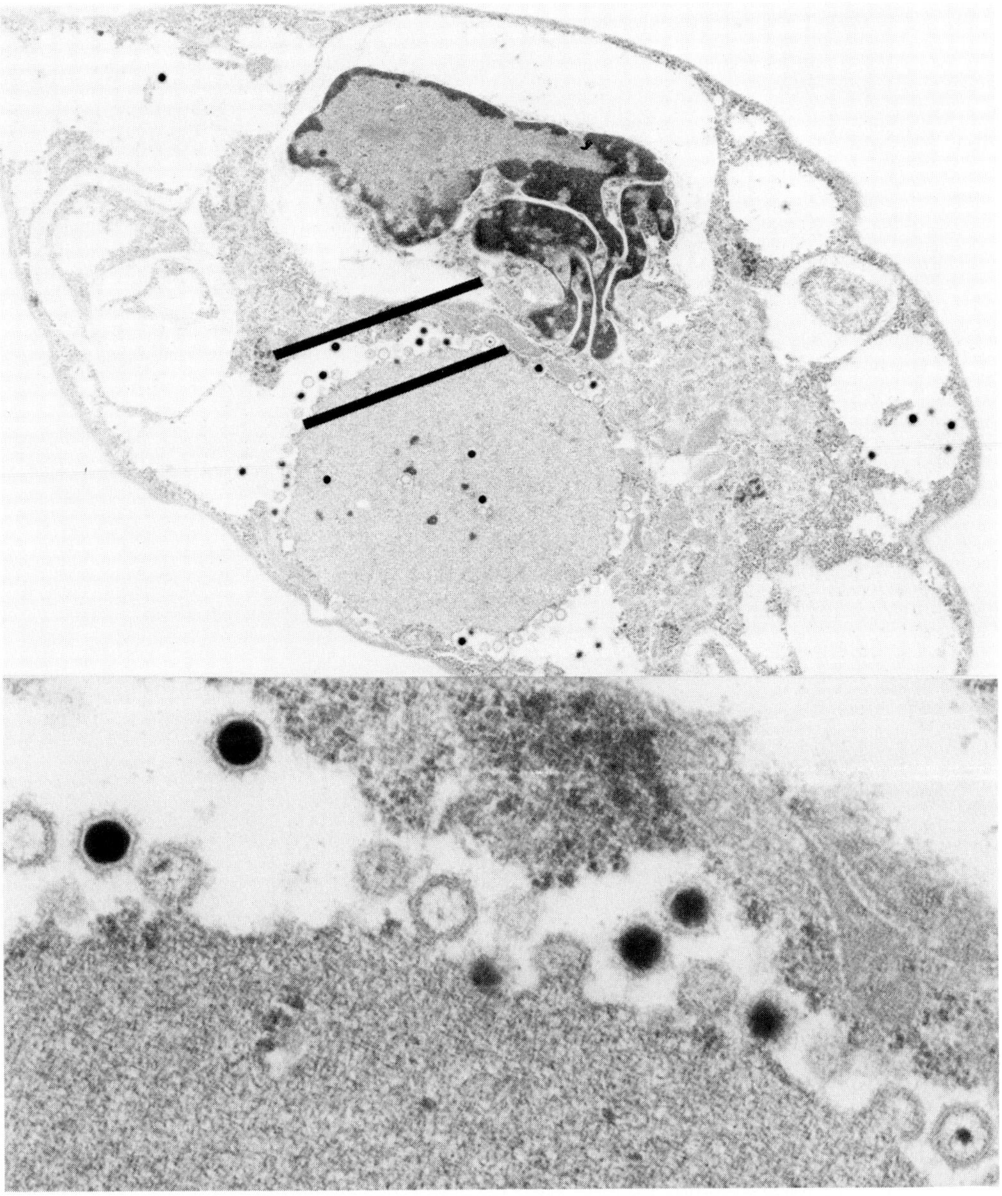

Figure 1. Electron micrographs of a cell of the CAR line infected with the goldfish iridovirus. *Top.* Prominent vacuolation is present in the cytoplasm and marked margination of chromatin is evident in the nucleus. Circular viroplasm (inclusion body) shows internal and external stages of virus replication. *Bottom* (inset from top photo). Virions in the various stages of development occur on the edge of viroplasm. Courtesy of E. S. Berry and T. B. Shea.

Host and Geographic Range

Virus was found only in goldfish from the Boston area; however, the virus probably will be found elsewhere and possibly in other cyprinid species.

References

Berry, E. S., T. B. Shea, and J. Gabliks. 1983. Two iridovirus isolates from *Carassius auratus* (L.). J. Fish Dis. 6:501–510.

Shea, T. B. 1983a. Enhancement of goldfish virus-2 in vitro replication by the pesticides carbaryl and toxaphene. Appl. Environ. Microbiol. 45:1859–1864.

Shea, T. B. 1983b. Enhancement of goldfish virus type 2 replication by 1-naphthol, the major hydrolysis product of the pesticide carbaryl. Appl. Environ. Microbiol. 46:1230–1231.

34 Opaleye Calicivirus

Synonyms: San Miguel sea lion virus, SMSV-6, SMSV-7

Definition

The opaleye calicivirus is a well-recognized pathogen of marine pinniped mammals, but is better known as San Miguel sea lion virus (SMSV). It has been shown to produce a disease condition "identical to vesicular exanthema in experimentally infected swine" (Smith, Skilling, Dardiri, and Latham 1980). Although the virus is replicated in the opaleye (*Girella nigricans*), its role in producing disease in the fish is unknown. The fish is, however, circumstantially involved as a natural reservoir for the virus, and a complex web is believed to link the virus to marine mammals by way of a nematode lung worm (Smith, Skilling, and Brown 1980). More important, SMSV is considered to be the etiologic agent of the calicivirus-caused vesicular exanthema that swept across swine herds in the United States during the 1930s to 1950s. Bankowski (1981), who reviewed the vesicular exanthema–San Miguel sea lion viral infection, considers the same virus to be involved.

History

Viral vesicular exanthema of swine first broke out in California in 1932 and subsequently spread across the United States. Control was effected by late 1959 by slaughtering affected herds and prohibiting the feeding of uncooked garbage.

In 1971, abortion among California sea lions rose to sufficient magnitude to cause a multidisciplinary team to be organized, and a detailed investigation to be carried out. Within a few months, virus was isolated from several dead sea lions and soon shown to have the size and morphology of what was known as the virus of vesicular exanthema of swine. The SMSV was later isolated from a cetacean and five species of pinnipeds. It was postulated that garbage contaminated with raw fish scraps in coastal California communities was a source of the virus infecting swine, and SMSV was subsequently isolated from the opaleye and from a nematode lung worm of marine mammals that has the opaleye as an intermediate host. Serologically, the opaleye calicivirus isolants have not been shown to be identical to serotypes of the agents of swine vesicular exanthema. The opaleye virus, however, has been used to infect swine and produce disease identical to vesicular exanthema.

The exciting story of discovery of marine mammal, fish, and nematode involvement with calicivirus and the role of key investigators in the history was told by Smith (1981), who played a leading role in the epizootiologic research.

Signs and Pathologic Changes

Experimental infections have clearly shown that a calicivirus isolant from the northern fur seal (*Callorhinus ursinus*), designated SMSV-5, replicates in the opaleye (Smith et al. 1981). Four fish (8 to 15 g) were injected intraperitoneally with 10^6 $TCID_{50}$ virus, then held in seawater at 15°C. Neither behavior changes nor signs of disease were observed, but virologic assay of individual organs and tissues of two fish at 14 days and two fish at 28 days showed that virus was present in most tissues. At 14 days, virus was present in all specimens of liver, kidneys, muscle, gills, and gut, but the greatest levels—$10^{7.1}$ to $10^{7.6}$ $TCID_{50}/g$—were in spleens. Lesser amounts—$10^{3.4}$ to $10^{4.2}$ $TCID_{50}/g$—were in splenic tissue at 28 days, and other tissues showed either less virus or none. Histologic examination was not carried out.

Etiology

Members of the genus *Calicivirus* belong to the family Picornaviridae and have a genome of single-stranded RNA. The general size is 35 to 40 nm, and the morphology is probably icosahedral. The opaleye isolants have been identified as SMSV serotypes 6 and 7. In vitro replication of SMSV-5 occurs at 15 and 37°C, and peak titers have reached 10^8 $TCID_{50}/mL$.

Isolation

Isolations were made from homogenates of whole viscera or individual internal organs of opaleye specimens. Preparation included dilution and clarification by centrifugation and then adsorption for an hour on monolayers of Vero (African green monkey kidney) cells grown in tube cultures. Incubation was at 37°C. Cytopathic effects were found after three passages (Smith, Skilling, Dardiri, and Latham 1980). Titers of original organ and visceral homogenates are not known and neither is the effect of the virus on any cell line of fish origin.

Host and Geographic Range

Calicivirus serotypes SMSV-6 and SMSV-7 have been isolated from the opaleye and from one or another of the following: the northern elephant seal (*Mirounga agusti-*

rostris), the California sea lion (*Zalophus californianus*), the northern fur seal, and a liver trematode of the genus *Zalophatrema* taken from a California sea lion (Smith, Skilling, Dardiri, and Latham 1980).

The opaleye calicivirus has thus far been isolated from fish taken from tidal pools on San Miguel and San Nicholas Islands, which are about 50 miles off the coast of southern California. The range of the opaleye extends from Point Conception, California, south along the coast into Mexican waters. Other fish hosts might well be found in other regions, for the virus has been isolated from feces of walrus (*Odobenus rosmarus*) taken off the ice in the Chukchi Sea, and SMSV-neutralizing antibodies have been found in bowhead whales (*Balaena mysticetus*), another boreal mammal (Smith et al. 1983).

References

Bankowski, R. A. 1981. Vesicular exanthema of swine: San Miguel sea lion viral infection. Pages 176–182 in G. W. Beran, ed. CRC handbook series in zoonoses. Section B: viral zoonoses, Vol. II. CRC Press, Boca Raton, Florida.

Smith, A. W. 1981. Marine reservoirs for caliciviruses. Pages 182–190 in G. W. Beran, ed. CRC handbook series in zoonoses. Section B: viral zoonoses, Vol. II. CRC Press, Boca Raton, Florida.

Smith, A. W., D. E. Skilling, and R. J. Brown. 1980. Preliminary investigation of a possible lung worm (*Parafilaroides decorus*), fish (*Girella nigricans*), and marine mammal (*Callorhinus ursinus*) cycle for San Miguel sea lion virus type 5. Am. J. Vet. Res. 41:1846–1850.

Smith, A. W., D. E. Skilling, A. H. Dardiri, and A. B. Latham. 1980. Calicivirus pathogenic for swine: a new serotype isolated from opaleye *Girella nigricans*, an ocean fish. Science 209:940–941.

Smith, A. W., D. E. Skilling, C. M. Prato, and H. L. Bray. 1981. Calicivirus (SMSV-5) infection in experimentally inoculated opaleye fish (*Girella nigricans*). Arch. Virol 67:165–168.

Smith, A. W., D. G. Ritter, G. C. Ray, D. E. Skilling, and D. Wartzok. 1983. New calicivirus isolates from feces of walrus (*Odobenus rosmarus*). J. Wildl. Dis. 19:86–89.

PART III

VIRUSES VISUALIZED BUT NOT YET ISOLATED

35 Atlantic Cod Adenovirus

Definition

The Atlantic cod adenovirus is an icosahedral form seen in nuclei of cells of hyperplastic epidermal plaques on the body surfaces of *Gadus morhua* from the Baltic Sea off Denmark (Jensen and Bloch 1980).

History

A few cod with characteristic raised hyperplastic lesions in their skin were taken during a study of fish having the better known cod ulcus syndrome. The brief report of Jensen and Bloch (1980) is the only published account of the adenovirus-like forms, and the first report of a member of the group from a fish.

Signs and Pathologic Changes

The skin lesions are slightly raised, flat, transparent plaques occurring at various places on the body surface but most numerous in the caudal regions. Diameters range from 3 to 20 mm. Histologically, the epidermis is four times the normal thickness and contains few or no mucous glands. Although the cells themselves appeared to be normal, those of the germinal layer are shorter than normal and vascularity is greatly increased beneath the plaques. No other details have been reported.

Etiology

In the absence of evidence to the contrary, the intranuclear forms in cells of the epidermal plaques are presumed to be causal (Fig. 1). Virus particles are not universally present, however, being found only in a few cells scattered among the outermost cell layers of the epidermis. The particles measure 77 nm from corner to corner and have hexagonal or pentagonal profiles. Fibers 20 to 25 nm long some-

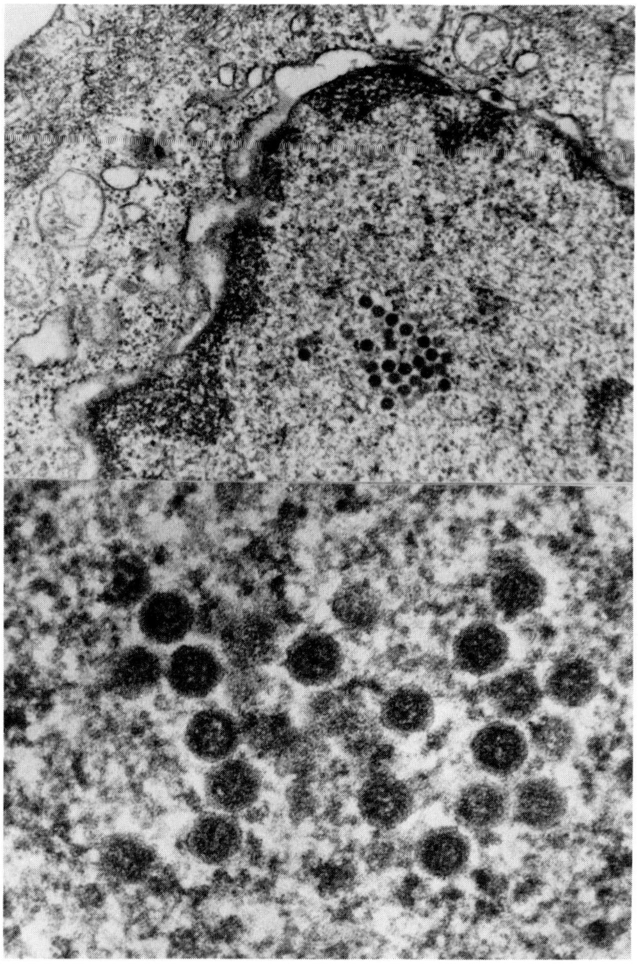

Figure 1. Atlantic cod adenovirus. *Top.* A hyperplastic epidermal cell showing margination of chromatin and a cluster of intranuclear virions. ×2800. *Bottom.* Icosahedral morphology of 77-nm virions. ×90,000. From Jensen and Bloch (1980). Reprinted with the permission of Danske Dyrlaegforening Alhambravej, Copenhagen.

times connect corners to the surrounding nucleoplasm. A 10-nm outer layer has been interpreted as a capsid surrounding a more densely stained nucleoid. Compared with sections of human adenovirus type 7, the agent in cod cells has a less dense and more weakly contrasted nucleoid.

It is worth noting that similar-size adenovirus-like particles have been found in epithelial neoplasms in the dab (*Limanda limanda*) from the adjacent North Sea (Bloch et al. 1986).

Reference

Bloch, B., S. Mellergaard, and E. Nielsen. 1986. Adenovirus-like particles associated with epithelial hyperplasias in dab, *Limanda limanda* (L.). J. Fish Dis. 9:281–285.

Jensen, N. J., and B. Bloch. 1980. Adenovirus-like particles associated with epidermal hyperplasia in cod (*Gadus morhua*). Nord. Vet. Med. 32:173–175.

36 Atlantic Salmon Swim Bladder Sarcoma Virus

Definition

The Atlantic salmon swim bladder sarcoma virus is a C-type retrovirus abundantly present in massive swim bladder tumors that were associated with mortality among subadult *Salmo salar* reared in cages in a Scottish loch.

History

In 1976, mortality in 1-year-old smolts and 2-year-old Atlantic salmon in commercial cage culture was brought to the attention of investigators at the Marine Laboratory, Aberdeen, Scotland. McKnight (1978) reported on the gross and microscopic pathology of swim bladder tumors associated with some of the dead fish and arrived at a definitive identification of leiomyosarcoma. Electron microscopy was also used, and descriptions and illustrations of associated C-type retrovirus were provided by Duncan (1978).

Signs and Pathologic Changes

Mortality attributed to the sarcoma was 3.4% in a population of about 500 fish, but the prevalence might have been greater (living fish could not be examined).

Affected fish of both sexes are sluggish and in poor physical condition. There are no external lesions. Internally, in the most severely afflicted specimens, hard nodular tumors are extensively distributed along the length of the swim bladder (Fig. 1). Tumors are 15 to 30 mm in diameter, protrude from the organ, and occupy much of the body cavity. Adjacent nodules appear to have fused. The anterior swim bladder is most heavily involved, but neither infiltration of adjacent organs nor metastases have been observed. In fish least affected, nodules on the swim bladder are barely visible.

Histologically, the tumors arise at the junction of the circular smooth muscle

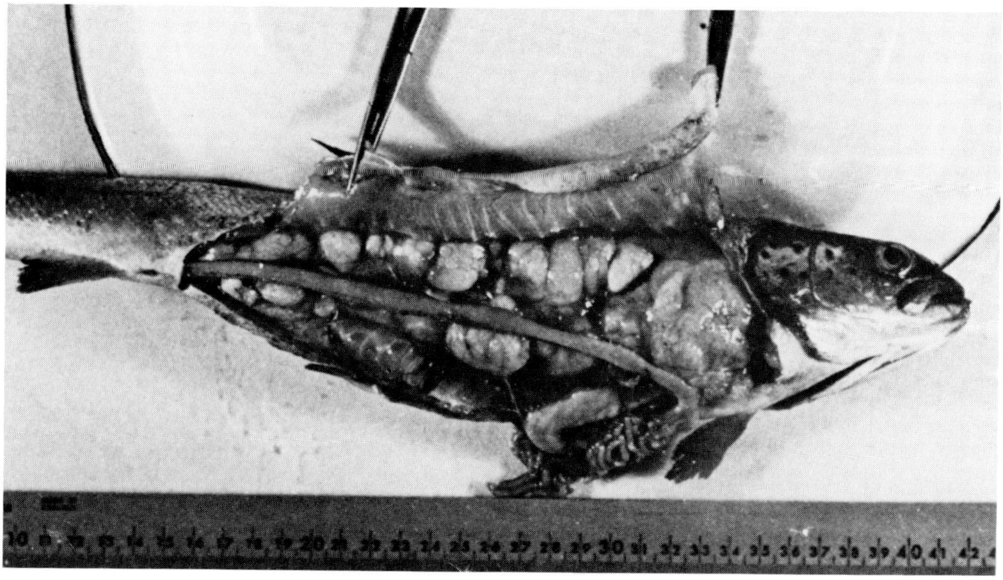

Figure 1. A subadult Atlantic salmon with massive sarcomatous nodules along the entire length of the swim bladder. From McKnight (1978). Reprinted with the permission of Elsevier Science Publishers.

layer and the area of loose areolar tissue. They consist of well-differentiated, interlacing bundles of fibroblastic cells, mostly with elongated nuclei and cytoplasmic processes. The peripheries of the nodules are well vascularized but the centers are not, and consequently show ischemic necrosis. Mitoses are fairly common, but little collagen is present. Histopathologically, the tumors are classified as leiomyosarcomas (McKnight 1978).

Etiology

The virus has not been isolated, nor has experimental transmission been demonstrated. Nevertheless, the presence of C-type retrovirus in neoplastic tissue is widely considered as evidence of probable causal association.

The virus particles are about 120 nm in diameter and present only in cytoplasm, where they show all indications of budding from membranes. In section, the particles show a crescent-shaped nucleoid. Membrane-enclosed aggregates of other electron-dense spheroidal bodies in the cytoplasm are about 110 nm in diameter. The smaller particles are accompanied by collagenous fibrils about 20 nm thick, produced by adjacent cells. Duncan (1978) suggested that the smaller particles might represent earlier maturing virus that was somehow trapped.

Host and Geographic Range

The virus has been found only in Atlantic salmon in Scotland.

References

Duncan, I. B. 1978. Evidence for an oncovirus in swim bladder fibrosarcoma of Atlantic salmon *Salmo salar* L. J. Fish Dis. 1:127–131.

McKnight, I. J. 1978. Sarcoma of the swim bladder of Atlantic salmon (*Salmo salar* L.). Aquaculture 13:55–60.

37 Carp Coronavirus

Definition

The carp coronavirus is an intracellular particle that has the morphology appropriate to the group and was discovered in electron micrographs of pathologically altered kidney tissue from specimen fish, presumably *Cyprinus carpio*, suspected of having a viral disease. The fish were from the state of Morelos, Republic of Mexico.

History

Published reports have not appeared, and the only available information is in an explanatory letter (P. C. Giron, personal communication, 1979) that accompanied electron micrographs sent to the National Fish Health Research Laboratory in 1979.

Signs and Pathologic Changes

Affected fish were described as having "clinical signs and lesions suspicious of spring viremia of carp." Liver and kidney tissues were taken for electron microscopy, but pathologic changes were found only in the kidneys and consisted of mononuclear infiltration "in the intertubular area." Further details were not provided.

Etiology

It is not known if the virus was the cause of the disease condition. Isolation was attempted with inocula prepared from kidneys and liver and the following cell cultures: BB, FHM, and RTG-2 lines, primary cultures of carp and trout ovaries, the mammalian lines BHK-21 and PK-15, and primary cultures of bovine and porcine testicles. Temperature of incubation, duration of observation, and possible blind passage were not mentioned—only that CPE was not seen. Micrographs were referred to a consulting electron microscopist, who concurred in the opinion that the particles resembled coronavirus (R. W. Darlington, personal communication, 1979). Attempts to obtain additional information from the investigator were unsuccessful.

38 Dab Adenovirus

Definition

The dab adenovirus is an icosahedral particle visualized in nuclei of hyperplastic epithelial cells that occur in discrete plaquelike growths on the pigmented skin of the pleuronectid dab *Limanda limanda* in eastern waters of the North Sea (Bloch et al. 1986).

History

Small papillomata or patches of epithelial hyperplasia were observed on 1 to 2% of the dabs taken during stock assessment surveys made in 1983 and 1984. Representative tumors were fixed in formalin for histologic examination and postfixed in osmium tetroxide for electron microscopy. Adenovirus-like particles found in growths from four representative fish were examined by electron microscopy; additional reports on the condition and on attempts at isolation are anticipated (Bloch et al. 1986).

Signs and Pathologic Changes

The growths occur on otherwise normal-looking 3- and 4-year-old fish. The usual tumor is a smooth, somewhat transparent, creamy-white patch, 2 to 10 mm in diameter (Fig. 1). Less frequently, larger rough-surfaced and vascularized patches as large as 15 mm are also found.

Histologically, the normal epithelium of the dab is only 5 to 10 cell layers thick, but the growths may have as many as 50 layers. The smaller areas of hyperplasia are believed to be early or initial stages that develop into larger papillomata. Basal cells of prismatic to cuboidal form are arrayed vertically, whereas deeper more centrally located cells are larger and polyhedral and in horizontal arrangement. The outermost cells are elongated or flattened, and arrayed horizontally. Some eosinophilic granulated cells are present, but mucous cells are rare.

Figure 1. Partly transparent focus of epithelial hyperplasia on the pigmented side of a dab. ×2.3. From Bloch et al. (1986). Reprinted with the permission of Blackwell Scientific Publications Ltd.

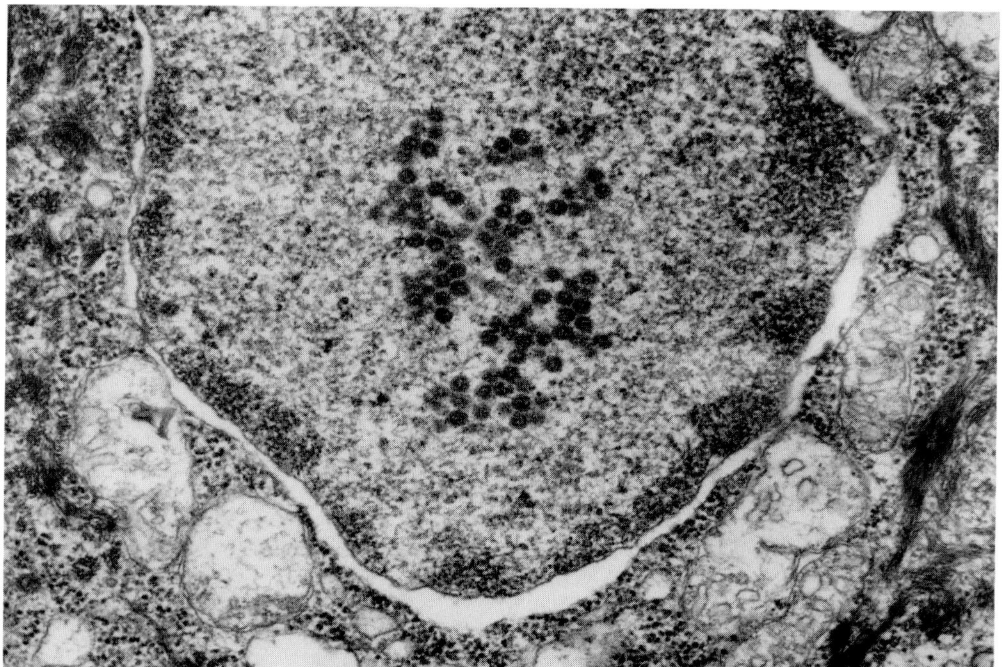

Figure 2. Adenovirus-like particles and marginated chromatin in the nucleus of an epithelial cell from the dab. ×37,600. From Bloch et al. (1986). Reprinted with the permission of Blackwell Scientific Publications Ltd.

Etiology

Because the virus is present in some of the cells that compose the tumors, it is assumed to have a causal role, but the agent has not yet been isolated, nor transmission attempted.

Virions are only in the nucleus—usually in cells or cell debris near the surface, or at least in the upper third of the affected tissue. The particles have profiles of an icosahedron, are 70 nm in mean diameter, and have a capsid 7-nm thick surrounding a core that stains nonhomogeneously with uranyl acetate and lead citrate (Fig. 2).

The viruslike particles in lesions of the dab are much like those described and illustrated in a report on epidermal hyperplasia in Atlantic cod by Jensen and Bloch (1980). Moreover, the cod were taken in the Baltic Sea, which adjoins the eastern arm of the North Sea, where the affected dab were found. Accordingly, and even though the cods (Gadidae) and dabs (Pleuronectidae) are in widely separated families, the similarities of the respective neoplasms and of the intracellular viruslike particles strongly suggest that only one virus is involved. Isolation and serologic identification should resolve the matter of identification.

References

Bloch, B., S. Mellergaard, and E. Nielsen. 1986. Adenovirus-like particles associated with epithelial hyperplasias in dab, *Limanda limanda* (L.). J. Fish Dis. 9:281–285.

Jensen, N. J., and B. Bloch. 1980. Adenovirus-like particles associated with epidermal hyperplasia in cod (*Gadus morhua*). Nord. Vet. Med. 32:173–175.

39 Pacific Cod Herpesvirus

Definition

The Pacific cod herpesvirus is an agent visualized in giant cells peculiar to certain skin lesions of *Gadus macrocephalus* from the Bering Sea.

History

The virus was first reported in an abstract by McArn et al. (1978), who noted that nearly 4% of some 2000 Pacific cod examined were found to have either ulcers or pale, raised, circular skin lesions. The raised lesions contained hypertrophied cells with intracellular herpesvirus-like particles (Fig. 1).

In a report on the prevalence of tumors and diseases in 37,000 fish from the Bering Sea, McCain et al. (1979) included descriptions and illustrations of the skin lesions of cod—presumably the same condition described in 1978. Although McCain was second author of the 1978 report and first author of the 1979 paper, the descriptions of the lesions and their photomicrographic details differed.

Signs and Pathologic Changes

McArn et al. (1978) described the skin lesions as being circular, pale, and 1 to 15 mm in diameter. The epithelium of the lesions contained giant cells that were surrounded by a hyaline extracellular coat of PAS-positive material. McCain et al. (1979) described the lesions as being "ringlike" and surrounding a circular disk of normal, or sometimes hemorrhagic, skin. The diameter of such lesions was 10 to 50 mm.

Histologic features described in the two references also differ, but that could be a matter of interpretation. McCain et al. (1979) described and illustrated "cystlike bodies" within the epidermis, which had a vacuolated basophilic center surround-

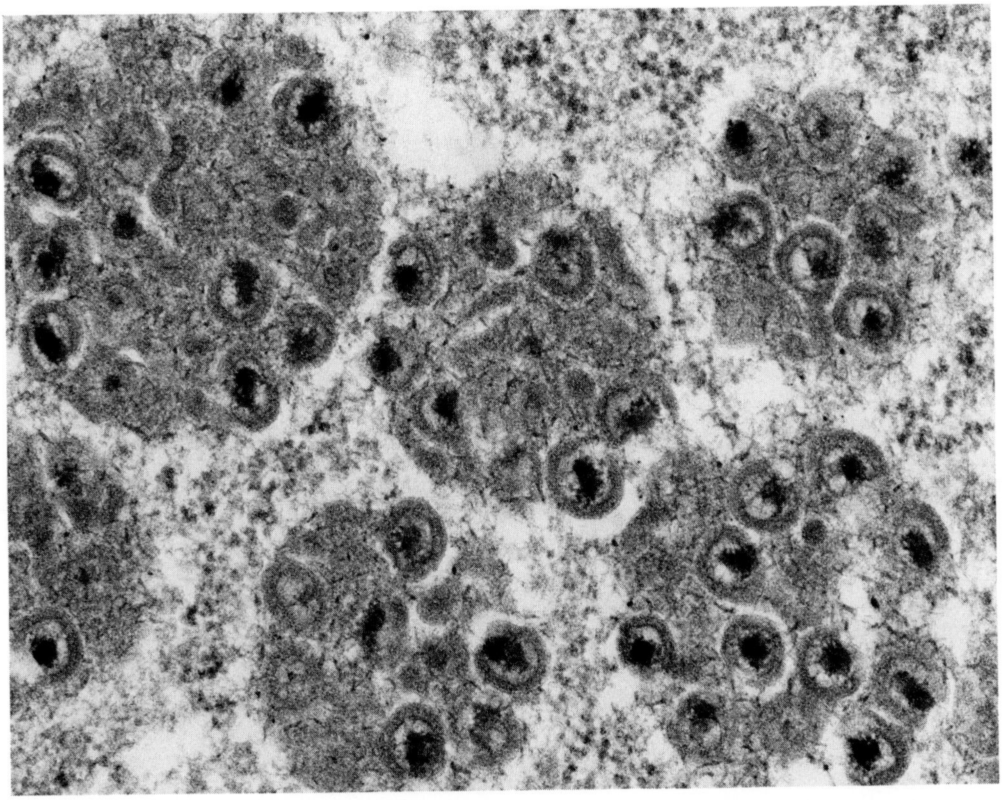

Figure 1. Aggregates of enveloped herpesvirus-like particles occur in the cytoplasm of hypertrophied skin cells of a Pacific cod. ×68,500. Courtesy of G. E. McArn.

ed by an eosinophilic margin (Fig. 2). Both the margin and the central portion contained eosinophilic bodies.

McArn et al. (1978) described the objects as giant hypertrophied cells, rather than as "cystlike bodies."

The skin lesions of Atlantic cod (*Gadus morhua*) illustrated by Jensen and Bloch (1980) look much like those of Pacific cod, but there the similarity ends; histologic sections of Atlantic cod did not contain giant cells or herpesvirus.

Etiology

Intranuclear particles in various stages of development measure 80 × 110 nm and cytoplasmic forms are 120 × 170 nm (Fig. 1).

It seems reasonable to consider the virus as the cause of cell hypertrophy and associated hyperplasia. The virus is yet to be isolated and characterized.

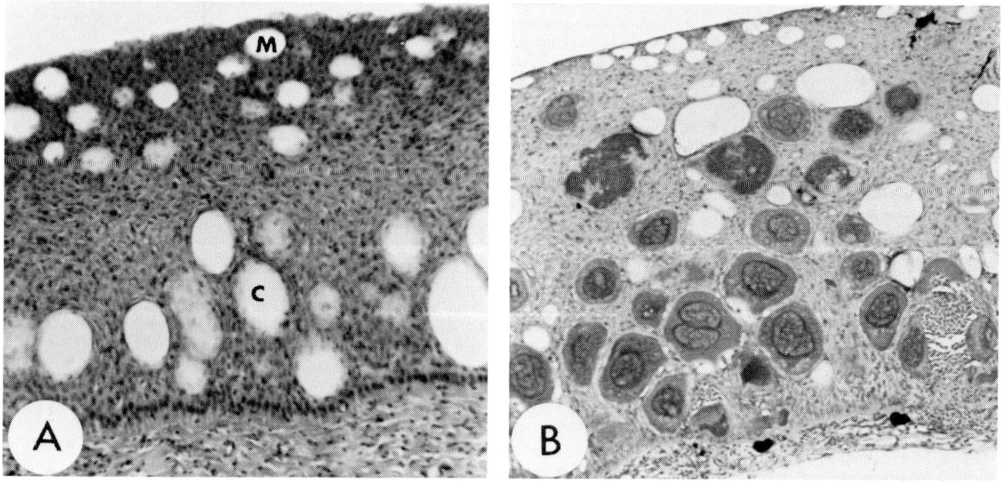

Figure 2. Comparison of normal and abnormal skin of Pacific cod. *A.* Normal epidermis contains mucous cells (m) and cystic bodies (c) at greater depth. The dermis is normal. *B.* The epidermis of infected skin shows abnormal basophilic cells with vacuolated or reticulated nucleus. The dermis is infiltrated with inflammatory cells. ×155. H & E. From McCain et al. (1979). Reprinted with the permission of Blackwell Scientific Publications Ltd.

Host and Geographic Range

This herpesvirus infection is thus far known only in Pacific cod in the Bering Sea.

References

Jensen, N. J., and B. Bloch. 1980. Adenovirus-like particles associated with epidermal hyperplasia in cod (*Gadus morhua*). Nord. Vet. Med. 32:173–175.

McArn, G. E., B. McCain, and S. R. Wellings. 1978. Skin lesions and associated virus in Pacific cod (*Gadus macrocephalus*) in the Bering Sea. Fed. Proc. 37:937.

McCain, B. B., W. D. Gronlund, M. S. Meyers, and S. R. Wellings. 1979. Tumors and microbial diseases of marine fishes in Alaskan waters. J. Fish Dis. 2:111–130.

40 Pacific Salmon Anemia Virus

Definition

Pacific salmon anemia virus is a transmittable 80-nm particle found in the cytoplasm of erythrocytes of young *Oncorhynchus* salmon showing signs of severe anemia and sustaining an associated mortality.

History

Two apparently new kinds of salmon anemia—one of which had been observed since 1980 in juvenile and yearling coho salmon (*Oncorhynchus kisutch*) during autumn—were reported by Holt and Rohovec (1984). The first form was distinguished by a precipitous mortality that occurred about 15 minutes after feeding. The etiology was under investigation. The second form occurred during spring in yearling coho salmon. The fish did not die after feeding, but blood films showed erythrocytic inclusions and intracytoplasmic virions morphologically distinct from the iridovirus of viral erythrocytic necrosis (Fig. 1).

The next development was a report by Leek (1985) of an anemia in fingerling spring chinook salmon (*O. tshawytscha*). They found that transmission of the intra-erythrocytic inclusions and virus could be effected by gastric intubation of healthy young fall chinook salmon with unspecified tissue homogenates from naturally infected spring chinook salmon. A full report of the virus-associated anemia is in preparation (J. R. Winton, personal communication, 1985).

Signs and Pathologic Changes

Salmon that were anemic, and whose erythrocytes showed inclusions and virus, commonly also had external fungal lesions or systemic bacterial infection with *Yersinia ruckeri* or *Renibacterium salmoninarum*. No behavior changes, external or internal signs, or histopathologic findings were described.

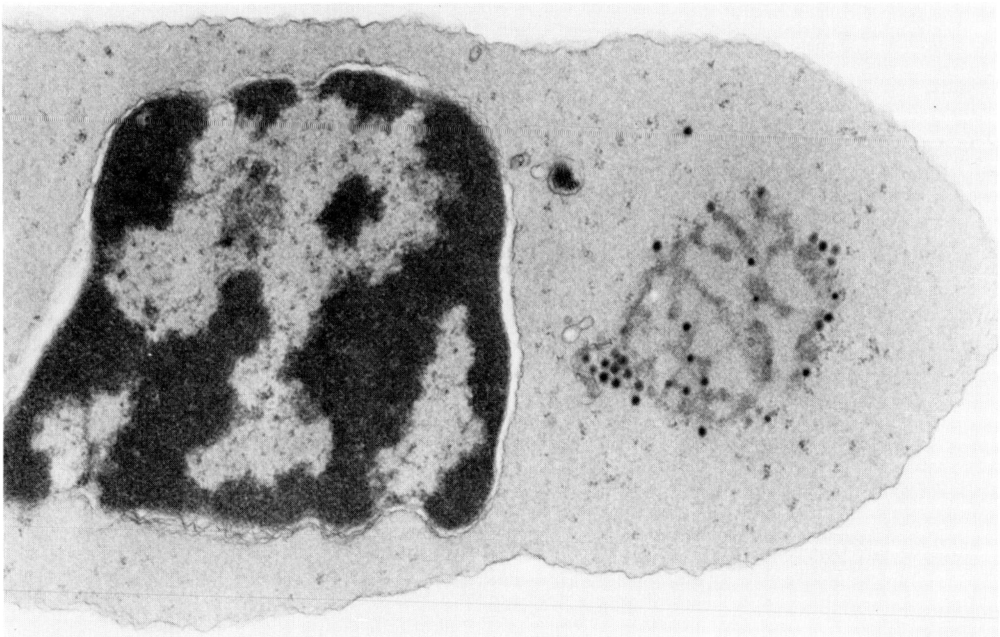

Figure 1. Electron micrograph of a salmon erythrocyte infected with anemia virus, showing cytoplasmic inclusion and associated virus particles. Courtesy of J. R. Winton.

Etiology

Holt and Rohovec (1984) indicated a particle size of about 80 nm, but J. R. Winton (personal communication) reported a size of about 60 nm and a core of 40 nm (Fig. 2). The virus is known to be chloroform sensitive, but has thus far eluded efforts at isolation. Considering the general size, lipid solvent sensitivity, and cytoplasmic replication, the agent could be a togavirus. Whether it is the cause of the anemia remains to be determined.

Diagnosis

A presumptive diagnosis can be based on the finding of anemia and the presence of erythrocytic inclusions and accompanying cytoplasmic virions of the size range noted.

Isolation

No cell line has yet been found susceptible. As suggested for the agent of viral erythrocytic necrosis, primary cultures of hematopoietic tissue should be tried.

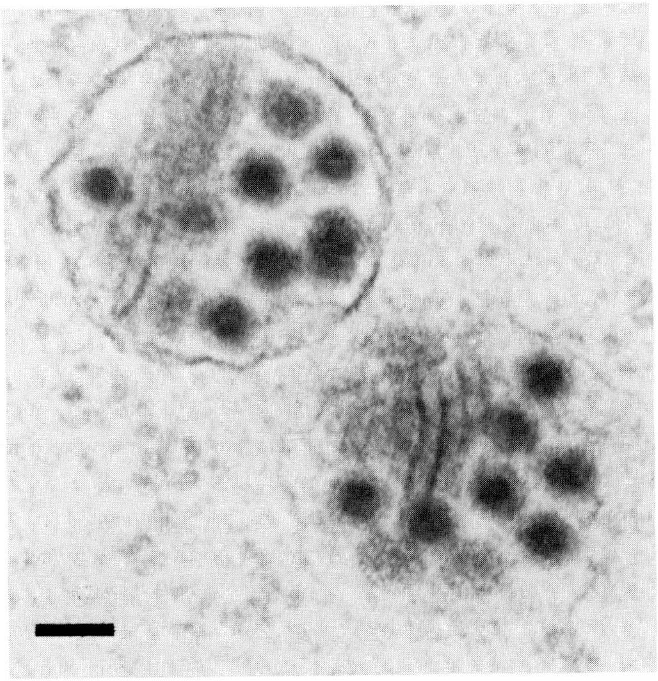

Figure 2. Membrane-enclosed aggregates of 60-nm virus particles in the cytoplasm of an erythrocyte infected with Pacific salmon anemia virus. Bar = 100 nm. Courtesy of J. R. Winton.

Identification

Identification is based solely on size and shape, and the cytoplasmic presence of virus in association with intraerythrocytic inclusion substance. A possible relationship with the intraerythrocytic virus of rainbow trout merits investigation.

Transmission and Incubation

Whether anemia was induced with transmission of the virus is not known. Inclusions were transiently evident 4 days after the gastric intubation of fingerling chinook salmon and again on day 28, when virus was found (temperature was not given).

Source

The source is considered to be a salmonid.

Host and Geographic Range

The anemia and virus are known only among young chinook and coho salmon in the Pacific Northwest and more specifically in the Columbia River system.

References

Holt, R., and J. Rohovec. 1984. Anemia of coho salmon in Oregon. Am. Fish. Soc. Fish Health Sect. Newsl. 12:4.

Leek, S. L. 1985. Artificial transmission of erythrocytic virus. Am. Fish. Soc. Fish Health Sect. Newsl. 13:4.

41 Pike Epidermal Proliferation Retrovirus

Definition

Pike epidermal proliferation retrovirus is a C-type retrovirus particle that is within and between cells of, and presumably the cause of, benign plaquelike areas of neoplasia in mature *Esox lucius*, the northern pike.

History

Investigations of dermal sarcomas of northern pike in Sweden in 1966 and 1967 revealed a low incidence of epidermal hyperplasia, and specimens were collected for closer examination. Electron microscopy revealed C-type virus particles in four of five fish studied (Winqvist et al. 1968). In later work, similar particles were found in two cases of sarcoma and the Swedish authors made a point of noting that epithelial proliferations were abundant in the same population of northern pike.

The next report of epidermal hyperplasia in the northern pike, a species of circumpolar distribution, came from Canada (Yamamoto et al. 1984). Two distinctly different kinds of epidermal lesions were noted: one had an associated herpesvirus and the other a retrovirus. A few fish showed both kinds of lesions, but efforts to isolate virus were not successful.

In the absence of evidence to the contrary, it is cautiously assumed that the C-type particles seen in the American and the European pike were identical.

Signs and Pathologic Changes

The lesions of American and European specimens are described as smooth, transparent or translucent, and about 2 mm thick. The range of diameters is 5 to 20 mm, and adjacent areas sometimes coalesce. In some cases in Canadian northern pike, the lesions damaged the dermis and hemorrhage resulted.

Histologically, the Swedish and Canadian lesions are similar in being composed of undistinguished, undifferentiated normal-size homogeneous cells in random arrangement that extend from the basal layer to the surface. The normal stratification of epidermis is lacking.

Etiology

Maximum diameter of the putative causal virus is uniformly about 150 nm and the nucleoid is 80 nm. Maturation occurs by budding of the cytoplasmic membrane, with resulting aggregates of virus in intercellular spaces.

Diagnosis

A presumptive diagnosis can be made by finding the described smooth plaquelike lesions. The presence of retrovirus is confirmed by electron microscopy.

Host and Geographic Range

Particles of C-type virus occur in external neoplasms of several kinds of fishes, but involvement in the epithelium of northern pike is thus far known only from Sweden and central Canada. The condition might also occur in other members of the family Esocidae.

References

Winqvist, G., O. Ljungberg, and B. Hellstroem. 1968. Skin tumors of northern pike (*Esox lucius* L.). II. Viral particles in epidermal proliferations. Bull. Off. Int. Epizoot. 69:1023–1031.

Yamamoto, T., R. K. Kelly, and O. Nielsen. 1984. Epidermal hyperplasias of northern pike (*Esox lucius*) associated with herpesvirus and C-type particles. Arch. Virol. 79:255–272.

42 Pike Herpesvirus

Synonym: esocid herpesvirus 1

Definition

Pike herpesvirus is an agent associated with and most likely the cause of unusual hypertrophy of some epithelial cells that are embedded in plaques of hyperplastic skin epithelium of *Esox lucius* (Yamamoto et al. 1983).

History

The discovery of the pike herpesvirus, or esocid herpesvirus 1, was made incidental to an investigation of neoplastic skin lesions in another species, the walleye (*Stizostedion vitreum vitreum*). Because some northern pike were trapped with the walleyes and because the pike also had hyerplastic lesions of skin tissue, the pike lesions received the closer scrutiny. Some lesions of pike contained a herpesvirus, and others—which were morphologically distinct—contained a retrovirus. The year of first discovery was 1978, and additional infected pike were collected in 1981 and 1982. The prevalence ranged from less than 1% to about 7% (Yamamoto et al. 1983).

Signs and Pathologic Changes

The appearance of the two kinds of virus-associated growths differ. Herpesvirus lesions are thin (0.25 mm), flat, bluish white areas of granular texture 3 to 10 mm in diameter (Fig. 1A). The growths are generally on the dorsal surface of the fish, but also occur on fins.

In section, affected tissue is greatly thickened and shows the presence of numerous unusual and hypertrophied cells up to 150 mm in diameter that are surrounded by normal-size epithelial cells (Fig. 1B). The giant cells are not a result of syncytium

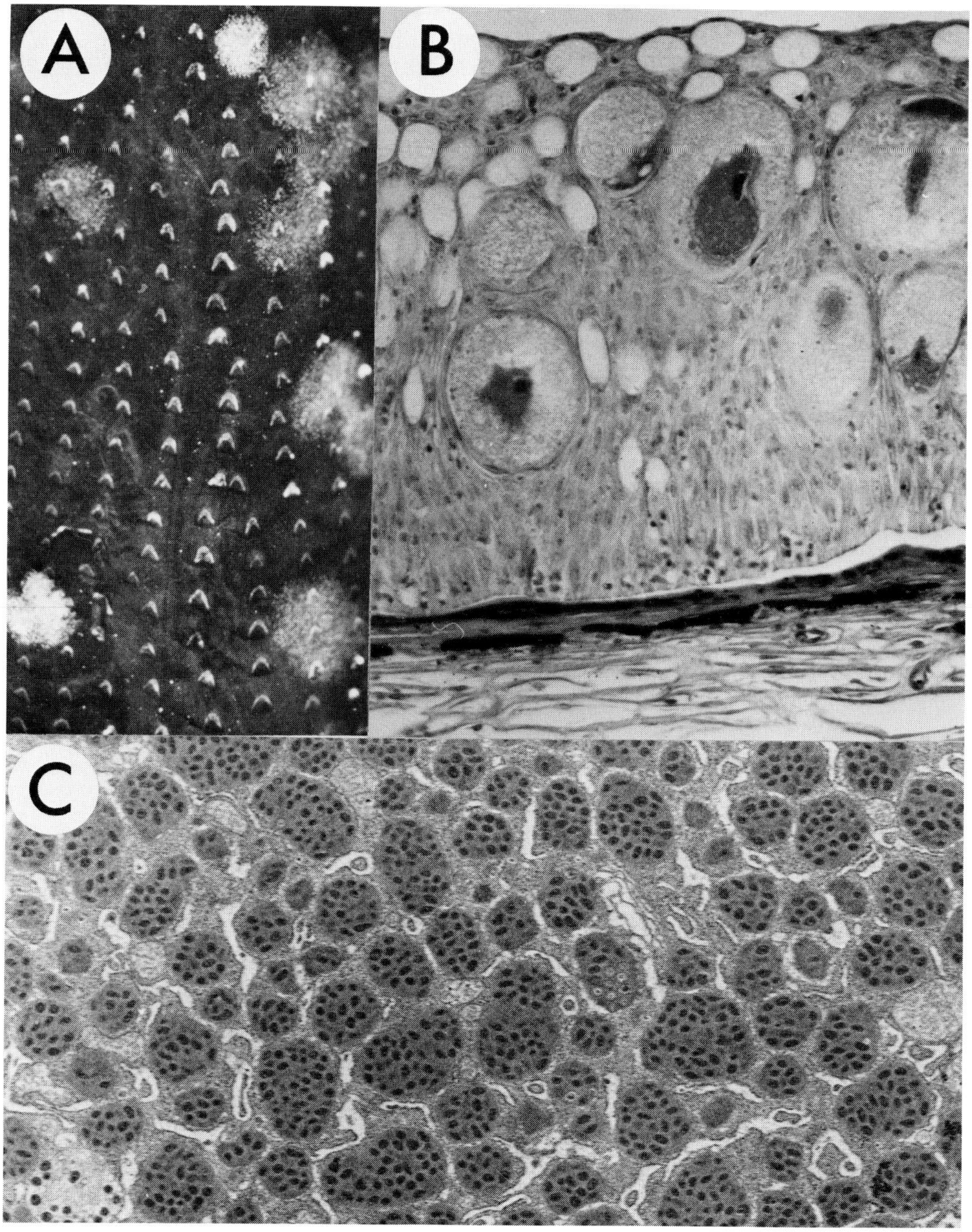

Figure 1. Features of pike herpesvirus infection. *A.* Granular lesions on the skin show variation in density; in fresh specimens they are pale bluish white. *B.* Histologic section through a lesion, showing hypertrophied infected cells with darkly stained nucleus and granular cytoplasm. *C.* Electron micrograph of infected cells containing abundant and unusual cytoplasmic inclusions, within which are virions in various stages of condensation. Virion envelopes are indistinct in mature and presumably older inclusions. From Yamamoto et al. (1983). Reprinted with the permission of Springer-Verlag.

formation; instead, they originate as single cells at the basal layer and enlarge during outward movement to the surface, where they are lysed. The giant cells are also unusual in that they contain dark-staining granular material in the cytoplasm. The material gives a positive reaction for DNA when cells are stained with the fluorochrome bisbenzamide. The hypertrophied herpesvirus cells could be confused with cells infected with lymphocystis iridovirus if it were not for the presence of the thick hyaline capsule that envelops lymphocystis cells. Also, the pike is a relatively primitive fish and lymphocystis disease is known to occur only among more highly evolved orders of fishes.

Electron micrographs of infected cells show herpesvirus in various stages of development, but a unique feature of the esocid herpesvirus is the presence of aggregates of up to 50 virions within numerous cytoplasmic inclusions (Fig. 1C).

The giant cells in pike skin somewhat resemble those caused by the turbot herpesvirus, but the etiology cannot be the same: herpesvirus specificity is too exclusive—and geographic separation and the difference in environments (freshwater and marine) are other factors.

Etiology

By virtue of its presence in the hypertrophied cells, the herpesvirus is the etiologic agent. In typical herpesvirus maturation, capsids with and without a nucleoid are present in enlarged cell nuclei. Virion size is about 100 nm. Envelopment occurs as virions pass through membranes. The mature and presumably infectious forms measure about 200 nm.

The esocid herpesvirus has not been isolated, although Yamamato et al. (1983) attempted to do so using percid cells. Prospects for isolation would be improved if epithelial cells from esocid skin or fins were used.

Diagnosis

Lesions of pike herpesvirus can be diagnosed by finding the described hypertrophied cells in sections of granular plaquelike skin lesions. Confirmation requires electron microscopy.

Host and Geographic Range

Esocid herpesvirus 1 is known from sites in Saskatchewan and Manitoba, Canada, where viral lesions have been found on adult pike during the spring spawning season. The effect of the virus on young pike is unknown, and the susceptibility of other esocid species is conjectural. The presence of the virus in other geographic regions is considered likely.

Reference

Yamamoto, T., R. K. Kelly, and O. Nielsen. 1983. Epidermal hyperplasias of northern pike (*Esox lucius*) associated with herpesvirus and C-type particles. Arch. Virol. 79:255–272.

43 Shark (Smooth Dogfish) Herpesvirus

Definition

The herpesvirus of the smooth dogfish *Mustelus canis* (a shark) is found in cells and intercellular spaces of mild localized and necrotic skin lesions. Termed a viral dermatitis for its presumed cause, the condition is remarkable in that lesions are temporary and do not involve obvious cell proliferation (Leibovitz and Lebouitz 1985).

History

The condition came to light as the result of daily surveillance of newly captured and exhibit-type elasmobranchs, a program carried out at the Laboratory for Marine Animal Health, Marine Biological Laboratory, Woods Hole, Massachusetts. The lesions were separable from those caused by trauma, and their prevalence seemingly increased among confined animals. Indications of an infectious process prompted investigation that led to the discovery (Leibovitz and Lebouitz 1985).

Signs and Pathologic Changes

Behavioral and internal and clinical changes in infected sharks were apparently unremarkable; they were not reported.

Externally, the smallest lesions are round to oval elevated areas of grayish white depigmented skin less than 1 mm in diameter. Lesions occur anywhere on the body but predominantly on the tail and fins. Large lesions 1 cm or more in diameter display three zones: a small, dark, depressed center; a broad grayish white midzone; and an elevated granular, shaggy perimeter. Areas of apparent healed skin are melanized and level.

Histologically, small lesions have foci of degenerating basal cells and show intracellular edema, balloon degeneration, acantholysis, reticular colliquation, and the formation of eosinophilic intranuclear and cytoplasmic inclusion bodies (Fig. 1).

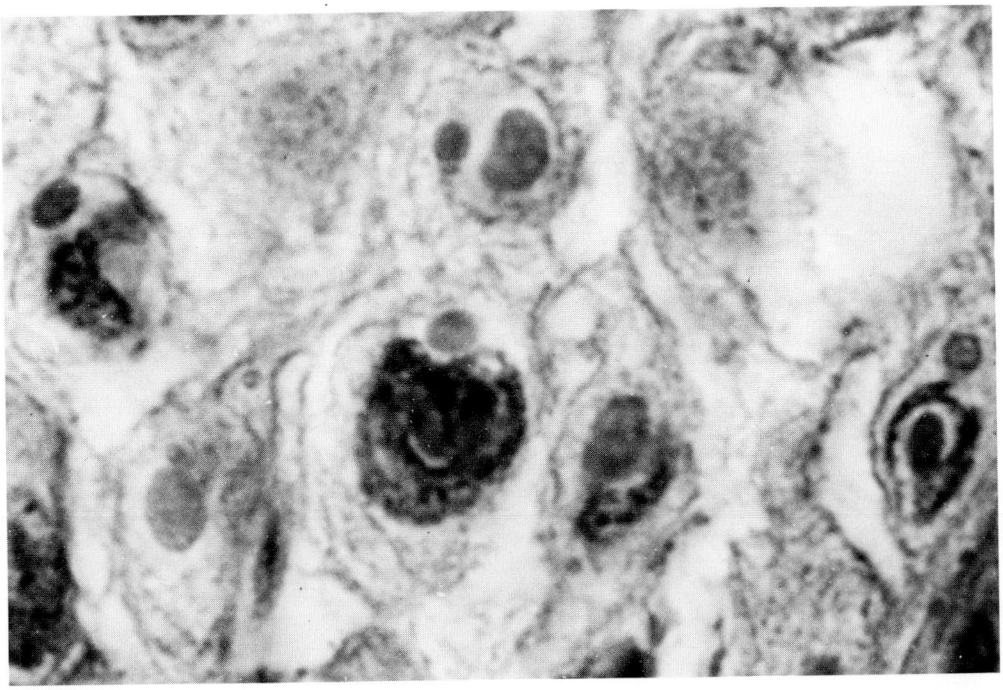

Figure 1. Swollen degenerating and necrotic epidermal cells, showing eosinophilic inclusion bodies in polar proximity to the nuclei. From Leibovitz and Lebouitz (1985). Reprinted with permission of Blackwell Scientific Publications Ltd.

Lesions enlarge and form vesicles by lysis and separation of the lower cells. Cells at the surface slough and capillary rupture results in hemorrhages within lesions and at their surfaces.

Electron microscopy shows margination of chromatin, nuclear and nucleolar hypertrophy, intracellular edema, and apparent sequential stages of virus replication. Accumulations of enveloped virions are found between the inner and outer lamellae of nuclear membranes. Such accumulations, with accompanying membrane fragments and other nuclear debris, rupture into the cytoplasm and constitute the cytoplasmic inclusion bodies seen by light microscopy.

Etiology

Mature particles are 150 to 185 nm in diameter, are enveloped, and have a capsid that is either hexagonal or pentagonal in profile. No attempt has been made to isolate the virus, but its presence is considered reasonable evidence for a causal role.

Diagnosis

Diagnosis is based on finding nontraumatic skin lesions in which histologic sections show nuclear hypertrophy with eosinophilic intranuclear and cytoplasmic inclusions. More specific determination requires demonstration of herpesvirus particles.

Transmission and Incubation

No attempt was made to transmit the virus to other animals. The incidence of lesions was 1% in newly caught smooth dogfish, but 5 to 8% in animals on exhibit or held in the laboratory. That observation can be interpreted as evidence of fish-to-fish transmission, or as a *Herpes simplex* kind of response to stress.

Host and Geographic Range

As a result of the surveillance program, the infection is known from waters off Woods Hole. The host shark, however, ranges extensively in the Atlantic. If lesions are sought in fish from other populations, they will undoubtedly be found.

Reference

Leibovitz, L., and S. S. Lebouitz. 1985. A viral dermatitis of the smooth dogfish, *Mustelus canis* (Mitchill). J. Fish Dis. 8:273–279.

44 Sheatfish Herpesvirus

Definition

The sheatfish herpesvirus has been visualized in and is presumably the cause of localized areas of epidermal hyperplasia (a fish poxlike condition) in the sheatfish *Silurus glanis* (also known as the European catfish).

(Until shown otherwise, the sheatfish herpesvirus is considered distinct from the carp pox herpesvirus, because most herpesviruses show a high degree of host specificity and the common carp and sheatfish are members of different families. Depending on the scheme of classification, the two fishes are also members of different orders or suborders.)

History

Although Lucký (1970) cited earlier publications on fish pox as it occurred in common carp and other species, he was the first to describe and illustrate a comparable epidermal hyperplasia that occurred in sheatfish. Lucký's interpretation of the condition as being viral was confirmed by the work of Békési et al. (1981, 1984), whose illustrated report first appeared in a paperback proceedings and 3 years later in a hardcover volume. The presence of virus in the lesions was clearly established.

Signs and Pathologic Changes

Specimens described by Békési et al. (1984) were from a population that, in the winter of the second year, developed serious skin lesions and sustained a 50% mortality. Behavior changes were not reported, nor was the condition of internal organs described. Externally, affected specimens have a generally slimy appearance. Lucký (1970) did not mention mortality associated with the infection.

Lesions are raised above normal skin, grayish, and range in area from 1 to 100 mm^2 or larger (Fig. 1). Both sources described the lesions as elastic.

Figure 1. Hyperplastic epidermal lesions on the head and barbel of a sheatfish with herpesvirus infection. From Békési et al. (1984). Reprinted with the permission of Akadémiai Kiadó, Budapest.

Histologic sections of the lesions show hyperplasia of the malpighian layer or deeper germinal portion of the epidermis. Mucous cells are greatly diminished in number and the abnormal tissue is infiltrated with numerous foci of melanin. Some cells have an enlarged nucleus and Cowdry type A inclusions.

Electron microscopy reveals margination of nuclear chromatin, intranuclear capsids, and enveloped particles in the cytoplasm (Fig. 2).

Etiology

By virtue of their herpesvirus size, shape, and apparent mode of replication in abnormally proliferating tissue, the virus is presumptively considered to be causal. Intranuclear capsids measure 85 to 90 nm. Enveloped virions of 145 to 160 nm are present in the cytoplasm. The role of virus in the mortality reported by Békési et al. (1981, 1984) is not known.

Diagnosis

The tissue lesions are recognized on the basis of the appearance of host species and the aforementioned histologic findings.

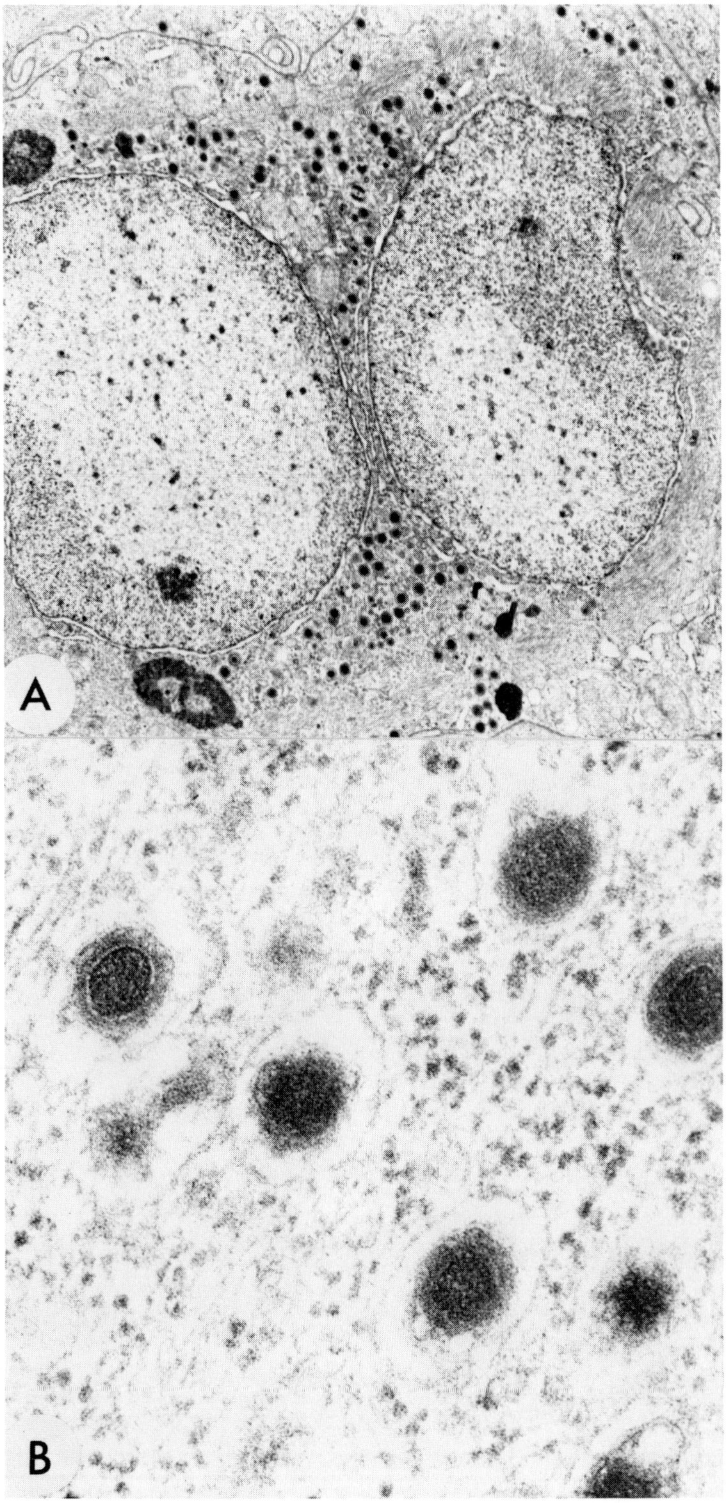

Figure 2. Electron micrographs of sections through cells infected with sheatfish herpesvirus. *A.* Infected cells show mild margination of chromatin, intranuclear inclusion substance, and nonenveloped particles in the nuclei. Enveloped particles are prominent in the cytoplasm. ×9,575. *B.* Enveloped particles show a dense capsid. ×14,700. From Békési et al. (1984). Reprinted with the permission of Akadémiai Kiadó, Budapest.

Identification

The virus is identified solely by electron microscopy. Although attempts have been made, the virus has not yet been isolated.

Host and Geographic Range

The sheatfish is native to Europe and (with the qualifications previously noted) is the only known host.

References

Békési, L., E. Kovács-Gayer, F. Ratz, and O. Turkovics. 1981. Skin infection of the sheatfish (*Silurus glanis* L.) caused by a herpes virus. Pages 59–69 in Proceedings of an international seminar on fish, pathogens and environment in European polyculture, Fisheries Research Institute, Szarvas, Hungary, June 23–27, 1981.

Békési, L., E. Kovács-Gayer, F. Rátz, and O. Turkovics. 1984. Skin infection of the sheatfish (*Silurus glanis* L.) caused by a herpes virus. Pages 25–30 in J. Olah, ed. Symposia biologica Hungarica 23, Akadémiai Kiadó, Budapest.

Lucký, Z. 1970. Pathological changes with pox (*Epithelioma papillosum*) in the sheat fish (*Silurus glanis*). Acta Vet. Brno, Suppl. 1:81–86.

45 Smelt Papilloma Herpesvirus

Definition

The smelt papilloma herpesvirus is an unequivocal viral agent that is associated with and presumptively the cause of seasonally transient skin neoplasms of *Osmerus eperlanus* in Europe.

History

The first account of the papillomatosis was a brief mention by Möller and Anders (1984) that the condition occurred seasonally, in spring. The following year, Anders and Möller published a short illustrated communication that included most of what is known of the condition.

Signs and Pathologic Changes

About 5% of the spawning smelts bore one to several smooth white external growths, mostly on the head and fins (Fig. 1). The growths were flat and some covered extensive areas but were easily dislodged—particularly near the end of the spawning season.

Histology showed papillomatous growths of epithelial cells with enlarged and polymorphic nuclei and Cowdry type A inclusions.

Etiology

The putative causal agent is present in the nucleus as 95-nm polygonal particles, some of which show a dense central nucleoid. Enveloped particles measure about 135 nm and are found only in the cytoplasm; some have tail-like projections (Fig. 2).

Anders and Möller (1985) recognized the similarity of the smelt lesions to those of

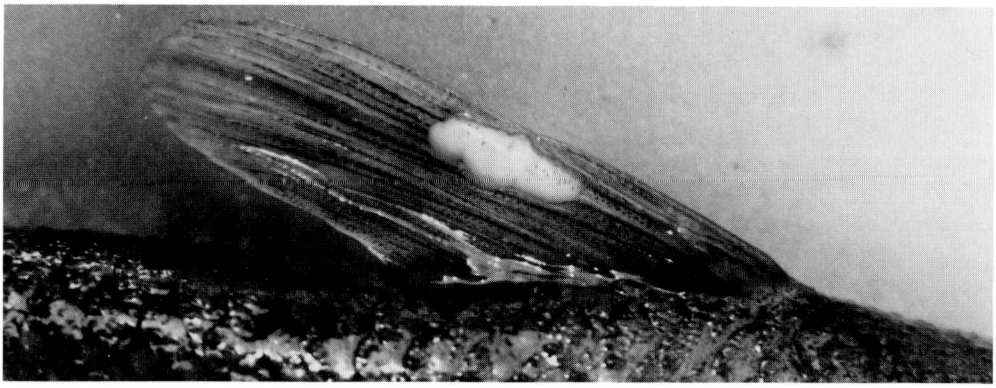

Figure 1. Papilloma on the dorsal fin of a smelt. From Anders and Möller (1985). Reprinted with the permission of Blackwell Scientific Publications Ltd.

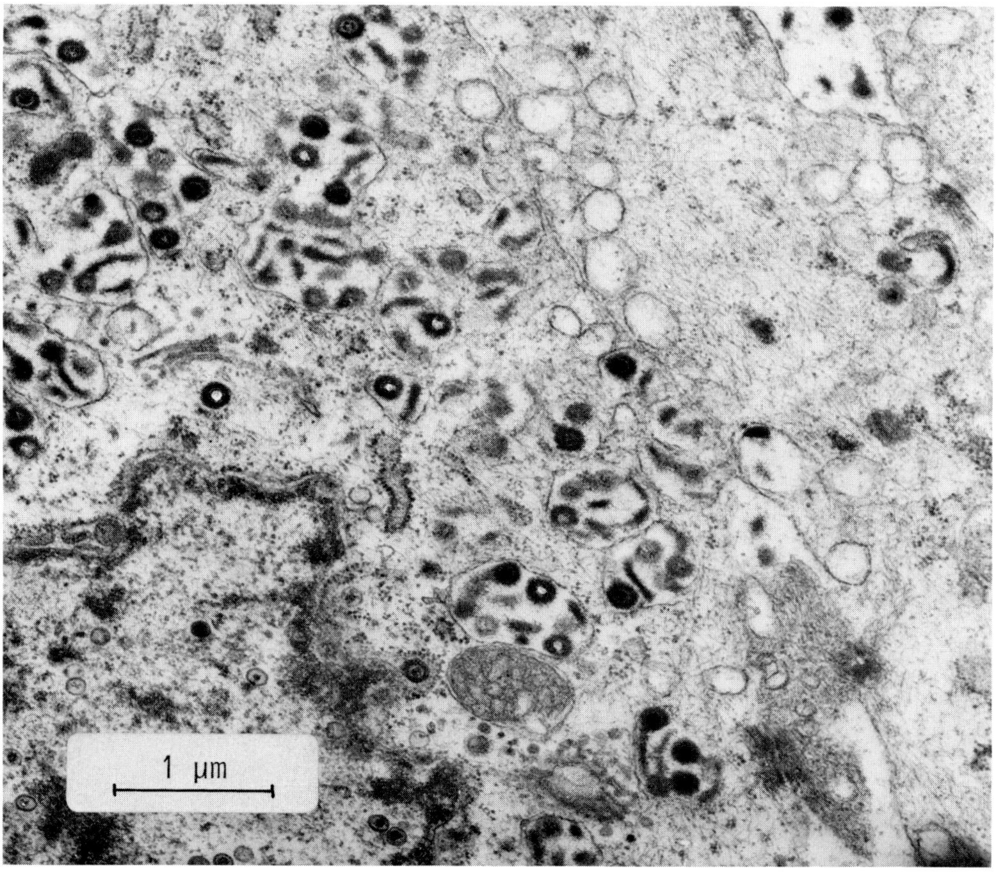

Figure 2. Virions in smelt papilloma cells. Capsids with and without nucleoids are present in the nucleus (lower left). Enveloped particles—some with a prominent tail-like extension—are abundant in the cytoplasm. From Anders and Möller (1985). Reprinted with the permission of Blackwell Scientific Publications Ltd.

the so-called fish pox. That matter can best be resolved when the respective agents are isolated and compared serologically.

Host and Geographic Range

The condition is known only from smelt taken in the Elbe estuary. However, other areas of the North Sea could also be expected to yield affected specimens.

References

Anders, K., and H. Möller. 1985. Spawning papillomatosis of smelt, *Osmerus eperlanus* L., from the Elbe estuary. J. Fish Dis. 8:233–235.

Möller, H., and K. Anders. 1984. A compilation of viruses from marine and brackish environments. Fisch Umwelt 13:59–70.

46 Sturgeon Adenovirus

Definition

The sturgeon adenovirus is a doubly encapsidated icosahedron visualized in pathologically hypertrophied nuclei of epithelial cells of the gastrointestinal tract of white sturgeons (*Acipenser transmontanus*) afflicted with a debilitating condition associated with chronic mortality.

History

During 1984, white sturgeon fry, the progeny of adults taken from California's Sacramento River, sustained severe mortality that was associated with flexibacterial infection. Examination of specimens from one farm showed no evidence of bacterial involvement; instead, a virus was found—the first to be documented from a sturgeon (Hedrick et al. 1985).

Signs and Pathologic Changes

Specimen fry are lethargic, anorexic, and emaciated. The liver is pale and the alimentary tract is devoid of food. Histologic examination showed a marked hypertrophy of nuclei of mucosal cells of the straight intestine and spiral valve (Figs. 1 and 2). That change was also found in hepatic cells of one specimen.

Etiology

The marked involvement of mucosal cells of the digestive tract, in association with anorexia and emaciation, appears to be causally related to the virus. The agent is icosahedral and has an inner capsid of 47 nm and an outer capsid of 74 nm. Attempts to isolate the virus in BB, BF-2, CCO, and CHSE-214 cell lines, and by co-

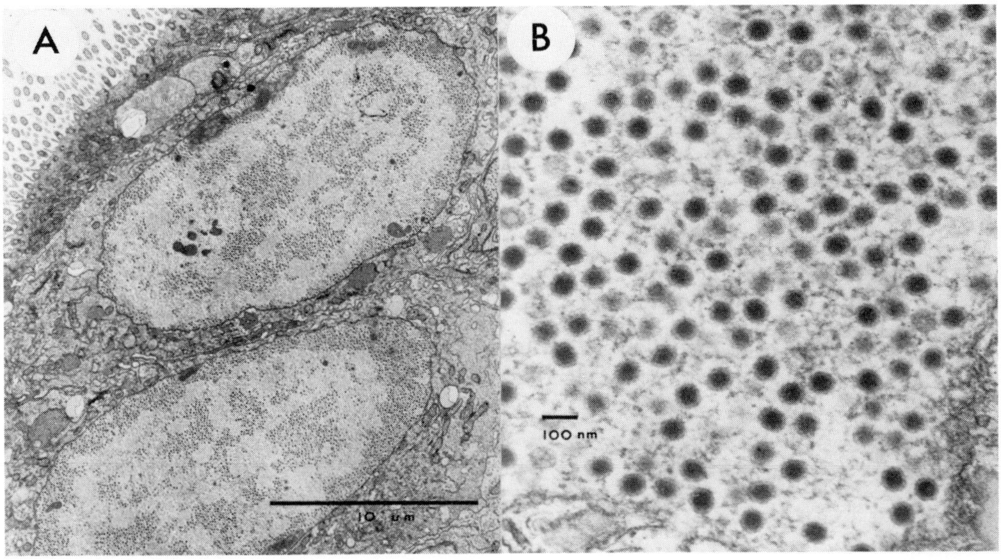

Figure 1. A. Mucosal epithelial cells of the spiral valve, showing grossly enlarged nuclei and aggregates of intranuclear virions. *B.* At higher magnification most virions show double encapsidation. From Hedrick et al. (1985). Reprinted with the permission of the *Canadian Journal of Fisheries and Aquatic Sciences.*

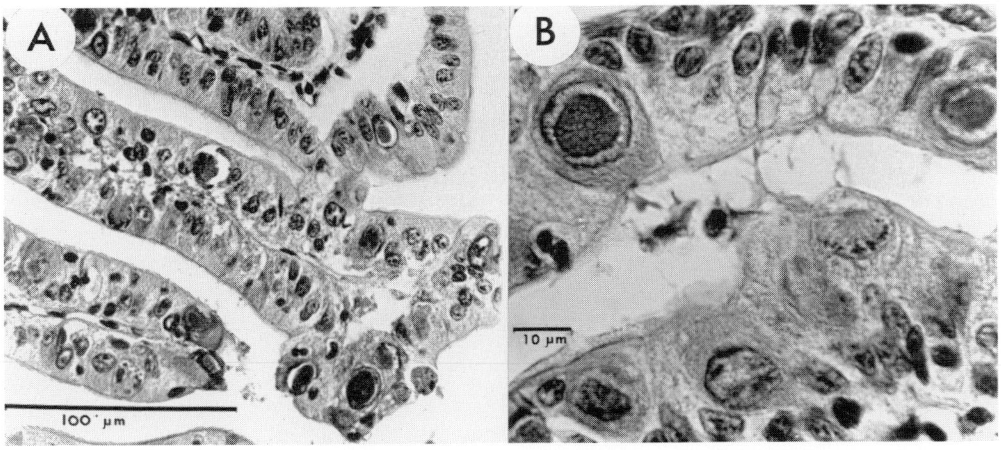

Figure 2. A. Histological section of straight intestine, showing enlarged nuclei. *B.* At higher magnification, enlarged nuclei are shown in the epithelium of the spiral valve. H & E. From Hedrick et al. (1985). Reprinted with permission of the *Canadian Journal of Fisheries and Aquatic Sciences.*

cultivation in the sturgeon cell lines SH-1 (heart) and SS-2 (spleen), were not successful. Also, primary cultures initiated from affected tissues failed to show CPE.

Transmission and Incubation

Fish of 10 g that were inoculated intraperitoneally with homogenates of affected visceral tissues and held at 16°C were examined histologically after 4 weeks. Gut epithelium showed nuclear hypertrophy identical to that found in naturally infected fry; however, the intensity was much less than in natural infection.

Host and Geographic Range

The white sturgeon in husbandry in California is the only known host. Presumably, the adult is the source animal.

The relationship, if any, to the adenovirus-like agent found in skin lesions of the dab (*Limanda limanda*) and the cod (*Gadus morhua*) is not known. Infected cod and dab cells, although epithelial, are not hypertrophied.

Reference

Hedrick, R. P., J. Speas, M. L. Kent, and T. McDowell. 1985. Adenovirus-like particles associated with a disease of cultured white sturgeon, *Acipenser transmontanus*. Can. J. Fish. Aquat. Sci. 42:1321–1325.

47 Sucker Papilloma Retrovirus

Definition

The sucker papilloma retrovirus is a C-type retrovirus found in papillomatous epidermal neoplasms on *Catostomus commersoni* in North American waters, but particularly in certain polluted locations.

History

In a survey intended to compile data on the occurrence of neoplasms in fish and other poikilothermic vertebrates, Sonstegard (1977a) distributed more than 6000 questionnaires to biologists and fisheries personnel in various parts of the world. Although it is not clear whether the information was obtained from the survey or from Sonstegard's own investigations, his report stated that a 7% incidence of papillomas was discovered in white suckers in a moderately polluted watershed in North America. He stated further that electron microscopy of the tumors showed the presence of C-type particles, but that studies of cell culture transformation and experimental transmission were not successful. Greater detail on the virus, the tumors, and the epizootiology appeared in a second report (Sonstegard 1977b).

Signs and Pathologic Changes

Behavior changes were not noted in the fish, but they bore multiple benign cauliflower-like epidermal growths. The neoplasms were white to pinkish white, and firm in consistency. In Lakes Huron and Superior—relatively lightly or nonpolluted watersheds, where the incidence of tumors was less than 1%—the growths were found on fin, flanks, eyes, opercula, and lips. In a heavily polluted bay of Lake Ontario, where the incidence was about 30%, the tumors were exclusively on the lips. Sonstegard (1977b) interpreted his data as showing a gradient of occurrence, with the incidence being highest ("clustered") in the most heavily polluted areas.

The tumors are not invasive, nor do they occur internally or metastasize. Their histologic features have not been described. Fish first show the growths at sexual maturity, at the age of 4 years. Thereafter, the neoplasms increase in size, and frequencies of 50 to 70% are found in 6- to 9-year-old fish. Growths are believed to persist for life.

Etiology

The cause of the tumors has not been determined, but the virus—alone or in concert with extrinsic environmental factors—is suspected. The virus is a C-type particle measuring about 100 nm. It buds from cytoplasmic membranes, and aggregates of virions are present in intercellular spaces. Sonstegard (1977b) found reverse transcriptase associated with fractions of particulate papilloma material that banded at 1.15–1.16 g/mL on sucrose gradients. The presence of reverse transcriptase supports the hypothesis that the C-type particle is involved in tumor induction. Attempts to show transformation of RTG-2, FHM, BB, and BF-2 cell lines were not successful. Primary epithelial cell cultures of catostomid fish origin were not tried.

Transmission and Incubation

All attempts at experimental transmission—cohabitation, application of tumor tissue to abraded lips, implantation of tumor fragments, and percutaneous inoculation of cell-free tumor extracts—have failed. Sonstegard did not indicate how long the fish were held for observation. The chronic nature of the tumors suggests that considerable time is required for development. Also, as in all studies of infectious diseases of wild fishes, health history and possible prior exposure were unknown.

Host and Geographic Range

The white sucker is a species native to North America, and the tumors, at least grossly, have been found in areas separated by as much as 3000 miles. It is not known if the virus is common to all tumors.

References

Sonstegard, R. A. 1977a. The potential utility of fishes as indicator organisms for environmental carcinogens. Pages 561–577 in F. M. D'Itri, ed. Wastewater renovation and reuse. Marcel Dekker, New York.

Sonstegard, R. A. 1977b. Environmental carcinogenesis studies in fishes of the Great Lakes of North America. Ann. N.Y. Acad. Sci. 298:261–269.

48 Turbot Herpesvirus

Synonym: *Herpesvirus scophthalmi*

Definition

The herpesvirus infection of turbot (*Scophthalmus maximus*) is an enzootic and specific giant cell condition in skin and gill epithelium afflicting young-of-the-year fish. The disease occurs both in the wild and in populations in husbandry.

History

Heavy mortality in 1978 among young turbot in a fish farm whose water source was the warm effluent from a nuclear power plant in Scotland prompted a search for an infectious agent (Buchanan et al. 1978). Histologic examination showed numerous giant cells in the epithelium of the skin and gill (Figs. 1 and 2). Scrapings and ultrathin sections of affected tissues, examined by electron microscopy, showed a herpesvirus-like agent.

A second and third report appeared during the year of discovery. Richards and Buchanan (1978) briefly described clinical signs of affected fish and gave details and illustrations of their histopathologic findings; and Buchanan and Madeley (1978) followed with an ultrastructural study. Bucke (1983) encountered what he presumptively identified as cells that were hypertrophied as a result of the turbot herpesvirus. Neither disease nor loss was associated with the case. Confirmation by electron microscopy was lacking.

Signs and Pathologic Changes

Obvious external lesions are not to be found, neither are there internal signs. The virus seen in the giant cells has not been isolated; therefore, classical induction of

Figure 1. Differential phase-contrast micrograph of giant cells in the skin of an infected turbot. The black dendritic forms are melanophores. Bar = 50 μm. From Buchanan et al. (1978). Reprinted with the permission of the British Veterinary Association.

experimental infections in healthy stock has not been carried out. Accordingly, aspects of the disease presumed to be viral must be considered provisional.

Behavior

Affected fish are anorexic and lethargic, and offer little resistance to netting. Their posture is abnormal for a flatfish: in fish lying on the bottom, the head and tail are raised.

Histopathologic Findings

Pathologic changes are found only in the epithelium of the skin and gills. Giant cells ranging from 9 × 15 μm to 70 × 130 μm are present individually in light cases and to a depth of several layers in heavy infections. Infected cells are invariably more numerous on the upper surface than on the side in contact with the bottom of the tank. By light microscopy the giant cells usually have only one nucleus, which occupies up to 90% of the cell (Richards and Buchanan 1978). Cells sometimes

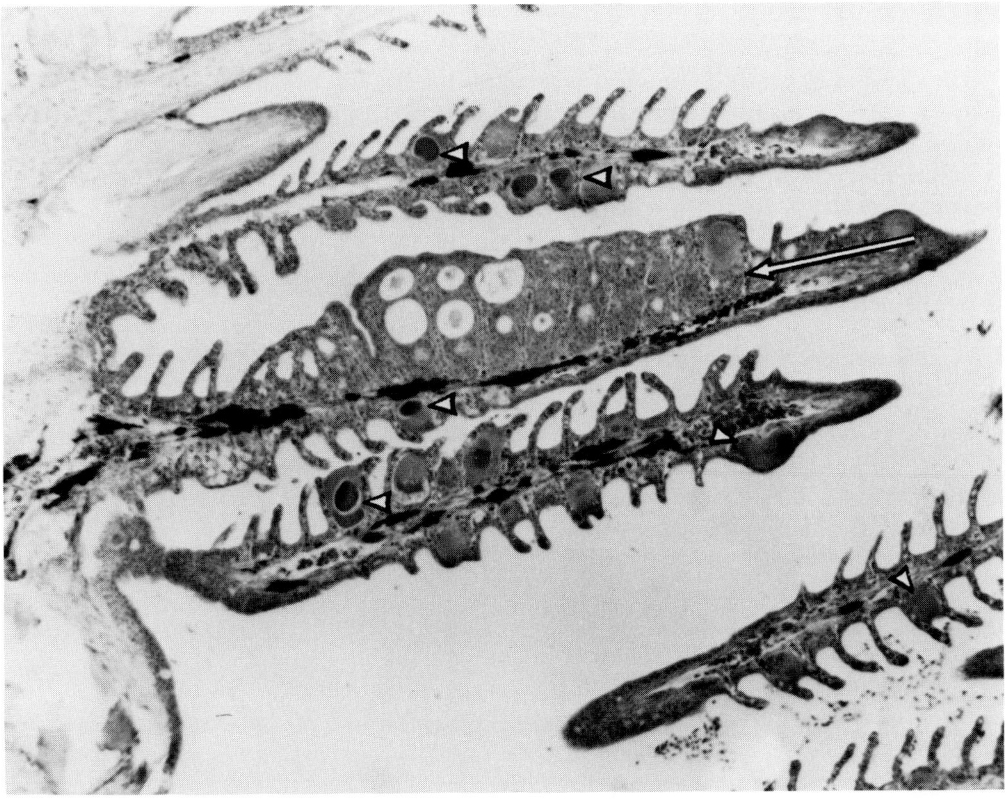

Figure 2. Section through the gills of an infected turbot showing giant cells (points) and hyperplasia and fusion of lamellae. H & E. From Richards and Buchanan (1978). Reprinted with the permission of Blackwell Scientific Publications Ltd.

contain several nuclei having the size and shape of the nuclei in surrounding malpighian cells. It is believed that such cells had been formed by fusion but that their nuclei had not yet coalesced to form giant nuclei. Giant cell nuclei measure 45 to 75 μm, are frequently fenestrated, and show a densely basophilic Feulgen-positive, PAS-negative area. In some cells a Feulgen-negative inclusion or a dense eosinophilic area surrounds a less eosinophilic central area of the nucleus. The cytoplasm is basophilic but contains PAS-negative, Feulgen-positive granules. Some areas of the epidermis from moribund fish show cryptlike voids where giant cells presumably had been lost or destroyed.

In gill tissue of light cases, the giant cells occur singly between adjacent lamellae and additional host response is light. Tips of the involved lamellae are occasionally fused. In heavy infections, several lamellae fuse and the surrounding malpighian cells are hyperplastic. In such areas there is vascular stasis and fibrinous thrombosis, and considerable reduction of the functional respiratory surface.

Fish without clinical signs of disease have fewer giant cells than those that are obviously affected. Pathognomonic cells are absent in some populations, but are found in some fish taken directly from the wild.

Etiology

The giant cells contain abundant particles having the general characteristics of a herpesvirus. Parasites are not involved and bacteriologic culture showed no evidence of microorganisms. Accordingly, and even though the virus has not been isolated, it is presumed to be the etiologic agent of the giant cells and the binomial *Herpesvirus scophthalmi* has been proposed as a working name.

Size and Shape

Naked capsids, some empty and some containing a dense nucleoid, occur in the nucleus (Fig. 3). Enveloped particles most often occur in membrane-enclosed aggregates in the cytoplasm (Buchanan and Madeley 1978). Nuclear capsids occur in paracrystalline arrays; the individual particles have a circular or hexagonal profile and are about 100 nm in diameter. In ultrathin section, cores or nucleoids appear as bars or rings of about 25 to 30 nm.

Capsomeres are hollow and are 9 nm in diameter—a size that is compatible with that of herpesviruses. However, the resolution of published electron micrographs is not sufficient to show unequivocally a tubelike form; neither can one confidently count capsomeres along an edge and calculate the total number.

Enveloped particles occur only in the cytoplasm, are often pyriform or tailed, and are 200 to 220 nm in diameter. The envelope is trilaminar and shows an outer fringe of petal-like spikes (18 nm long) resembling those of a coronavirus. Particles have not been seen budding from membranes, although this is postulated to occur. Replication apparently begins in the nucleus, and the particles move into the cytoplasm at points where the nuclear membrane has deteriorated. Virions are believed to acquire an envelope as they leave the nucleus or (more commonly) bud into cytoplasmic vacuoles. Mature particles make their way out of cells through perinuclear cisternae and modified endoplasmic reticulum that forms channels through the cytoplasm.

Diagnosis

Infection with *Herpesvirus scophthalmi* can be suspected in young turbot showing mortality accompanied by the described clinical signs. At present, the best diagnosis is the demonstration of the pathognomonic giant cells containing virus.

Transmission and Incubation

Because the giant cells have been found in wild fish, it is assumed that the infection is spread horizontally. The condition is probably persistent and health problems arise as a result of the stress of transport, handling, temperature fluctuation, or high chlorine levels when water conduits are treated for removal of fouling sessile marine invertebrates (Richards and Buchanan 1978).

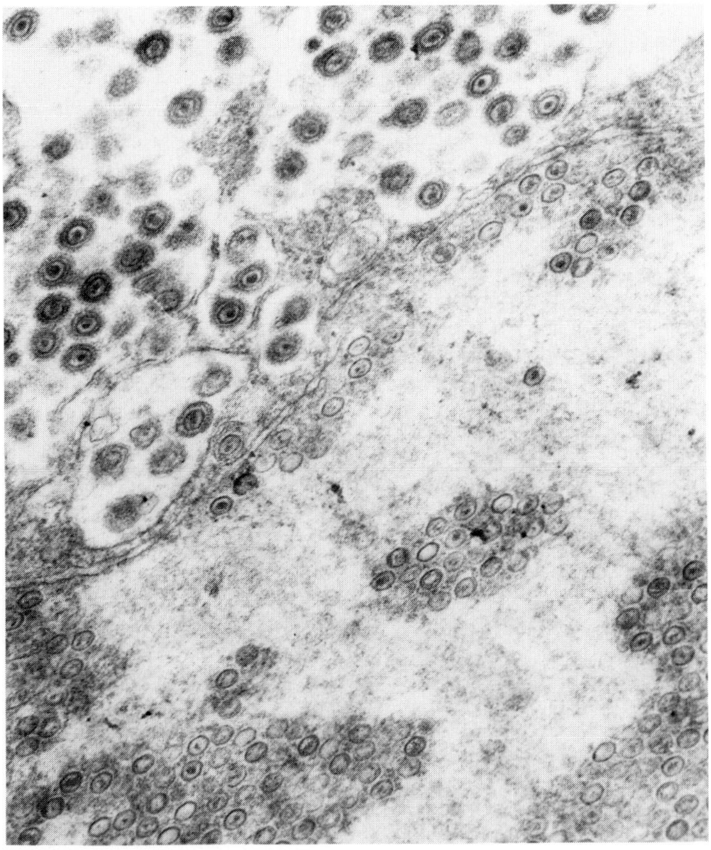

Figure 3. Electron micrograph of an infected cell showing intranuclear particles—some with and others without a nucleoid—and enveloped cytoplasmic virions. Courtesy of J. S. Buchanan.

Source

Predicated on its presumed membership in the herpesvirus group and the typical host specificity of those viruses, the source is most likely the turbot or a closely related pleuronectiform fish.

Host and Geographic Range

The only fish known to be susceptible is the young turbot, and the only affected populations are in Scotland and Wales (J. Buchanan, personal communication). The turbot, however, is circumpolar in distribution and additional populations harboring the infection may be discovered.

Control

Richards and Buchanan (1978) suggested that the only means of reducing losses is to apply judicious principles of nutrition, handling, and husbandry.

References

Buchanan, J. S., and C. R. Madeley. 1978. Studies on *Herpesvirus scophthalmi* infection of turbot *Scophthalmus maximus* (L.) ultrastructural observations. J. Fish Dis. 1:283–295.

Buchanan, J. S., R. H. Richards, C. Sommerville, and C. R. Madeley. 1978. A herpes-type virus from turbot (*Scophthalmus maximus* L). Vet. Rec. 102:527–528.

Bucke, D. 1983. A preliminary histological study of normal and pathological tissues in cultivated turbot (*Scophthalmus maximus* L.). Rapp. P.-V. Reun. Cons. Int. Explor. Mer 182:111–115.

Richards, R. H., and J. S. Buchanan. 1978. Studies on *Herpesvirus scophthalmi* infection of turbot *Scophthalmus maximus* (L.): histopathological observations. J. Fish Dis. 1:251–258.

49 Viral Erythrocytic Necrosis

Synonym: piscine erythrocytic necrosis (PEN)

Definition

Viral erythrocytic necrosis (VEN) is a pathologic condition of certain poikilothermic vertebrates from elasmobranchs to reptiles. The disease is recognized by the finding of altered red blood cells, which sometimes have an obviously damaged nucleus and always have a DNA-positive cytoplasmic inclusion and associated (and presumably causal) iridovirus-like particles.

History

The history of VEN was traced back to parasitologic reports of the late 1800s and early 1900s in reviews by Johnston (1975) and Marquardt and Yaeger (1967). As in the story of lymphocystis disease, the cytoplasmic inclusions and otherwise pathologically altered erythrocytes from certain reptiles and amphibians were for many years thought to be protozoan parasites. Also as in lymphocystis, binomial nomenclature was coined for the presumed hematozoa; the more common names of genera were *Pirhemocyton*, *Toddia*, *Cytamoeba*, and *Immanoplasma*.

The first suggestion that the altered erythrocytes involved virus was made by Blanc and Ascione (1958). Interestingly, Blanc had earlier coauthored a paper in which the name *Pirhemocyton* was coined (Chatton and Blanc 1914). Stehbens and Johnston (1966) provided the first concrete evidence of *Pirhemocyton*'s true nature: their micrographs of reptilian erythrocytes showed cytoplasmic icosahedra like those of the iridescent viruses of insects. Marquardt and Yaeger (1967) concluded, from their study and their review of the literature, that *Toddia* and *Pirhemocyton* were both "viruses of the DNA type." Soon thereafter, Bernard et al. (1968) demonstrated cytoplasmic icosahedra in amphibian erythrocytes and noted a feature to be found later among fish—namely, virus size differences; particles in the frog were larger than those in the reptile.

Erythrocytic virus in fish came to light with the report of Laird and Bullock (1969) on their 1958–1961 surveys of marine species taken in the Atlantic off North America. In addition to those blood films showing bona fide hematozoa, they found four blood films that they described as having "piscine erythrocytic necrosis (PEN)." Although the authors were limited to light microscopy, they described "viruslike particles" in affected cells and illustrated them with camera lucida drawings.

Although virologic evidence was accumulating, parasitologic interpretations of the pathologic erythrocytes persisted (Fletch and Karstad 1968; Pessôa 1969; Pessôa and Campos 1966). Interestingly, Pessôa et al. (1974) eventually acknowledged that *Toddia* was probably a virus. As recently as 1971, Arcay de Peraza and McLure described *Toddia piscium* as a new parasite, and even though they cited Marquardt and Yaeger (1967), they considered *Toddia* to be "endoerythrocytic protozoan parasites."

The actual virologic confirmation of Laird and Bullock's interpretation was reported by Walker (1971), who briefly described the virions and correlated Laird and Bullock's descriptions of nuclear lesions with cytoplasmic viroplasm and virions. Six years elapsed, however, before the full report of that work—electron micrographs, elegant diagrammatic drawings, and tabular data on new teleost hosts—was published (Walker and Sherburne 1977). In the meantime, other micrographs of the virus in teleosts had been published (Appy et al. 1976; Doyle 1971; Johnston 1975; Johnston and Davies 1973). Among these, the report by Johnston (1975) merits special attention because it constituted the most extensive review of *Pirhemocyton* and related infections (74 references), and illustrated unequivocal virions in an elasmobranch. Surprisingly, he stated "they cannot be considered to be an ICDV [icosahedral cytoplasmic DNA virus]."

Sherburne (1973) reported degenerating erythrocytes in Atlantic herring (*Clupea harengus harengus*) and considered virus as one of the possible causes. That work had interesting ramifications, for it led to three notable reports in 1977: his authorship of a report on PEN in the alewife (*Alosa pseudoharengus*), and two coauthorships (Philippon et al. 1977; Walker and Sherburne 1977). Significant size differences between the fish viruses were noted by Philippon et al.: the herring virus was about half the size of that reported earlier.

Most of what was then known about PEN in fishes concerned host species from the Atlantic, but Bell and Margolis (1976) reported that the infection occurred among Pacific salmon in Canada and, more important, that their colleagues had found it to be transmissible by intraperitoneal inoculation.

At the Pacific Biological Station (British Columbia, Canada), Evelyn and Traxler (1978) gave details of serial transmission in Pacific salmon and stated that the condition had been seen there as early as 1964. Considering that the infection occurred in reptiles and amphibians, as well as in fish, they took issue with the restrictive term *piscine* and proposed the now accepted viral erythrocytic necrosis.

Thus far, VEN had been found only among marine or anadromous species, but Sherburne and Bean (1979) discovered the condition among landlocked American smelt (*Osmerus mordax*). Workers at the National Fisheries Research Center (Seattle) of the U.S. Fish and Wildlife Service showed that many species of salmonids were susceptible to experimental VEN in freshwater, but also that incidences were highest near seawater (MacMillan and Mulcahy 1979). A year later, VEN was shown to adversely affect chum salmon (*Oncorhynchus keta*) by increasing their susceptibility to vibriosis and oxygen depletion, and by decreasing their ability to regulate sodium and potassium (MacMillan et al. 1980).

The prevalence of VEN in blennies and cod from waters near the United Kingdom was reported by Smail and Egglestone (1980a). These authors also introduced a method of electron microscopy that gave documentation of virus from erythrocytes in as little as 1 to 2 hours (Smail and Egglestone 1980b). Reno and Nicholson gave details of in vitro metabolic studies in 1980, and in 1981 reported results of a survey of VEN among about 1100 Atlantic cod (*Gadus morhua*) taken in waters off Newfoundland and Greenland.

Awareness of VEN was growing in North America's Pacific Northwest. A survey of freshwater hatchery stocks of salmonids in Oregon by Rohovec and Amandi (1981) revealed the infection in one of every four stocks sampled. Moreover, the investigators found indications that a natural outbreak of VEN could result in mortality. Although they did not comment on the possibility, their description can be interpreted as evidence for vertical transmission.

Smail (1982) published a review of VEN. In three other reports, the presence of virus was not documented, but Kahn and Newman (1982) found presumptive evidence of the infection in a second species of elasmobranch; Eiras (1984) described the disease in two species of bony fishes in coastal waters of Portugal; and Sherburne (1984) reported PEN in 50% of 142 sea lampreys (*Petromyzon marinus*) from Maine.

Signs and Pathologic Changes

The virus has yet to be isolated and used for experimental infections. Accordingly, information about its effects is both limited and somewhat conjectural. It is likely that effects vary with degree of infection and with host species and age. As an example, some evidence suggests that the virus is virulent in the young. Because hematopoiesis is adversely affected, heavy infections cause anemia. Other than the occurrence of pale gills and viscera in infected Pacific salmon, noteworthy changes are not evident in external or internal appearance or in behavior.

Blood films stained with Wright's or Wright-Giemsa show pathognomonic VEN cells: erythrocytes that have a small, round, eosinophilic inclusion of 1 to 4 μm in the cytoplasm or budding from the nucleus (Fig. 1). The nucleus of VEN cells may appear normal or show signs of alteration that can vary from vacuolation or blebbing to advanced necrosis, or even to complete dissolution. Apparent fragmentation of erythrocyte nuclei is common in cods. The percentage of infected erythrocytes can range from less than 1% to 100%, but leukocytes are unaffected.

Greater than normal numbers of immature erythrocytes occur in peripheral blood, and hematocrit values (where they have been determined) are depressed. Evelyn and Traxler (1978) stated that hematocrits of 2 to 10% are common in Pacific salmon, and that in addition blood from fish with VEN either clots abnormally slowly or not at all. MacMillan and Mulcahy (1979) showed that when chum salmon

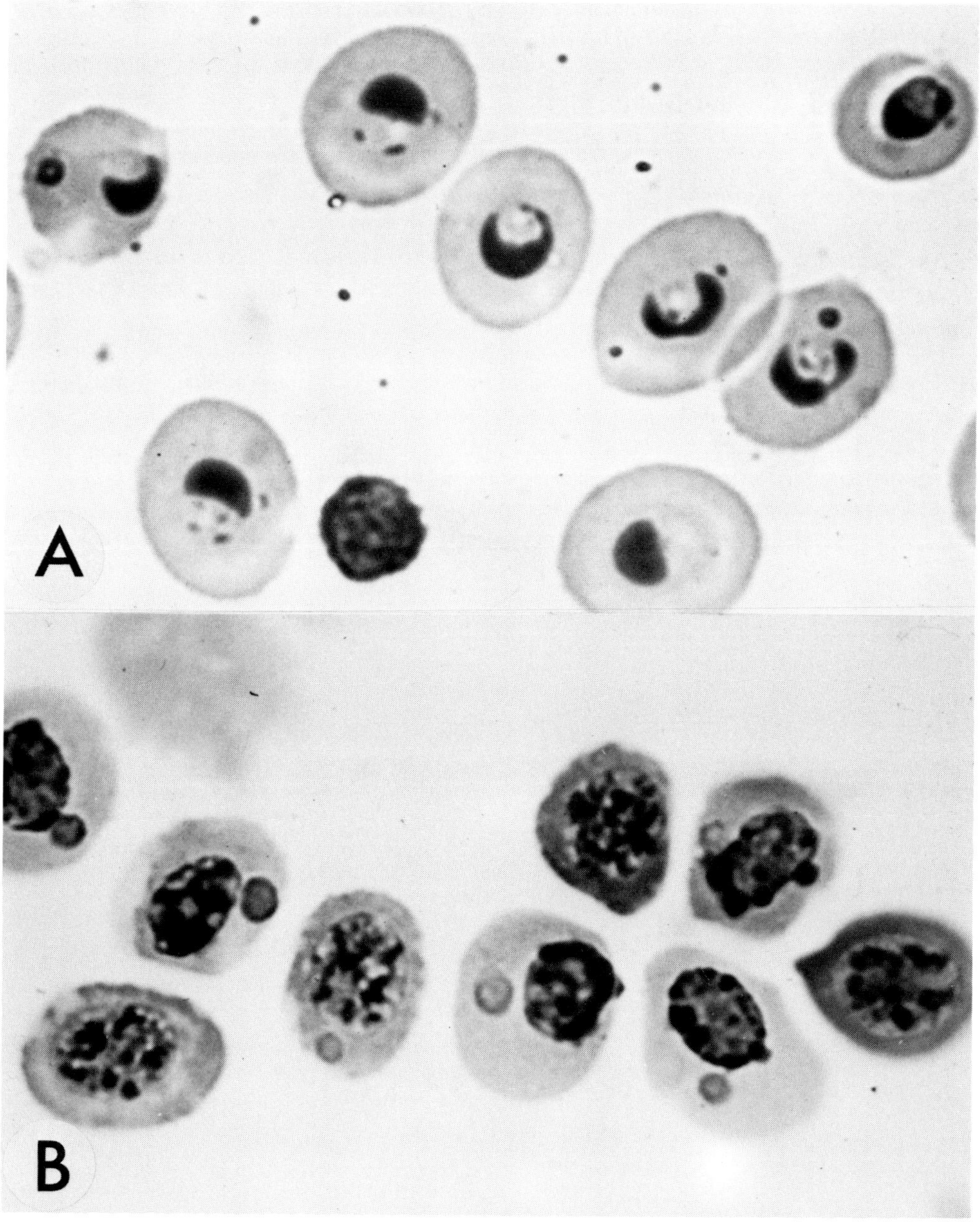

Figure 1. Cells infected with viral erythrocytic necrosis show a small, discrete, generally circular, DNA-positive cytoplasmic inclusion. *A.* Cod and blenny cells have nuclei that appear to be partly lysed or eroded. (Photograph of specimen slide of *Gadus morhua* blood film, Cat. No. 39499, U.S. National Museum, Washington, D.C.) *B.* Salmon cells have intact but commonly immature nuclei. Inclusions may be cytoplasmic or seem to bud from the nucleus (Photograph of specimen slide of *Oncorhynchus keta* blood film.) Courtesy of G. E. Hoskins.

Table 1. Sizes of viral eyrthrocytic necrosis virions in thin section from different fishes

Host and reference	Size (nm)
Atlantic herring (*Clupea harengus*)	
Reno et al. (1978)	146
Pacific salmon (*Oncorhynchus* spp.)	
Evelyn and Traxler (1978)	190
Blenny (*Blennius pholis*)	
Johnston and Davies (1973)	200–300
Smail and Egglestone (1980b)	250–300
Atlantic cod (*Gadus morhua*)	
Reno and Nicholson (1981)	260–325
Walker and Sherburne (1977)	310–350
Appy et al. (1976)	310–360

fry are injected with virus, the average hematocrit value of about 40% drops to a low of about 10% in 4 weeks.

Etiology

Cytoplasmic icosahedral virus has been found universally in the pathognomonic VEN cells, and the disease has been serially transmitted with filtrates of infected kidney and spleen homogenate (Evelyn and Traxler 1978) and with filtrate (0.22 μm) of sonicated infected whole blood (MacMillan and Mulcahy 1979). Efforts to isolate and grow the virus in fish cell lines have not been successful; accordingly, Rivers' postulates have yet to be fulfilled.

Size and Shape

Although virions seen in thin sections of cells from different species of fish all have the same shape, they vary considerably in size—from 150 to 360 nm (Table 1). Virion size in the shark (*Scyliorhinus canicula*) is given as 450 to 500 nm (Johnston 1975), but the micrographs show obvious compression, and the true size is probably closer to 400 nm. Marked differences in size have also been found in another iridovirus, the agent of lymphocystis disease. Reno and Nicholson (1981) suggested that minor size differences (their data suggested values on the order of 10%) can be due to methods of fixation, and that the use of isotonic instead of hypotonic salines probably contributes to more accurate determinations.

Virions of VEN from the cod and blenny differ in size and ultrastructural detail from those of herring, and their respective cells also show differences. As described and illustrated by Appy et al. (1976), Johnston and Davies (1973), and Walker and Sherburne (1977), the cytoplasmic inclusion body of cod and blenny cells is viroplasm, and adjacent granular material consists of virions in various stages of assembly. Complete virions have a spheroidal electron-dense nucleoid and an outer surcoat of fibrillar material, clearly distinguishing the particles from the cytoplasm. The seemingly fragmented nucleus of cod and blenny VEN cells shows an intact nuclear membrane with regionally dense clumps of chromatin and, in some cases, internal areas devoid of chromatin (pseudovacuoles) having a stellate body.

Reno et al. (1978), who studied VEN cells from herring, agreed that inclusion bodies are viroplasm, but there the similarities end, because the virions have a doughnut-shaped nucleoid with an electron-lucent center. In addition, virions show no outer region or surcoat; instead, they are in direct contact with cytoplasm.

Biophysical Properties

By its presumed membership in the iridovirus group, the VEN agent has a DNA genome, ether sensitivity, heat lability, and other group attributes. Heat lability has been determined for VEN: 15 minutes at 60°C effected complete inactivation.

Reno and Nicholson (1980), working with radiolabeled precursors and infected cod erythrocytes in vitro, found impressive evidence that the virus has a DNA genome. Tritium-labeled thymidine uptake by infected cells exceeded that of normal erythrocytes by as much as 10-fold, and the uptake of uridine was comparable. Infected cells also incorporated labeled amino acids at a maximum rate twice that of normal cells.

Diagnosis

Diagnosis of VEN is made by finding the pathognomonic erythrocytes in blood films treated with a hematologic stain. Demonstration that the cytoplasmic inclusions stain for DNA with Feulgen or bisbenzamide reagents reinforces the diagnosis. Confirmation to the extent now possible may be obtained by electron microscopy; the rapid method of Smail and Egglestone (1980b), in which lysed erythrocytes are stained with molybdate or phosphotungstate, is suggested.

Isolation

Virus isolation has been tried in cultures of commonly used fish cell lines (RTG-2, CHSE-214, FHM), but all attempts have failed (Evelyn and Traxler 1978). Considering the cell tropism of the virus, the results are not surprising. New effort should be made with primary cultures of renal hematopoietic tissues and cells—cultures that will provide the normal target for the virus.

Identification

The only current means of identifying the virus is by electron microscopy. Size differences of the virus in different hosts suggest that antigenic differences might also occur.

Transmission and Incubation

Nothing is known about transmission in the marine environment, except that it occurs, and that incubation time probably involves weeks. Sherburne (1973) noted that among captive Atlantic herring the incidence increased from an initial 7% to 96% over a 2-week period. Similar data were obtained by Philippon et al. (1977): at the time of capture 11% of their Atlantic herring showed VEN, and 13 days after capture the incidence was 90%. In neither case was it possible to determine whether transmission occurred or whether stress exacerbated existing subclinical infection.

Among salmonids in freshwater, VEN has been transmitted by intraperitoneal injection and, to a lesser extent, by waterborne exposure. Evelyn and Traxler (1978) were the first to report experimental transmission: administering a filtrate of homogenized kidney and spleen from an infected chum salmon by intraperitoneal injection, they infected chum and pink salmon, but not coho, chinook, or sockeye salmon. It is worth noting that specimens of the refractory species were fairly large, weighing 10 to 24 g. MacMillan and Mulcahy (1979), who used a filtrate of sonicated whole blood from chum salmon, successfully infected nine species of salmonids: five species of *Oncorhynchus*, Atlantic salmon, and brook, brown, and rainbow trout. Their greater success was probably due to the smaller size (0.5 g) of the fish tested.

Although MacMillan and Mulcahy (1979) provided no details, they reported that 50% of 50-g chum salmon and an unspecified percentage of brook trout—but no pink salmon or rainbow trout—were infected by exposure to waterborne infectivity. One month elapsed before VEN was found in the chum salmon, but in brook trout the first infection was found at 7 days and the last apparently after 3 months.

The shortest incubation times after injection of virus into fry ranged from 2 days in chum salmon and brook and brown trout to 7 days in rainbow trout.

MacMillan and Mulcahy (1979) were able to infect chum salmon with virus obtained from Pacific herring.

Reno et al. (1986) infected Atlantic cod with intraperitoneal or intravascular injection of infected erythrocytes or cell-free homogenates of erythrocytes and noted an incubation time of 1 to 2 months at 5 to 10°C.

Source

Within susceptible species, older carrier fish are probably the source of virus that infects the young. Successful interfamily transmission from herring to salmon under experimental conditions indicates a possibility of a marine source for virus among anadromous fish in husbandry.

Host and Geographic Range

Most VEN occurs in marine species and secondarily in anadromous species. Experimental infections have been initiated in salmonids that are usually considered to be freshwater species—brook, brown, and rainbow trout. However, all three species are anadromous in certain populations or waters.

Table 2. Families of fishes in which at least one species is known to have viral erythrocytic necrosis

Family number[a]	Class and family
	Myxini (hagfishes and lampreys)
2	Petromyzontidae
	Chondrichthyes (sharks, skates, and rays)
13	Scylliorhinidae
22	Rajidae
	Osteichthyes (bony fishes)
49	Anguillidae
69	Clupeidae
137	Osmeridae
143	Salmonidae
174	Gadidae
175	Merluciidae
177	Ophidiidae
276	Cottidae
281	Cyclopteridae
326	Sciaenidae
358	Labridae
362	Zoarcidae
395	Blenniidae
434	Bothidae

Source: Adapted and expanded from Kahn and Newman (1982) and Walker and Sherburne (1977).
[a]In phylogenetic sequence from Nelson (1984).

The host range of VEN is broad, extending from cyclostomes and elasmobranchs through 14 families of bony fishes (Table 2). All indications from results of experimental infections are that the young are more susceptible than older fish of the same species. Representative hosts range from the relatively primitive clupeids and salmonids to the advanced bleniids. In that respect, the virus of VEN differs from lymphocystis—an iridovirus that occurs only in the more highly evolved fishes.

Geographic regions where VEN is known to occur include the Atlantic and Pacific Oceans off North America, the Atlantic off Greenland and the United Kingdom, and the Mediterranean.

If, as may be the case, the electric eel (*Electrophorus electricus*) had VEN and not the "endoerythrocytic protozoan" parasite described by Arcay de Peraza and McLure (1971), the geographic range is extended to Venezuela and the host range to a freshwater species of the gymnotid family.

Immunity

Because some fish have been known to recover from VEN, an immune response is presumably evoked.

References

Appy, R. G., M. D. B. Burt, and T. J. Morris. 1976. Viral nature of piscine erythrocytic necrosis (PEN) in the blood of Atlantic cod (*Gadus morhu*). J. Fish. Res. Board Can. 33:1380–1385.

Arcay de Peraza, L., and M. T. McLure. 1971. The paranuclear corpuscles in poikilothermal vertebrates. II. Description of a new species of *Toddia* in *Electrophorus electricus* (electric eel), with an expansion of the key to the species of *Toddia* in poikilothermal vertebrates. Acta Biol. Venez. 7:201–209.

Bell, G. R., and L. Margolis. 1976. The fish health program and the occurrence of fish diseases in the Pacific region of Canada. Fish Pathol. 10:115–122.

Bernard, G. W., E. L. Cooper, and M. L. Mandell. 1968. Lamellar membrane encircled viruses in the erythrocytes of *Rana pipiens*. J. Ultrastruct. Res. 26:8–16.

Blanc, G., and L. Ascione. 1958. Sur un parasite endoglobulaire du lézard *Eremias guttatulus olivieri* Audoin de la région Marrakech Maroc. Bull. Soc. Pathol. Exot. Filiales 51:508–511.

Chatton, E., and G. Blanc. 1914. Sur un hématozoaire nouveau, *Pirhemocyton tarentole* de gecko *Tarentola mauritanica*. C. R. Soc. Biol. (Paris) 77:496–498.

Doyle, W. L. 1971. Occurrence of a virus in erythrocytes of the eel, *Anguilla*. Bull. Mt. Desert Isl. Biol. Lab. 10:11. Abstract.

Eiras, J. C. 1984. Virus infection of marine fish: prevalence of viral erythrocytic necrosis (VEN) in *Mugil cephalus* L., *Blennius pholis* L. and *Platichthys flesus* L. in coastal waters of Portugal. Bull. Eur. Assoc. Fish Pathol. 4:52–56.

Evelyn, T. P. T., and G. S. Traxler. 1978. Viral erythrocytic necrosis: natural occurrence in Pacific salmon and experimental transmission. J. Fish. Res. Board Can. 35:903–907.

Fletch, A. L., and L. Karstad. 1968. An intraerythrocytic protozoan parasite of the garter snake *Thamnophis sirtalis*. Bull. Wildl. Dis. Assoc. 4:9–11.

Johnston, M. R. L. 1975. Distribution of *Pirhemocyton* Chatton and Blanc and other, possibly related, infections of poikilotherms. J. Protozool. 22:529–535.

Johnston, M. R. L., and A. J. Davies. 1973. A *Pirhemocyton*-like parasite of the blenny *Blennius pholis* L. (Teleostei; Blenniidae) and its relationship to *Immunoplasma* Neumann, 1909. Int. J. Parasitol. 3:235–241.

Kahn, R. A., and M. W. Newman. 1982. Blood parasites from fish of the Gulf of Maine to Cape Hatteras, Northwest Atlantic Ocean, with notes on the distribution of fish hematozoa. Can. J. Zool. 60:396–402.

Laird, M., and W. L. Bullock. 1969. Marine fish hematozoa from New Brunswick and New England. J. Fish. Res. Board Can. 26:1075–1102.

MacMillan, J. R., and D. Mulcahy. 1979. Artificial transmission to and susceptibility of Puget Sound fish to viral erythrocytic necrosis (VEN). J. Fish. Res. Board Can. 36:1097–1101.

MacMillan, J. R., D. Mulcahy, and M. Landolt. 1980. Viral erythrocytic necrosis: some physiological consequences of infection in chum salmon (*Oncorhynchus keta*). Can. J. Fish. Aquat. Sci. 37:799–804.

Marquardt, W. C., and R. G. Yaeger. 1967. The structure and taxonomic status of *Toddia* from the cottonmouth snake *Agkistrodon piscivorous leucostoma*. J. Protozool. 14:726–731.

Nelson, J. S. 1984. Fishes of the world. 2nd ed. Wiley, New York.

Pessôa, S. B. 1969. Experiências sôbre a transmissão de hemoparasitas a animais de sangue frio. An. Fac. Med. Univ. Fed. Parana 11-12:21–24.

Pessôa, S. B., and E. P. Campos. 1966. Sôbre um hemoparasita do gênero "Pirhaemocyton" (organismo de afinidades incertas) encontrado em serpentes Brasileiras. Rev. Bras. Biol. 26:417–423.

Pessôa, S. B., P. deBiasi, and G. Puorto. 1974. Nota sobre a freqüência de hemoparasitas em serpentes do Brasil. Mem. Inst. Butantan (São Paulo) 38:69–118.

Philippon, M., B. L. Nicholson, and S. W. Sherburne. 1977. Piscine erythrocytic necrosis (PEN) in the Atlantic herring (*Clupea harengus harengus*): evidence for a viral infection. Fish Health News 6:6–10.

Reno, P. W., and B. L. Nicholson. 1980. Viral erythrocytic necrosis (VEN) in Atlantic cod (*Gadus morhua*): in vitro studies. Can. J. Fish. Aquat. Sci. 37:2276–2281.

Reno, P. W., and B. L. Nicholson. 1981. Ultrastructure and prevalence of viral erythrocytic necrosis (VEN) virus in Atlantic cod, *Gadus morhua* L., from the northern Atlantic Ocean. J. Fish Dis. 4:361–370.

Reno, P. W. , M. Philippon-Fried, B. L. Nicholson, and S. W. Sherburne. 1978. Ultrastructural studies of piscine erythrocytic necrosis (PEN) in Atlantic herring (*Clupea harengus harengus*). J. Fish. Res. Board Can. 35:148–154.

Reno, P. W., K. Kleftis, S. W. Sherburne, and B. L. Nicholson. 1986. Experimental infection and pathogenesis of viral erythrocytic necrosis (VEN) in Atlantic cod, *Gadus morhua*. Can. J. Fish. Aquat. Sci. 43:945–951.

Rohovec, J. S., and A. Amandi. 1981. Incidence of viral erythrocytic necrosis among hatchery reared salmonids of Oregon. Fish Pathol. 15:135–141.

Sherburne, S. W. 1973. Erythrocyte degeneration in the Atlantic herring, *Clupea harengus harengus* L. U.S. Natl. Mar. Fish. Serv. Fish. Bull. 71:125–134.

Sherburne, S. W. 1977. Occurrence of piscine erythrocytic necrosis (PEN) in the blood of the anadromous alewife, *Alosa pseudoharengus* from Maine coastal streams. J. Fish. Res. Board Can. 34:281–286.

Sherburne, S. W. 1984. The occurrence of piscine erythrocytic necrosis (PEN), in the sea lamprey *Petromyzon marinus*, from several Maine localities. U.S. Dep. Commer. Fish. Bull. 82:541–544.

Sherburne, S. W., and L. L. Bean. 1979. Incidence and distribution of piscine erythrocytic necrosis and the microsporidian, *Glugea hertwigi*, in rainbow smelt, *Osmerus mordax*, from Massachusetts to the Canadian Maritimes. U.S. Natl. Mar. Fish. Serv. Fish. Bull. 77:503–509.

Smail, D. A. 1982. Viral erythrocytic necrosis in fish: a review. Proc. R. Soc. Edinburgh [B] 81:169–175.

Smail, D. A., and S. I. Egglestone. 1980a. Virus infections of marine fish erythrocytes: prevalence of piscine erythrocytic necrosis in cod *Gadus morhua* L. and blenny *Blennius pholis* L. in coastal and offshore waters of the United Kingdom. J. Fish Dis. 3:41–46.

Smail, D. A., and S. I. Egglestone. 1980b. Virus infections of marine fish erythrocytes: electron microscopical studies of the blenny virus. J. Fish Dis. 3:47–54.

Stehbens, W. E., and M. R. L. Johnston. 1966. The viral nature of *Pirhemocyton tarentolae*. J. Ultrastruct. Res. 15:543–554.

Walker, R. 1971. PEN, a viral lesion of erythrocytes. Am. Zool. 11:707. Abstract.

Walker, R., and S. W. Sherburne. 1977. Piscine erythrocytic necrosis virus in Atlantic cod, *Gadus morhua*, and other fish: ultrastructure and distribution. J. Fish. Res. Board Can. 34:1188–1195.

50 Walleye Dermal Sarcoma Virus

Definition

Walleye dermal sarcoma virus is a retrovirus or leukovirus-like particle found in abundance in benign sarcomatous tissue of adult walleyes, *Stizostedion vitreum vitreum*, in North America.

History

The discovery of walleye dermal sarcoma is credited to Walker (1947), who reported that adult fish migrating from Lake Oneida, New York, in addition to having lymphocystis nodules, also bore "hyperplastic lumps resembling sarcoma." He apparently had access to specimens preserved 20 years previously, and noted that they too had sarcomas. A popular article appeared next, in which sarcomas were described in greater detail, and a statement was made that they were probably not a response to lymphocystis virus (Walker 1957). This article preceded by 5 years Walker's report on the documentation of lymphocystis virus. He first cautiously broached the possibility of a separate viral etiology (Walker 1958), and then briefly reported the presence of 100-nm virions in the sarcomas (Walker 1961). The first micrographs of the sarcoma-associated virus were provided by Walker for inclusion in a review of fish viruses by Wolf (1966). Subsequently, spawning walleyes were found to have lesions of epidermal hyperplasia and an associated virus, but Walker (1969) specifically noted that the virions in sarcomas were larger. Seven years later, Yamamoto et al. (1976) reported sarcomas with an associated leukovirus in spawning walleyes taken in Saskatchewan, Canada. The virus-associated skin tumors of walleyes were reviewed by Yamamoto et al. (1984).

Signs and Pathologic Changes

Superficially, the lesions can be confused with those of lymphocystis and presumably also with the plaques of epidermal hyperplasia. Examination with a hand lens or dissecting microscope provides easy recognition of the hypertrophied lymphocystis cells, but histologic sections are probably the most certain way of separating the two kinds of skin neoplasms.

The sarcomas are fine in texture, variably firm, and vascularized. Their color is pink to white, and their shape tends toward the spherical or hemispherical. The location can be anywhere on the skin; Walker (1969) noted that the nodules are often in clusters, occasionally arrayed closely or on scale tips as if initiated after an abrasion. The size can range from 1 to 10 mm or more, but Yamamoto et al. (1976) gave the mean diameter as 5 mm.

Histologically, the architecture is a range of highly cellular sarcoma, ranging from fibrosarcoma to densely fibrous hard fibroma. The nodules do not contain giant cells, and mucous cells of normal epidermis are generally lacking. Lymphocystis cells have occasionally been found superficially embedded in a sarcomatous nodule.

Electron microscopy shows the cells of the sarcoma have interdigitating surfaces, making it difficult to distinguish cell membranes from endoplasmic reticulum. Nuclei are normal (Yamamoto et al. 1976).

Etiology

The virus particles have yet to be causally related to the sarcoma.

According to Walker (1961, 1969), the particles have a regular form, measure about 100 nm, and are fairly common in intercellular spaces but seldom found in masses. Complete virions were not found intracellularly; instead they appeared as budding particles, and envelopes were continuous with the cell membrane.

Yamamoto et al. (1976) confirmed the presence of virus, but gave its size as 135 nm and described and illustrated an estimated 162 studlike surface features. They concluded that the viruses from the two widely separated regions are possibly the same or related and added that, by its form and apparent mode of maturation, the virus has an RNA genome and most closely resembles leukoviruses.

Transmission and Incubation

Neither transmission nor transplantation has been reported. In the early 1960s, Walker froze fresh sarcomatous tissue in liquid nitrogen and then lyophilized the material. It was later reconstituted and inoculated subdermally into fingerling walleyes (obtained from the midwestern United States, where the disease has not been found) and bluegills (*Lepomis macrochirus*). The fish were held for 4 months at 12°C, but no tumors were induced (Wolf 1966).

Host and Geographic Range

In addition to the Lake Oneida fish that showed a 5% incidence of sarcomas, Walker (1969) reported the condition in Lake Champlain and the eastern Great

Lakes. For Canadian fish, Yamamoto et al. (1976) gave an average incidence of 11% but noted both an age (or size) and sex differential: growths were found on 30% of the small females, but only 9% of the small males; no growths were found on large fish of either sex.

References

Walker, R. 1947. Lymphocystis disease and neoplasia in fish. Anat. Rec. 99:559–560.

Walker, R. 1957. Warty walleyes. N.Y. State Conserv. 12:28–29.

Walker, R. 1958. Lymphocystis warts and skin tumors of walleyed pike. Rensselaer Rev. Grad. Stud. 14:1-5.

Walker, R. 1961. Fine structure of a virus tumor of fish. Am. Zool. 1:395–396. Abstract.

Walker, R. 1969. Virus associated with epidermal hyperplasia in fish. Natl. Cancer Inst. Monogr. 31:195–207.

Wolf, K. 1966. The fish viruses. Adv. Virus Res. 12: 35–101.

Yamamoto, T., R. D. Macdonald, D. C. Gillèspie, and R. K. Kelly. 1976. Viruses associated with lymphocystis disease and dermal sarcoma of walleye (*Stizostedion vitreum vitreum*). J. Fish. Res. Board Can. 33:2408–2419.

Yamamoto, T., R. K. Kelly, and O. Nielsen. 1984. Morphological differentiation of virus associated skin tumors of walleye (*Stizostedion vitreum vitreum*). Page 53 in International seminar on fish pathology for the 20th Anniversary of the Japanese Society of Fish Pathology, Tokyo, Japan, September 8–10, 1984. Abstract.

51 Walleye Epidermal Hyperplasia Retrovirus

Definition

The walleye epidermal hyperplasia retrovirus is an agent visualized in abnormal skin plaques on walleyes, *Stizostedion vitreum vitreum* (Walker 1969a; Yamamoto et al. 1985).

History

Lesions of virus-associated epithelial hyperplasia were first reported by Walker (1969a), who noted that their presence had been recognized for many years but that attention had been concentrated instead on lesions of lymphocystis and dermal sarcoma. Epithelial hyperplasia in walleyes came under scientific scrutiny in the late 1960s in New York, but no further reports from that location have appeared. Instead, the peculiar neoplasms and their contained C-type virus particles have been found in walleyes from two provinces in central Canada (Yamamoto et al. 1985).

Signs and Pathologic Changes

The incidence was about 5% in the New York population and as high as 20% in the several walleye populations in Canada.

Several lesions may occur anywhere on the skin or fins, but particularly on the caudal peduncle and tail. Shape ranges from broad and plaquelike (New York) to nodular (Canada), but margins are always raised (Fig. 1). Size varies from a few millimeters to several centimeters, and coloration, although translucent, is that of the background skin pigmentation.

Histologically, hyperplasia is confined to the epidermis and consists predominantly of cuboidal cells. Mucous cells may be present but are generally at the surface of the lesions and not deep. The presence or absence of mitoses depends on the stage of development; few are found in mature lesions.

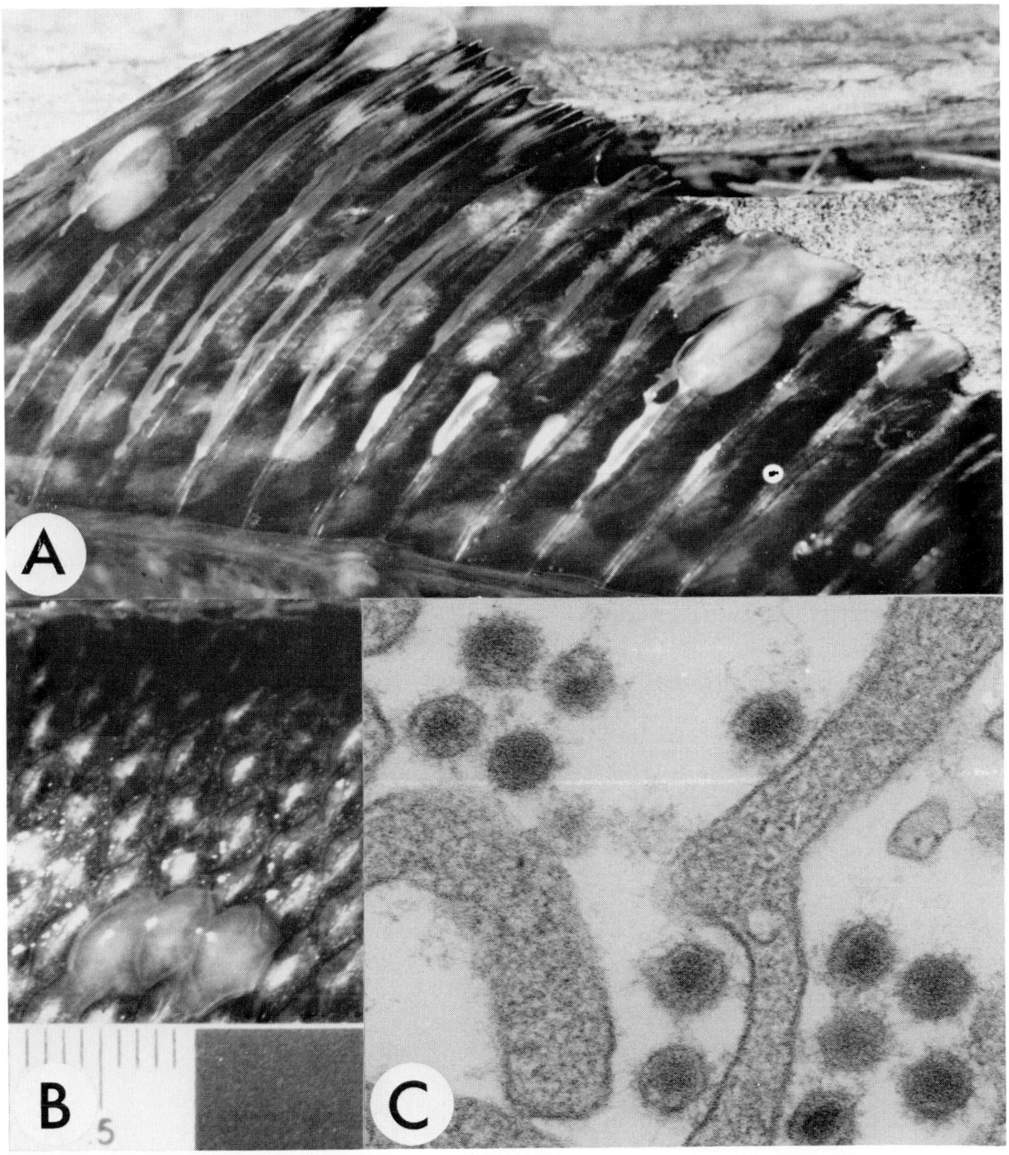

Figure 1. A. Dorsal fin of a walleye showing translucent raised nodular lesions of hyperplastic epidermis. *B.* Trio of nodules on the lateral body surface. Scale is in millimeters. *C.* Electron micrograph of 120-nm retrovirus-like particles in an intercellular space. From Yamamoto et al. (1985). Reprinted with the permission of Blackwell Scientific Publications Ltd.

Walker (1969a) reported that the virus-associated epithelial hyperplasia is sometimes present on fish with either of two other viral infections—lymphocystis or dermal sarcoma. Yamamoto et al. (1985) found that two kinds of epidermal hyperplasias occur in Canadian walleyes: a superficial lesion with the contained retrovirus, and a more diffuse growth with a herpesvirus infection.

Electron micrographs showed that lesions consist of unremarkable, simple, undifferentiated epithelial cells with long cytoplasmic processes and C-type particles budding into and collecting in intercellular spaces (Fig. 1C).

Etiology

The virus is postulated to be causal, but experimental transmission and tumor induction has not been attempted. Yamamoto et al. (1985) attempted to isolate the agent, but were not successful. Accordingly, the virus is known only from electron micrographs. Walker (1969a) gave the size as 80 nm. In contrast, Yamamoto et al. (1985) reported the size as 120 nm, but attributed no significance to the difference because different parts of the virus were used for size determinations.

Host and Geographic Range

The lesions and accompanying retrovirus have been found only in Lake Oneida, New York, and in populations in Manitoba and Saskatchewan, Canada. However, Yamamoto et al. (1985) believe that the infection might be widely distributed in North America. Walker (1969b) looked at similar lesions in the related yellow perch, *Perca flavescens*, but found no virus. Abnormal skin proliferations in America's other percids—particularly the sauger, *Stizostedion canadense*—merit study.

References

Walker R. 1969a. Virus associated with epidermal hyperplasia in fish. Natl. Cancer Inst. Monogr. 31:195–207.

Walker R. 1969b. Epidermal hyperplasia in fish: two types without visible virus. Natl. Cancer Inst. Monogr. 31:209–213.

Yamamoto, T., R. K. Kelly, and O. Nielsen. 1985. Epidermal hyperplasia of walleye *Stizostedion vitreum vitreum* (Mitchell), associated with retrovirus-like type-C particles: prevalence, histologic and electron microscopic observations. J. Fish Dis. 8:425–436.

PART IV

VIRUSLIKE PARTICLES

52 Atlantic Salmon Papillomatosis

Synonym: salmon wart disease

Definition

Atlantic salmon papillomatosis is a chronic, benign, plaquelike proliferation of epidermis—usually at multiple sites—that occurs seasonally in parr and occasionally in young adult *Salmo salar*, particularly in Scandinavia. In Scotland, viruslike particles have recently been found in some of the neoplastic cells (Carlisle 1977).

History

It has not been possible to determine exactly when salmon papillomatosis was first recognized. B. Svenonius, a visitor to the U.S. Fish and Wildlife Service Eastern Fish Disease Laboratory in 1954, showed photographs of afflicted young Atlantic salmon and spoke of the condition as being common among hatchery fish in Sweden (personal communication, 1954). Ljungberg (1963) briefly mentioned the condition in a report on fish diseases in Sweden and noted that papillomatosis was possibly infectious. Wiren (1971), who studied salmon papillomatosis with both light and electron microscopy, noted intracellular forms that were suggestive of virus. Chronwall (1976), who conducted a similar but more detailed study, found no evidence of virus. Johansson (1977) made brief mention of seasonality and duration of the condition. Workers at Sterling University, Scotland, explored aspects of epizootiology, pathology, and immunology; ultrastructural findings showed evidence of viruslike particles (Carlisle 1977; Carlisle and Roberts 1977). Two episodes of similar lesions were reported among rainbow trout under conditions of marine husbandry in Scotland by Roberts and Bullock (1979), and afflicted Atlantic salmon were found in Finland by Bylund et al. (1980).

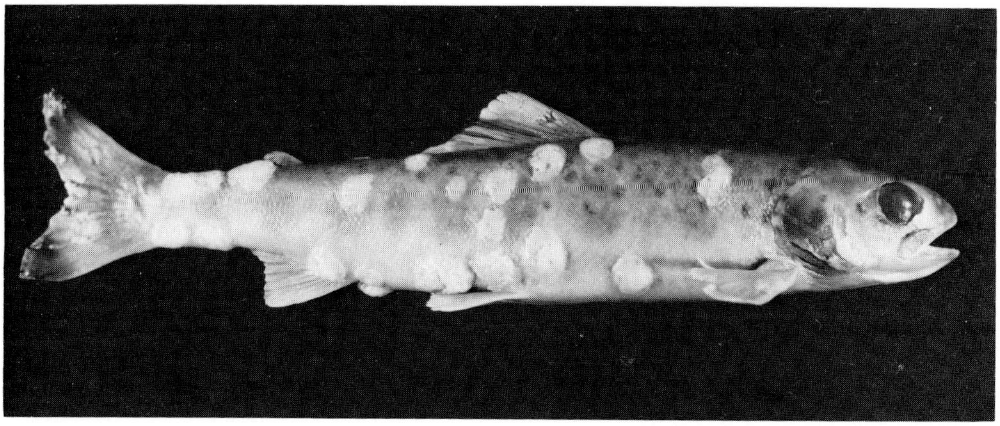

Figure 1. Young Atlantic salmon from Sweden showing multiple skin lesions of papillomatosis. Specimen courtesy of B. Svenonius.

Signs and Pathologic Changes

Outbreaks usually occur among salmon in freshwater; lesions begin during July or August, peak in September or October, and disappear in winter (Johansson 1977). The incidence sometimes reaches 90%. The condition may also afflict parr in seawater; the involvement is usually less than 1%, but lesions sometimes persist for a year.

External Signs

Affected fish behave normally but usually have multiple skin lesions anywhere on the body. The growths are smooth to rough in texture, several millimeters thick, and as large as 40 mm (Fig. 1). Some lesions show areas of hyperemia. No internal involvement has been found.

The papillomas themselves are usually of little significance, but they are sometimes invaded secondarily, and such infections may result in mortality. Bylund et al. (1980) reported an incidence of 80 to 90%, and one outbreak in which mortality (aggravated by seasonal temperature that fell rapidly to 1 to 3°C) reached 50%.

Histopathologic Findings

Descriptions of the several histologic studies differ somewhat, but there is general agreement that the neoplastic tissue is readily distinguishable from adjacent normal epithelium. The normal architecture is deranged, mucous cells are infrequent to absent, and mitoses are common and may be tripolar. Cords of fibrovascular cells generally occupy the central or thicker portions of the lesions. Eosinophilic cytoplasmic inclusions are present in some growths, particularly in degenerating cells. Host response varies from localized hyperemia involving a few leukocytes to massive lymphocytic infiltration.

At the ultrastructural level, Chronwall (1976) and Wirén (1971) found that normal interdigitation of neoplastic epithelial cells was lacking and that desmosomes seemed not to have contact with other portions of cell membranes. Neoplastic tissue examined by Carlisle (1977) showed the normal condition for these features, but cell nuclei were more nearly spherical than normal and the chromatin was obviously clumped and marginated.

Comments on the presence or absence of virus or viruslike particles differed among the several investigators: Chronwall (1976) found none in his material; Wirén (1971) noted inclusions that were believed to be virus; and Carlisle (1977) found both intracellular and extracellular viruslike particles.

Etiology

Although the papillomas on Atlantic salmon and rainbow trout are histologically similar, there is no assurance that they share a common etiology in the two species and in different parts of the world. The cause (or possibly the causes) remains to be demonstrated, but the localization of the lesions supports the hypothesis of a viral etiology.

Size and Shape

Somewhat ambiguously, Carlisle (1977) noted viruslike particles in "five blocks of tissue from three fish." The particles had an electron-dense outer coat and were 125 to 150 nm in diameter. An electron-lucent layer between the outer coat and a core or nucleoid was 70 to 95 nm wide. Particles were found in degenerating cells or associated with plasma membranes of intact cells.

Diagnosis

Diagnosis can be made on the basis of the presence of the macroscopic lesions described and their confirming histopathologic features.

Isolation

Carlisle (1977) assayed papillomatous tissue and internal organs in cell cultures but found no CPE or other indication of virus. Fresh or frozen tissues were homogenized, decontaminated by filtration, and inoculated into the AS cell line from Atlantic salmon. Cultures were incubated at 12°C for 1 week and, in the absence of CPE, the material was blind passaged five times at weekly intervals. No viral effects were observed. Primary cultures of papillomatous tissue that were initiated survived for 1 month, but cell numbers were never sufficient to permit subculturing.

In the United States, R. Dexter (personal communication) inoculated RTG-2 cells with homogenates of salmon papillomas but found no evidence of virus.

Considering the equivocal evidence thus far obtained for virus, further attempts at isolation should be made with salmon epithelial cells—preferably of skin or fin origin—from young donors. The AS and RTG-2 cell lines are both fibroblastic and, considering the chronic nature of the lesions, incubation of cultures should be prolonged, and cell transformation rather than lytic effects should be sought.

Transmission and Incubation

Both Wirén (1971) and R. Dexter (personal communication) were unsuccesful in attempts to transmit the disease experimentally.

Host and Geographic Range

One-year-old and, more commonly, 2-year-old parr of *S. salar* in hatcheries in Sweden, Norway, Scotland, Finland, and Great Britain have had the disease. Infrequently, the condition has been found in young salmon in North American and European streams. Morphologically indistinguishable lesions have also been reported in rainbow trout from Scotland.

Immunity

Carlisle and Roberts (1977) claimed evidence for a cell-mediated immunity when the migration of leukocytes from afflicted Scottish salmon was reacted with papilloma antigen, also from Scotland. In contrast, migration occurred in the presence of antigen from Norwegian salmon. No reactions were observed in immunoprecipitin tests in which similar sera and tumor antigens were used.

Control

Bylund et al. (1980) reported that papillomatosis did not occur among yearling Atlantic salmon that had been treated prophylactically with sodium chloride and formalin for about 2.5 months. However, the authors noted that the lesions appeared only among fish in two tanks not included in the first several weeks of the treatment program.

References

Bylund, G., E. T. Valtonen, and E. Niemelä. 1980. Observations on epidermal papillomata in wild and cultured Atlantic salmon *Salmo salar* L. in Finland. J. Fish Dis. 3:525–528.

Carlisle, J. C. 1977. An epidermal papilloma of the Atlantic salmon II: ultrastructure and etiology. J. Wildl. Dis. 13:235–239.

Carlisle, J. C., and R. J. Roberts. 1977. An epidermal papilloma of the Atlantic salmon. I: epizootiology, pathology and immunology. J. Wildl. Dis. 13:230–234.

Chronwall, B. 1976. Epidermal papillomas in *Salmo salar* L.: a histological description. Zoon 4:109–114.

Johansson, N. 1977. Studies on diseases in hatchery-reared Atlantic salmon (*Salmo salar* L.) and sea trout (*Salmo trutta* L.) in Sweden. Abstracts of Uppsala dissertations from the Faculty of Science, 401, Acta Univ. Upsaliensis, Sweden.

Ljungberg, O. 1963. Report on fish diseases and inspection of fish products in Sweden. Bull. Off. Int. Epizoot. 59:111–120.

Roberts, R. J., and A. M. Bullock. 1979. Papillomatosis in marine cultured rainbow trout *Salmo gairdneri* Richardson. J. Fish Dis. 2:75–77.

Wirén, B. 1971. Wart disease in Atlantic salmon (*Salmo salar* L.): histological studies of epidermal papilloma in reared salmon. (Vartsjuka hos lax [*Salmo salar* L.] histologiska studier över epidermala papillom hos odlad lax.) Swedish Salmon Res. Inst. Rep. LFI MEDD 7/1971.

53 Brown Bullhead Papilloma Virus

Definition

The brown bullhead papilloma virus is a small, regular, intracellular form found in oral papillomatous tissue from a single specimen of *Ictalurus nebulosus* taken from a lake in New York (Edwards and Samsonoff 1977; Edwards et al. 1977).

History

In the northeastern United States, some populations of brown bullheads have long been known to have a noticeable incidence of neoplasms on or about the mouth. Lucké and Schlumberger (1941), who investigated such tumors from brown bullheads in Pennsylvania, found them to be benign slow-growing epitheliomas. Of special interest was the finding that the tumors were transplantable to the anterior chamber of the eye of other brown bullheads. R. Walker (personal communication), who examined thin sections of the tumors, found no evidence of virus. Tumors from specimens taken in Pennsylvania (Fig. 1) were next investigated by Sullivan et al. (1965), who reported the isolation of a DNA virus; however, that claim subsequently proved to be unfounded.

Steeves (1969), who carried out a histologic study of a single epithelial papilloma on a brown bullhead taken in Virginia, noted that the neoplasm did not differ from the Pennsylvania specimens described by Lucké and Schlumberger (1941). He offered no evidence that suggested a virus.

At the Smithsonian Institution Registry of Tumors in Lower Animals, Harshbarger (1972) described and illustrated papillomas and carcinomas occurring on the head and particularly about the mouth of brown bullheads from lakes in the southeastern United States. Although a collaborative study was conducted with the National Fish Health Research Laboratory, neither electron microscopy nor virologic examination showed evidence of virus.

The most recent investigation of brown bullhead papilloma was that of Edwards and Samsonoff (1977) and Edwards et al. (1977), whose brief reports covered essen-

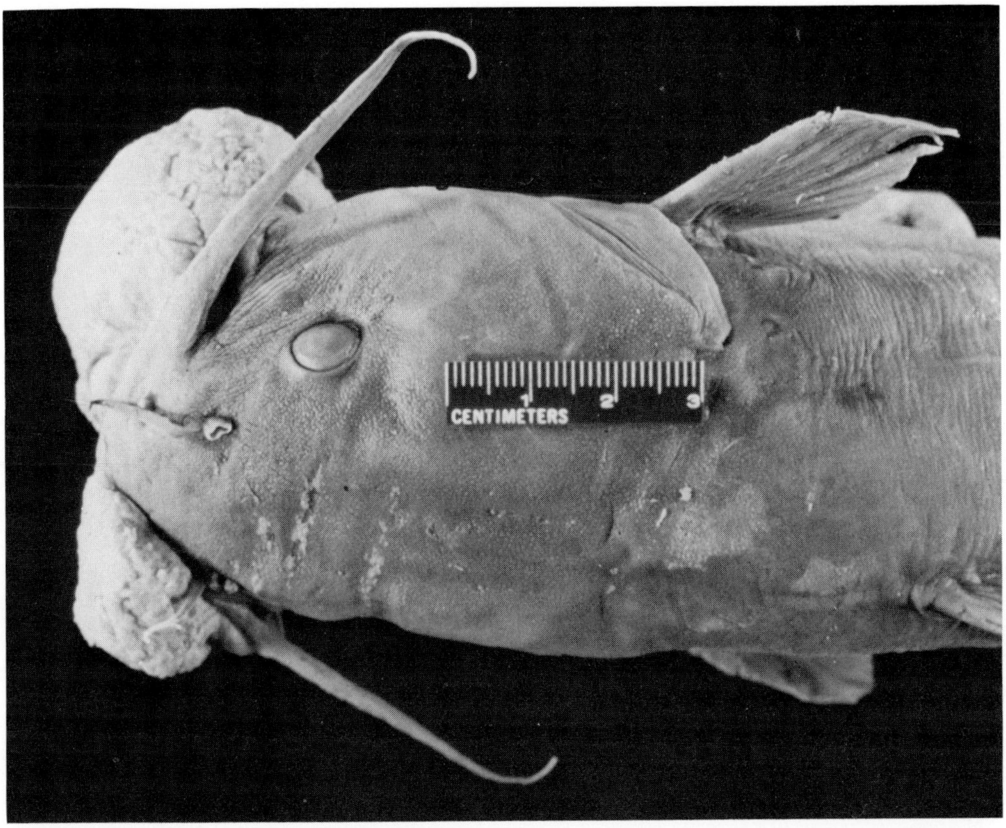

Figure 1. Two fleshy and vascularized papillomas on the mandible of an adult brown bullhead from a river in Pennsylvania. Courtesy of R. G. Malsberger.

tially the same information about viruslike forms found in tumors from a specimen in New York.

Signs and Pathologic Changes

The various cases of papilloma share generalized common morphology and anatomic location on or about the mouth, but other aspects of possible relatedness are unknown.

The tissue examined by Edwards et al. (1977) was taken from one of four tumor-bearing fish that were submitted frozen, then thawed and fixed in glutaraldehyde and osmium by routine methods. Histologic features are those found solely by electron microscopy. The neoplasm consisted largely of fusiform epidermal cells with pleomorphic nuclei having irregular nuclear membranes and large lacunae. Cytoplasmic vesiculation was prominent and attributed to apparent dilation or proliferation of the endoplasmic reticulum. Proliferation of tonofibrils was extensive.

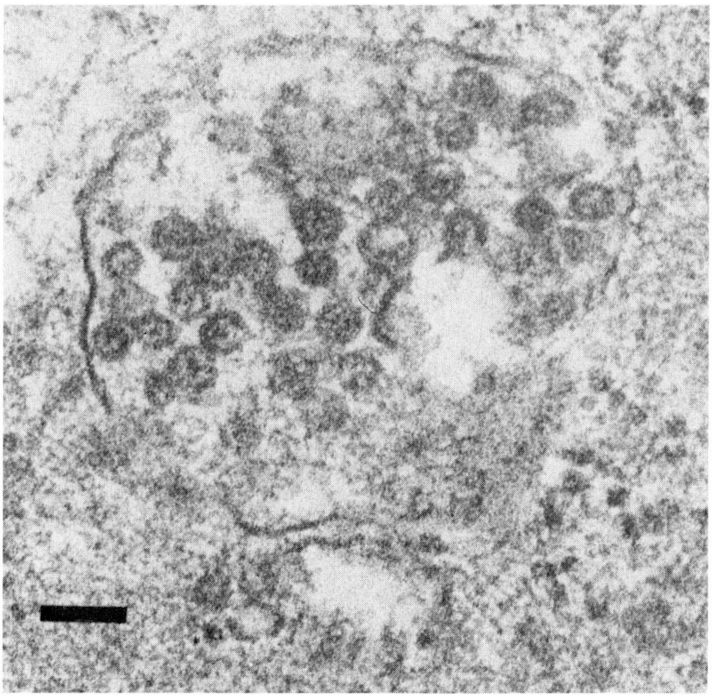

Figure 2. Aggregated viruslike particles within a cytoplasmic vesicle of a neoplastic epidermal cell from a brown bullhead papilloma. From Edwards and Samsonoff (1977). Reprinted with the permission of the Electron Microscopy Society of America.

Other features of histology were not reported, nor was the gross appearance of the tumor described.

Etiology

Viruslike forms have been found only by Edwards and Samsonoff (1977) and Edwards et al. (1977); moreover, the true nature of the forms, i.e., whether virus or artifact, remains to be determined. Accordingly, the cause of the neoplasms remains unknown.

The particles have a diameter of about 50 nm, an electron-dense core, and an enveloping double membrane interpreted to be the capsid (Fig. 2). Particles are abundant in the dilated cisternae of the nuclear membrane, but budding from the membrane has not been observed. Particles are also observed in dilated cisternae of the endoplasmic reticulum, particularly at the cell periphery. The authors postulated that the particles originate in the nucleus and then pass into the cytoplasm; from this pattern and the size and shape, they believed the particles to be similar to papilloma viruses.

References

Edwards, M. R., and W. A. Samsonoff. 1977. Electron microscopic observation on virus-like particles of a catfish papilloma. Pages 394–395 in G. W. Bailey, ed. 35th annual proceedings of the Electron Microscopy Society of America, Boston, Massachusetts.

Edwards, M. R., W. A. Samsonoff, and E. J. Kuzia. 1977. Papilloma-like viruses from catfish. Fish Health News 6:94–95.

Harshbarger, J. C. 1972. Work of the registry of tumors in lower animals with emphasis on fish neoplasms. Pages 285–303 in L. E. Mawdesley-Thomas, ed. Symposium of the Zoological Society of London. Academic Press, London.

Lucké, B., and H. G. Schlumberger. 1941. Transplantable epithelioma of the lip and mouth of catfish. I. Pathology. Transplantation to anterior chamber of eye and into cornea. J. Exp. Med. 74:397–408.

Steeves, H. R. 1969. An epithelial papilloma of the brown bullhead, *Ictalurus nebulosus*. Natl. Cancer Inst. Monogr. 31:215–216.

Sullivan, R., R. G. Malsberger, and E. J. Benz. 1965. Isolation of a virus from papillomas of the catfish, *Ictalurus nebulosus*. Bacteriol. Proc. 18:100. Abstract.

54 Gilthead Bream Viruslike Particles

Definition

The gilthead bream viruslike particles are forms of 35 to 65 nm visualized by electron microscopy in the cytoplasm of fibroepithelial papilloma tissue composing benign maxillary neoplasms in *Sparus aurata* taken in saline water near Cadiz, Spain.

History

The available information consists of a single report of seven affected specimens among 39 examined; lengths and weights of the tumor-bearing fish were 155 to 297 mm and 50 to 248 g (Gutierrez et al. 1977).

Signs and Pathologic Changes

The tumors occur principally on the maxillae of the fish and are slow growing and benign. No metastases have been found. The neoplasms apparently reached sizes that interfered with feeding and, presumably, as a result the growth of the fish was retarded.

Histologically, the tumor shows no necrosis, nor does it invade adjacent tissue. The basal membrane remains intact. Further details have not been described.

Etiology

The cause of the tumors has not been determined, but the viruslike forms must certainly be considered causal because of their presence and abundance.

The agent has not been isolated. Viruslike particles were found singly and in groups and in some instances appeared to bud from cytoplasmic membranes.

Morphologically, although the authors did not comment on shape, the forms generally conformed to spheres of C-type virions and were never found in cell nuclei. Some of the particles displayed electron-dense nucleoids and bore regular surface projections.

Host and Geographic Range

Thus far, the viruslike forms have been found only in the dorada or gilthead bream in Spanish waters. Since this species is found elsewhere, however, the tumors may be encountered in fish from other places.

Reference

Gutierrez, M., J. Perez Crespo, and A. Arias. 1977. Particulas virus-like en un tumor en boca de dorado, *Sparus aurata* L. Invest. Pesq. 41:331–336.

55 Platyfish Viruslike Particles

Definition

Platyfish viruslike particles are an assemblage of intracellular forms found in neoplastic tissues of certain hybrid xiphophorine fish. The particles occur in two sizes that differ in cellular location and develop after dissimilar manipulations. The subject is presented here because a genus of tropical aquarium fish is involved, and the possibility exists that one or another of the particles might in time be confirmed as virus.

History

The history of the platyfish particles is both recent and limited. Hybrids of *Xiphophorus maculatus* and *X. helleri* develop melanomas spontaneously, and a different neoplasm—neuroblastoma—arises after the hybrids are treated with *N*-methyl-*N*-nitrosourea. The neoplasms are believed to be mediated by a tumor gene, which is additionally regulated by other genes. The tumor gene might be related to a virus, but no candidate agent was found until Kollinger et al. (1979) reported the presence of three kinds of cytoplasmic particles.

Confirmation of similar viruslike particles in melanoma tissue was reported by Esaka et al. (1981), who conducted an ultrastructural study and found particles of a size reported by Kollinger et al. (1979), but in the nucleus instead of the cytoplasm. The nature of the particles remains to be determined.

Signs and Pathologic Changes

Descriptions and illustrations of the subject neoplastic cells are to be found in the two references cited above.

Etiology

Just as their nature is unknown, the role of viruslike particles in the neoplastic tissue is wholly uncertain. The particles could be artifacts that are merely associated with and not a causal factor in the neoplasia.

Particles of three size groups have been found after chemical treatment. Melanomas in 2-g hybrid specimens that had been injected with 200 μg of bromodeoxyuridine 10 days earlier showed abundant cytoplasmic viruslike particles 25 to 30 nm in diameter, and other particles that measured 40 to 50 nm. Clusters of the larger particles were frequently seen within a membranous envelope. Kollinger et al. (1979) suggested that the smaller particles resembled papovavirus. They induced neuroblastomas with *N*-methyl-*N*-nitrosourea, and after injection of bromodeoxyuridine the neoplasms showed infrequent solitary cytoplasmic particles about 100 nm in diameter. The particles were said to resemble B- and C-type forms of what is now termed retrovirus. Neoplastic tissues of control specimen fish injected with saline were devoid of viruslike particles.

Invasive melanotic and amelanotic melanomas developed in hybrids of *Xiphophorus* species; Esaka et al. (1981), who examined such tissues, found viruslike particles 40 to 50 nm in diameter in cell nuclei.

The relationship of the cytoplasmic forms to the 40- to 50-nm nuclear forms is not known, and no efforts to isolate the forms or to show infectivity have been reported.

References

Esaka, T., M. Asada, Y. Wakamatsu, and K. Ozato. 1981. Differentiation of melanomas occurring in platyfish-swordtail hybrids of different ages: an ultrastructural study. J. Exp. Zool. 215:133–142.

Kollinger, G., M. Schwab, and F. Anders. 1979. Virus-like particles induced by bromodeoxyuridine in melanoma and neuroblastoma of *Xiphophorus*. J. Cancer Res. Clin. Oncol. 95:239–246.

56 Pleuronectid Papilloma

Definition

Pleuronectid papilloma is a collective term used for morphologically and histologically similar growths that occur singly or in multiples in epizootic proportions on the skin of several species of Pacific flatfishes. Although the tumors have been studied by pathologists and other specialists, their etiology has proved elusive. Different causes have been proposed—parasite, pollution, and virus. However, the evidence for a viral etiology is meager. Electron micrographs of some tumor cells show various objects that several investigators have termed viruslike particles. However, the viruslike particles lack uniformity and do not conform in size and shape with any presently recognized virus group.

History

In North America, the story of pleuronectid papillomas began with unpublished field notes made in 1922 that described abundant dermal growths on young English sole (*Parophrys vetulus*) in San Francisco Bay (Cooper and Keller 1969; Nigrelli et al. 1965). Little interest was shown in the neoplasms of flatfishes until the 1960s, when several investigations were begun. Much work was done at the University of Oregon School of Medicine, where Wellings and his collaborators reported epizootiologic and histologic findings of various affected species taken in waters off Washington State (Wellings and Chuinard 1964; Wellings et al. 1964, 1965, 1966). At the New York Aquarium, Nigrelli et al. (1965) worked with affected species taken off British Columbia. Knowledge of the neoplasms was extended to even more distant places and to additional species when Honma and Kon (1968) and Kimura et al. (1967) reported growths on flatfishes in Japan.

During the late 1960s, reports from the University of Oregon appeared on the comparative histology of tumors in various species and on ultrastructural details of the so-called X cells that had been found common to certain tumors of species of

Pacific flatfishes (McArn et al. 1968). At the University of California, Cooper and Keller (1969) reported on the epizootiology of tumors on the English sole in San Francisco Bay.

Little evidence was found to support statements regarding findings of virus; nevertheless, nearly two-thirds of the reports that appeared during the 1960s mentioned that virus was seen in cells of the papillomas. The theme of virus involvement was reiterated in more than half of the reports published during the 1970s, but additional evidence was not forthcoming.

During the 1970s, the case for a viral cause was only slightly strengthened, whereas the possible role of the X cell, postulated to be a parasite, was significantly enhanced. Reviews of papillomas were published by Wellings (1970) and Wellings et al. (1976). Wellings and new collaborators at the University of California described the development of tumors, their prevalence, and host survival (Miller and Wellings 1971), and seasonal change in prevalence (Angell et al. 1975).

At the University of British Columbia, the use of the papillomas and other neoplasms as a system of monitoring the environment for carcinogens was proposed by Stich and Acton (1976) and Stich et al. (1976, 1977).

In Japan, Ito et al. (1976) briefly reviewed earlier work in that country, particularly that of Kimura et al. (1967). Wellings et al. (1977) reported on the prevalence, distribution, and pathology of three common diseases of fishes of the Bering Sea, including papillomas. In Germany, Peters et al. (1978) reported on differences in skin tumors of Atlantic and Pacific flatfishes, and noted that papillomas were rare in the Atlantic species. Interest in the X cells continued: Yamazaki et al. (1978) described developmental changes in X cell morphology and origin. McCain et al. (1979) reported on an extensive survey of tumors and diseases of demersal fishes of the Bering Sea: some 36,000 specimens were examined and geographic distribution of the tumors and diseases given. Peters and Watermann (1979) reviewed three kinds of papillomas that occur on the skin of flatfishes, describing the growths and discussing the probable causes; they considered the complex X cell–associated tumors of Pacific flatfishes to be of probable viral etiology.

Signs and Pathologic Changes

The papillomas are external and occur anywhere on the body, but preferentially on the eyed or pigmented side. Terminology for the tumors was supplied by Wellings et al. (1964). The youngest growths, angioepithelial nodules, are smooth, raised, and pink areas 1 to 2 mm in diameter. With time, the growths enlarge to several centimeters and develop a pebbled surface that may or may not be pigmented; they are then termed epidermal papillomas (Fig. 1). The assumed last stage in tumor development is termed an angioepithelial polyp.

The growths are not invasive, only infrequently show necrosis, have a low level of mitoses, and seemingly do not slough spontaneously.

The details of histopathologic findings were described and illustrated by Nigrelli et al. (1965), Peters et al. (1978), Wellings et al. (1965, 1976), and others. A common finding among the lesions of all species is the presence of a foreign cell type, the so-

Figure 1. Subyearling flathead sole bearing multiple tumors, ranging from angioepithelial nodules to epidermal papillomas. Plaquelike spread at margins is evident. From Wellings et al. (1964).

called X cell, presumed to be protozoal. Dawe et al. (1979) found evidence to indicate that X cells are amebic and not of fish origin.

Etiology

Cytoplasmic viruslike particles have been reported in tumor cells of only two species of host fishes. Wellings and Chuinard (1964) described and illustrated 44-nm particles having a 6-nm internal granule present in papilloma cells of flathead sole (*Hippoglossoides elassodon*). McArn et al. (1968) found numerous electron-dense particles of uniform 22.5- to 30-nm size in cells of an angioepithelial polyp from a starry flounder (*Platichthys stellatus*).

Wellings et al. (1976) attempted homotransplantation of tumors, but that effort was not successful; normal flatfish cell cultures were similarly unresponsive to homogenates of tumor cells. When young English sole (*Parophrys vetulus*) were inoculated with preparations of tumor cells or normal cells, all groups developed tumors and the prevalence was essentially the same.

Diagnosis

The papillomas are diagnosed on the basis of their occurrence in a flatfish species in Pacific waters, morphology of the tumor, and histologic demonstration of the presence of the X cells.

Brooks et al. (1969) defined the X cells as having membrane-bound, pleomorphic cytoplasmic granules, large mitochondria, and prominent nucleoli. Wellings et al. (1976) considered X cells to be pathognomonic.

Transmission and Incubation

In general, the pattern of occurrence is greatest in the youngest specimens; prevalence declines as fish increase in age. The inference is that the condition ultimately results in mortality (Peters and Watermann 1979; Wellings et al. 1976).

Host and Geographic Range

Stich et al. (1977), who conducted a global survey of the literature on the occurrence of pleuronectid papillomas, found 18 species on which the growths were known. The regions involved were the Northern Pacific Ocean, Bering Sea, Sea of Okhotsk, and Sea of Japan. Hard et al. (1979) found no papillomas among pleuronectid fishes during a 17-month survey of demersal fishes in waters near Melbourne, Australia.

References

Angell, C. L., B. S. Miller, and S. R. Wellings. 1975. Epizootiology of tumors in a population of juvenile English sole (*Parophrys vetulus*) from Puget Sound, Washington. J. Fish. Res. Board Can. 32:1723–1732.

Brooks, R. E., G. E. McArn, and S. R. Wellings. 1969. Ultrastructural observations on an unidentified cell type found in epidermal tumors of flounders. J. Natl. Cancer Inst. 43:97–109.

Cooper, R. C., and C. A. Keller. 1969. Epizootiology of papillomas in English sole, *Parophrys vetulus*. Pages 173–185 in C. J. Dawe and J. C. Harshbarger, eds. Neoplasms and related disorders of invertebrate and lower vertebrate animals. Natl. Cancer Inst. Monogr. 31.

Dawe, C. J., J. Bagshaw, and C. M. Poore. 1979. Amebic pseudotumors in pseudobranchs of Pacific cod, *Gadus macrocephalus*. Proc. Am. Assoc. Cancer Res. 20:245. Abstract 990.

Hard, G. C., R. Williams, and J. Lee. 1979. Survey of demersal fish in Port Phillip Bay for incidence of neoplasia. Aust. J. Mar. Freshwater Res. 30:73–79.

Honma, Y., and T. Kon. 1968. A case of the epidermal papilloma in the witchflounder from the Sea of Japan. Bull. Jpn. Soc. Sci. Fish. 34:1–5.

Ito, Y., I. Kimura, and T. Miyake. 1976. Histopathological and virological investigations of papillomas in soles and gobies in coastal waters of Japan. Prog. Exp. Tumor Res. 20:86–93.

Kimura, I., T. Sugiyama, and Y. Ito. 1967. Papillomatous growth in sole from Wakasa Bay area. Proc. Soc. Exp. Biol. Med. 125:175–177.

McArn, G. E., R. G. Chuinard, B. S. Miller, R. E. Brooks, and S. R. Wellings. 1968. Pathology of skin tumors found on English sole and starry flounder from Puget Sound, Washington. J. Natl. Cancer Inst. 41:229–242.

McCain, B. B., W. D. Gronlund, M. S. Myers, and S. R. Wellings. 1979. Tumors and microbial diseases of marine fishes in Alaskan waters. J. Fish Dis. 2:111–130.

Miller, B. S., and S. R. Wellings. 1971. Epizootiology of tumors on flathead sole (*Hippoglossoides elassodon*) in East Sound, Orcas Island, Washington. Trans. Am. Fish. Soc. 100:247–266.

Nigrelli, R. F., K. S. Ketchen, and G. D. Ruggiori. 1965. Studies on virus diseases of fishes. Epizootiology of epithelial tumors in the skin of flatfishes of the Pacific coast, with special reference to the sand sole (*Psettichthys melanosticus*) from northern Hecate Strait, British Columbia, Canada. Zoologica 50:115–122.

Peters, N., and B. Watermann. 1979. Three types of skin papillomas of flatfishes and their causes. Mar. Ecol. Prog. Ser. 1:269–276.

Peters, N., G. Peters, H. F. Stich, A. B. Acton, and G. Bresching. 1978. On differences in skin tumors of Pacific and Atlantic flatfish. J. Fish Dis. 1:3–25.

Stich, H. F., and A. B. Acton. 1976. The possible use of fish tumors in monitoring for carcinogens in the marine environment. Prog. Exp. Tumor Res. 20:44–54.

Stich, H. F., A. B. Acton, and C. R. Forrester. 1976. Fish tumors and sublethal effects of pollutants. J. Fish. Res. Board Can. 33:1993–2001.

Stich, H. F., A. B. Acton, B. P. Dunn, K. Oishi, F. Yamazaki, T. Harada, G. Peters, and N. Peters. 1977. Geographic variations in tumor prevalence among marine fish populations. Int. J. Cancer 20:780–791.

Wellings, S. R. 1970. Biology of some virus diseases of marine fish. Pages 296–306 in S. F. Snieszko, ed. A symposium on diseases of fishes and shellfishes. Am. Fish. Soc., Spec. Publ. 5.

Wellings, S. R., and R. G. Chuinard. 1964. Epidermal papillomas with virus-like particles in flathead sole, *Hippoglossoides elassodon*. Science 146:932–934.

Wellings, S. R., R. G. Chuinard, R. T. Gourley, and R. A. Cooper. 1964. Epidermal papillomas in the flathead sole, *Hippoglossoides elassodon*, with notes on the occurrence of similar neoplasms in other pleuronectids. J. Natl. Cancer Inst. 33:991–1004.

Wellings, S. R., R. G. Chuinard, and M. Bens. 1965. A comparative study of skin neoplasms in four species of pleuronectid fishes. Ann. N.Y. Acad. Sci. 126:479–501.

Wellings, S. R., R. A. Cooper, and R. G. Chuinard. 1966. Skin tumors of pleuronectid fishes in Puget Sound, Washington. Bull. Wildl. Dis. Assoc. 2:66.

Wellings, S. R., B. B. McCain, and B. S. Miller. 1976. Epidermal papillomas in Pleuronectidae of Puget Sound, Washington. Prog. Exp. Tumor Res. 20:55–74.

Wellings, S. R., C. E. Alpers, B. B. McCain, and M. S. Myers. 1977. Fish disease in the Bering Sea. Ann. N.Y. Acad. Sci. 298:290–304.

Yamazaki, F., T. Hibino, K. Oishi, and T. Harada. 1978. Morphological changes in the X-cells in epidermal papillomas of the flatfish, *Limanda schrenki*. Bull. Jpn. Soc. Sci. Fish. 44:723– 727.

57 Puffer Viruslike Particles

Definition

Puffer viruslike particles are electron-dense cytoplasmic bodies found in degenerated cells of small necrotic foci of hematopoietic spleen tissue of *Fugu pardale*.

History

The source fish was an 80-g male caught off Iwate, Japan, in late 1981; it was held for 2 months before its tissues were fixed and stained for ultrastructural study of what was assumed to be normal hematopoietic tissue. Small necrotic foci, each consisting of a large macrophage and two or three cells showing pyknotic nuclei and degenerate cytoplasm were found. Macrophage granules and degenerate cell cytoplasm contained many viruslike particles. The diameter of the particles was 60 to 75 nm. Electron-dense cores were 35 to 45 nm and an outer envelope or array of projections was 10 to 15 nm thick. The particles had morphologic features of reovirus, orbivirus, or coronavirus (Watanabe and Yamamori 1983).

The nature and role of the particle is not known.

Reference

Watanabe, T., and K. Yamamori. 1983. Virus-like particles in the spleen hematopoietic tissue of puffer *Fugu pardale*. Bull. Jpn. Soc. Sci. Fish. 49:307.

58 Rainbow Trout Intraerythrocytic Virus

Definition

The rainbow trout intraerythrocytic virus is a regular cytoplasmic particle found in erythrocytes of *Salmo gairdneri* afflicted with a hemorrhagic condition. The subject trout are the inbred Donaldson strain at the University of Washington and, more specifically, are predominantly adult females.

History

The particles were detected by Landolt et al. (1977) during an investigation of sudden deaths among the captive brood stock. Constraints of funding and accessibility to specimens have precluded further work (M. Landolt, personal communication, 1985).

Signs and Pathologic Changes

Affected specimens show exsanguinating hemorrhage and evidence of hypoxia, and die suddenly. Massive subcapsular hematomas in the liver or free blood in the body cavity are found on necropsy. Histologic findings consist of diffuse necrotic hepatitis, eosinophilic cytoplasmic inclusions in renal tubule and liver cells, scattered areas of vascular coagulation, and dystrophy of skeletal muscle. Afflicted fish have abnormally high numbers of immature erythrocytes that frequently contain basophilic cytoplasmic inclusions about 1000 nm in diameter and spherical vesicles of about 900 nm. Electron microscopy has shown the cytoplasmic inclusions to be membrane-bound structures with an amorphous inner matrix but without virus. The vesicles, however, contain viruslike particles.

Etiology

The cause of the diseased condition is unknown, but in the absence of other recognizable pathogens and on the premise that the disease is contagious, a viral etiology is suspected.

The viruslike forms in the vesicles have a uniform diameter of 80 to 90 nm. Their electron-dense outer margins enclose a somewhat less electron-dense reticulated interior. Attempts at isolation in RTG-2 and FHM cells incubated at three temperatures have been unsuccessful.

The authors noted that the viruslike forms do not resemble the iridovirus of viral erythrocytic necrosis.

Host and Geographic Range

The sole report involved the Donaldson rainbow trout at the University of Washington, but eggs and progeny of these fish have been widely distributed to various parts of the world. Accordingly, the appearance of hemorrhagic disease and sudden mortality in that strain or in other rainbow trout stocks suggests that a renewed search for the viruslike forms might be fruitful.

Reference

Landolt, M. L., J. R. MacMillan, and M. Patterson. 1977. Detection of an intraerythrocytic virus in rainbow trout (*Salmo gairdneri*). Fish Health News 6:4–6.

59 Winter Flounder Papilloma

Definition

The winter flounder papilloma is a benign neoplasm of the skin of *Pseudopleuronectes americanus*. The abnormal cells of the neoplasm contain abundant cytoplasmic viruslike particles 30 nm in size that appear to consist of DNA.

History

The sole report of the lesion with viruslike particles is the well-illustrated publication of Emerson et al. (1985).

Signs and Patholgic Changes

The neoplasms were found on only 2 of more than 100 winter flounders. The lesions are distinct areas of raised tissue that occur only on pigmented skin and superficially resemble lymphocystis (Fig. 1). However, the authors carefully noted that the papillomas are "nonlymphocystis"—a fact confirmed by histologic sections that show extreme distention of intercellular spaces and no cellular hypertrophy (Fig. 2).

Cell nuclei are typically lobate and have a swollen envelope. Mitochondria are numerous, enlarged, and swollen. Cisternae of the endoplasmic reticulum are distended, and the cytoplasm shows many vacuoles and prominent aggregations of 30-nm electron-dense particles (Fig. 3).

Etiology

The particles are found only in cells of the lesion and not in normal epithelial cells. Some particles show polyhedral morphology, and treatment with DNAse

Figure 1. Specimen winter flounder with prominent papillomas. From Emerson et al. (1985). Reprinted with the permission of Blackwell Scientific Publications Ltd.

markedly reduces electron density. Reactivity for glycogen does not increase electron density. Claims for a viral nature are not adequately documented.

Diagnosis

Lesions are recognized by the lack of cell hypertrophy (lymphocystis) and histologic demonstration of the peculiar tissue structure and the presence of the 30-nm cytoplasmic particles.

Transmission and Incubation

Specimen flounders have been taken only in an estuary that receives high levels of hydrocarbon and sewage pollution. A causal relationship is inferred.

Host and Geographic Range

Winter flounders in Conception Bay, Newfoundland, Canada, are the only known victims.

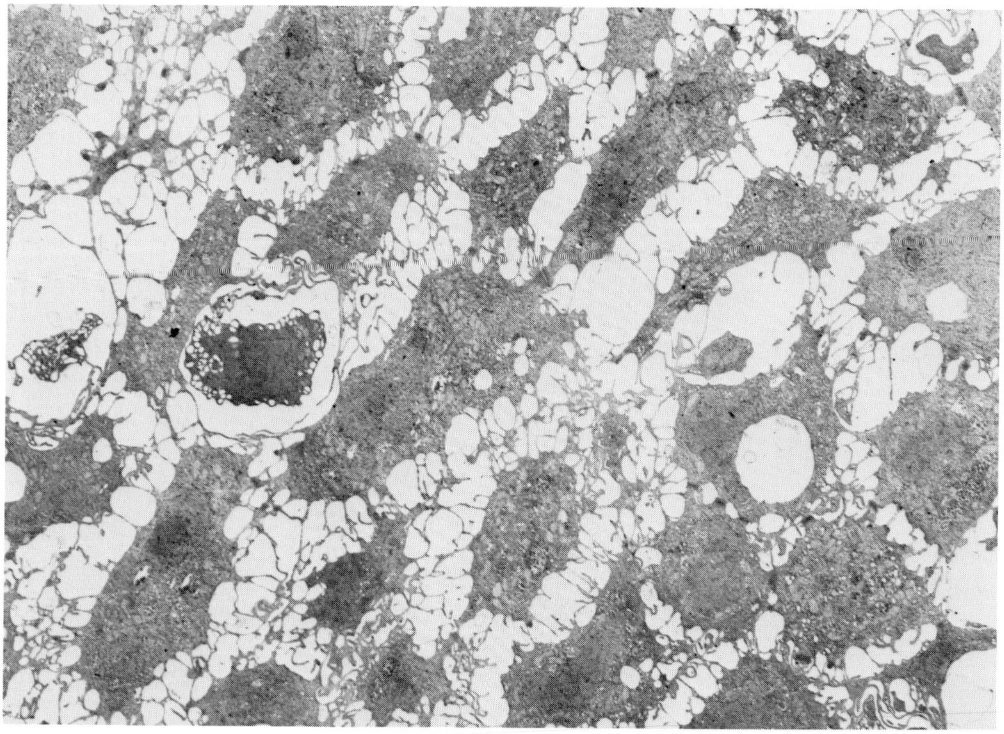

Figure 2. Histologic section of a lesion showing extreme separation of cells and enlarged intercellular space. From Emerson et al. (1985). reprinted with the permission of Blackwell Scientific Publications Ltd.

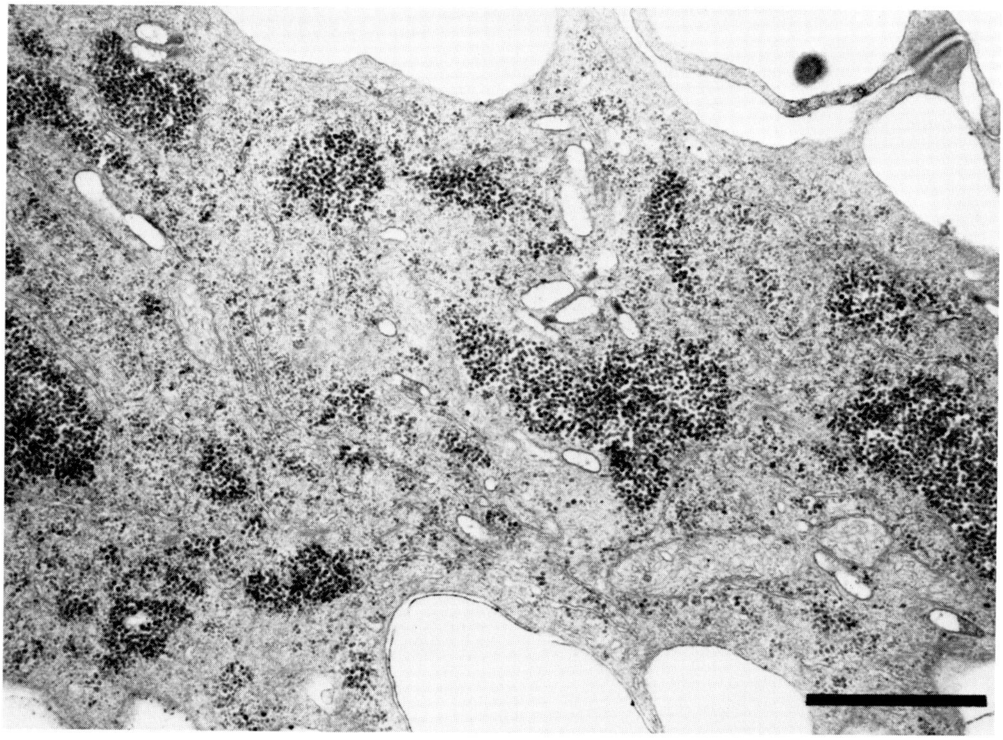

Figure 3. Electron micrograph showing cytoplasmic aggregates of 30-nm viruslike particles. From Emerson et al. (1985). Reprinted with the permission of Blackwell Scientific Publications Ltd.

Reference

Emerson, C. J., J. F. Payne, and A. K. Bal. 1985. Evidence for the presence of a viral non-lymphocystis type disease in winter flounder, *Pseudopleuronectes americanus* (Walbaum), from the north-west Atlantic. J. Fish Dis. 8:91–102.

PART V

CHLAMYDIAL INFECTIONS

60 Epitheliocystis

Definition

Epitheliocystis is a chronic and unique infection by a chlamydial organism that results in hypertrophied epithelial cells—typically of gills but sometimes also skin—of certain freshwater and marine fishes. Epitheliocystis cells alone are superficial and benign. Their presence in gills, however, can stimulate hyperplasia and fusion of adjacent lamellae. If the infection is extensive, respiration is impaired and death may result.

History

Much like the history of lymphocystis, the story of epitheliocystis opens with the discovery of the disease and an erroneous attribution of cause. In Germany, Plehn (1920) described and illustrated a newly recognized disease afflicting the gills of common carp (*Cyprinus carpio*). Multiple small "cysts," sometimes surrounded by proliferated epithelial cells, were described as present in gill epithelium. When the cysts were abundant, respiration was impaired and mortality resulted. The causal agent of the cysts was considered to be a unicellular alga that, because it stimulated mucus production, was named *Mucophilus cyprini*; accordingly, the disease bore the name mucophilosus. Plehn's 1924 text on fish diseases included an illustration and brief discussion of the disease that were later also included in Schäperclaus's (1954) book, but under the misnomer that was to be continued even in the 1979 revision of the book and into the European literature of the 1980s.

Meanwhile, in America, Hoffman found abundant small white "cysts" not only in the gills but also in the fins and skin of young bluegills (*Lepomis macrochirus*), believed them to be protozoan parasites, and termed the condition epitheliocystis. Dunbar then sectioned the tissues and found that the cysts were unusual hypertrophied epithelial cells. He believed that the uniformly granular intracellular material was viral. Material was sent to Zwillenberg, in Switzerland, for electron microscopy, where it was found that epitheliocystis cells uniformly contained myriads of minute coccuslike cells. The microorganisms were considered to be bedsonia and were so reported by Hoffman et al. (1969). Bedsonia are now termed chlamydia.

Epitheliocystis was next reported in gills of striped bass (*Morone saxatilis*) and white perch (*M. americanus*), anadromous North American species that were taken in estuarine waters off Connecticut (Wolke et al. 1970). In a comparison of epitheliocystis and lymphocystis cells, Wolf (1974) described and illustrated key features shared by both kinds of hypertrophied cells and other features that were unique to each. The known range of epitheliocystis was extended when it was found in striped bass in the Chesapeake Bay (Paperna and Zwerner 1976); Zachary and Paperna (1977) published a detailed light and electron microscopic study of the disease in striped bass from that source. The epitheliocystis cells contained a microorganism with both coccoid and rod forms that was considered to be rickettsia-like. From Israel, Paperna (1977) reported epitheliocystis again in gills of two marine species—the gray mullet (*Liza ramada*) and the sea bream (*Sparus aurata*). Severe infections among young hatchery-propagated sea bream were described as resulting in mortality—the first indication that epitheliocystis could be a serious disease. Electron microscopy of material from the two Israeli fishes temporarily presented the possibility that epitheliocystis cells might in some disease cases be individual hypertrophied cells and in other cases result from fusion, and also that two kinds of organisms might be involved (Paperna et al. 1978).

Epitheliocystis was next reported in France and in the gills of another marine species—the sea bass, *Dicentrarchus labrax* (Paperna and Baudin-Laurencin 1979).

As a contribution to the third session of the Cooperative Program of Research on Aquaculture (COPRAQ) in Munich, Paperna and Sabnai (1980) reviewed the disease and added new species to the list of known hosts. They suggested that the several morphologic forms of the organism in cells of the sea bream might be successive developmental stages such as are known in chlamydia. They did not, however, rule out a possible rickettsial nature.

The year 1981 was particularly significant because it saw the unification of the older and relatively obscure disease termed mucophilosus with the newer condition termed epitheliocystis. Clarification came from Hungary, where Molnar and Boros (1981) noted the striking similarity of epitheliocystis with mucophilosus cysts and undertook studies with light and electron microscopy that unequivocally showed chlamydial cells in the carp cysts. Their review of the literature was enlightening, for it revealed that the European reports on mucophilosus "generally centered on the prevalence, aetiology, and possible control . . . without closer investigations into the systematic position and biological properties of *M. cyprini* itself." (Molnar and Boros, however, chose to retain the less descriptive term mucophilosus in the title of their report.)

Paperna et al. (1981) described the apparent developmental cycle of the epitheliocystis agent in a representative of each of four families of fishes. They noted and illustrated similarities and differences, and concluded that a pleomorphic development cycle was characteristic of epitheliocystis, but that the organism was specific for each fish family. Opinion favoring a chlamydial nature was growing, and Wolf (1981) noted in a review that rickettsiae were transmitted in terrestrial animals by hematophagous arachnoids and insects—invertebrate vectors that are rare in aquatic environments.

Awareness of epitheliocystis was growing: Morrison and Shum (1983) found it in a marine flatfish, Rourke et al. (1984) considered it to be a cause of mortality in a salmonid, and Zimmer et al. (1984) reported a high incidence of benign infection in an ictalurid. In a report on epitheliocystis in common carp, Paperna and Alves de Matos (1984) described the apparent cycle of development and confirmed the earlier claim that the disease was what had been reported as mucophilosus.

The interesting story of epitheliocystis is still incomplete: yet to be added are reports of isolation and graphs of growth cycles and developmental forms as they occur in vivo or in vitro in studies of known-age material.

Signs and Pathologic Changes

Two degrees of severity are recognized—from benign light cases to heavy infections involving skin epithelium. Light infection can also occur in gill epithelium, but heavy infection of gills sometimes evokes a proliferative condition that greatly reduces the respiratory area and is accompanied by mortality.

Epitheliocystis can be confused with lymphocystis, but that infection typically results in a larger cell and seldom if ever involves the gills. Also, lymphocystis does not occur in salmonids, ictalurids, or cyprinids, all of which are susceptible to epitheliocystis. Heavy infection of skin has been confused with infestation by the flagellated protozoan parasite *Ichtyobodo necatrix*.

Behavior

Lightly infected fish behave normally. Terminally ill fish with the proliferative form are lethargic, their gill covers are flared, and respiration is rapid.

External Signs

Light infections can be wholly inapparent to the unaided eye, whereas heavy infections can be dramatically obvious (Fig. 1A). If infected gills or skin tissues are examined closely, but preferably with some degree of magnification, epitheliocystis cells—termed "cysts" by some authors—appear as small white ovals or spheres (Fig. 1B).

The size of epitheliocystis cells ranges from several tens of micrometers in young cells to several hundred micrometers in mature ones. Brownian movement is evident in wet mounts of epitheliocystis cells; when the cells are ruptured, the outpouring cell contents appear as a swarmlike mass of minute coccoid forms.

Internal Signs

Specific internal pathologic changes have not been reported but, perhaps as in occasional cases of lymphocystis, an involvement of internal organs may eventually be found.

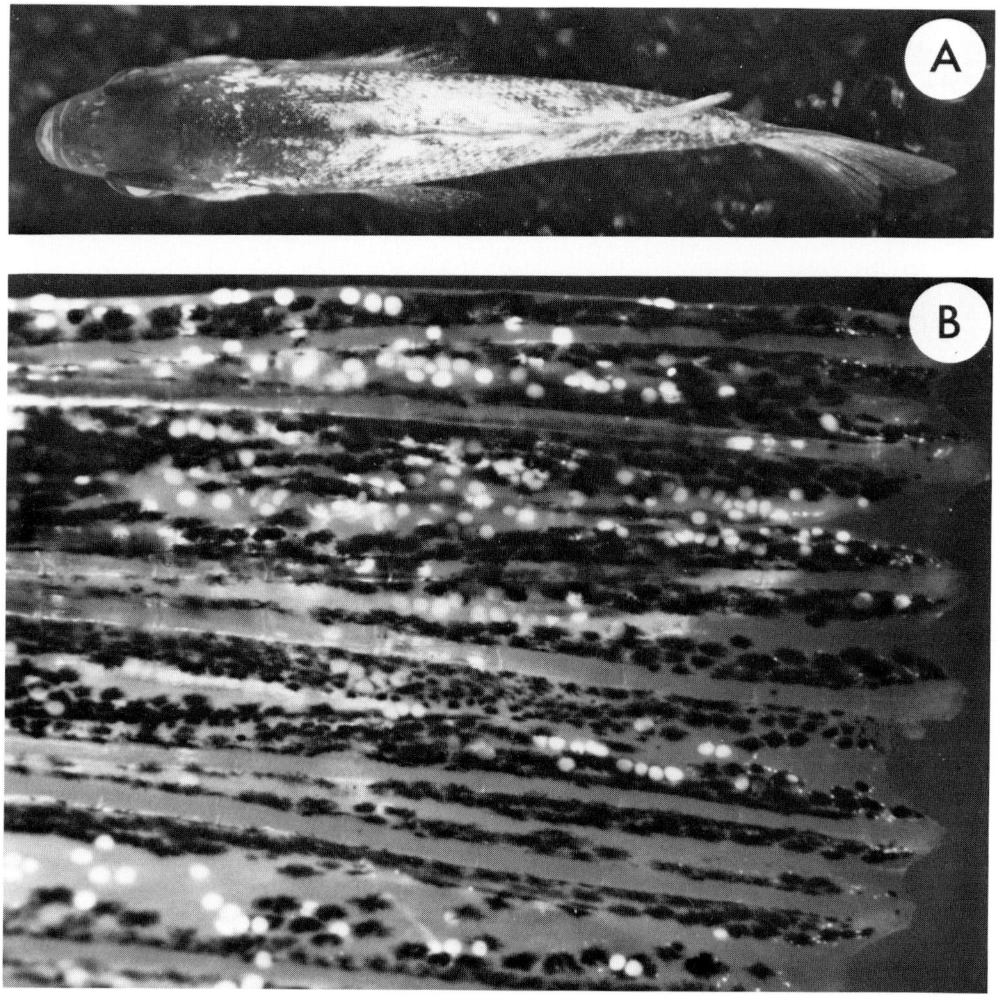

Figure 1. A. Heavy but benign epitheliocystis on the skin of a small centrarchid. Individual infected cells, termed "cysts" by some authors, are barely visible. *B.* Under low magnification individual epitheliocystis cells appear as round to ovate white bodies. From Wolf (1980). Reprinted with the permission of The Written Word.

Histopathologic Findings

Molnar and Boros (1981) described and illustrated sequential changes in infected cells of carp gills. The youngest cells (10 to 15 μm in diameter) have a small centrally located inclusion surrounded by a considerable volume of foamy cytoplasm. As the cell enlarges (20 to 30 μm in diameter) the inclusion enlarges, chlamydia are evident, and the foamy cytoplasm contracts. The cytoplasm stains red with PAS, and electron micrographs show the inclusion to be surrounded by mucine droplets 1 to 3 μm in diameter. Mature cells (60 to 70 μm in diameter) retain only a thin rim of

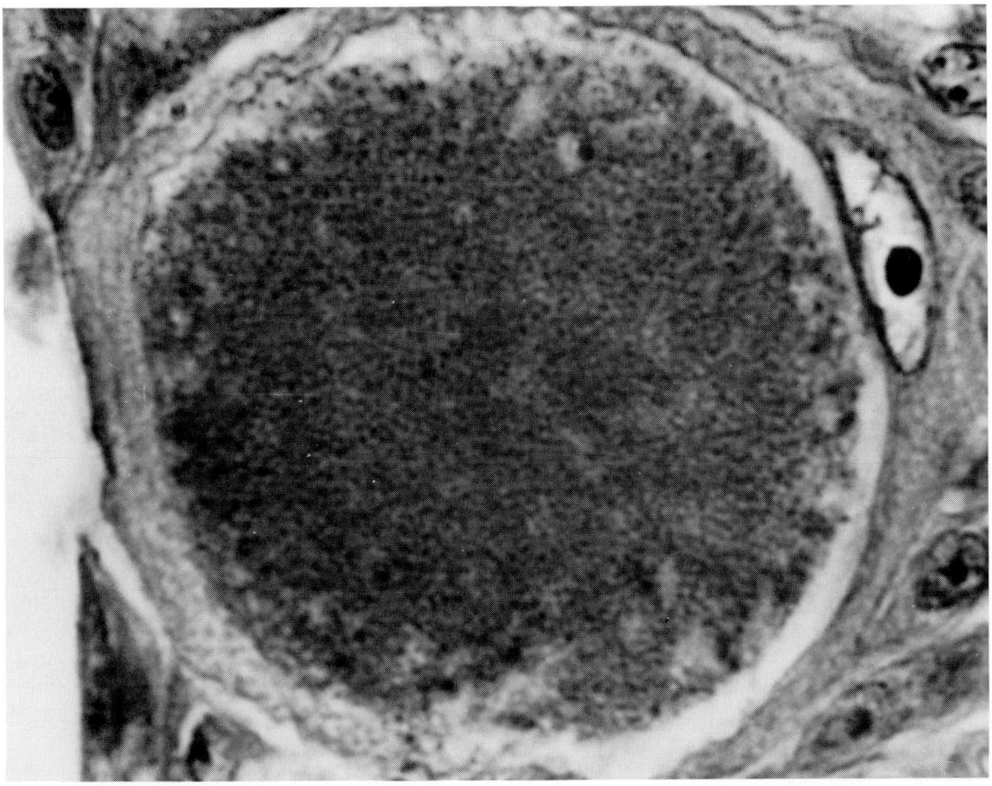

Figure 2. Histologic section of a mature epitheliocystis cell. The much enlarged nucleus (upper right) is within a thin rim of cytoplasm. The great homogeneously granular central mass consists of uniformly sized chlamydial cells. From Hoffman et al. (1969). Reprinted with the permission of Swets Publishing Services.

cytoplasm; the internal inclusion with its contained mass of chlamydia constitutes the bulk of the cell's volume (Fig. 2).

At maturity, epitheliocystis cells are obviously enlarged and surrounded by a thick hyaline capsule (Fig. 2). The cytoplasmic space contains the large basophilic membrane-bound inclusion that may show either uniform granularity or a hyaline condition with or without vacuolation. Such variation is probably due to cell age or to fixation methods. Sections that include the nucleus show both it and the nucleolus to be enlarged several-fold. The nucleus is typically located eccentrically and retains a normal intact configuration (Fig. 3). (In contrast, lymphocystis cell nuclei are typically located centrally and have grossly irregular outlines.)

Epitheliocystis cells are often surrounded by one or more layers of normal epithelial cells showing some evidence of proliferation. The hypertrophied cells themselves do not divide. Proliferation of surrounding epithelium is particularly marked in cases where the disease occurs in gills (Fig. 4). Paperna (1977) noted that in advanced cases in the sea bream, the proliferated tissue was infiltrated by macrophages and eosinophils and the capillaries were sometimes blocked.

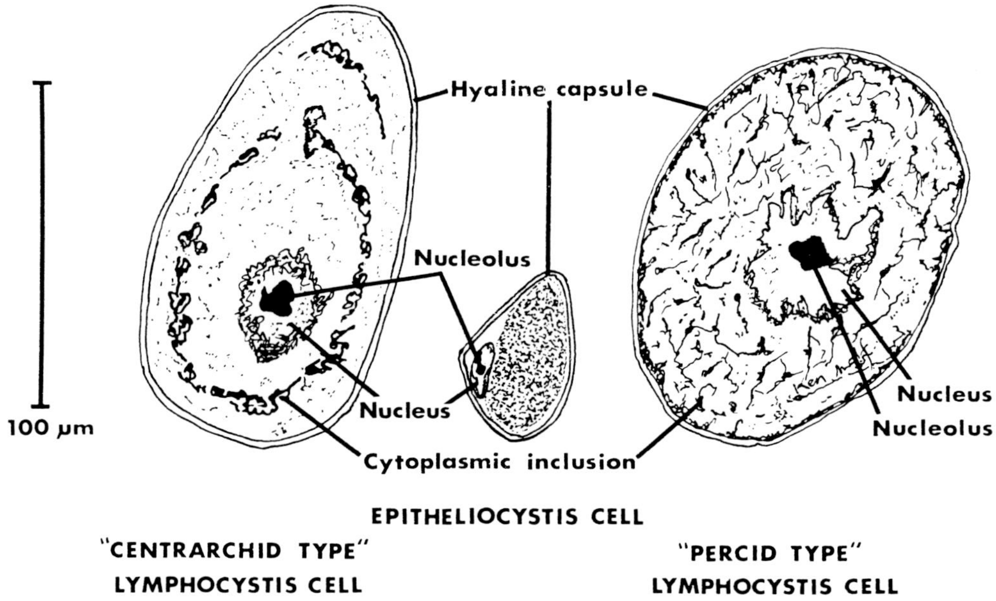

Figure 3. Comparative size and morphologic features of a representative epitheliocystis cell of chlamydial etiology (center) and two kinds of lymphocystis cells of viral etiology (left and right). Note the eccentric location of the nucleus in the epitheliocystis cell and the more centrally located nuclei of the lymphocystis cells; note also that the cytoplasmic inclusion substance of the epitheliocystis cell is coherently granular, whereas that of the lymphocystis cells is disseminated in a cordlike or dendritic array.

Some histochemical determinations were reported by Wolke et al. (1970). The cell inclusion is bordered by the hyaline lamina, which stains acidophilic. The cytoplasmic or inclusion granules give a positive reaction with Macchiavello's stain—a feature that is presumptively diagnostic for chlamydia.

Etiology

The cause of epitheliocystis disease has not been formally proven. However, chlamydial cells are universally present in the pathologically hypertrophied cells and are tacitly accepted to be causal.

Size and Shape

Considering the diversity of hosts and the fact that extant studies have shown natural infections of undetermined stages of development, the size and shape of the epitheliocystis cell is perhaps best described as being pleomorphic.

Ultrastructural studies of the epitheliocystis chlamydia have revealed forms ranging from elongated hyphal-like cells to small round or ovate forms and even to cells that are tailed and look somewhat like a tadpole. Names have been given to the various forms of cells and interpretations made of the apparent sequence or cycle of development (Paperna and Alves de Matos 1984; Paperna et al. 1978, 1981).

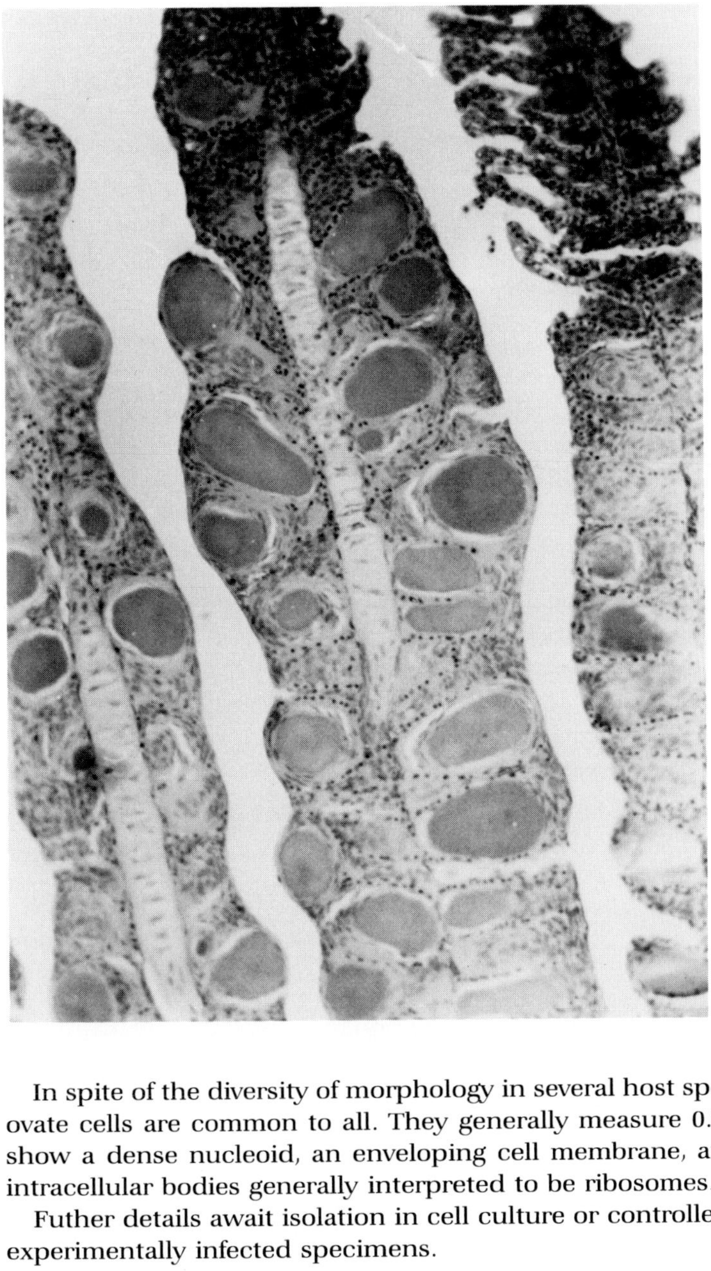

Figure 4. Section through infected gills showing proliferative condition that resulted in fusion of adjacent lamellae and a consequent reduction in the respiratory exchange surface. From Paperna (1977). Reprinted with the permission of Elsevier Science Publishers.

In spite of the diversity of morphology in several host species, the small round to ovate cells are common to all. They generally measure 0.3–0.5 × 0.5–1.0 μm and show a dense nucleoid, an enveloping cell membrane, and (in some specimens) intracellular bodies generally interpreted to be ribosomes.

Futher details await isolation in cell culture or controlled time-course studies of experimentally infected specimens.

Biophysical Properties

Until the agents are isolated and tested, nothing can be said with certainty except that chlamydia are sensitive to penicillin, chloramphenicol, and the tetracycline antibiotics. That sensitivity should be considered when in vitro isolation is attempted.

Diagnosis

Epitheliocystis may be diagnosed by the finding of numerous small white spherical forms in skin or gills. Wet mounts of such material shows the thick capsule and the uniformly granular contents. Histologic examination yields great accuracy, but a definitive diagnosis is possible now only after electron microscopy has shown the presence of the intracellular forms described and illustrated herein.

Isolation

As reported by Hoffman et al. (1969), isolation might have been achieved in BF-2 cell cultures, but the agent was not successfully transferred. D. S. Wyand (personal communication) unsuccessfully attempted isolation in embryonate fowl eggs and neonatal mice. Wolf (1981) used BF-2 cells that were irradiated under a cobalt-60 source and then subcultured to remove dead cells. Inocula were prepared from a clinical case of epitheliocystis in a bluegill and centrifuged at 1300 × g onto the irradiated BF-2 cells. Although the method is close to the standard procedures used to isolate chlamydia, the work was not successful in isolating the fish agent. Perhaps the fibroblastic morphology of BF-2 cells was the limiting factor.

Identification

At present, the only method of identification of the epitheliocystis chlamydia is by electron microscopy.

Transmission and Incubation

Epitheliocystis apparently is easily transmitted horizontally, but the usual portal of entry has yet to be identified. Paperna (1977) provided some evidence for transmission by nets and other fish culture equipment.

G. L. Hoffman (personal communication) effected experimental transmission to bluegills by placing healthy fish in an aquarium with gills that had been removed from infected specimens. The temperature was 20 to 26°C and infections were established by the 25th day (earlier examinations were not made). Unexposed control fish remained healthy.

When D. S. Wyand (personal communication) added a crude emulsion of gill material from *Morone* to a tank of water with goldfish (*Carassius auratus*), several of the goldfish developed epitheliocystis gill lesions in 3 to 4 weeks, and the condition was confirmed histologically. Again, control fish remained free of the lesions.

Zachary and Paperna (1977) mentioned using a specimen of striped bass that was infected in the laboratory—presumably by cohabitation—but details were not provided.

Source

The biology of chlamydia among homeotherms is such that it seems unquestionable that the epitheliocystis agent is exclusively a pathogen of fish, although invertebrate vectors or transport hosts might conceivably be involved.

Host and Geographic Range

The infection is known among fishes in eastern North America, Eastern and Western Europe, South Africa, and the Middle East. P. Cheung (personal communication) discovered the condition at the New York Aquarium among tropical marine species imported from the Phillipines. Collectively, susceptible fishes are divided among 11 families: Centrarchidae, Chaetodontidae, Cichlidae, Cyprinidae, Hippoglossidae, Ictaluridae, Moronidae, Mullidae, Salmonidae, Sparidae, and Zanclidae. Infections occur in marine and in freshwater environments and in coldwater as well as tropical species.

Control

If possible, avoidance should be practiced where susceptible species are to be propagated under conditions of husbandry.

Paperna (1977) implied that a measure of control was achieved under hatchery conditions when young sea bream suspected of having epitheliocystis were fed for 3 days with ground liver that had been treated with 1% chloramphenicol. Although the response of untreated fish was not mentioned, the results are consistent with the known sensitivity of chlamydia to antibiotics.

References

Hoffman, G. L., C. E. Dunbar, K. Wolf, and L. O. Zwillenberg. 1969. Epitheliocystis, a new infectious disease of bluegill (*Lepomis macrochirus*). Antonie Van Leeuwenhoek J. Microbiol. Serol. 35:146–158.

Molnar, K., and G. Boros. 1981. A light and electron microscopic study of the agent of carp mucophilosus. J. Fish Dis. 4:325–334.

Morrison, C., and G. Shum. 1983. Epitheliocystis in American plaice *Hippoglossoides platessoides* (Fabricius). J. Fish Dis. 6:303–308.

Paperna, I. 1977. Epitheliocystis infection in wild and cultured sea bream (*Sparus aurata*, Sparidae) and grey mullets (*Liza ramada*, Mugilidae). Aquaculture 10:169–176.

Paperna, I., and A. P. Alves de Matos. 1984. The developmental cycle of epitheliocystis in carp, *Cyrpinus carpio* L. J. Fish Dis. 7:137–147.

Paperna, I., and F. Baudin-Laurencin. 1979. Parasitic infections of sea bass, *Dicentrarchus labrax*, and gilt head sea bream, *Sparus aurata*, in mariculture facilitites in France. Aquaculture 16:173–175.

Paperna, I., and I. Sabnai. 1980. Epitheliocystis disease in fishes. Pages 228–234 in W. Ahne, ed. Fish diseases. Springer-Verlag, Berlin.

Paperna, I., and D. E. Zwerner. 1976. Parasites and diseases of striped bass, *Morone saxatilis* (Walbaum), from the lower Chesapeake Bay. J. Fish Biol. 9:267–287.

Paperna, I., I. Sabnai, and M. Castel. 1978. Ultrastructural study of epitheliocystis organisms from gill epithelium of the fish *Sparus aurata* (L.) and *Liza ramada* (Risso) and their relation to the host cell. J. Fish Dis. 1:181–189.

Paperna, I., I. Sabnai, and A. Zachary. 1981. Ultrastructural studies in piscine epitheliocystis: evidence for a pleomorphic development cycle. J. Fish Dis. 4:459–472.

Plehn, M. 1920. Neue Parasiten in Haut und Kiemen von Fischen. *Ichthyochytrium* und *Mucophilus*. Zentralbl. Bakteriol. Parasitenkd. I Orig. 85:275–281.

Plehn, M. 1924. Praktikum der Fischkrankheiten. E. Schweizerbart'sche Verlag, Stuttgart.

Rourke, A. W., R. W. Davis, and T. M. Bradley. 1984. A light and electron microscopic study of epitheliocysits in juvenile steelhead trout, *Salmo gairdneri* Richardson. J. Fish Dis. 7:301–309.

Schäperclaus, W. 1954. Fischkrankheiten. Akademie-Verlag, Berlin.

Wolf, K. 1974. Epitheliocystis and lymphocystis—a comparison of hypertrophied fish cells. Fish Health News 3:13–15.

Wolf, K. 1980. Virology of aquarium fishes, a review—1980. J. Aquaric. 1:58–62.

Wolf, K. 1981. Chlamydia and rickettsia of fish. Fish Health News 10(3):i–v.

Wolke, R. E., D. S. Wyand, and L. H. Khairallah. 1970. A light and electron microscopic study of epitheliocystis disease in the gills of Connecticut striped bass (*Morone saxatilis*) and white perch (*Morone americanus*). J. Comp. Pathol. 80:559–563.

Zachary, A., and I. Paperna. 1977. Epitheliocystis disease in the striped bass *Morone saxatilis* from the Chesapeake Bay. Can. J. Microbiol. 23:1404–1414.

Zimmer. M. A., M. S. Ewing, and K. M. Kocan. 1984. Epitheliocystis disease in the channel catfish, *Ictalurus punctatus* (Rafinesque). J. Fish Dis. 7:407–410.

PART VI

NONVIRAL CONDITIONS, AGENTS, AND ARTIFACTS

61 Eel Stomatopapilloma

Synonyms: Blumenkohlkrankheit, cauliflower disease, orocutaneous papillomatosis, papillomatosis

Definition

Stomatopapilloma or cauliflower disease of eels is a villous to papular, chronic, and benign epidermal neoplasm that develops preferentially about the mouth, nares, and head of *Anguilla anguilla* (Fig. 1) in some tributaries and brackish waters of the Baltic and North seas. The tumors have yet to be transmitted experimentally and their cause has yet to be recognized, but virus has long been suggested as being the etiologic agent. Indeed, viruses have been isolated from tumor-bearing eels, but none have been demonstrably oncogenic. The possible role of virus is now highly questionable.

The first virus to be isolated from a tumor-bearing eel was assumed to have a causal role in tumorigenesis. Preliminary results with that agent, named EV (Berlin), indicated that it belonged to the papovavirus group; later work, however, clearly showed it to be a birnavirus. Also, other isolants having the size, shape, and major features of EV (Berlin) have been reported from tumor-bearing and, equally important, from wholly normal-looking eels. Those isolants proved to be simply strains of the cosmopolitan infectious pancreatic necrosis virus (IPNV). Accordingly, it is a considered judgment that eel stomatopapilloma should no longer be regarded as a viral disease. The rationale for that decision is supported by highlights of the relevant history.

History

The peculiar eel tumors were noted in the Baltic as early as 1910 (Koops et al. 1970), but it remained for Christiansen and Jensen (1950) to provide an illustrated scientific report of the condition, in which the tumors were described as cauli-

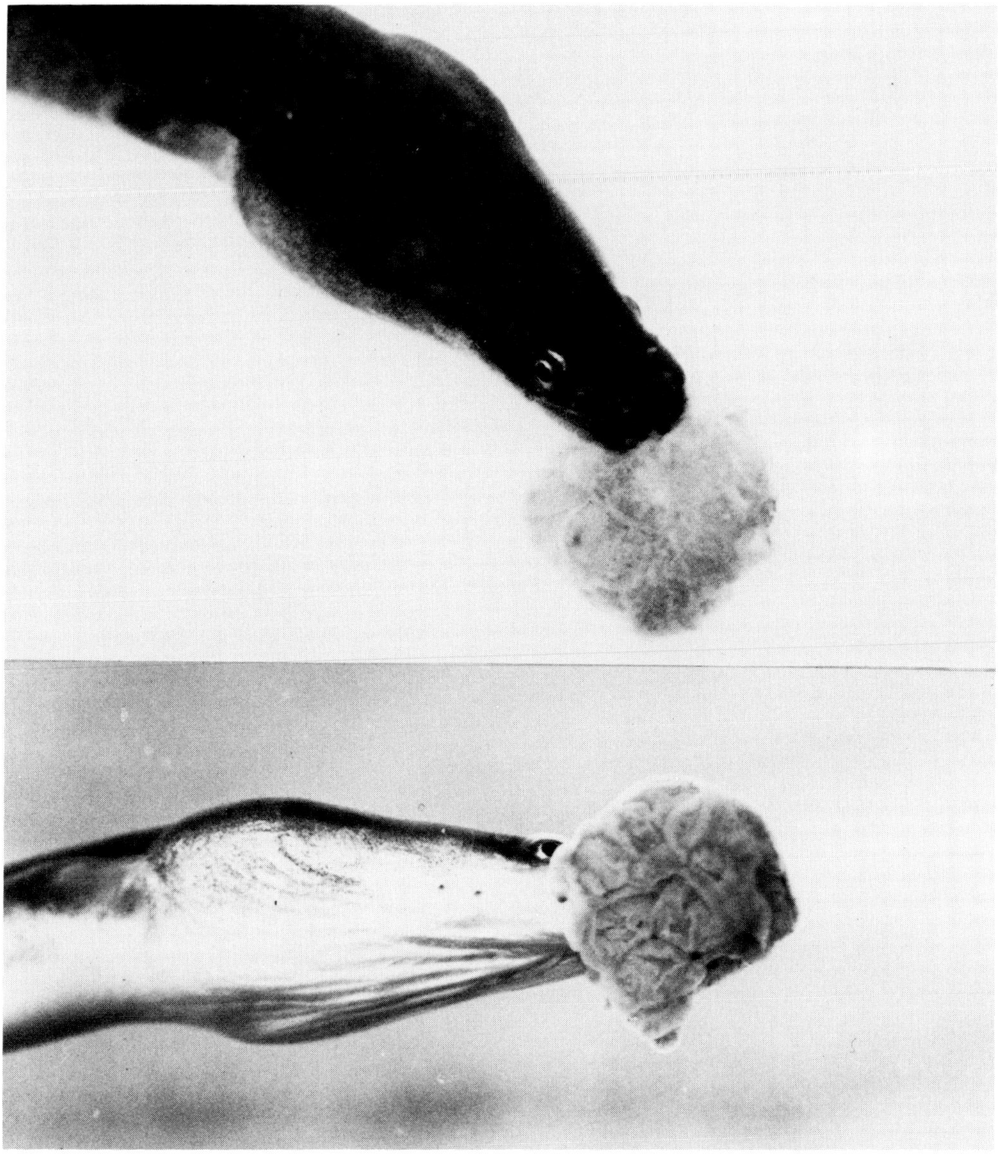

Figure 1. Dorsal and ventral views of a specimen *Anguilla anguilla* with stomatopapilloma.

flower-like. Soon afterward, Schäperclaus (1953) published his account on Blumenkohlkrankheit (cauliflower disease) of eels, and suggested that virus might be involved. Lühmann and Mann (1956) reviewed earlier reports and presented their findings on physical characteristics of normal and tumorous specimens. Ten years later, that work was reviewed in an English-language account by Koops and Mann (1966). Early on, the eel tumors were included in reviews of fish viruses (Wolf 1966).

The presence of any virus in tumorous eels was reported first by Pfitzner and

Schubert (1969), who investigated eels from the Hamburg area and isolated a 55-nm polyhedral particle that occurred in cytoplasmic vesicles of infected cell cultures. A special supplement to Volume 20 of the *Archiv für Fischereiwissenschaft* consisted of four relevant papers: Pfitzner (1969) gave more details of the virus—particularly its effects on cell cultures, but also noted that culture-grown virus had no visible effect on healthy eels; Schubert (1969) described and illustrated findings by electron microscopy; Koops and Mann (1969) reviewed the occurrence and distribution of the tumors; and Schmid (1969) discussed the histology and etiology, and concluded that virus was probably involved. On a different theme, Peters and Peters (1970) began continuing investigations of environmental and metabolic factors of tumor development in the Hamburg area.

Koops et al. (1970), in a brief review of the subject of tumors in eels, maintained that virus was the probable cause, but conceded that it was possible that virus played no role in tumor development. That possibility was also evident in a report by Schwanz-Pfitzner (1976), who was unable to induce tumors with culture-grown virus. Schwanz-Pfitzner et al. (1984) showed that EV (Berlin) was a bisegmented double-stranded RNA agent and so belonged to the birnavirus group. They claimed that, although EV (Berlin) was a birnavirus, it differed from IPNV. However, critical serologic and electrophoretic comparisons with known IPNV were not made; instead, EV (Berlin) was compared with, and found to be different from, reovirus types 1 and 3.

Ahne and Thomson (1985), working with tumorous eels from Hamburg, isolated three different viruses from a single specimen: a rhabdovirus, a syncytium-inducing virus, and a birnavirus that was identified serologically as IPNV strain Ab.

Eel virus (Berlin) appears to be simply an IPNV strain Ab virus that has been found in eels that happen to have stomatopapilloma. The virus is not demonstrably oncogenic, and strain Ab has been isolated from normal-looking *A. anguilla* taken from waters elsewhere.

References

Ahne, W., and I. Thomson. 1985. The existence of three different viral agents in a tumor bearing European eel (*Anguilla anguilla*). Zentralbl. Veterinärmed. [B] 32:228–235.

Christiansen, M., and A. J. C. Jensen. 1950. On a recent and frequently occurring tumor disease in eel. Rep. Dan. Biol. Stn. 1947 50:31–44.

Koops, H., and H. Mann. 1966. The cauliflower disease of eels in Germany. Bull. Off. Int. Epizoot. 65:991–998.

Koops, H., and H. Mann. 1969. Die Blumenkohlkrankheit der Aale: Vorkommen und Verbreitung der Krankheit. Arch. Fischereiwiss. 20(Suppl. 1):5–15.

Koops, H., H. Mann, I. Pfitzner, O. J. Schmid, and G. Schubert. 1970. The cauliflower disease of eels. Pages 291–295 in S. F. Snieszko, ed. A symposium on diseases of fishes and shellfishes. Am. Fish. Soc., Spec. Publ. 5.

Lühmann, M., and H. Mann. 1956. Beobachtungen über die Blumenkohlkrankheit der Aale. Arch. Fischereiwiss. 7:229–230.

Peters, N., and G. Peters. 1970. Tumorgenese, ein Energieproblem der Zelle? Untersuchungen an Papillomen des europäischen Aals, *Anguilla anguilla* (L.). Arch. Fischereiwiss. 20:238–257.

Pfitzner, I. 1969. Zur Aetiologie der Blumenkohlkrankheit der Aale. Arch. Fischereiwiss. 20(Suppl. 1):24–35.

Pfitzner, I., and G. Schubert. 1969. Ein Virus aus dem Blut mit Blumenkohlkrankheit behafteter Aale. Z. Naturforsch. 24b:790.

Schäperclaus, W. 1953. Die Blumenkohlkrankheit der Aale und anderer Fische der Ostsee. Z. Fisch. Hilfswiss. 2:105–124.

Schmid, O. J. 1969. Beitrag zur Histologie und Ätiologie der Blumenkhohlkrankheit der Aale. Arch. Fischereiwiss. 20(Suppl. 1):16–23.

Schubert, G. 1969. Elektronenmikroskopische Untersuchungen an der Haut mit Blumenkohlkrankheit behafteter Aale. Arch. Fischereiwiss. 20(Suppl. 1):36–49.

Schwanz-Pfitzner, I. 1976. Further studies of eel virus (Berlin) isolated from the blood of eels (*Anguilla anguilla*) with skin papilloma. Prog. Exp. Tumor Res. 20:101–107.

Schwanz-Pfitzner, I., M. Özel, G. Darai, and H. Gelderblom. 1984. Morphogenesis and fine structure of eel virus (Berlin), a member of the proposed birnavirus group. Arch. Virol. 81:151–162.

Wolf, K. 1966. The fish viruses. Adv. Virus Res. 12:35–101.

62 Grunt Fin Agent

Definition

The grunt fin agent (GFA) is a microbial entity of undetermined nature that appeared unexpectedly after many passages of a cell line that originated from fins of a marine fish. Although the GFA has some attributes of virus, it has not been convincingly visualized, nor has it been shown to infect or to evoke disease in the source fish. Earlier claims of a viral nature have yet to be substantiated.

History

A cell line, designated GF, was initiated from fins of the bluestriped grunt (*Haemulon sciurus*) by Clem et al. (1961). Two years later, one culture in 65th passage showed a focus of necrosis that eventually enveloped the entire cell sheet and proved to be serially transferable. The causal agent was considered to be an orphan virus (Clem et al. 1965). Beasley et al. (1966), at the same laboratory, reported the establishment of a subpopulation of GF cells that was persistently infected with GFA. In a review of interferon production in poikilothermic vertebrates, Beasley and Sigel (1968) reported that ability in GF cells that carried GFA. No later reports on GFA have appeared.

The agent is available from the American Type Culture Collection as VR-683, but requests are infrequent—e.g., only one during 1972–1975, none in 1983, and two in 1984 (D. Stevens, personal communication).

Signs and Pathologic Changes

As yet, GFA has no known role in fish disease. Adult grunts inoculated with the agent and held at 25 to 28°C for 1 month showed no signs of disease. In fact, assays showed that virus was not recoverable after the second day. Suckling mice and chick embryos were similarly unresponsive, and repeated injection of GFA into margate (*Haemulon album*) did not produce detectable neutralizing antibody (Clem et al. 1965).

Size and Shape

Electron micrographs of pelleted material show a few oval particles that measure 120 to 140 nm and have been termed viruslike. However, filtration has shown that 0.45-μm membranes trap 99% or more of the infectivity. Micrography of thin sections has not been enlightening.

Biophysical Characteristics

The agent is ether sensitive. It does not hemagglutinate, and its replication is not inhibited by halogenated uridine. Replication in GF cells at 20°C results in CPE and a yield of about 10^6 $TCID_{50}$/mL. Comparable CPE and yield are obtained in CAR cells at 20 and 25°C. The RTG-2, KB, and HeLa lines are refractory. Infectivity is not reduced by three cycles of freezing and thawing; at −70°C infectivity persists for at least 4 months, but is completely destroyed in 15 minutes at 45°C.

Cytopathic effect has been illustrated by pyknosis, with only slight involvement of the nucleus. There was no evidence of cytoplasmic or intranuclear inclusions (Clem et al. 1965).

Grunt fin cells infected with GFA yield survivors that are serially cultivable. Designated G1A, the carrier line continued to yield about $10^{4.5}$ $TCID_{50}$/mL over a period of 2 years and 125 transfers. A culture of the G1A line challenged with infectious pancreatic necrosis virus (IPNV) yielded cells that were doubly infected with GFA and IPNV (Beasley et al. 1966).

Identification

Identification could be made with rabbit antiserum. Clem et al. (1965) reported neutralizing titers of 1 : 125 (against 100 $TCID_{50}$) within a period as short as 16 days.

Source

By reason of its initial appearance in a fish cell line, its temperature of replication, and its ability to grow only in certain fish cells and not in the cells of higher vertebrates, Clem et al. (1965) favored the hypothesis that GFA had its origin in fish and that it was carried in a latent condition. They did not rule out, however, the possibility of its being a chance laboratory contaminant. They reported that they had inoculated cell cultures with tissue extracts, especially intestinal, from more than 400 grunts and had found no evidence of virus.

Host and Geographic Range

The agent is known only by its association with the GF cell line. The fish originally came from south Florida waters.

Immunity

Grunt fin agent proved to be weakly antigenic in grunts and other fishes. If prior exposure was a factor in the failure to evoke disease, one could correctly expect GFA to stimulate an amnestic reaction in the grunt; clearly, it did not.

References

Beasley, A. R., and M. M. Sigel. 1968. Interferon production in cold-blooded vertebrates. In Vitro 3:154–165.

Beasley, A. R., M. M. Sigel, and L. W. Clem. 1966. Latent infection in marine fish tissue cultures. Proc. Soc. Exp. Biol. Med. 121:1169–1174.

Clem, L. W., L. Moewus, and M. M. Sigel. 1961. Studies with cells from marine fish in tissue culture. Proc. Soc. Exp. Biol. Med. 108:762–766.

Clem, L. W., M. M. Sigel, and R. R. Friis. 1965. An orphan virus isolated in marine fish cell tissue culture. Ann. N.Y. Acad. Sci. 126:343–361.

63 Kidney Tumor Virus

Definition

The kidney tumor virus is the name applied to the postulated etiologic agent of transmissible malignant tumors of tropical aquarium fish. Although it is not possible to reexamine the original material, there is a considerable likelihood that incomplete pathologic examination and less than critical interpretation of experimental data led to an erroneous conclusion of viral etiology for what were probably mycobacterial granulomas.

History

In Germany in 1956, a specimen of the characin *Pristella riddlei* held in an aquarium developed tumorlike growths in the body cavity and musculature. The condition was contagious and spread to other specimens. Investigation to determine the cause was undertaken, but since the application of fish tissue culture to virologic examination was then in its infancy, research on the tumors was carried out solely with susceptible fish as the assay system (Wessing 1959; Wessing and von Bargen 1959).

Wessing and von Bargen (1959) transmitted the tumors with cell-free high-speed centrifugates of tumor homogenate administered in food or by intramuscular injection. Infectivity of such preparations was inactivated by drying and in 50% glycerol. Four of nine species of tropical aquarium fish proved susceptible to experimental infection, but their very young fry were resistant. Induction time was generally chronic and 3 to 16 weeks were required for injections to result in mortality. Kidneys were the primary target organs, and there was evidence for infiltration and metastases.

All experimental fish were destroyed accidentally, and a diligent search of *Citation Index* failed to show additional reports by the authors.

At Lehigh University, Pennsylvania, Beckwith and Malsberger (1980) reopened the search for the presumed causal virus; in response to an advertisement, they received a specimen of *Brachydanio rerio* with an abnormal mass protruding from its abdomen. Histologic sections showed multiple tumors that were "compatible with the description of Wessing and von Bargen."

The initial work with cell-free extracts of tumors from *B. rerio* gave evidence of seeming infectiousness for *B. albolineatus*. Again, the experimentally transmitted tumors were found to be compatible with those described in Germany.

The tumors consisted of distinct granulomatous masses of mononuclear epithelioid cells, which showed occasional mitoses, surrounded by connective tissue. Tumor size ranged from 20 to nearly 400 μm, and the centers of the more advanced lesions were necrotic. Kidneys and, secondarily, submucosal connective tissue of the intestine were most frequently involved.

In contrast, however, Wessing and von Bargen had found *B. albolineatus* to be resistant to their agent. Of greater significance was the fact that the work in Pennsylvania showed evidence neither for serially transferable cell culture infectivity nor for virus in tumors examined by electron microscopy.

Failing to find evidence for virus involvement, Beckwith and Malsberger looked to other possible etiologies. The application of acid-fast staining and bacteriologic culture revealed the presence of *Mycobacterium fortuitum* in the granulomas. Acid-fast bacilli were similarly found in unfiltered tumor extracts but not in those that had been filtered through 0.45-μm membranes.

Beckwith and Malsberger specifically noted that the German investigators did not apply acid-fast staining to their material. They further noted that, among homeotherms with low-grade mycobacterial infection, cell-free fractions of acid-fast bacilli are capable of evoking the formation of tubercles. That fact is consistent with the induction of tumors by the high-speed centrifugates of tumor extracts prepared by Wessing and von Bargen.

The inference by Beckwith and Malsberger that the kidney tumor virus is in reality a case of mycobacterial granuloma seems well justified. Accordingly, Wolf (1982) proposed that the kidney tumor agent no longer be considered a virus.

References

Beckwith, D. G., and R. G. Malsberger. 1980. Kidney tumour virus—tumour or mycobacterial tubercle? J. Fish Dis. 3:339–348.

Wessing, A. 1959. Über einen bösartigen, virusbedingten Tumor bei tropischen Zierfischen. Naturwissenschaften 46:517–518.

Wessing, A., and G. von Bargen. 1959. Untersuchungen über einen virusbedingten Tumor bei Fischen. Arch. Gesamte Virusforsch. 9:521–536.

Wolf, K. 1982. Newly discovered viruses and viral diseases of fishes, 1977–1981. Pages 59–90 in R. J. Roberts, ed. Microbial diseases of fishes. Academic Press, London. (Soc. Gen. Microbiol. Spec. Publ. 9.)

Appendix 1 Common and Scientific Names of Fishes

Alewife	*Alosa pseudoharangus*
Amago	*Oncorhynchus rhodurus*
Ayu	*Plecoglossus altivelis*
Barramundi	*Lates calcarifer*
Bass, largemouth	*Micropterus salmoides*
Bass, sea	*Dicentrarchus labrax*
Bass, striped	*Morone saxatilis*
Bluegill	*Lepomis macrochirus*
Bream	*Abramis abrama*
Bream, sea	*Sparus aurata*
Bullhead, brown	*Ictalurus nebulosus*
Bullhead, yellow	*Ictalurus natalis*
Carp, bighead	*Aristichthys nobilis*
Carp, black	*Mylopharyngodon piceus*
Carp, common	*Cyprinus carpio*
Carp, crucian	*Carassius carassius*
Carp, grass	*Ctenopharyngodon idella*
Carp, silver	*Hypothalmichthys molitrix*
Catfish, blue	*Ictalurus furcatus*
Catfish, channel	*Ictalurus punctatus*
Catfish, walking	*Clarias batrachus*
Char, Arctic	*Salvelinus alpinus*
Chebachek	*Pseudorasbora parva*
Cichlid, Ramirez' dwarf	*Apistogramma ramirezi*
Cod, Atlantic	*Gadus morhua*
Cod, Pacific	*Gadus macrocephalus*
Dab	*Limanda limanda*
Danio, pearl	*Brachydanio albolineatus*
Danio, zebra	*Brachydanio rerio*
Eel, American	*Anguilla rostrata*
Eel, ditch	*Fluta alba*
Eel, electric	*Electrophorus electricus*
Eel, European	*Anguilla anguilla*
Eel, Japanese	*Anguilla japonica*
Eel, swamp	*Fluta alba*
Flounder, starry	*Platichthys stellatus*
Flounder, winter	*Pseudopleuronectes americanus*
Goldfish	*Carassius auratus*
Grayling	*Thymallus thymallus*
Grunt, bluestriped	*Haemulon sciurus*
Guppy	*Poecilia reticulata*
Halibut, bastard	*Paralichthys olivaceus*
Herring, Atlantic	*Clupea harengus harengus*
Himemasu	*Oncorhynchus nerka*
Hirame	*Paralichthys olivaceus*
Ide	*Leuciscus idus*
Kokanee	*Oncorhynchus nerka*
Lamprey, sea	*Petromyzon marinus*
Loach, cobitid	*Misgurnus anguillicaubulatus*
Margate	*Haemulon album*
Menhaden, Atlantic	*Brevoortia tyrannus*
Minnow, fathead	*Pimephales promelas*
Mullet, gray	*Liza ramada*
Muskellunge	*Esox masquinongy*
Opaleye	*Girella nigricans*
Perch, Eurasian	*Perca fluviatilis*
Perch, redfin	*Perca fluviatilis*
Perch, Rio Grande	*Cichlasoma cyanoguttatum*
Perch, white	*Morone americanus*
Perch, yellow	*Perca flavescens*
Pickerel, chain	*Esox niger*
Pickerel, grass	*Esox americanus vermiculatus*
Pickerel, redfin	*Esox americanus americanus*
Pike, northern	*Esox lucius*
Plaice, American	*Hippoglossoides platessoides*

Platyfish, southern	*Xiphophorus maculatus*
Puffer	*Fugu pardale*
Roach	*Rutilus rutilis*
Salmon, Atlantic	*Salmo salar*
Salmon, chinook	*Oncorhynchus tshawytscha*
Salmon, chum	*Oncorhynchus keta*
Salmon, coho	*Oncorhynchus kisutch*
Salmon, Danube	*Hucho hucho*
Salmon, pink	*Oncorhynchus gorbuscha*
Salmon, sockeye	*Oncorhynchus nerka*
Sauger	*Stizostedion canadense*
Seahorse, lined	*Hippocampus erectus*
Shad, American	*Alosa sapidissima*
Shark, lesser spotted dogfish	*Scyliorhinus caniculus*
Shark, smooth dogfish	*Mustelus canis*
Sheatfish	*Silurus glanis*
Shiner, golden	*Notemigonus crysoleucas*
Smelt, European	*Osmerus eperlanus*
Smelt, rainbow	*Osmerus mordax*
Snakehead, striped	*Ophiocephalus striatus*
Sole, English	*Parophrys vetulus*
Sole, flathead	*Hippoglossoides elassodon*
Sole, sand	*Psettichthys melanosticus*
Splake	*Salvelinus fontinalis* × *S. namaycush*
Sturgeon, white	*Acipenser transmontanus*
Sucker, white	*Catostomus commersoni*
Swordtail, green	*Xiphophorus helleri*
Tench	*Tinca tinca*
Tilapia, blue	*Tilapia aurea* and *Sarotherodon aureus*
Tilapia, redbelly	*Tilapia zilli*
Trout, brook	*Salvelinus fontinalis*
Trout, brown	*Salmo trutta*
Trout, cutthroat	*Salmo clarki*
Trout, golden	*Salmo aguabonita*
Trout, lake	*Salvelinus namaycush*
Trout, rainbow	*Salmo gairdneri*
Trout, steelhead	*Salmo gairdneri*
Turbot	*Scophthalmus maximus*
Walleye	*Stizostedion vitreum vitreum*
Whitefish, mountain	*Prosopium williamsoni*
X-ray fish	*Pristella riddlei*
Yamamae	*Oncorhynchus masou*
Yellowtail	*Seriola quinqueradiata*

Appendix 2 Fish Cell and Tissue Culture

Tissue culture is the principal biological tool of fish virology because it is the best and most practical approach for research, diagnostics, virus isolations, and serologic identification tests.

The information presented here is intended to be a guide for those with prior tissue culture experience. Comprehensive treatments of fish cell and tissue culture were published by Wolf and Ahne (1982) and Wolf and Quimby (1969). The basic requirements for cultivation of cells from fish, and also from amphibians and reptiles, were given in an authoritative and scholarly review by Clark (1972), who stressed comparative aspects.

Short discussions of fish tissue trypsinization were presented by McKenzie and Stephenson (1973), Sigel and Beasley (1973a, b), and Wolf (1973). Ahne and Bachmann (1974) described in considerable detail the procedures for preparation of standardized cell cultures by trypsinization of selected tissues from a warm-water fish (common carp) and a cold-water species (rainbow trout). The culture of fish leukocytes was described by Etlinger et al. (1976), Sigel et al. (1973), and Wolf and Quimby (1976a). Stepwise procedures for initiating fish cell lines from minced tissues or by trypsinization, and for passaging the resulting cultures and established cell lines were also given by Wolf and Quimby (1976b, c, d).

For those with only limited experience in animal cell or tissue culture, textbooks or references—e.g., Paul (1975)—will prove helpful. The handbook and laboratory guide by Merchant et al. (1964) is a valuable addition to the fish health specialist's library. One of the most comprehensive works on plant and animal tissue culture is that by Kruse and Patterson (1973); it is particularly valuable because it includes selected aspects of fish cell and fish tissue culture.

Today, the only practical approach to fish cell and tissue culture is with tested and proved media and materials from commercial sources. Collectively, catalogs of the biological supply houses also provide valuable reference material. In addition to product listings, several of the catalogs contain formulations and other useful information.

Table 1. Composition (g/L) of physiologic salines for use in fish tissue culture

Component	Earle's BSS	Hanks' BSS	Phosphate-buffered saline[a]
NaCl	6.80	8.00	8.00
$CaCl_2 \cdot 2H_2O$	0.27	0.19	0.132
KCl	0.40	0.40	0.20
$NaH_2PO_4 \cdot H_2O$	0.14	—	—
$Na_2HPO_4 \cdot 2H_2O$	—	0.045	1.15
$NaHCO_3$	2.20	0.35	—
KH_2PO_4	—	0.06	0.20
$MgCl_2 \cdot 6H_2O$	—	0.10	0.10
$MgSO_4 \cdot 7H_2O$	0.20	0.10	—
Glucose	1.00	1.00	—
Phenol red, 1% soln.[b]	1 mL	1 mL	1 mL

Note: Use pyrogen-free water—triple glass distilled or equivalent.
[a]In preparing phosphate-buffered saline, add chemicals individually in order given. Dissolve each completely before adding the next.
[b]Value for phenol red is 1 mL/L.

Physiologic Salines

Physiologic salt solutions are essential to most work in tissue culture. Several formulations are available commercially, but Hanks' balanced salt solution (BSS) is one of the most useful. It contains a small amount of sodium bicarbonate plus a phosphate buffer system, and exposure to air does not induce high pH values as quickly as it does in Earle's BSS. To obtain a truly stable pH, one should use a saline having good buffering capacity, such as phosphate-buffered saline (PBS; Table 1). For emergency use, for physiologic washing, or for reduction of tissue to homogenates, "normal" saline (0.9 g/100 NaCl) may be used. Commercial parenteral isotonic saline solution, which is available from hospitals or pharmacies, is economical, sterile, and pyrogen free.

Culture Medium

Although it might occasionally prove inadequate, Eagle's minimal essential medium (MEM), supplemented with 10% fetal bovine serum, comes close to being an all-purpose culture medium for cells of mammals, birds, reptiles, amphibians, and bony fishes (teleosts),

For use with cells of anadromous and freshwater teleosts, Eagle's MEM has the proper osmolarity (a freezing point of about −0.6°C) and requires no adjustment. Inasmuch as the freezing point of the blood of strictly marine teleosts is about −0.75°C, the culture medium for marine teleost cells should be adjusted by adding 0.45 g/100 mL NaCl. One can do this easily by adding 20.6 mL/L sterile 3.4M (20% w/v) stock solution of NaCl.

Eagle's MEM can be purchased in liquid form (both in working strength and as concentrates), or in two powder forms—one of which is heat stable and may be sterilized by autoclaving, and the other of which is heat labile and is intended for decontamination by filtration. Powdered MEM must be dissolved in water of high purity or of tissue culture quality. High-purity water can be prepared by demineralization or distillation, or both (Pumper 1973). Alternatively, USP water for injection (obtainable at pharmacies or hospital supply houses) is totally satisfactory and generally economical.

If fish cell cultures grow more slowly than desired, the addition of 5% extra serum or 5 to 7% whole-egg ultrafiltrate should stimulate renewed vigor. Such measures should be considered on a temporary basis only; if slow growth persists, other causes should be sought.

The use of antibiotics in a culture medium should be governed by the expected risks of contamination. For routine purposes, the addition of 100 IU sodium penicillin G, 100 μg dihydrostreptomycin, and 25 IU nystatin per milliliter yields what is probably the most widely used combination in fish virology. Penicillin and streptomycin are being supplanted by the bactericidal antibiotic gentamicin, at 100 μg/mL; it can be autoclaved, is bactericidal, and is effective against many mycoplasmas.

Control of pH

Management of pH is absolutely essential, not only for good cell growth but also for the replication of fish viruses by the cells. A ph of 7.6 to 7.8 supports good growth of BB, BF-2, CHSE-214, EPC, FHM, and RTG-2 cell lines (Table 2). Collectively these cell lines are able to replicate many of the fish viruses thus far isolated.

Table 2. Useful cell lines for fish virology

Cell line designations						
Common	ATCC[a]	Animal	Tissue	Morphology	Some fish viruses replicated[b]	Temperature tolerance (°C)
BB	CCL-59	Brown bullhead	Caudad trunk	Fibroblast	CCV, IPNV	4–34
BF-2	CCL-91	Bluegill	Caudad trunk	Fibroblast	EVA, IPNV, LV	10–30
CAR	CCL-71	Goldfish	Fin	Fibroblast	GFV, GSV	15–30
CHSE-214	—	Chinook salmon	Embryo	Fibroblast	*H. salmonis*, IHNV, IPVN, OMV, VHSV	4–27
EPC	—	Common carp	Epithelioma	Epithelial	PFR, SVC, VHSV	15–30
FHM	CCL-42	Fathead minnow	Caudad trunk	Epithelial	EV-2, GSV, IHNV, IPNV, PFR, SVCV	0–36
RTG-2	CCL-55	Rainbow trout	Gonad	Fibroblast	*H. salmonis*, IHNV, IPNV, OMV, PFR, VHSV	4–26

Note: For a longer listing of available fish cell lines, see Wolf and Mann (1980).

[a]American Type Culture Collection designation for Certified Cell Line.

[b]CCV, channel catfish virus; EV-2, eel virus 2; EVA, eel virus, American; GFV, goldfish virus; GSV, golden shiner virus; *H. salmonis, Herpesvirus salmonis;* IHNV, infectious hematopoietic necrosis virus; IPNV, infectious pancreatic necrosis virus; LV, lymphocystis virus; OMV, *Oncorhynchus masou* virus; PFR, pike fry rhabdovirus; SVCV, spring viremia of carp virus; VHSV, viral hemorrhagic septicemia virus.

Most, if not all, established lines of fish cells have a requirement for carbon dioxide, but that need can also be met with bicarbonate. Some sodium bicarbonate is found in virtually all cell culture media. An added function of sodium bicarbonate is that of providing buffering capacity. Buffer function is provided by the bicarbonate ion in equilibrium with carbon dioxide and with the carbonate ion, but this fact is often misunderstood and can present problems in pH control. For example, there is a prevalant misconception that newly seeded animal cells alkalize the medium, with a resulting high pH. In reality, carbon dioxide initially in the medium diffuses into the atmosphere of the culture vessel, causing the pH to rise. Some workers use dilute hydrochloric acid to lower the pH, but such an adjustment reduces the buffering capacity as well. The belief that newly seeded cultures alkalize the medium can be disproved by adding the medium alone to culture vessels; the pH will rise to even higher levels than if cells are present. If one uses a medium such as MEM with Earle's BSS in sealed vessels, a nearly ideal initial pH of 7.6 to 7.8 can be achieved by using carbon dioxide to lower the pH to about 7.2 before the medium is seeded. An alternative method of achieving pH 7.6 to 7.8, which was tailored for culture and virus plaque assay in open-dish cultures, is to use organic buffers. One such buffer system consists of 9 mM sodium bicarbonate ($NaHCO_3$) and 16 mM tris (hydroxymethyl) aminomethane. Alternatively, 14 mM *N*-2-hydroxyethylpiperazine-*N*′-2-ethane sulfonic acid (HEPES) can be used, again with 9 mM $NaHCO_3$. The organic buffers are added from sterile stock solutions that have been set to pH 7.8 at 25°C.

Cultures of Tissue Explants

Tissue culture—literally the methodology by which small fragments of organs are grown in vitro—gave rise to today's more efficient and versatile cell culture. Tissues such as heart, fin, skin, ovary, cornea, gas bladder, or kidneys are cut into fragments no larger than 2 mm and explanted to culture vessels. To keep them in place for viewing microscopically, one should allow the explants to adhere to the substrate for 30 minutes or more before the medium is added. Alternatively, the explants may be positioned in chicken plasma that is made to clot with embryo extract, thus providing a transparent physiologic adhesive. One to two explants per square centimeter of growth area usually produce cell sheets that are sufficiently confluent for many virologic needs.

Fin, skin, and corneal explants usually show migration of macrophages and epithelial cells within a matter of a few hours. Although mitosis is negligible within such sheets, confluent sheets are developed rather quickly by the outward-moving epithelial cells. When sheets have reached adequate size, the culture can be inoculated. The circumscribed sheet of migrating epithelium usually has a finite area beyond which extension does not occur; in the meantime, however, fibroblast-like cells will have appeared that show mitotic activity and are usually long-lived. Explant cultures of fish gas bladder have shown rhythmic muscle contractility and beating cilia for many months, and cardiac tissues have been kept alive and pulsating for years.

Preparation of tissue explant cultures is somewhat time consuming and seldom necessary today. However, explant cultures are appropriate when only small amounts of tissue are available. The methodology has much more realistic applications when cells lines are not immediately available or when there is reason to believe that host specificity is criticial for virus replication.

Primary Monolayer Cell Cultures

Established cell lines offer many advantages in fish virology, but there are times when primary or original monolayer cultures are either the only available susceptible system or the most appropriate in vitro system. Because primary cultures usually offer several kinds of cells, their sensitivity to viruses is generally considered to exceed that of established lines. Then too, host specificity is a factor—there simply may not be a line of fish cells from the animal that one wishes to work with. As shown by the survey by Wolf and Mann (1980), there are conspicuous gaps in availability of representative fish cell lines from important families—the Serranidae (sea basses) and Pleuronectidae (right eye flounders), for example. Some fish viruses apparently have a high specificity for cells of a single morphologic type. Lymphocystis virus and the bluegill virus, for example, replicate only in cells having fibroblast-like morphology. All of these limitations of tissue culture point to the need for primary monolayer cultures, at least for some applications.

Two generalized methods of preparing monolayer cultures of fish cells are presented here. The first is strictly by mechanical disaggregation, and the second (and more commonly used) is by controlled tryptic enzyme digestion. Both methods require that tissues be obtained by procedures that eliminate or greatly reduce the risk of microbial contamination. Guidelines for obtaining tissues follow.

In contrast to the skin of terrestrial vertebrates, the skin of healthy fishes generally supports only a meager microbial flora, although ectoparasites may be present and can present problems.

Fin, barbel, or corneal tissues are readily obtained for culture and generally grow well. Preparatory cleaning of such tissues should include thorough rinsing in cool to cold tap water (preferably chlorinated). Precautions in obtaining fin or other external tissue should include avoidance of contamination and fouling from regurgitated food and feces. Well-planned work should include withdrawal of food from the fish for several days before the tissue samples are collected.

Several common antibiotics are bactericidal and efficiently decontaminate fish tissues. One hour of exposure of external tissues to concentrations of polymyxin B at 1000–2000 IU/mL have been useful in reducing contamination before cells are prepared for culture. Similarly, the commercial preparation Neosporin (Burroughs Wellcome Co.), a mixture of three bactericidal antibiotics, has been used as a bath at a concentration giving 500 IU polymyxin B, 500 μg neomycin, and 40 IU of zinc bacitracin per milliliter. Treatment in cold water (4 to 10°C) of tissues or whole nonfeeding fry or larvae for 1 to 2 hours should be followed by thorough rinsing in sterile water or BSS.

For internal tissues, a strictly asceptic technique should be used to obtain the organs of choice. Topical disinfection should be carried out, and complete immersion in disinfectant is often practical. A level of 500 ppm available chlorine, obtained

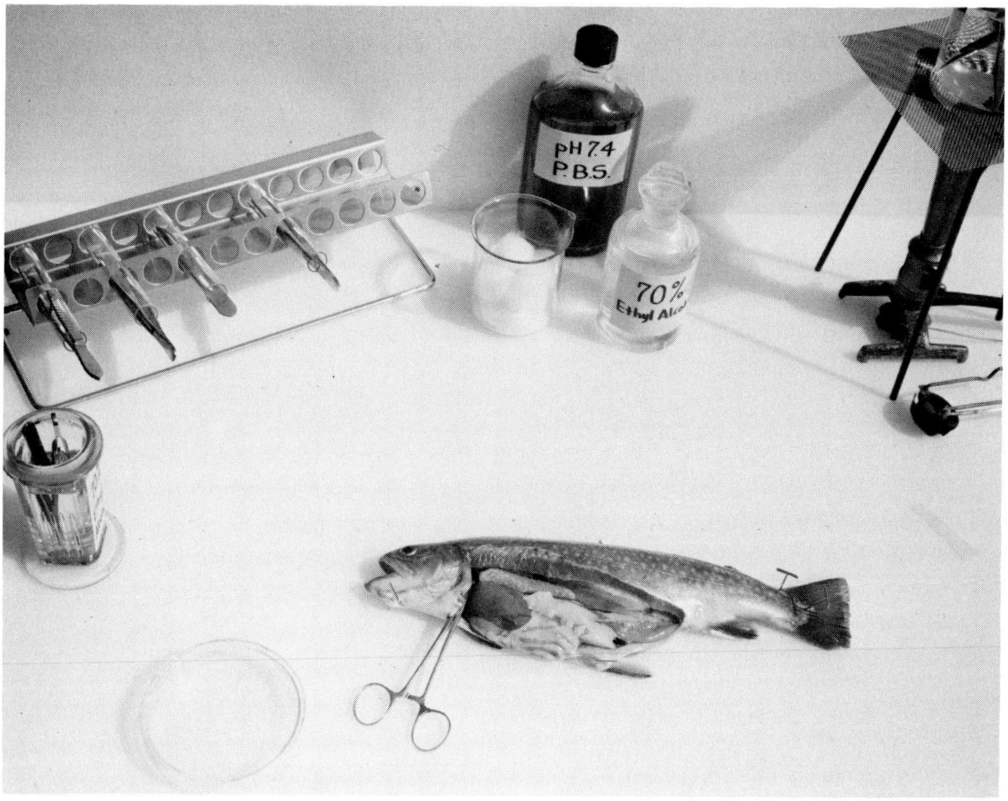

Figure 1. Adult brook trout (*Salvelinus fontinalis*) opened for aseptic removal of internal organs. Hemostat is clamped on anterior attachment of left ovary. Nearby rack holds sterile instruments shielded in sterile tubes. Boiling water, sterile saline, and 70% ethanol are used for cleaning and disinfecting contaminated instruments. From Wolf and Quimby (1969). Reprinted with the permission of Academic Press, Inc.

with calcium or sodium hypochlorite solution (household bleach), is effective as a bath of several minutes' duration. A 1 : 1000 solution of benzalkonium chloride (Roccal or Zephiran) has been used with equally satisfactory results. Both quartenary ammonium compounds are toxic to living tissues; therefore excess disinfectant should be removed by thorough rinsing. One may use 70% ethanol or isopropanol for rinsing, but if no other disinfectant is available, alcohol may be used as the sole decontaminating liquid. Alcohols do, however, cause rapid coagulation of the skin components, and probably are more effective if disinfection is first carried out with hypochlorite solution.

After external decontamination of donor fish, aseptic technique is used to open the body cavity laterally (Fig. 1) and expose the internal organs and tissues. Immature gonads or ovaries at almost all stages of egg development have been the most easily and consistently cultivable of all fish tissues. Ovaries provide vigorous cells, even when the eggs are nearly ripe. One should, however, remove as many mature eggs as possible before the ovaries are cultured. For best results with tissues of almost all fishes, one should keep the tissues cool throughout handling and preparation. Kidney tissue usually does well in culture, as do the tissues of the heart, gas bladder, mesentery, liver, and spleen. The liver and spleen, especially when taken from adult fish, usually give rise to much debris and contain a large proportion of differentiated cells.

Young tissues often provide better material for cultures than do the same kinds of tissues from older animals. Embryos from live-bearing fishes or from eggs, sac fry (with the yolk sacs removed), or tissues from fry stages have performed well in vitro. To minimize gut flora, one should take such tissues from fry that have not been fed for at least 1 or 2 days. Very young fry may be used in their entirety, but only selected portions of large fry are recommended. After the disinfectant has been removed by thorough rinsing, the caudal portion of the body should be amputated along a diagonal line extending from the dorsal fin to a point posterior to the vent. Considerable tissue can be obtained in this way, with little or no contamination.

The procedure for preparation of cultures was described by Wolf and Quimby (1976b). Several grams of tissue are usually preferred, but smaller amounts can be cultured if necessary. Tissues are placed in a small, sterile beaker (25 to 50 mL) and enough PBS is added to keep the tissues barely covered. The tissues are then minced with sterile, fine scissors until the pieces to be used are no larger than 2 mm^3. It is essential that the pieces be small. A suitable mince is usually obtained in 3 to 10 minutes.

The finely minced tissues are then transferred to plastic culture flasks (or other culture vessels) at the rate of 0.1–0.2 mL/25 cm^2 of the area available for growth. The tissue fragments are spread uniformly over the available area, and the flasks are then capped or stoppered securely. At a temperature appropriate for the fish, the flasks are tilted or otherwise positioned so that fluids drain from around fragments. Tissues are left to adhere for about 1 hour before the collected fluid is removed. Flasks are then positioned so that adherent tissues are uppermost, 5 to 7 mL of culture medium is added per 25 cm^2 of area, and the flasks are moved to an incubator. There the flasks are oriented to normal culture position and the medium is allowed to flow gently over the adherent fragments. After one to several days of incubation, the cultures may be moved for examination.

These procedures have consistently provided confluent primary monolayer cultures; in addition the cultures are easily and quickly set up—a dozen or more flasks can be prepared in about 2 hours.

Minced tissues can be further reduced by enzymatic digestion, and trypsinization is undoubtedly the most popular method of preparing primary monolayer cultures. Detailed procedures were described by Wolf and Quimby (1976c). Fluted trypsinizing flasks are ideal, but ordinary Erlenmeyer flasks or covered beakers are also satisfactory. The digestion fluid is composed of PBS (Table 1) containing 0.25% trypsin and optional serum and antibiotics (Table 3). Minced tissues are washed free of debris, and then treated for 30 to 60 minutes with about 10 volumes of digeston fluid. The controlled digestion may be carried out with intermittent manual shaking, but continual agitation (e.g., with a magnetic stirrer) is preferred. Debris and cells released during the first period of digestion should be discarded and replaced with an equal volume of fresh dispersant.

Digestion may be carried out at 15 to 20°C, and the progress checked at 20- to 30-minute intervals. The goal is a mixed suspension of individual cells and small

Table 3. Composition of fish tissue digestion fluid

Ingredient	No. of volumes
Phosphate-buffered saline (PBS)	34
Trypsin or pancreating solution (2.5% in normal saline)	4
Serum (fetal bovine)	1 (optional)
Antibiotic solution[a]	1 (optional)

[a]Stock solution of penicillin and streptomycin (10,000 IU and 10,000 μg/mL), or stock solution of gentamicin (2000 μg/mL).

clusters of incompletely disaggregated tissues. The cell and tissue dispersion shown in Figure 2 is an example of an ideal digestion. A preferred method is to carry out digestions overnight at 4 to 6°C—a procedure that has consistently yielded good results.

Precautions with Primary Monolayer Cultures

One disadvantage of primary cultures is that they might harbor resident pathogens. As reviewed and carefully documented by Bullock and Snieszko (1969), signifi-

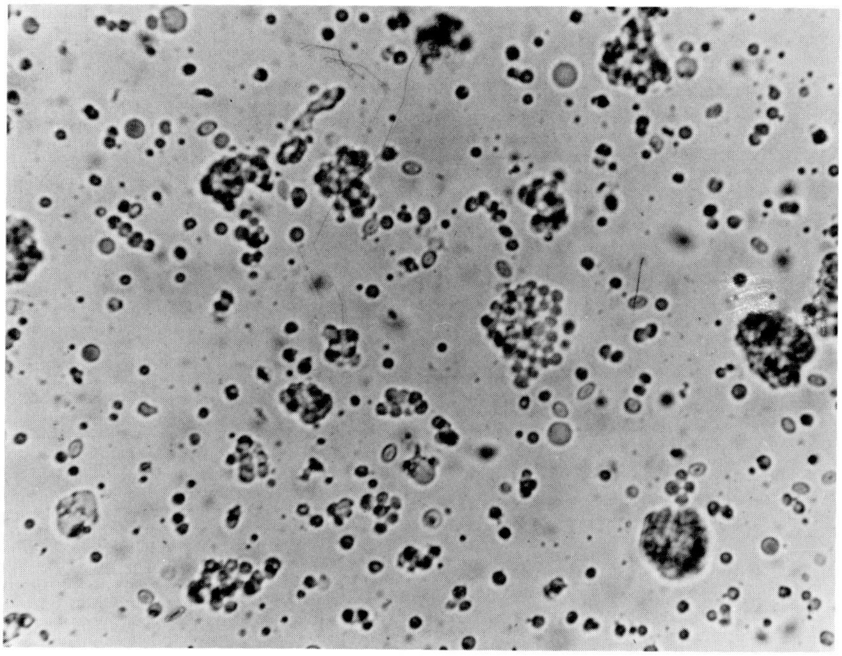

Figure 2. A properly harvested trypsinized tissue preparation, showing a mixture of individual cells and fragments that are incompletely dispersed. Trypan blue dye exclusion should indicate viability in excess of 95% in such a preparation. From Wolf (1973). Reprinted with the permission of Academic Press, Inc.

cant numbers (up to 26% of some populations) of apparently healthy hatchery fishes sometimes have a low-level bacteremia. Antibiotics used in the medium help suppress such flora. Viruses, however, present a different problem, and if agents such as infectious pancreatic necrosis virus (IPNV) are present, they can destroy primary cultures. The prudent worker attempts to learn the health history of the fish from which tissues are to be taken for initiating primary monolayer cell cultures. Populations known to have had IPN or any other viral disease should be avoided.

Fish Cell Lines

With few exceptions, fish viruses can be replicated only in fish or in fish cell cultures. However, BHK-21 cells, originating from the kidneys of baby hamster, are an exception: they can be used to grow the rhabdoviruses of fishes, as well as those of mammals, birds, and amphibians. Although many different lines of fish cells are available (Wolf and Mann 1980), seven dominate the literature and suffice for much of today's fish virology (culture of some herpesviruses excepted): BB, BF-2, CAR, CHSE-214, EPC, FHM, and RTG-2 (Table 2). Further details of the characteristics of Certified Cell Lines are given in *The American Type Culture Collection Catalogue of Strains II* (Hay et al. 1983).

Permanent or established lines of cells are those that have been passaged 70 or more times and meet the requirements of Federoff (1975), who additionally provided the current standard for tissue culture terminology. The RTG-2 cell line of Wolf and Quimby (1962) was the first fish cell line to be developed and established and is the most thoroughly studied. Depending on incubation temperature and initial seeding density, Plumb and Wolf (1971) found that the time required for production of standard tube cultures varies widely—from only 1.5 days to as long as 67 days (Table 4). At the optimum incubation temperature of 20°C, the uniform population doubling time for RTG-2 cells is 2 days (Table 5). Times required for population doubling increase at higher or lower temperatures.

Although comparable studies have not been done with other fish cell lines, these general relations probably hold. Cell lines such as BB, BF-2, EPC, and FHM, which have higher optimum temperatures, have shorter population doubling times. Recommended seeding densities for major fish cell lines are given in Table 6.

Procedures for subculturing fish cells and propagating fish cell lines were described by Wolf and Quimby (1967d).

Precautions with Fish Cell Lines

In general, cultures to be used in fish virology should be young enough to show obviously dividing cells throughout the area of growth. At the same time, the available growth area should be well covered by the cells—as a rule of thumb, about 80 to 90 or 95%. Dense cultures or cultures that have become old and crowded and show no mitoses are likely to perform poorly for virology.

One should also monitor the cultures to ensure that they retain susceptibility to the virus to be studied. The FHM and BF-2 cells have a history of recurrent loss of some or all of their susceptibility to one or more viruses. Some lineages of these two cell lines have become refractory to almost all fish viruses.

One solution to the problem of refractory cells is to obtain a susceptible lineage by using cells of lower passage, either kept as frozen stocks or obtained from other laboratories. An alternative is to use any of several cell cloning procedures to derive subpopulations. When clones are established, they may be individually tested and susceptible stocks selected for propagation.

Today, most diagnostic virology and virologic research is done with fish cell lines. The day-to-day culture of animal cell lines has built-in hazards—microbial contamination, toxicity of components of the medium, finite cell line life, temperature extremes, mixed cultures, and accidents. Wolf and Quimby (1973, 1978) summarized much of their experience in establishing and managing continuous laboratory propagation of animal cell lines. The flow chart (Fig. 3A, B) and listing of 13 precepts (Table 7) to be followed have been taken from that work—a system that was developed during 17 years of experience in establishing and maintaining fish and amphibian cell lines.

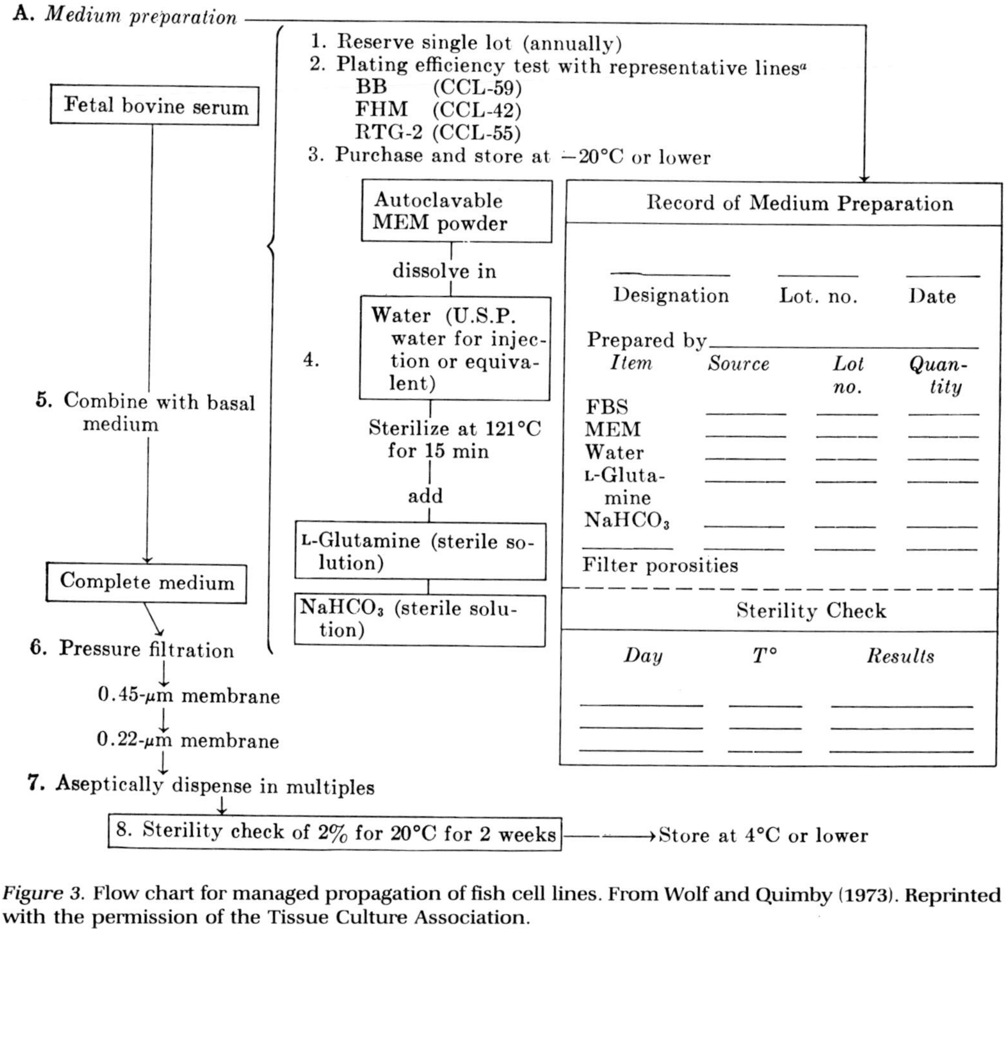

Figure 3. Flow chart for managed propagation of fish cell lines. From Wolf and Quimby (1973). Reprinted with the permission of the Tissue Culture Association.

B. *Stock culture propagation protocol*

1. Certified cell line[b]
2. Remove medium, rinse with Versene solution, treat for 5 to 15 min, pipette vigorously to release cells, "neutralize" with medium, centrifuge at 200 to 500 × *g* for 10 min at 4°C, resuspend in multiples in two lots of sterile antibiotic-free media

Versene (1:5000) in Ca- and Mg-free salt solution at 121°C for 15 min

Series I (one lot of medium and/ or one person) → Multiple cultures

0 0 0 0 0

Series II (second lot of medium and/ or second person) → Multiple cultures

0 0 0 0 0

Sample label

Cell line →	RTG-2
Passage →	176 s̄
Medium →	Lot 105 MEM 10
Date →	1-27-70

Each daughter culture labeled (example as above)

3. Incubate to confluency and hold for 1 to 2 months
4. Inspect for pH, clarity, freedom from contamination
5. Examine microscopically for morphology, cell quality, and cleanliness
6. Select best for subcultivation
7. Retain unused cultures as reserve

   0 0 0 0 ← Multiple reserve cultures → 0 0 0 0
8. Subculture to multiple daughter cultures

   0′ 0′ 0′ 0′ 0′ 0′   0′ 0′ 0′ 0′ 0′ 0′
9. Record all relevant data

Record of______ cell line   Page______

Date	*Pass*	*Source*	*Split ratio*	*Medium*	*Versene*

10. Check annually for mycoplasmas
11. Expand appropriate stock cultures annually and use with antibiotics as source of fresh working stocks. Older working stocks to be discarded.

[a] For virological application, each lot of serum is also tested for freedom from neutralizing activity against four fish viruses.

[b] Or equivalent. Certified cell lines available from American Type Culture Collection, Rockville, Maryland 20852.

Figure 3. (continued)

Table 4. Approximate time (days) for different seeding densities of RTG-2 cells to attain confluency (275,000 cells/mL)

Initial seeding of cells (no./mL)	Temperature (°C)				
	25	20	15	10	5
200,000	1.8	1.5	2.4	5.0	8.0
100,000	3.3	3.1	4.8	9.3	18.0
50,000	6.2	4.8	7.3	14.0	37.0
25,000	9.3	7.5	14.0	34.0	53.0
12,500	13.0	10.5	19.8	45.0	67.0

Source: From Plumb and Wolf (1971).

Table 5. Population doubling times (days) of RTG-2 cells at different seeding densities and temperatures

Initial seeding of cells (no./mL)	Temperature (°C)				
	25	20	15	10	5
100,000	2.4	2.0	3.8	7.0	10.0
50,000	2.5	2.0	3.5	5.5	14.5
25,000	2.5	2.0	4.0	8.0	16.0
12,500	2.7	2.0	4.1	8.8	13.5
Average	2.6	2.0	3.8	7.6	13.8

Source: From Plumb and Wolf (1971).

Table 6. Suggested seeding density and incubation temperature for several established fish cell lines

Cell designation[a]	Suggested seeding density (no./mL)	Incubation temperature (°C)	Time to confluency (days)
BB (CCL-59)	200,000	30	3[b]
BF-2 (CCL-91)	250,000	25	3
CHSE-214	300,000	20	3
EPC	500,000	25	3
FHM (CCL-42)	500,000	25	3
RTG-2 (CCL-55)	200,000	20	3

[a]CCL, Certified Cell Line of the American Type Culture Collection.

[b]In my experience, BB cells are most sensitive to unidentified factors in certain lots of fetal bovine serum. Some lots of serum have proved wholly unsatisfactory for sustaining fish cell growth, whereas others have been of marginal quality and have required higher seeding densities (300,000 cells/mL). Serum should be tested before purchase.

Table 7. An outline of precepts to be followed in maintaining fish and other animal cell lines

1. Determine the degree of discipline required for cell culture needs.
2. Maintain stock cultures separate from working cultures.
3. Employ a binary or redundant approach for maintaining stock cultures.
4. Employ a testing program for each year's serum supply.
5. Use high-purity or tissue culture quality water for all solutions.
6. Test the sterility of all locally prepared solutions before use.
7. Avoid the use of any antibiotics in maintaining stock cultures.
8. Disperse stock cultures with autoclaved versene solution instead of versene-trypsin solution.
9. Handle only one cell line at a time.
10. Prepare and keep multiple daughter cultures.
11. Subculture stock cultures infrequently and keep at low temperature.
12. Keep detailed records of solution preparation and cell histories.
13. Test cultures at least annually for mycoplasmas (see Barile 1977).

Source: Adapted from Wolf and Quimby (1973, 1978).

References

Ahne, W., and P. A. Bachmann. 1974. Versuch zur Standardisierung der Methodik bei der Herstellung primärer Zellkulturen aus Süsswasserteleosteergeweben. Zentrabl. Bakteriol. Hyg. I Abt. Orig A. 228:447–464.

Barile, M. P. 1977. Mycoplasma contamination of cell cultures: a status report. Pages 291–334 in R. Acton, ed. Cell culture and its application. Academic Press, New York.

Bullock, G. L., and S. F. Snieszko. 1969. Bacteria in blood and kidney of apparently healthy trout. Trans. Am. Fish. Soc. 98:268–271.

Clark, H. F. 1972. Cultivation of cells from poikilothermic vertebrates. Pages 287–325 in G. H. Rothblatt and V. J. Cristofalo, eds. Growth, nutrition, and metabolism of cells in culture. Academic Press, New York.

Etlinger, H. M., H. O. Hodgins, and J. M. Chiller. 1976. Rainbow trout leucocyte culture: a simplified method. In Vitro 12:599–601.

Federoff, S. 1975. Proposed usage of animal tissue culture terms: Procedure 81169. Pages 53–57 in V. J. Evans, V. P. Perry, and M. M Vincent, eds. TCA manual 2. Tissue Culture Association, Rockville, Maryland.

Hay, R., M. Macy, A. Hamburger, A. Weinblatt, T. R. Chen, D. Stevens, R. L. Gherna, P. Pienta, S.-C. Jong, H.-T. Hsu, B. Brandon, and M. T. Alexander, editors. 1983. The American Type Culture Collection catalogue of strains II, 4th ed. Rockville, Maryland.

Kruse, P. F., Jr., and M. K. Patterson, Jr., editors. 1973. Tissue culture methods and applications. Academic Press, New York.

McKenzie, L. S., and N. G. Stephenson. 1973. Goldfish tissues. Pages 143–146 in P. F. Kruse, Jr., and M. K. Patterson, Jr., eds. Tissue culture methods and applications. Academic Press, New York.

Merchant, D. J., R. H. Kahn, and W. H. Murphy, Jr. 1964. Handbook of cell and organ culture. Burgess, Minneapolis.

Paul, J. 1975. Cell and tissue culture, 5th ed. Churchill Livingstone, New York.

Plumb, J. A., and K. Wolf. 1971. Fish cell growth rates. Quantitative comparison of RTG-2 cell growth at 5–25 C. In Vitro 7:42–45.

Pumper, R. W. 1973. Purification and standardization of water for tissue culture. Pages 674–677 in P. F. Kruse, Jr., and M. K. Patterson, Jr., eds. Tissue culture methods and applications. Academic Press, New York.

Sigel, M. M., and A. R. Beasley. 1973a. Trypsin. C. Marine teleost fish tissues. Pages 12–14 in P. F. Kruse, Jr., and M. K. Patterson, Jr., eds. Tissue culture methods and applications. Academic Press, New York.

Sigel, M. M., and A. R. Beasley. 1973b. Marine teleost fish tissues. Pages 133–135 in P. F. Kruse, Jr., and M. K. Patterson, Jr., eds. Tissue culture methods and applications. Academic Press, New York.

Sigel, M. M., E. C. McKinney, and J. C. Lee. 1973. Fish lymphocytes and blastogenesis. Pages 135–138 in P. F. Kruse, Jr., and M. K. Patterson, Jr., eds. Tissue culture methods and applications. Academic Press, New York.

Wolf, K. 1973. Freshwater fishes. Pages 138–143 in P. F. Kruse, Jr., and M. K. Patterson, Jr., eds. Tissue culture methods and applications. Academic Press, New York.

Wolf, K., and W. Ahne. 1982. Fish cell culture. Adv. Cell Cult. 2:305–328.

Wolf, K., and J. A. Mann. 1980. Poikilotherm vertebrate cell lines and viruses: a current listing for fishes. In Vitro 16:168–179.

Wolf, K., and M. C. Quimby. 1962. Established eurythermic line of fish cells in vitro. Science 135:1065–1066.

Wolf, K., and M. C. Quimby. 1969. Fish cell and tissue culture. Pages 253–305 in W. S. Hoar, and D. J. Randall, eds. Fish physiology, Vol. 3. Academic Press, New York.

Wolf, K., and M. C. Quimby. 1973. Towards a practical fail-safe system of managing vertebrate cell lines in culture. In Vitro 8:316–321.

Wolf, K., and M. C. Quimby. 1976a. Culture of leucocytes from teleost (bony) fishes: Procedure 41105. Pages 441–444 in V. J. Evans, V. P. Perry, and M. M. Vincent, eds. TCA manual 2. Tissue Culture Association, Rockville, Maryland.

Wolf, K., and M. C. Quimby. 1976b. Primary monolayer culture of fish cells initiated from minced tissues: Procedure 41125. Pages 445–448 in V. J. Evans, V. P. Perry, and M. M. Vincent, eds. TCA manual 2. Tissue Culture Association, Rockville, Maryland.

Wolf, K., and M. C. Quimby. 1976c. Primary monolayer culture of fish cells initiated from trypsinized tissues: Procedure 41541. Pages 453–456 in V. J. Evans, V. P. Perry, and M. M. Vincent, eds. TCA manual 2. Tissue Culture Association, Rockville, Maryland.

Wolf, K., and M. C. Quimby. 1976d. Procedures for subculturing fish cells and propagating fish cell lines: Procedure 49145. Pages 471–474 in V. J. Evans, V. P. Perry, and M. M. Vincent, eds. TCA manual 2. Tissue Culture Association, Rockville, Maryland.

Wolf, K., and M. C. Quimby. 1978. Systematic management of animal cell lines: Procedure 40091. Pages 741–744 in V. J. Evans, V. P. Perry, and M. M. Vincent, eds. TCA manual 2. Tissue Culture Association, Rockville, Maryland.

Index